AF864215

Theory of Elastic Waves

Peijun Wei

Theory of Elastic Waves

Science Press
Beijing

Springer

Peijun Wei
Department of Applied Mechanics
School of Mathematics and Physics
University of Science and Technology
Beijing
Beijing, China

ISBN 978-981-19-5661-4   ISBN 978-981-19-5662-1 (eBook)
https://doi.org/10.1007/978-981-19-5662-1

Jointly published with Science Press
The print edition is not for sale in China (Mainland). Customers from China (Mainland) please order the print book from: Science Press.

© Science Press 2022
This work is subject to copyright. All rights are solely and exclusively licensed by the Publisher, whether the whole or part of the material is concerned, specifically the rights of reprinting, reuse of illustrations, recitation, broadcasting, reproduction on microfilms or in any other physical way, and transmission or information storage and retrieval, electronic adaptation, computer software, or by similar or dissimilar methodology now known or hereafter developed.
The use of general descriptive names, registered names, trademarks, service marks, etc. in this publication does not imply, even in the absence of a specific statement, that such names are exempt from the relevant protective laws and regulations and therefore free for general use.
The publishers, the authors, and the editors are safe to assume that the advice and information in this book are believed to be true and accurate at the date of publication. Neither the publishers nor the authors or the editors give a warranty, expressed or implied, with respect to the material contained herein or for any errors or omissions that may have been made. The publishers remain neutral with regard to jurisdictional claims in published maps and institutional affiliations.

This Springer imprint is published by the registered company Springer Nature Singapore Pte Ltd.
The registered company address is: 152 Beach Road, #21-01/04 Gateway East, Singapore 189721, Singapore

Preface

The theory of elastic wave propagation in complex media is of great importance. For example, it is desired to understand the mechanism of propagation and attenuation of seismic wave in the stratum for the seismic exploration. Based on the propagation and attenuation law, the morphology and physical parameters can be obtained after inverse analysis of the wave motion signals collected by seismic detector in the wild field, and thus provide useful information about the distribution and the location of underground oil and gas deposit. In the area of medical science, the ultrasound inspection imaging also needs to understand the patterns of propagation and dissipation of elastic wave in the bio-tissue. These laws can be used to design various imaging algorithm to reconstruct the colored images of bio-tissue and organs based on the inverse analysis of the reflection and attenuation information and thus help clinical doctors to acquire the scientific evidence of disease diagnosis. In the area of industrial manufacturing, the knowledge regarding elastic wave propagation characteristics can be used in non-destructive testing (NDT) of materials and structures. The information about the propagation constants in the interior, the reflection and transmission at interface, the scattering at the inclusions and voids can be collected by the sensor arranged at surface or interior of material. Further performing the signal processing such as the de-noising and spectrum analysis by computer, the location, distribution, geometry and size of the flaws inside the materials would be available. This is crucial to maintain good quality of products and to guarantee the security service of structures. No matter the seismic exploration, medical ultrasound imaging and non-destructive testing of material, the basic scientific problem is the law of elastic wave propagation in complex media. This book focuses on the basic scientific problem. The contents are arranged by consideration of the completeness of theory and research methods and the necessary involvement of latest research results. Under such guideline, this book is finally finished and named *Theory of Elastic Waves.*

Among all the books on elastic wave propagation theory, the *Wave Propagation in Elastic Solids* (by J. D. Achenbach, Elsevier Science Publishers, 1973), *Diffraction of Elastic Waves and Dynamic Stress Concentrations* (by Pao Y. H., Mao C. C. Grane, Russak & Company Inc, 1973) and *Ultrasonic Waves in Solid Media* (by J. L. Rose,

Cambridge University Press, 1999) are of great influence. However, these books were published early. Many new scientific problems and new analytic and numerical methods have emerged during the past decades. It becomes necessary to recall and summarize these analytic and numerical methods systematically which dispersed in massive literatures. This is one motivation of writing this book. On the other hand, the teaching experience of giving the lecture for graduate students for about ten years makes me more and more feel that the advanced textbooks which are more suited to the graduate students are rare and become the desideratum to accommodate the research-type teaching. This is another motivation to write this book. Based on my research efforts on the elastic wave propagation in complex media, I strive to incorporate the latest research results into the contents of book. This book is finally finished based on the lecture notes and after several addition and amendment.

This book focuses on the elastic wave propagation in isotropic media; no contents of special subjects are included. If the basic theory and methods about elastic wave propagation in the isotropic solid are grasped, then a firm and strong foundation is established to further study the elastic wave propagation in anisotropic solid, viscoelastic solid, porous solid, etc. With the consideration of systematicness and completeness, the bulk wave, the surface wave and the guided wave are all involved in this book. Firstly, the dispersion and attenuation features of elastic wave propagating in elastic media of infinite extension are addressed. Many important conceptions about elastic wave propagation are provided in this chapter. Then, the reflection and transmission problems at interfaces are addressed. The reflection and transmission of single interfaces is first discussed. Apart from flat interfaces, the periodic undulation interfaces are also involved. Moreover, the various imperfect interfaces are also involved beside the perfect interface. The layered media are often met in actual engineering fields as a common type. Then, the elastic wave propagation through layered structure is addressed based on the reflection and transmission investigation at single interface. Consider the fact that the reflecting and transmitting features of elastic wave in layered media can be embodied in the sandwiched structure with two interfaces; we addressed the reflection and transmission of sandwiched structure with two interfaces with great interest. The simultaneous interface condition method, the transfer matrix method, the stiffness matrix method, the multiple reflection/transmission method, the super-interface method and the state transition matrix method, basically all mainstream research methods, are all included. These methods can easily be extended to the laminated structure with arbitrary N layers with appropriate modification. In the chapter of surface wave, not only the classic Rayleigh wave, Love wave and Stanley wave, but also the rotating surface wave is also addressed. The rotation surface wave is a natural result in the cylindrical coordinate system of the surface waves studied in the rectangular coordinates. In the chapter of guided wave, the guided wave propagation in bar, pipe, beam, plate, cylindrical shell and spherical shell is all involved. In particular, the elastic wave propagation in spherical shells is rarely mentioned in the existing published books and literatures. The guided waves propagation in spherical shell is addressed in this chapter, and the comparison with the vibration mode is also made. Moreover, the leaky waves due to the liquid loads are also addressed in this chapter. These contents make a distinguishing feature of this book.

The main focus of this book is the basic theory and analytic methods of elastic wave propagation problem. This book is suitable for those who work in the field of seismic survey, the material characterizing and non-destructive testing, the medical ultrasound imaging, the phononic crystal/metamaterial and the structure health monitoring, especially for high-grade undergraduate and postgraduate students as textbook for the systematic study of elastic wave propagation theory.

During the process of writing this book, my Ph.D. students, including Guo Xiao, Zhang Peng, Li Yueqiu, Li Li, Wei Zibo, Xu Chunyu, Ma Zhanchun, Xu Yuqian, Zhao Lingkang, Wang Ziwei, Zhao Lina, etc., did a great deal of work, including but not limited to the typing and checking of manuscript, text layout and drawing illustrations. Here, I'll express my loyal thanks for their helpful works. Moreover, I'll also express my thanks for the financial support from "National Natural Science Foundation of China (No. 11872105, No. 12072022)" and "Project of Graduate Textbook Construction of University of Science and Technology Beijing."

There are inevitably some mistakes existing in the book due to the author's limited level and careless omissions, and I'm pleased and encourage whoever to give comments and correct mistakes.

Beijing, China Peijun Wei
September 2020

Introduction

The theory of elastic wave propagation in complex media is widely used in many fields, such as geophysical exploration, seismic survey, medical ultrasound imaging and non-destructive testing of material and structure. However, the books which systematically introduce the theory of elastic wave propagation are rare. This book systematically introduced the basic theory of elastic wave propagation in isotropic solid media, including elastic wave propagation in infinite media, reflection and transmission of elastic wave at interfaces, reflection and transmission of elastic wave through layered structure with finite thickness, Rayleigh wave and Love wave propagating along the surface of semi-infinite solid and covering layer, the guided waves and leaky waves in flat plates and in cylindrical rods. The propagation patterns and features of guided waves in cylindrical shells and spherical shells are also introduced. The single scattering and multiple scattering of elastic waves, although very important also, but are not included due to the limitation of length. The author has been teaching the course of *Theory of Elastic Wave* for graduate students for over ten years. At the same time, the author also has been conducting the research works on the elastic wave propagation in complex media and the actual applications for over two decades. Hence, this book is written based on the lecture notes of "elastic wave theory" and has combined with the related research results. The entire book is divided into six chapters and is mainly focused on the basic theory and the systematicness of analytic methods.

This book is suitable to those who work in the fields of geophysical exploration, non-destructive testing, medical ultrasound imaging, phononic crystal, metamaterial, structure health monitoring and so on. Especially, it is suited to the high-grade undergraduate and graduate students to study the elastic wave theory systematically as textbook.

Contents

Chapter 1
Fundamentals of Elastodynamics

1.1 Basic Hypothesis of Elastodynamics

1.1.1 Continuity Hypothesis

The continuity hypothesis holds that the research object of elastic mechanics is the elastic deformable body which is a continuous medium filled with material points without any voids inside. In fact, all matter is composed of atoms and molecules, and matter is not continuous at the microscopic level. Even at the macroscopic level, the existence of internal cavities and cracks cannot be avoided. Continuity is only an idealized model. When studying the macroscopic phenomena and motion laws of objects, the continuity assumption makes it easier to deal with the problem. For example, the physical quantities of stress, strain and displacement are all continuous functions of coordinates, so that mathematical tools such as calculus can be used to establish and solve mathematical models of dynamic problems.

1.1.2 Elasticity Hypothesis

Under the action of external load, the object will generate stress field and strain field inside. When the amplitude of the stress field does not exceed the elastic limit of the material, after the external load is removed, the stress field and the strain field will disappear accordingly. This property is called the elastic property of the material. When the external load is large enough, for example, the amplitude of the stress field generated inside the object exceeds the yield limit of the material, there will be residual deformation (i.e. plastic deformation) existing inside the object and cannot be recovered when the external load is removed. The properties are called the plastic properties of the material. The elasticity hypothesis assumes that the internal stress field of the object is always in the elastic range under the action of external load.

© Science Press 2022

P. Wei, *Theory of Elastic Waves*,
https://doi.org/10.1007/978-981-19-5662-1_1

1.1.3 Small Deformation Hypothesis

Small deformation means that the deformation at various points within the object due to external load is small relative to the size of the object. In other words, the strain components (including line strain and shear strain) are all quantities much smaller than 1. Their second powers and products are small quantities of higher order relative to the first-order quantity and can be ignored without large precision loss. Due to the small deformation assumption, the stress field and strain field can be thought to satisfy the generalized Hooke's law. Moreover, when establishing the equilibrium equation, the geometry before deformation (initial configuration) can be used instead of the geometry after deformation (current configuration).

1.1.4 Homogeneous Hypothesis

Homogeneous hypothesis means that all points inside the object have the same elastic properties, namely the material elastic parameters do not change with the spatial coordinates. For non-homogeneous materials, the elastic parameters of the material are functions of coordinates, such as functional gradient materials, where the material parameters are continuous functions of coordinates. Another example is the fiber or particle reinforced composites, where the material parameters are discontinuous functions of coordinates or piecewise continuous functions. The assumption of uniformity makes the mechanical properties of the material not depend on the location, but does not guarantee that the material properties do not depend on the direction. The direction-dependent character of the material properties is described by isotropy or anisotropy. Therefore, the homogeneous assumption is not the same thing as the isotropic assumption.

1.1.5 Isotropic Hypothesis

The isotropic means that each point inside object has the same elastic properties along different directions, namely the material elastic parameters do not change with the change of direction. For isotropic materials, only two independent elastic parameters are needed to describe the elastic properties of the material. The commonly used elastic parameters are: modulus of elasticity E, shear modulus G, Lamé constants λ and μ, Poisson's ratio ν. However, they are not independent of each other, and there are only 2 independent parameters. For completely anisotropic materials, where the material has different properties along different directions, 21 independent parameters are needed to describe the elastic properties of the material. Usually, materials have certain symmetry, are not completely anisotropic, and their independent material parameters are between 2 and 21. For example, transverse isotropic materials have five independent elastic parameters; Orthotropic anisotropic materials have nine independent elastic parameters; Cubic crystalline systems have three independent elastic parameters; triangular crystalline systems have seven independent elastic parameters, etc.

1.1.6 Zero Initial Stress Hypothesis

The zero initial stress assumes that the object is in its natural state before the external load is applied and that there is no stress field inside the object. When the initial stress field exists, the equations of motion, the constitutive equation and the boundary conditions should be modified, and in general, all three sets of equations contain information about the initial stress. The initial stress field can be homogeneous or non-homogeneous. For the initial stress problem, there are three deformation configurations, i.e. no stress configuration, initial stress configuration (the configuration generated under the action of initial stress) and current configuration (the configuration generated under the action of both initial stress and external load). Deformations generated by initial stresses are usually relatively large and are taken into account as finite deformations. In the case that the deformation produced by the external load is not large, it can be considered as a small deformation. Therefore, the problem of the existence of initial stress is attributed to the superposition of small deformations on finite deformations. The incremental stress method is a common method for dealing with initial stress problems. The equations of motion, the constitutive equations and the boundary conditions satisfied by the incremental stresses are all related to the initial stresses. Usually, initial stresses lead to the anisotropy of the material as well as the change of equivalent modulus.

1.2 Basic Conservation Laws of Elastodynamics

1.2.1 Law of Mass Conservation

According to the assumption of continuity of objects, an object is a continuous collection of moving particles. If there is no "source" that generates new matter inside the object, and there also is no mass flux at the boundary of the object, the mass of the object is constant before and after the deformation, which is the law of mass conservation, i.e.

$$\int_{V} \rho \mathrm{d}V = \int_{V_0} \rho_0 \mathrm{d}V_0, \tag{1.2.1}$$

where V_0 and V are the volumes of the space occupied by the object before and after deformation; ρ_0 and ρ are the mass density of the object before and after deformation.

Considering

$$\mathrm{d}V = J\mathrm{d}V_0, \tag{1.2.2}$$

where $J = \det(x_i, X_j) = \left|\frac{\partial x_i}{\partial X_j}\right|$ (x_i is the spatial coordinate of the mass point, i.e. Eulerian coordinates; X_i is the co-moving coordinate or body-fitted coordinates of the mass point, i.e. Lagrangian coordinate), is called Jacobi determinant. The differential form of the conservation law, i.e. Eq. (1.2.1), can be expressed as

$$\rho J - \rho_0 = 0. \tag{1.2.3}$$

When there is a mass flux on the surface of an object, the law of conservation of mass is expressed as

$$\frac{\mathrm{d}}{\mathrm{d}t}\int\limits_V \rho \mathrm{d}V = \int\limits_V \frac{\partial \rho}{\partial t}\mathrm{d}V + \int\limits_S \rho \dot{\boldsymbol{u}} \cdot \boldsymbol{n}\mathrm{d}S = 0. \tag{1.2.4}$$

Using the divergence theorem, we can obtain its differential form

$$\frac{\partial \rho}{\partial t} + \nabla \cdot (\rho \dot{\boldsymbol{u}}) = 0, \tag{1.2.5}$$

where $\dot{\boldsymbol{u}}$ denotes the velocity of motion of the material point and $\boldsymbol{I} = \rho \dot{\boldsymbol{u}}$ denotes **flow density of mass**, i.e. the mass passing through a unit area per unit time. Equation (1.2.5) can also be written as

$$\frac{\partial \rho}{\partial t} + \nabla \rho \cdot \dot{\boldsymbol{u}} + \rho \nabla \cdot \dot{\boldsymbol{u}} = 0. \tag{1.2.6}$$

For solid materials, $\nabla \rho$ is very small at small deformation. Equation (1.2.6) can be approximated as

$$\frac{\partial \rho}{\partial t} + \rho \nabla \cdot \dot{\boldsymbol{u}} = 0. \tag{1.2.7}$$

The above equation indicates that when the volume of the micro-element of the object expands ($\nabla \cdot \boldsymbol{u} > 0$), the mass density of the micro-element decreases; conversely, when the volume of the micro-element contracts ($\nabla \cdot \boldsymbol{u} < 0$), the mass density of the micro-element increases.

1.2.2 Law of Conservation of Momentum

Defining the force acting on the boundary S of the object with volume V as t_i, the volume force and the body couple acting on the unit mass of the object as $\boldsymbol{b}(\boldsymbol{r}, t)$ and $\boldsymbol{m}(\boldsymbol{r}, t)$ respectively, the resultant force acting on the object is

$$\boldsymbol{F} = \int_S \boldsymbol{t}\mathrm{d}S + \int_V \rho\boldsymbol{b}\mathrm{d}V. \tag{1.2.8}$$

The resultant moment of the body forces, the boundary traction and the body couple about the origin of the coordinates is [1]

$$\boldsymbol{M} = \int_S \boldsymbol{r} \times \boldsymbol{t}\mathrm{d}S + \int_V \rho\boldsymbol{r} \times \boldsymbol{b}\mathrm{d}V + \int_V \rho\boldsymbol{m}\mathrm{d}V. \tag{1.2.9}$$

Defining $\rho\dot{\boldsymbol{u}}$ as the linear momentum density per unit volume and $\boldsymbol{r} \times \rho\dot{\boldsymbol{u}}$ as the angular momentum density about the coordinate origin, the law of conservation of linear momentum and the law of conservation of angular momentum can be expressed as, respectively,

$$\frac{\mathrm{d}}{\mathrm{d}t}\int_V \rho\dot{\boldsymbol{u}}\mathrm{d}V = \boldsymbol{F}, \tag{1.2.10}$$

$$\frac{\mathrm{d}}{\mathrm{d}t}\int_V (\boldsymbol{r} \times \rho\dot{\boldsymbol{u}})\mathrm{d}V = \boldsymbol{M}. \tag{1.2.11}$$

Considering $t_i = \sigma_{ij}n_j$ (σ_{ij} is the Cauchy stress tensor), inserting Eq. (1.2.8) into Eq. (1.2.10) and using Gauss' law yields the equation of conservation of momentum.

$$\sigma_{ij,j} + \rho b_i = \rho\ddot{u}_i. \tag{1.2.12}$$

The component form of the law of conservation of angular momentum, i.e. Eq. (1.2.11), is

$$\frac{\mathrm{d}}{\mathrm{d}t}\int_V e_{ijk}x_j\rho\dot{u}_k\mathrm{d}V = \int_S e_{ijk}x_jt_k\mathrm{d}S + \rho\int_V e_{ijk}x_jb_k\mathrm{d}V + \rho\int_V m_i dV. \tag{1.2.13}$$

After using the divergence theorem, we get

$$\int_V e_{ijk}\left[\frac{\partial}{\partial x_l}(x_j\sigma_{lk}) + \rho x_jb_k - \rho\frac{\mathrm{d}}{\mathrm{d}t}(x_j\dot{u}_k)\right]\mathrm{d}V + \int_V \rho m_i\mathrm{d}V = 0. \tag{1.2.14}$$

Considering

$$e_{ijk}\frac{\partial}{\partial x_l}(x_j\sigma_{lk}) = e_{ijk}(\delta_{jl}\sigma_{lk} + x_j\sigma_{lk,l}),$$

$$e_{ijk}\rho\frac{\mathrm{d}}{\mathrm{d}t}(x_j\dot{u}_k) = \rho e_{ijk}(\dot{x}_j\dot{u}_k + x_j\ddot{u}_k) = \rho e_{ijk}x_j\ddot{u}_k,$$

and using Eq. (1.2.12), Eq. (1.2.14) becomes

$$\int_V (e_{ijk}\sigma_{jk} + \rho m_i)\mathrm{d}V = 0. \tag{1.2.15}$$

The differential form of the law of conservation of angular momentum can be expressed as

$$e_{ijk}\sigma_{jk} + \rho m_i = 0. \tag{1.2.16}$$

This means that the Cauchy stress tensor is not symmetrical anymore when the body couple is not zero. When the body couple is not included, the law of conservation of angular momentum requires that the stress tensor must be a symmetric tensor.

1.2.3 The Law of Conservation of Energy

Defining the kinetic energy per unit volume of the object as $T = \frac{1}{2}\rho\dot{u}_i\dot{u}_i$ and the deformation energy per unit volume as $A = \frac{1}{2}\sigma_{ij}\varepsilon_{ij}$, the law of conservation of energy can be expressed as

$$\frac{\mathrm{d}}{\mathrm{d}t}\int_V (T + A)\mathrm{d}V = \int_S t_i\dot{u}_i\mathrm{d}S + \int_V \rho b_i\dot{u}_i\mathrm{d}V. \tag{1.2.17}$$

The left side of the equals sign represents the rate of the time change rate of the total mechanical energy (kinetic energy plus deformation energy) of the object; the right side of the equals sign represents the work power done by the boundary traction and the volume force on the object. Using the divergence theorem, the differential form of the energy conservation theorem can be obtained

$$\frac{\mathrm{d}}{\mathrm{d}t}(T + A) = (\sigma_{ij}\dot{u}_i)_{,j} + \rho b_i\dot{u}_i. \tag{1.2.18}$$

The $\sigma_{ij}\dot{u}_i = I_j$ in the above equation is generally referred to as the energy flow density vector, which represents the inflow or outflow energy from the boundary of the micro-volume element. Therefore, the first term at the right side of the equal sign of Eq. (1.2.18) represents the energy exchange between the micro-volume element and the surrounding medium through the boundary, or the "source" of the energy flow density vector field.

1.3 Variational Principle of Elastodynamics

We know that the value of univariate or multivariate functions depends on the values of the independent variables. If the value of a function depends on the selection of one or more functions, the function is called functional, and the function on which the functional depends (corresponding to the independent variable of the ordinary function) is called the argument function of the functional. Mathematically, the argument functions that satisfy certain continuity conditions, boundary conditions and certain constraints are called admissible functions. The variational method is to find a specific function in a certain allowable function family, so that the given functional takes a stationary value. For example, selecting a curve so that its length is the shortest among all the curves that connect two given points is a variational problem. All curves connecting two given points here are the family of admissible functions, and the length of any curve is functional, which takes values that depend on the particular curve selected.

The state of a physical system can usually be described by some kind of scalar field, vector field or tensor field. A functional dependent on these physical fields can be established, so that in a series of possible states, the real state (corresponding to the equilibrium state of the physical system) makes the functional take a stationary value. Therefore, the governing equation describing a certain physical process can be obtained from the stationary value conditions of the functional. The principle of minimum potential energy and the principle of minimum complementary energy in elastic mechanics are the successful applications of the variational principle in solid mechanics. For elastodynamics, in addition to considering deformation energy and external force potential energy, the kinetic energy of the object should also be considered when establishing an energy functionals. In Sect. 1.2, we derive the governing equation Eq. (1.2.12) of the elastodynamic problem through the law of conservation of linear momentum. Below we use the variational principle to clarify that this governing equation can also be derived from the stationary value conditions of the Lagrangian energy functional.

The displacement field is used to express the state of a certain mechanical system under the action of external load, and the displacement fields that meets the following conditions are called the allowable displacement fields

(1) Boundary conditions

$$u_i(x, t) = u_i^s(x, t), x \in S_u, t > 0, \tag{1.3.1}$$

(2) Strain–displacement relationship

$$\varepsilon_{ij} = \frac{1}{2}(u_{i,j} + u_{j,i}), \tag{1.3.2}$$

(3) Stress–strain relationship

$$\sigma_{ij} = \lambda\varepsilon_{kk}\delta_{ij} + 2\mu\varepsilon_{ij}, \tag{1.3.3}$$

(4) Time node conditions

$$u_i(x,t)|_{t=0} = u_i^0(x), \tag{1.3.4a}$$

$$u_i(x,t)|_{t=t_1} = u_i^1(x), \tag{1.3.4b}$$

where $x \in S_u$ in Eq. (1.3.1) denotes the mass points on the boundary with the given displacement. The symbol $x \in S_\sigma$ will also be used below, which represents the mass points on the boundary with the given surface force.

The kinetic energy, deformation energy and external force work corresponding to any allowable displacement field can be expressed as, respectively [2],

$$T(u_i) = \int\limits_V \left[\frac{1}{2}\rho\dot{u}_i\dot{u}_i\right] dV, \tag{1.3.5}$$

$$A(u_i) = \int\limits_V \frac{1}{2}\sigma_{ij}\varepsilon_{ij} dV, \tag{1.3.6}$$

$$W(u_i) = \int\limits_{S_\sigma} \bar{t}_i u_i \mathrm{d}S + \int\limits_V \rho b_i u_i \mathrm{d}V. \tag{1.3.7}$$

Hamilton's variational principle can be stated as, in all allowable displacement fields, the actual displacement field $u_i^*(x,t)$ makes the functional which is expressed as

$$\Pi = \int\limits_0^{t_1} [T(u_i) - A(u_i) + W(u_i)]\mathrm{d}t, \tag{1.3.8}$$

take the stationary value, i.e.

$$\delta\Pi = 0. \tag{1.3.9}$$

The explicit expressions of the stationary value condition of $\delta\Pi = 0$ are derived as follow

$$\delta\Pi = \delta\int_0^{t_1}\int_V \left[\frac{1}{2}\rho\dot{u}_i\dot{u}_i - \frac{1}{2}\sigma_{ij}\varepsilon_{ij} + \rho b_i\right] dV dt + \delta\int_0^{t_1}\int_{S_\sigma} \bar{t}_i u_i \mathrm{d}S\mathrm{d}t, \tag{1.3.10}$$

where

$$\begin{aligned}
\delta\int_0^{t_1}\int_V \frac{1}{2}\rho\dot{u}_i\dot{u}_i \mathrm{d}V\mathrm{d}t &= \int_0^{t_1}\int_V \rho\dot{u}_i\delta\dot{u}_i \mathrm{d}V\mathrm{d}t \\
&= \int_V \left[\int_0^{t_1} \rho\frac{\mathrm{d}u_i}{\mathrm{d}t}\cdot\frac{\mathrm{d}\delta u_i}{\mathrm{d}t}\mathrm{d}t\right]\mathrm{d}V \\
&= \int_V \left\{\left[\rho\frac{\mathrm{d}u_i}{\mathrm{d}t}\delta u_i\right]_0^{t_1} - \int_0^{t_1}\rho\frac{\mathrm{d}^2u_i}{\mathrm{d}t^2}\delta u_i \mathrm{d}t\right\}\mathrm{d}V \\
&= -\int_0^{t_1}\int_V (\rho\ddot{u}_i\delta u_i)\mathrm{d}V\mathrm{d}t,
\end{aligned} \tag{1.3.11}$$

(The time node conditions of the allowable displacement field are used here.)

$$\begin{aligned}
\delta\int_0^{t_1}\int_V \left(\frac{1}{2}\sigma_{ij}\varepsilon_{ij}\right)\mathrm{d}V\mathrm{d}t &= \int_0^{t_1}\int_V (\sigma_{ij}\delta\varepsilon_{ij})\mathrm{d}V\mathrm{d}t \\
&= \int_0^{t_1}\int_V \left[\sigma_{ij}\cdot\frac{1}{2}(\delta u_{i,j} + \delta u_{j,i})\right]\mathrm{d}V\mathrm{d}t \\
&= \int_0^{t_1}\int_V \sigma_{ij}\delta u_{i,j}\mathrm{d}V\mathrm{d}t
\end{aligned}$$

(The symmetry of the stress tensor is used here.)

$$\begin{aligned}
&= \int_0^{t_1}\int_V \left[(\sigma_{ij}\delta u_i)_{,j} - \sigma_{ij,j}\delta u_i\right]\mathrm{d}V\mathrm{d}t \\
&= \int_0^{t_1}\left(\int_S \sigma_{ij}\delta u_i n_j \mathrm{d}S\right)\mathrm{d}t - \int_0^{t_1}\int_V \sigma_{ij,j}\delta u_i \mathrm{d}V\mathrm{d}t
\end{aligned}$$

(The divergence theorem is used here.)

$$= \int_0^{t_1}\int_{S_\sigma} \sigma_{ij}n_j\delta u_i \mathrm{d}S\mathrm{d}t - \int_0^{t_1}\int_V \sigma_{ij,j}\delta u_i \mathrm{d}V\mathrm{d}t$$

$$(\delta u_i = 0 \,\text{on}\, x \in S_u)$$
$$= \int_0^{t_1}\int_{S_\sigma} t_i \delta u_i \mathrm{d}S \mathrm{d}t - \int_0^{t_1}\int_V \sigma_{ij,j} \delta u_i \mathrm{d}V \mathrm{d}t. \quad (1.3.12)$$

Substituting Eq. (1.3.11) and Eq. (1.3.12) into Eq. (1.3.10) leads to

$$\delta\Pi = \int_0^{t_1}\int_V [\sigma_{ij,j} + \rho b_i - \rho \ddot{u}_i]\delta u_i \mathrm{d}V \mathrm{d}t + \int_0^{t_1}\int_{S_\sigma} (\bar{t}_i - t_i)\delta u_i \mathrm{d}S \mathrm{d}t = 0. \quad (1.3.13)$$

Taking into account the arbitrariness of δu_i, Eq. (1.3.13) requires

$$\sigma_{ij,j} + \rho b_i = \rho \ddot{u}_i, \text{ (inside the object } \boldsymbol{x} \in S_u) \quad (1.3.14)$$

$$t_i = \bar{t}_i, \text{ (on the boundary } \boldsymbol{x} \in S_\sigma \text{ with the given stress).} \quad (1.3.15)$$

Equation (1.3.14) is exactly the governing equation of the physical system to be sought.

1.4 The Initial Boundary Value Problem of Elastodynamics

Using the law of conservation of linear momentum or the variational principle of energy, the governing equations of elastodynamics can be obtained. The governing equations together with the constitutive equation reflecting the properties of the material and the initial and the boundary conditions of the physical problem constitutes the initial boundary value definite solution problem of elastodynamics.

Governing equation

$$\sigma_{ij,j} + \rho b_i = \rho \ddot{u}_i. \quad (1.4.1)$$

Constitutive equation

$$\sigma_{ij} = \lambda \varepsilon_{kk} \delta_{ij} + 2\mu \varepsilon_{ij}. \quad (1.4.2)$$

Geometric equation

$$\varepsilon_{ij} = \frac{1}{2}(u_{i,j} + u_{j,i}). \quad (1.4.3)$$

Initial conditions

$$u_i(x, t)|_{t=0} = u_i^0(x), \tag{1.4.4a}$$

$$\dot{u}_i(x, t)|_{t=0} = \dot{u}_i^0(x). \tag{1.4.4b}$$

Boundary conditions

$$u_i(x, t)|_{x \in S_u} = \overline{u}_i(x, t), \tag{1.4.5}$$

$$t_i(x, t)|_{x \in S_\sigma} = \overline{t}_i(x, t). \tag{1.4.6}$$

In the above formula, we adopt the Einstein summation convention: when a certain index appears twice in an item, it means that the index is summed from 1 to 3 (in the case of a three-dimensional index). The index that appear repeatedly is called **dummy index**. The index that does not appear repeatedly in one item is called the **free index**. The dummy index and free index have the following properties [3].

1. The occurrence of a duplicate index (dummy index) in one term of the expressions means that the index is summed. The index is not repeated in this item, which means 1, 2, 3 can be taken, respectively. In the example Eq. (1.4.1), $\sigma_{ij,j}$ means $\sum_{j=1}^{3} \sigma_{ij,j}$ (i can be arbitrarily selected as 1, 2, 3);
2. For different terms in the same equation, the number and signs of free indexes must be the same. For example, three items in Eq. (1.4.1) all have a free index i;
3. Replacing the dummy index symbol with another symbol does not affect the result. For example, changing j in Eq. (1.4.1) to k does not change the result of this item, that is, $\sigma_{ij,j} = \sigma_{ik,k}$;
4. Changing all the free indices in the same equation to other symbols does not affect the result. For example, changing all i in Eq. (1.4.1) to k does not change the meaning of the formula.

According to the nature of the dummy index and the free index, it can be known that the governing equation Eq. (1.4.1) is equivalent to three coupled differential equations; the constitutive equation Eq. (1.4.2) is equivalent to nine differential equations; the geometric equation Eq. (1.4.3) is equivalent to nine differential equations. These 21 differential equations involve a total of 21 physical quantities, namely three displacement components, nine components of stress and nine components of strain (only six independent components due to symmetry). Using geometrical relations and constitutive relations, stress and strain can be eliminated, and a governing equation expressed only by displacement can be obtained. Substitute Eq. (1.4.3) into Eq. (1.4.2) leads to

$$\sigma_{ij} = \lambda u_{k,k}\delta_{ij} + \mu\left(u_{i,j} + u_{j,i}\right). \tag{1.4.7}$$

Substituting the above formula into Eq. (1.4.1), we get

$$\lambda u_{k,kj}\delta_{ij} + \mu\left(u_{i,jj} + u_{j,ij}\right) + \rho b_i = \rho \ddot{u}_i,$$

$$\lambda u_{k,ki} + \mu(u_{i,jj} + u_{j,ji}) + \rho b_i = \rho \ddot{u}_i,$$

$$\lambda u_{j,ji} + \mu(u_{i,jj} + u_{j,ji}) + \rho b_i = \rho \ddot{u}_i,$$

$$(\lambda + \mu)u_{j,ji} + \mu u_{i,jj} + \rho b_i = \rho \ddot{u}_i. \tag{1.4.8}$$

It also can be written in vector form

$$(\lambda + \mu)\nabla(\nabla \cdot \boldsymbol{u}) + \mu \nabla^2 \boldsymbol{u} + \rho \boldsymbol{b} = \rho \ddot{\boldsymbol{u}}, \tag{1.4.9}$$

where ∇ and ∇^2 are the Hamilton differential operator and Laplace differential operators, namely

$$\nabla = \frac{\partial}{\partial x_i}()\boldsymbol{e}_i, \tag{1.4.10}$$

$$\nabla^2 = \frac{\partial^2}{\partial x_i \partial x_i}(). \tag{1.4.11}$$

Further, using the vector formula [3]

$$\nabla^2 \boldsymbol{u} = \nabla(\nabla \cdot \boldsymbol{u}) - \nabla \times \nabla \times \boldsymbol{u}, \tag{1.4.12}$$

Eq. (1.4.9) can be written in other form

$$(\lambda + 2\mu)\nabla(\nabla \cdot \boldsymbol{u}) - \mu \nabla \times \nabla \times \boldsymbol{u} + \rho \boldsymbol{b} = \rho \ddot{\boldsymbol{u}}. \tag{1.4.13}$$

This is the famous Navier equation. The fundamental problem of elastodynamics is to solve the Navier equation under the given initial and boundary value conditions. Once the displacement field is known, the strain field and stress field inside the object can be obtained by further using geometric relations and constitutive equations.

1.5 Transient and Steady-State Problems

Elastodynamics studies the time-dependent mechanical behavior of materials or structures. According to the characteristics of mechanical behavior changing with time, elastodynamic problems can be further divided into two categories: transient problems and steady-state problems. The transient problem means that the change of displacement field or stress field is a non-periodic function of time (such as displacement field and stress field under impact load); steady-state problem means that the change of displacement field or stress field is a periodic function of time (such as free vibration problem and wave propagation problem). The steady-state problem can be divided into the standing wave problem and the traveling wave problem. The standing wave problem is the problem of vibration (the velocity of the wave packet is zero); the traveling wave problem is the wave propagation problem (the wave packet has a certain velocity). The solution to the standing wave problem generally has the form $u(x, y, z, t) = A(x, y, z)e^{-i\omega t}$. The solution to the wave propagation problem generally has the form $u(x, y, z, t) = Ae^{i(\boldsymbol{k}\cdot\boldsymbol{r}-\omega t)}$, that is, the displacement field is also a periodic function with respect to space coordinates. For transient problems, the displacement field can generally only be expressed as $u(x, y, z, t)$. The following are some examples of one-dimensional transient and steady-state problems.

Example 1 Suppose a soft and thin string with the length l and the density ρ. Both ends of the thin string are fixed. The initial displacement and initial velocity of the string are given. The self-weight of the string can be ignored. Solve the elasto dynamics problem of the thin string. The elastodynamic problem can be expressed as

$$c^2u_{xx} - u_{tt} = 0 \quad (0 < x < l, t > 0), \tag{1.5.1a}$$

$$u(0, t) = 0 \quad (t \geq 0), \tag{1.5.1b}$$

$$u(l, t) = 0 \quad (t \geq 0), \tag{1.5.1c}$$

$$u(x, 0) = u_0(x) \quad (0 < x < l), \tag{1.5.1d}$$

$$u_t(x, 0) = v_0(x) \quad (0 < x < l), \tag{1.5.1e}$$

where $c^2 = \frac{T}{\rho}$ (T is the tension of the string). Here $u_t = \dot{u}$, $u_{tt} = \ddot{u}$, $u_{xx} = \partial^2 u/\partial x^2$.

The solution to this problem will be given in the next chapter. Since the displacement is always zero at the two end points $x = 0$ and $x = l$, the displacement distribution between the two end points, that is, the initial disturbance, will not spread to both sides with the increase of time and is always limited between the two end points. This is the standing wave problem. In the standing wave problem, each mass particle maintains the same phase during the vibration process, that is, reaches the maximum and minimum at the same time.

Example 2 Solve the elastodynamic problem of an infinitely long soft thin string under given initial conditions. The elastodynamic problem can be expressed as

$$c^2u_{xx} - u_{tt} = 0 \quad (t > 0), \tag{1.5.2a}$$

$$u(x, 0) = u_0(x), \quad \mathrm{x} \in (-\mathrm{a}, \mathrm{a}), \tag{1.5.2b}$$

$$u_t(x, 0) = v_0(x) \quad \mathrm{x} \in (-\mathrm{a}, \mathrm{a}). \tag{1.5.2c}$$

The process to get specific solution of example 2 will also be given in the next chapter. It can be seen that the initial disturbance will spread to the left and right sides of interval (−a, a) at a certain speed with the increase of time. This is the traveling wave problem. In the traveling wave problem, the phases of the various mass particles are different from each other during the vibration process, and they will not reach the maximum or minimum displacement at the same time. This is an important feature that distinguishes the wave problem from the vibration problem.

Example 3 Suppose a soft and thin string with the length l and the density ρ. Both ends of the thin string are fixed. The thin string rests in a horizontal position at the initial time. Now a uniform load $p(t) = at$ is applied on the string. Find the displacement and velocity of the string before it breaks. The elastodynamic problem can be expressed as

$$Tu_{xx} - \rho u_{tt} = p(t) \quad (0 < x < l, t > 0), \tag{1.5.3a}$$

$$u(0, t) = 0 \quad (t \geq 0), \tag{1.5.3b}$$

$$u(l, t) = 0 \quad (t \geq 0), \tag{1.5.3c}$$

$$u(x, 0) = 0 \quad (0 < x < l), \tag{1.5.3d}$$

$$u_t(x, 0) = 0 \quad (0 < x < l). \tag{1.5.3e}$$

Obviously, the displacement distribution $u(x, t)$ of the soft thin string before breaking always satisfies the characteristic that the two ends are zero and the middle is the largest. Different from example 1 and example 2, the displacement $u(x, t)$ that changes with time is no longer a periodic function of time. This problem is neither a vibration problem nor a wave motion problem, but a transient problem. If $p(t)$ is a periodic function of time, obviously $u(x, t)$ is also a periodic function of time. Then, the problem is a forced vibration problem. If $p(t)$ suddenly withdraws at a certain time t_0, after this time, the movement of the string becomes a problem of free vibration.

Standing waves can be seen as the result of mutual interference of traveling waves with the same frequency moving in opposite directions. The solution of the transient problem can also be understood as the result of the mutual interference of countless traveling waves of different frequencies. Therefore, transient problems are different from the steady-state problems on the one hang but also related with the steady-state problems on the other hand. The frequency spectrum analysis of the time domain signal provides a good interpretation of the interrelationship between the transient problem and the steady-state problem.

Chapter 2
Elastic Waves in an Infinite Medium

2.1 Scalar Potential and Vector Potential

The Navier equation Eq. (1.4.13) is a coupled system of differential equations about three displacement components, i.e. u_x, u_y and u_z. It is very difficult to solve Navier equation directly. For the convenience of solving Navier equation, the scalar potential $\varphi(\boldsymbol{x}, t)$ and the vector potential $\boldsymbol{\Psi}(\boldsymbol{x}, t)$ are introduced and the displacement field is thus expressed as [4]

$$\boldsymbol{u}(\boldsymbol{x}, t) = \nabla\varphi(\boldsymbol{x}, t) + \nabla \times \boldsymbol{\Psi}(\boldsymbol{x}, t) = \boldsymbol{u}^{(1)}(\boldsymbol{x}, t) + \boldsymbol{u}^{(2)}(\boldsymbol{x}, t). \tag{2.1.1}$$

According to the decomposition theorem of vector field, any single-value and finite vector field can be decomposed into a field without divergence plus a field without curl. There must be a scalar potential function $\varphi(\boldsymbol{x}, t)$ existing for the field without curl, which can be expressed as $\nabla\varphi(\boldsymbol{x}, t)$. And there must be a vector potential function $\boldsymbol{\Psi}(\boldsymbol{x}, t)$ existing for the field without divergence, which can be expressed as $\nabla \times \boldsymbol{\Psi}(\boldsymbol{x}, t)$. The displacement field is a vector field, which can be decomposed naturally. Therefore, Eq. (2.1.1) always holds. In the same way, the volumetric force $\boldsymbol{b}(\boldsymbol{x}, t)$ is also a vector field and can be decomposed similarly

$$\boldsymbol{b}(\boldsymbol{x}, t) = \nabla q(\boldsymbol{x}, t) + \nabla \times \boldsymbol{Q}(\boldsymbol{x}, t) = \boldsymbol{b}^{(1)}(\boldsymbol{x}, t) + \boldsymbol{b}^{(2)}(\boldsymbol{x}, t). \tag{2.1.2}$$

By substituting Eqs. (2.1.1) and (2.1.2) into the Navier equation Eq. (1.4.9), the following equation can be obtained

$$(\lambda + \mu)\nabla[\nabla \cdot (\nabla\varphi + \nabla \times \boldsymbol{\Psi})] + \mu\nabla^2(\nabla\varphi + \nabla \times \boldsymbol{\Psi}) + \rho(\nabla q + \nabla \times \boldsymbol{Q})$$

$$= \rho\left(\nabla\ddot{\varphi} + \nabla \times \ddot{\boldsymbol{\Psi}}\right). \tag{2.1.3}$$

After merging similar terms, we obtain

© Science Press 2022
P. Wei, *Theory of Elastic Waves*,
https://doi.org/10.1007/978-981-19-5662-1_2

$$\nabla\left[(\lambda + 2\mu)\nabla^2\varphi + \rho q - \rho\ddot{\varphi}\right] + \nabla \times \left[\mu\nabla^2\boldsymbol{\Psi} + \rho\boldsymbol{Q} - \rho\ddot{\boldsymbol{\Psi}}\right] = 0. \tag{2.1.4}$$

In the derivation of the above formula, the condition without divergence is used, i.e.

$$\nabla \cdot (\nabla \times \boldsymbol{\Psi}) = 0. \tag{2.1.5}$$

The conditions that Eq. (2.1.4) holds are

$$(\lambda + 2\mu)\nabla^2\varphi + \rho q - \rho\ddot{\varphi} = c(t) \tag{2.1.6}$$

and

$$\mu\nabla^2\boldsymbol{\Psi} + \rho\boldsymbol{Q} - \rho\ddot{\boldsymbol{\Psi}} = \boldsymbol{A}(\boldsymbol{x}, t) = \nabla\phi(\boldsymbol{x}, t), \tag{2.1.7}$$

where $c(t)$ is a uniform field independent of coordinates, and $\boldsymbol{A}(\boldsymbol{x}, t)$ is a field without curl. It should be pointed out that the theorem of vector decomposition only emphasizes the existence of scalar potential $\varphi(\boldsymbol{x}, t)$ and vector potential $\boldsymbol{\Psi}(\boldsymbol{x}, t)$, but does not emphasize the uniqueness of $\varphi(\boldsymbol{x}, t)$ and $\boldsymbol{\Psi}(\boldsymbol{x}, t)$. In general, a vector field is determined completely by three scalar fields. However, Eq. (2.1.1) includes four scalar fields, i.e. $\varphi(\boldsymbol{x}, t)$ and three components of $\boldsymbol{\Psi}(\boldsymbol{x}, t)$. This implies that there is a large freedom for the selection of $\varphi(\boldsymbol{x}, t)$ and $\boldsymbol{\Psi}(\boldsymbol{x}, t)$ which satisfy the requirements of the equation. In order to eliminate a degree of freedom, the standard condition

$$\nabla \cdot \boldsymbol{\Psi} = 0, \tag{2.1.8}$$

is usually introduced. Similarly, there is

$$\nabla \cdot \boldsymbol{Q} = 0. \tag{2.1.9}$$

Performing divergence operation on both sides of Eq. (2.1.7) and considering the standard condition Eq. (2.1.8), we know that the field without curl, $\boldsymbol{A}(\boldsymbol{x}, t)$, in Eq. (2.1.7) also satisfies

$$\nabla \cdot \boldsymbol{A}(\boldsymbol{x}, t) = 0. \tag{2.1.10}$$

This indicates that the vector field $\boldsymbol{A}(\boldsymbol{x}, t)$ has neither curl nor divergence, so it is a harmonic field, which can be expressed as $\nabla\phi(\boldsymbol{x}, t)$ by the harmonic function $\phi(x, t)$, and the harmonic function $\phi(\boldsymbol{x}, t)$ satisfies

$$\nabla^2\phi(\boldsymbol{x}, t) = 0. \tag{2.1.11}$$

Even if the standard condition is introduced, due to the arbitrariness of $c(t)$ and $\phi(\boldsymbol{x}, t)$, the solutions of Eqs. (2.1.6) and (2.1.7) are still non-unique. According to the solution structure of differential equations, it can be generally expressed as

$$\varphi(\boldsymbol{x}, t) = \varphi^0(\boldsymbol{x}, t) + \varphi^*(\boldsymbol{x}, t), \tag{2.1.12}$$

and

$$\boldsymbol{\Psi}(\boldsymbol{x}, t) = \boldsymbol{\Psi}^0(\boldsymbol{x}, t) + \boldsymbol{\Psi}^*(\boldsymbol{x}, t), \tag{2.1.13}$$

where φ^0 and $\boldsymbol{\Psi}^0$ correspond to the general solution of the homogeneous equation, and φ^* and $\boldsymbol{\Psi}^*$ depend on the selection of functions $c(t)$ and $\varphi(\boldsymbol{x}, t)$, which are the particular solution of the inhomogeneous equation.

Considering

$$\boldsymbol{u}^{(1)} = \nabla\varphi^0 + \nabla\varphi^*, \tag{2.1.14}$$

$$\boldsymbol{u}^{(2)} = \nabla \times \boldsymbol{\psi}^0 + \nabla \times \boldsymbol{\psi}^*, \tag{2.1.15}$$

$\nabla\varphi^0$ and $\nabla\varphi^*$ satisfy the same equation, $\nabla \times \boldsymbol{\psi}^0$ and $\nabla \times \boldsymbol{\psi}^*$ satisfy the same equation. Therefore, considering φ^* and $\boldsymbol{\psi}^*$ doesn't expand the function space of $\boldsymbol{u}^{(1)}$ and $\boldsymbol{u}^{(2)}$, so we just have to figure out φ^0 and $\boldsymbol{\psi}^0$.

Performing the gradient and curl operations on Eqs. (2.1.6) and (2.1.7), respectively, leads to

$$(\lambda + 2\mu)\nabla^2\boldsymbol{u}^{(1)}(\boldsymbol{x}, t) + \rho\boldsymbol{b}^{(1)}(\boldsymbol{x}, t) - \rho\ddot{\boldsymbol{u}}^{(1)}(\boldsymbol{x}, t) = 0, \tag{2.1.16a}$$

and

$$\mu\nabla^2\boldsymbol{u}^{(2)}(\boldsymbol{x}, t) + \rho\boldsymbol{b}^{(2)}(\boldsymbol{x}, t) - \rho\ddot{\boldsymbol{u}}^{(2)}(\boldsymbol{x}, t) = 0. \tag{2.1.16b}$$

When the volumetric force is ignored, the above two expressions can be further simplified as

$$c_{\mathrm{p}}^2\nabla^2\boldsymbol{u}^{(1)}(\boldsymbol{x}, t) = \ddot{\boldsymbol{u}}^{(1)}(\boldsymbol{x}, t), \tag{2.1.17a}$$

$$c_{\mathrm{s}}^2\nabla^2\boldsymbol{u}^{(2)}(\boldsymbol{x}, t) = \ddot{\boldsymbol{u}}^{(2)}(\boldsymbol{x}, t), \tag{2.1.17b}$$

where

$$c_{\mathrm{p}}^2 = \frac{\lambda + 2\mu}{\rho}, \tag{2.1.18a}$$

$$c_{\mathrm{s}}^2 = \frac{\mu}{\rho}. \tag{2.1.18b}$$

Thus, it can be seen that each component of $\boldsymbol{u}^{(1)}(\boldsymbol{x}, t)$ and $\boldsymbol{u}^{(2)}(\boldsymbol{x}, t)$, as well as the scalar potential $\varphi(\boldsymbol{x}, t)$ and the vector potential $\boldsymbol{\Psi}(\boldsymbol{x}, t)$ satisfy the equation of the same form, i.e.

$$c^2\nabla^2\boldsymbol{F} = \ddot{\boldsymbol{F}}, \tag{2.1.19}$$

where $\boldsymbol{F}$ can be either a vector or a scalar. For steady-state problems, all mechanical quantities, including displacement field, strain field, stress field, etc., are harmonic functions of time; that is, they all contain a time factor $e^{-i\omega t}$. So the above equation can be further rewritten

$$\nabla^2\boldsymbol{F} + k^2\ddot{\boldsymbol{F}} = 0, \tag{2.1.20}$$

where $k^2 = \omega^2/c^2$. Equations (2.1.19) and (2.1.20) play an important role in elastodynamics and are basic field equations of elastodynamics. In particular, Eq. (2.1.20) is often referred to as the elastic wave equation. Performing divergence and curl operations on Eqs. (2.1.17a) and (2.1.17b), respectively, and introducing the volumetric strain and the angular displacement vector (the axis vector of rotational deformation) of volumetric element, i.e.

$$\theta = \nabla \cdot \boldsymbol{u}^{(1)}(\boldsymbol{x}, t), \tag{2.1.21a}$$

$$\boldsymbol{\Omega} = \frac{1}{2}\nabla \times \boldsymbol{u}^{(2)}(\boldsymbol{x}, t). \tag{2.1.21b}$$

It is noted that the volumetric strain θ and the angular displacement vector $\boldsymbol{\Omega}$ also satisfy the wave equation, i.e.

$$c_{\mathrm{p}}^2\nabla^2\theta = \ddot{\theta}(\boldsymbol{x}, t), \tag{2.1.22a}$$

$$c_{\mathrm{s}}^2\nabla^2\boldsymbol{\Omega} = \ddot{\boldsymbol{\Omega}}(\boldsymbol{x}, t). \tag{2.1.22b}$$

In the rectangular coordinate system, the general form of the solution of the wave equation Eq. (2.1.20) can be expressed as

$$\boldsymbol{F}(\boldsymbol{r}, t) = \boldsymbol{A}e^{i(\boldsymbol{k}\cdot\boldsymbol{r}+\omega t)}. \tag{2.1.23}$$

In the one-dimensional case, it can be expressed as

$$\boldsymbol{F}(\boldsymbol{x}, t) = \boldsymbol{A}e^{i(\boldsymbol{k}\cdot\boldsymbol{x}+\omega t)}, \tag{2.1.24}$$

where the parameters k (module of the vector $\boldsymbol{k}$) and ω satisfy the following relation, i.e.

$$k^2 = \frac{\omega^2}{c^2}. \tag{2.1.25}$$

This is often referred to as the **dispersion equation** of elastic wave. Given a series of values of time t = $t_0, t_1, t_2, \ldots$, corresponding curves for $f = ae^{i(\boldsymbol{k}\cdot\boldsymbol{x}-\omega t)}$ and $g = be^{i(\boldsymbol{k}\cdot\boldsymbol{x}+\omega t)}$ can be drawn. It can be seen that these curves are actually sinusoid and cosinusoid curves moving in opposite directions, commonly referred to as **left** and **right traveling waves**, where a and b represent the vibration amplitude of particle. $\boldsymbol{k}$ is called the **wave vector**, and its direction represents the propagation direction of the wave while its magnitude is called the **wave number**, and its relation with the wavelength λ is $k = \frac{2\pi}{\lambda}$. ω is called the **circular frequency**, which represents the vibration frequency of particle, and its relation to the period T is $\omega = \frac{2\pi}{T}$. For the displacement field $\boldsymbol{u}(\boldsymbol{r}, t) = \boldsymbol{a}e^{i(\boldsymbol{k}\cdot\boldsymbol{r}+\omega t)}$, consider that $\boldsymbol{u}^{(1)}(\boldsymbol{x}, t)$ is a field without curl, it is required

$$\nabla \times \boldsymbol{u}^{(1)} = i\,\boldsymbol{k} \times \left[\boldsymbol{a}e^{i(\boldsymbol{k}\cdot\boldsymbol{r}-\omega t)}\right] = \boldsymbol{0}. \tag{2.1.26}$$

Thus, $\boldsymbol{a}$ must be in the same direction as $\boldsymbol{k}$; that is, the vibration direction (also known as the direction of polarization) of particle is the same as the propagation direction of wave. The kind of wave that satisfies such condition is called **longitudinal wave**. Consider that $\boldsymbol{u}^{(2)}(\boldsymbol{x}, t)$ is without divergence, namely

$$\nabla \cdot \boldsymbol{u}^{(2)} = i\boldsymbol{k} \cdot \left[\boldsymbol{a}e^{i(\boldsymbol{k}\cdot\boldsymbol{r}-\omega t)}\right] = 0. \tag{2.1.27}$$

Therefore, $\boldsymbol{a}$ must be perpendicular to $\boldsymbol{k}$, and such a wave is called a **transverse wave.**

2.2 Solution of Wave Equation

Firstly, let's discuss the solution of one-dimensional wave equation, i.e.

$$\frac{\partial^2 \varphi}{\partial x^2} - \frac{1}{c^2}\frac{\partial^2 \varphi}{\partial t^2} = 0. \tag{2.2.1}$$

Rewrite Eq. (2.2.1) as

$$\left(\frac{\partial}{\partial x} - \frac{1}{c}\frac{\partial}{\partial t}\right)\left(\frac{\partial}{\partial x} + \frac{1}{c}\frac{\partial}{\partial t}\right)\varphi = 0. \tag{2.2.2}$$

Introducing new variables ξ and η, i.e.

$$\xi = x - ct, \tag{2.2.3a}$$

$$\eta = x + ct, \tag{2.2.3b}$$

and according to the derivative rule of function of functions, we can obtain [3]

$$\frac{\partial}{\partial x} = \frac{\partial}{\partial \xi}\frac{\partial \xi}{\partial x} + \frac{\partial}{\partial \eta}\frac{\partial \eta}{\partial x} = \frac{\partial}{\partial \xi} + \frac{\partial}{\partial \eta}, \tag{2.2.4a}$$

$$\frac{\partial}{\partial t} = \frac{\partial}{\partial \xi}\frac{\partial \xi}{\partial t} + \frac{\partial}{\partial \eta}\frac{\partial \eta}{\partial t} = c\left(\frac{\partial}{\partial \eta} - \frac{\partial}{\partial \xi}\right). \tag{2.2.4b}$$

From the above two equations, we can know

$$\frac{\partial}{\partial \xi} = \frac{1}{2}\left(\frac{\partial}{\partial x} - \frac{1}{c}\frac{\partial}{\partial t}\right), \tag{2.2.5a}$$

$$\frac{\partial}{\partial \eta} = \frac{1}{2}\left(\frac{\partial}{\partial x} + \frac{1}{c}\frac{\partial}{\partial t}\right). \tag{2.2.5b}$$

Using Eqs. (2.2.5a) and (2.2.5b), Eq. (2.2.2) can be written as

$$\frac{\partial}{\partial \xi}\left(\frac{\partial \varphi}{\partial \eta}\right) = \frac{\partial^2 \varphi}{\partial \xi \partial \eta} = 0. \tag{2.2.6}$$

The above equation shows that $\frac{\partial \varphi}{\partial \eta}$ is independent of ξ and is only a function of η. Let

$$\frac{\partial \varphi}{\partial \eta} = f(\eta). \tag{2.2.7}$$

Performing integral operation on the variable η on both sides of the above equation, we can obtain

$$\varphi(\xi, \eta) = \int f(\eta)\mathrm{d}\eta + \varphi_2(\xi) = \varphi_1(\eta) + \varphi_2(\xi). \tag{2.2.8}$$

After recovering variables ξ and η in the above equation to variables x and t, we get

$$\varphi(x, t) = \varphi_1(x + ct) + \varphi_2(x - ct), \tag{2.2.9}$$

where $\varphi_1(x + ct)$ and $\varphi_2(x - ct)$ are functions of arguments $x + ct$ and $x - ct$, respectively.

Similarly, the solution of the three-dimensional wave equation, i.e.

$$\frac{\partial^2 \varphi}{\partial x_j \partial x_j} - \frac{1}{c^2}\frac{\partial^2 \varphi}{\partial t^2} = 0, \tag{2.2.10}$$

can be discussed. The above equation can be written as

$$\left(\frac{\partial}{\partial x_j} - \frac{1}{c}\frac{\partial}{\partial t}\right)\left(\frac{\partial}{\partial x_j} + \frac{1}{c}\frac{\partial}{\partial t}\right)\varphi = 0. \tag{2.2.11}$$

Introducing new variables ξ and η, i.e.

$$\xi = n_j x_j - ct, \tag{2.2.12a}$$

$$\eta = n_j x_j + ct. \tag{2.2.12b}$$

According to the derivative rule of function of functions, we get

$$\frac{\partial}{\partial x_j} = \frac{\partial}{\partial \xi}\frac{\partial \xi}{\partial x_j} + \frac{\partial}{\partial \eta}\frac{\partial \eta}{\partial x_j} = n_j\frac{\partial}{\partial \xi} + n_j\frac{\partial}{\partial \eta}, \tag{2.2.13a}$$

$$\frac{\partial}{\partial t} = \frac{\partial}{\partial \xi}\frac{\partial \xi}{\partial t} + \frac{\partial}{\partial \eta}\frac{\partial \eta}{\partial t} = -c\frac{\partial}{\partial \xi} + c\frac{\partial}{\partial \eta}. \tag{2.2.13b}$$

From the above two equations, we further get

$$\left(\frac{\partial}{\partial x_j} - \frac{1}{c}\frac{\partial}{\partial t}\right) = n_j\frac{\partial}{\partial \xi} + n_j\frac{\partial}{\partial \eta} + \frac{\partial}{\partial \xi} - \frac{\partial}{\partial \eta}, \tag{2.2.14a}$$

$$\left(\frac{\partial}{\partial x_j} + \frac{1}{c}\frac{\partial}{\partial t}\right) = n_j\frac{\partial}{\partial \xi} + n_j\frac{\partial}{\partial \eta} - \frac{\partial}{\partial \xi} + \frac{\partial}{\partial \eta}. \tag{2.2.14b}$$

Thus,

$$\left(\frac{\partial}{\partial x_j} - \frac{1}{c}\frac{\partial}{\partial t}\right)\left(\frac{\partial}{\partial x_j} + \frac{1}{c}\frac{\partial}{\partial t}\right)\varphi = (n_j\frac{\partial \varphi}{\partial \xi} + n_j\frac{\partial \varphi}{\partial \eta})^2 - (\frac{\partial \varphi}{\partial \xi} - \frac{\partial \varphi}{\partial \eta})^2$$
$$= 4\frac{\partial^2 \varphi}{\partial \xi \partial \eta}. \tag{2.2.15}$$

Substituting the above equation into Eq. (2.2.11) leads to

$$\frac{\partial^2 \varphi}{\partial \xi \partial \eta} = 0. \tag{2.2.16}$$

The solution of the above equation is

$$\varphi(\xi, \eta) = \varphi_1(\xi) + \varphi_2(\eta). \tag{2.2.17}$$

After recovering variables ξ and η in the above equation to variables x and t, we get

$$\varphi(x,t) = \varphi_1(n_i x_i - ct) + \varphi_2(n_i x_i + ct). \tag{2.2.18}$$

Let

$$k = \frac{\omega}{c}. \tag{2.2.19}$$

Equation (2.2.18) can also be rewritten as

$$\begin{aligned}\varphi(x,t) &= \varphi_1(kn_i x_i - \omega t) + \varphi_2(kn_i x_i + \omega t)\\ &= \varphi_1(\boldsymbol{k}\cdot\boldsymbol{r} - \omega t) + \varphi_2(\boldsymbol{k}\cdot\boldsymbol{r} + \omega t),\end{aligned} \tag{2.2.20}$$

where

$$\boldsymbol{k} = kn_1\boldsymbol{e}_1 + kn_2\boldsymbol{e}_2 + kn_2\boldsymbol{e}_3, \tag{2.2.21}$$

and

$$\boldsymbol{r} = x_1\boldsymbol{e}_1 + x_2\boldsymbol{e}_2 + x_2\boldsymbol{e}_3. \tag{2.2.22}$$

$\boldsymbol{k}$ and $\boldsymbol{r}$ are, respectively, the wave vector and the position vector of any point in space. $(\boldsymbol{e}_1, \boldsymbol{e}_2, \boldsymbol{e}_3)$ is the unit basis vector in the spatial coordinate system.

Example 1 Let the length of the soft thin string be l, the density be ρ, and the two ends be fixed. The initial displacement and velocity of the string are given. The self-weight of the string is ignored. Solving the elastodynamic problem, i.e.

$$c^2 u_{xx} - u_{tt} = 0 \quad (0 < x < l, t > 0), \tag{2.2.23a}$$

$$u(0,t) = 0 \quad (t \geq 0), \tag{2.2.23b}$$

$$u(l,t) = 0 \quad (t \geq 0), \tag{2.2.23c}$$

$$u(x,0) = u_0(x) \quad (0 < x < l), \tag{2.2.23d}$$

$$u_t(x,0) = v_0(x) \quad (0 < x < l), \tag{2.2.23e}$$

where $c^2 = \frac{T}{\rho}$ (T is the tension of the string), u_{xx} and u_{tt} represent the twice partial derivatives with respect to coordinates x and time t, respectively.

Solution: Let the solution of the equation be

$$u(x,t) = X(x)T(t). \tag{2.2.24}$$

Substituting it into Eq. (2.2.23a), we can obtain

$$c^2 X''(x)T(t) - X(x)T''(t) = 0. \tag{2.2.25}$$

Divide both sides by $c^2 X(x)T(t)$, and we can obtain

$$\frac{X''(x)}{X(x)} = \frac{T''(t)}{c^2 T(t)}. \tag{2.2.26}$$

Note that the left-hand side of the equation is a function of the coordinate x and the right-hand side is a function of the time. Equation (2.2.26) can only be true if the left-hand side of the equation = the right-hand side of the equation = constant.

Let this constant be $-\lambda$, then Eq. (2.2.26) is decomposed into the following two equations, i.e.

$$X''(x) + \lambda X(x) = 0, \tag{2.2.27}$$

and

$$T''(t) + \lambda c^2 T(t) = 0. \tag{2.2.28}$$

Substituting the boundary condition into Eq. (2.2.24) leads to

$$X(0)T(t) = 0, \tag{2.2.29}$$

$$X(l)T(t) = 0. \tag{2.2.30}$$

Considering that the arbitrariness of $T(t)$, $X(x)$ has to satisfy

$$X(0) = 0, \tag{2.2.31a}$$

$$X(l) = 0. \tag{2.2.31b}$$

The combination of Eqs. (2.2.27) and (2.2.31a) constitutes the definite solution problem of ordinary differential equations. According to the knowledge of ordinary differential equations, the solution can be expressed as

$$X(x) = \begin{cases} Ae^{\sqrt{-\lambda}x} + Be^{-\sqrt{-\lambda}x} & (\lambda < 0) \\ Ax + B & (\lambda = 0) \\ Ae^{i\sqrt{\lambda}x} + Be^{-i\sqrt{\lambda}x} & (\lambda > 0) \end{cases}. \tag{2.2.32}$$

When $\lambda < 0$, the boundary condition requires

$$X(0) = A + B = 0, \tag{2.2.33}$$

and

$$X(l) = Ae^{\sqrt{-\lambda}l} + Be^{-\sqrt{-\lambda}l} = 0, \tag{2.2.34}$$

which results in

$$A = B = 0.$$

When $\lambda = 0$, the boundary condition requires

$$X(0) = B = 0, \tag{2.2.35}$$

and

$$X(l) = Al + B = 0, \tag{2.2.36}$$

which also results in

$$A = B = 0. \tag{2.2.37}$$

In both cases, $\lambda < 0$ and $\lambda = 0$, $X(x)$ can only be zero solution, thus

$$u(x, t) = X(x)T(t) \equiv 0. \tag{2.2.38}$$

In order to obtain a nontrivial solution to the elastodynamics problem, let's consider the third case.

When $\lambda > 0$, the boundary condition requires

$$X(0) = A + B = 0, \tag{2.2.39}$$

and

$$X(l) = Ae^{i\sqrt{\lambda}l} + Be^{-i\sqrt{\lambda}l} = 0. \tag{2.2.40}$$

Substituting Eq. (2.2.39) into Eq. (2.2.40) leads to

$$A\left(e^{i\sqrt{\lambda}l} - e^{-i\sqrt{\lambda}l}\right) = 0. \tag{2.2.41}$$

When $A \neq 0$, the above equation is equivalent to the requirement

$$\sin\sqrt{\lambda}l = 0. \tag{2.2.42}$$

The above equation is true only if λ has to take the discrete values, i.e.

$$\lambda = \left(\frac{n\pi}{l}\right)^2 \quad (n = \pm 1, \pm 2, \ldots). \tag{2.2.43}$$

Note that Eq. (2.2.27) has the same form as Eq. (2.2.28), thus

$$T(t) = Ce^{i\sqrt{\lambda}ct} + De^{-i\sqrt{\lambda}ct}. \tag{2.2.44}$$

Therefore, we have the solution of the elastodynamics problem of the thin string

$$\begin{aligned} u_n(x,t) = X_n(x)T_n(t) &= \left(A_n e^{i\frac{n\pi}{l}x} + B_n e^{-i\frac{n\pi}{l}x}\right)\left(C_n e^{i\frac{n\pi}{l}ct} + D_n e^{-i\frac{n\pi}{l}ct}\right) \\ &= A_n' e^{i\frac{n\pi}{l}(x+ct)} + B_n' e^{-i\frac{n\pi}{l}(x-ct)} \\ &\quad + C_n' e^{i\frac{n\pi}{l}(x-ct)} + D_n' e^{-i\frac{n\pi}{l}(x+ct)}. \end{aligned} \tag{2.2.45}$$

Since the governing equation of the elastodynamics problem Eq. (2.2.23) is linear, the superposition principle holds, and thus the general solution of the elastodynamics problem can be generally expressed as

$$u(x,t) = \sum_{n=\pm 1}^{\infty}\left[E_n e^{i\frac{n\pi}{l}(x+ct)} + F_n e^{-i\frac{n\pi}{l}(x-ct)}\right], \tag{2.2.46}$$

where E_n and F_n are combination coefficients whose values are determined by initial conditions, i.e.

$$u(x,0) = \sum_{n=\pm 1}^{\infty}\left(E_n e^{i\frac{n\pi}{l}x} + F_n e^{-i\frac{n\pi}{l}x}\right) = u_0(x), \tag{2.2.47}$$

and

$$u_t(x,0) = \sum_{n=\pm 1}^{\infty} i\frac{n\pi}{l}c\left(E_n e^{i\frac{n\pi}{l}x} + F_n e^{-i\frac{n\pi}{l}x}\right) = v_0(x). \tag{2.2.48}$$

Let's discuss the solution of the elastodynamics problem of a finite-length string with two fixed ends, which is expressed by Eq. (2.2.46).

(1) Let $f_n = E_n e^{i\frac{n\pi}{l}(x+ct)}$ and $g_n = F_n e^{-i\frac{n\pi}{l}(x-ct)}$, then both f_n and g_n are periodic functions of spatial coordinates x and time coordinates t, and their linear combination forms the general solution of the elastodynamics problem. So $u(x,t)$ is also a periodic function of x and t.

(2) Given different values of time t, and plot the curves of f_n and g_n. We can see that f_n and g_n are actually sine and cosine curves moving in opposite directions, and they are usually called **left** and **right traveling waves**, where E_n and F_n are called the **amplitude of the wave**, $k_n = \frac{n\pi}{l}$ is called the **wave number**, and its relationship with wavelength λ_n is $\lambda_n = \frac{2\pi}{k_n}$, $\omega_n = \frac{n\pi}{l}c$ is called the **circular frequency of the wave**, c is the moving speed of the left or right traveling wave, and is called the **propagation speed of the wave.**
(3) The general solution $u(x, t)$ of the elastodynamics problem is obtained by the superposition of the left travelling wave f_n and the right travelling wave g_n (usually called **monochromatic wave**) of different circular frequencies ω_n, wave numbers k_n and different amplitudes. $u(x, t)$ is called the **wave packet**. Its shape is the result of interference of these monochromatic waves, f_n and g_n, of different frequencies.
(4) For a bounded string, due to the limitation of the boundary condition,i.e. $u(0, t) = u(l, t) = 0$, the wave packet $u(x, t)$ has two nodes at $x = 0$ and $x = l$. With the increase of time t, the shape of the wave packet changes, but the position remains unchanged. Such kind of wave packet is called the **standing wave**. The standing wave problem is also called the **vibration problem**. In fact, Eq. (2.2.46) can also be written as

$$
\begin{aligned}
u(x,t) &= \sum_{n=\pm 1}^{\infty}\left[E_n e^{i\frac{n\pi}{l}x} + F_n e^{-i\frac{n\pi}{l}x}\right]e^{i\omega t} \\
&= e^{i\omega t}\sum_{n=\pm 1}^{\infty} G_n e^{i(k_n x+\theta_n)} \\
&= e^{i\omega t}F(x) \quad (k_n = n\pi/l)
\end{aligned},
$$

where $F(x)$ is the wave packet whose shape does not change with time but whose amplitude changes with time. For the soft thin string fixed at both ends, if the initial displacement $u_0(x) = 2\sin\frac{\pi x}{l}$ and the initial velocity $v_0(x) = -2\frac{\pi c}{l}\sin\frac{\pi x}{l}$, then, the change of the wave packet with time is shown in Fig. 2.1.

Example 2 Solving the elastodynamics problem of an infinitely long soft thin string under given initial conditions, i.e.

$$c^2 u_{xx} - u_{tt} = 0 \quad (t > 0), \tag{2.2.49a}$$

$$u(x, 0) = u_0(x), \tag{2.2.49b}$$

$$u_t(x, 0) = v_0(x). \tag{2.2.49c}$$

Solution: Let the solution of the equation be

$$u(x,t) = X(x)T(t). \tag{2.2.50}$$

Substituting the above equation into Eq. (2.2.49a), similar to Example 1, we get

$$X''(x) + \lambda X(x) = 0, \tag{2.2.51}$$

$$T''(t) + \lambda c^2 T(t) = 0. \tag{2.2.52}$$

Their solutions are, respectively,

$$X(x) = Ae^{i\sqrt{\lambda}x} + Be^{-i\sqrt{\lambda}x}, \tag{2.2.53}$$

and

$$T(t) = Ce^{i\sqrt{\lambda}ct} + De^{-i\sqrt{\lambda}ct}. \tag{2.2.54}$$

For infinite long and thin strings, since there is no boundary condition, the value of λ should be continuous, that is, any value within $(0, +\infty)$, not necessary to be discrete values. Thus, the solution of the elastodynamics problem can be expressed as [3].

$$\begin{aligned} u(x,t) &= \int_{-\infty}^{+\infty} \left[A(k)e^{ik(x+ct)} + B(k)e^{ik(x-ct)}\right]\mathrm{d}k \quad \left(k = \sqrt{\lambda}\right) \\ &= \int_{-\infty}^{+\infty} \left[A(\omega)e^{i(kx+\omega t)} + B(\omega)e^{i(kx-\omega t)}\right]\mathrm{d}\omega, \quad (\omega = kc) \end{aligned} \tag{2.2.55}$$

where the values of $A(\omega)$ and $B(\omega)$ are determined by the initial conditions which are

$$\int_{-\infty}^{+\infty} [A(k) + B(k)]e^{ikx}\mathrm{d}k = u_0(x), \tag{2.2.56}$$

and

$$\int_{-\infty}^{+\infty} [A(k) - B(k)]ikce^{ikx}\mathrm{d}k = v_0(x). \tag{2.2.57}$$

Let

$$\overline{u}_0(k) = \int_{-\infty}^{+\infty} u_0(x)e^{-ikx}\mathrm{d}x, \tag{2.2.58}$$

$$\overline{v}_0(k) = \int_{-\infty}^{+\infty} v_0(x)e^{-ikx}\mathrm{d}x, \tag{2.2.59}$$

where $\overline{u}_0(k)$ and $\overline{v}_0(k)$ are the Fourier transforms of $u_0(x)$ and $v_0(x)$, respectively, then

$$A(k) = \frac{1}{2}\overline{u}_0(k) + \frac{1}{2c}\cdot\frac{1}{ik}\overline{v}_0(k), \tag{2.2.60}$$

$$B(k) = \frac{1}{2}\overline{u}_0(k) - \frac{1}{2c}\cdot\frac{1}{ik}\overline{v}_0(k). \tag{2.2.61}$$

Consider the following properties of the Fourier transform

1) If $f(x) = \int g(x)dx$, then

$$\overline{f}(k) = \frac{1}{ik}\overline{g}(k). \tag{2.2.62}$$

2) If $\overline{f}(k) = \int_{-\infty}^{+\infty} f(x)e^{ikx}dx$, then

$$\overline{f}(k)e^{ikct} = \int_{-\infty}^{+\infty} f(x+ct)e^{-ikx}\mathrm{d}x. \tag{2.2.63}$$

Equation (2.2.55) can be further written as

$$\begin{aligned}
u(x,t) &= \int_{-\infty}^{+\infty}\left[A(k)e^{ik(x+ct)} + B(k)e^{ik(x-ct)}\right]\mathrm{d}k \\
&= \int_{-\infty}^{+\infty}\left[\frac{1}{2}\overline{u}_0(k)e^{ikct}\right]e^{ikx}\mathrm{d}k + \frac{1}{2c}\int_{-\infty}^{+\infty}\left[\frac{1}{ik}\overline{v}_0(k)e^{ikct}\right]e^{ikx}\mathrm{d}k \\
&\quad + \int_{-\infty}^{+\infty}\left[\frac{1}{2}\overline{u}_0(k)e^{-ikct}\right]e^{ikx}\mathrm{d}k - \frac{1}{2c}\int_{-\infty}^{+\infty}\left[\frac{1}{ik}\overline{v}_0(k)e^{-ikct}\right]e^{ikx}\mathrm{d}k \\
&= \frac{1}{2}[u_0(x+ct) + u_0(x-ct)] + \frac{1}{2c}\int_{x-ct}^{x+ct} v_0(\xi)d\xi.
\end{aligned} \tag{2.2.64}$$

This is known as **d' Alembert's formula.** If we take

$$u_0(x) = \begin{cases} a + x & x \in (-a, 0) \\ a - x & x \in (0, a) \\ 0 & x \notin (-a, a) \end{cases}, \tag{2.2.65}$$

and

$$v_0(x) = 0 \quad x \in (-\infty, +\infty). \tag{2.2.66}$$

The profile of $u(x, t)$ at different time, i.e. $= t_0, t_1, t_2 \ldots$, is as follows (Fig. 2.2).

The general elastodynamic solutions Eqs. (2.2.55) and (2.2.64) for infinite long and thin strings can be summarized as follows

1) Wave packet $u(x, t)$ is the result of superposition of left traveling wave $f = e^{ik(x+ct)}$ and right traveling wave $g = e^{ik(x-ct)}$ of different frequencies. Different from the bounded thin string, the wave packet of the infinite thin string contains wavelets of arbitrary frequency or wave number. In other words, the frequency or wave number of the wavelets is continuously distributed. However, the wavelet frequencies or wave numbers of the wave packets of bounded strings are discrete.
2) The shape of the wave packet depends on the initial conditions. When the initial velocity is 0, the wave packet keeps the same shape and moves to the left and right respectively with the increase of time t. Since there is no boundary, the

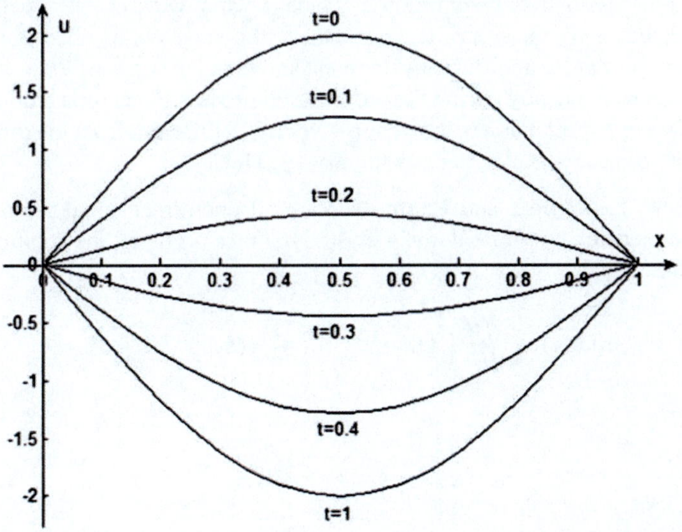

Fig. 2.1 Displacement waveform of flexible thin string fixed at both ends (bounded string) at different moments t_i

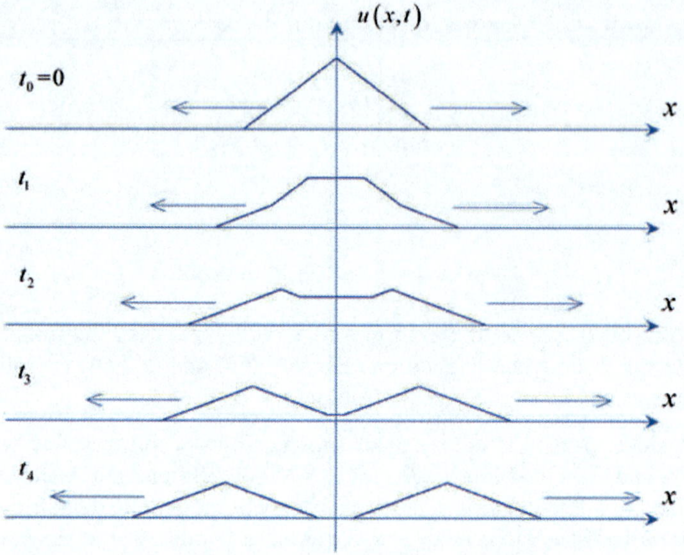

Fig. 2.2 Evolution process of initial perturbation with time in infinite long soft thin string, (movement of wave packet)

wave packet will propagate continuously without generating the reflected wave packet. The velocity of wave packet propagation is called **group velocity**.

3) In a non-dissipative medium, the shape of the wave packet will remain unchanged, while in a dissipative medium, the wavelets of different frequencies will propagate at different velocities, and thus the shape of the wave packet will change constantly during the propagation process. The propagation velocity of the wave packet, namely the **group velocity**, is different from the propagation velocity of each wavelet, namely the **phase velocity**.

The above **Example** 1 and **Example** 2 are all one-dimensional problems. For the three-dimensional infinite domain elastodynamics problem, the solution can be expressed as

$$\boldsymbol{u}(\boldsymbol{x},t)=\iiint\left[\boldsymbol{A}(\boldsymbol{k})e^{i(\boldsymbol{k}\cdot\boldsymbol{x}+ct)}+\boldsymbol{B}(\boldsymbol{k})e^{i(\boldsymbol{k}\cdot\boldsymbol{x}-ct)}\right]\mathrm{d}\boldsymbol{k}, \tag{2.2.67}$$

where

$$\boldsymbol{k}=k_1\boldsymbol{i}+k_2\boldsymbol{j}+k_3\boldsymbol{k},$$

$$\boldsymbol{x}=x_1\boldsymbol{i}+x_2\boldsymbol{j}+x_3\boldsymbol{k}.$$

For wavelets $f = e^{i(\boldsymbol{k}\cdot\boldsymbol{x}+\omega t)}$ and $g = e^{i(\boldsymbol{k}\cdot\boldsymbol{x}-\omega t)}$, the values depend on the argument $\boldsymbol{k}\cdot\boldsymbol{x} - \omega t$ which is often called the phase of the wave, and the plane represented by the equation is called the **equiphase plane**, i.e.

$$\boldsymbol{k}\cdot\boldsymbol{x} - \omega t = \text{const.} \tag{2.2.68}$$

It is assumed that the spatial positions of A point on the equiphase plane at times t_1 and t_2 are, respectively, x_1 and x_2, and from Eq. (2.2.68), we get

$$\boldsymbol{k}\cdot \mathrm{d}\boldsymbol{x} - \omega \mathrm{d}t = 0. \tag{2.2.69}$$

Notice that k goes in the same direction as dx, so

$$\frac{\mathrm{d}x}{\mathrm{d}t} = \frac{\omega}{k} = c. \tag{2.2.70}$$

This indicates that the moving velocity of the equiphase plane of the wavelet at a certain frequency is the velocity of the wave, which is usually called the **phase velocity**. Note that the phase velocity refers to the propagating velocity of the wavelet, while the group velocity refers to the propagating velocity of the wave packet. In a non-dissipative medium, the group velocity and the phase velocity are the same, but they are different in a dissipative medium.

A plane of equal phase may or may not be a plane. If the equiphase surface is flat, we generally call it a **plane wave**, as $f = e^{i(k\cdot x\pm\omega t)}$. However, there are also cases where the equal-phase plane is not a plane. For example, the equal-phase plane of the wave excited by a line source and a point source is a cylindrical surface and a spherical surface, respectively. Such waves are usually called **cylindrical waves** and **spherical waves** [4, 6].

2.3 Properties of Plane Waves

2.3.1 Propagation Mode of Plane Waves

Let the displacement component of a plane wave propagating in the direction of $\boldsymbol{n}$ be expressed as

$$u_i = a_i f\left(n_j x_j - ct\right), \tag{2.3.1}$$

where

$$n_j x_j - ct = \text{const.} \tag{2.3.2}$$

The above equation is an arbitrary plane in three-dimensional Euclidean space, which is called the equiphase plane of the plane wave. a_i represents the projection of the particle vibration amplitude on the coordinate axis. Let

$$\xi = n_j x_j - ct. \tag{2.3.3}$$

Then, we get from Eq. (2.3.1)

$$\frac{\partial u_i}{\partial x_j} = a_i \frac{\partial f(\xi)}{\partial x_j} = a_i \frac{df(\xi)}{d\xi} \frac{\partial \xi}{\partial x_j} = a_i n_j f'(\xi), \tag{2.3.4}$$

$$\frac{\partial^2 u_i}{\partial x_j^2} = a_i n_j n_j f''(\xi), \tag{2.3.5}$$

$$\frac{\partial^2 u_i}{\partial x_i \partial x_j} = a_i n_i n_j f''(\xi). \tag{2.3.6}$$

Substituting Eqs. (2.3.4) and (2.3.6) into the Navier equation (ignoring the volume force), i.e.

$$(\lambda + \mu)\frac{\partial \theta}{\partial x_i} + \mu \frac{\partial^2 u_i}{\partial x_j \partial x_j} = \rho \frac{\partial^2 u_i}{\partial t^2}, \tag{2.3.7}$$

we obtain

$$(\lambda + \mu) a_j n_j n_i + (\mu - \rho c^2) a_i = 0, \tag{2.3.8}$$

or

$$\left[(\lambda + \mu) n_j n_i + (\mu - \rho c^2)\delta_{ij}\right] a_j = 0. \tag{2.3.9}$$

The matrix form of Eq. (2.3.9) is

$$\begin{bmatrix} (\lambda+\mu)n_1 n_1 + (\mu - \rho c^2) & (\lambda+\mu)n_2 n_1 & (\lambda+\mu)n_3 n_1 \\ (\lambda+\mu)n_1 n_2 & (\lambda+\mu)n_2 n_2 + (\mu - \rho c^2) & (\lambda+\mu)n_3 n_2 \\ (\lambda+\mu)n_1 n_3 & (\lambda+\mu)n_2 n_3 & (\lambda+\mu)n_3 n_3 + (\mu - \rho c^2) \end{bmatrix} \begin{Bmatrix} a_1 \\ a_2 \\ a_3 \end{Bmatrix} = 0. \tag{2.3.10}$$

The condition that the above equation has a non-zero solution is that the determinant of the coefficient matrix is equal to zero. The condition reduces to

$$(\lambda + 2\mu - \rho c^2)(\mu - \rho c^2)^2 = 0. \tag{2.3.11}$$

This equation has two distinct real roots, i.e.

$$c_{\mathrm{p}} = \sqrt{\frac{\lambda + 2\mu}{\rho}}, \tag{2.3.12}$$

and

$$c_{\mathrm{s}} = \sqrt{\frac{\mu}{\rho}}. \tag{2.3.13}$$

These two eigenvalues physically represent two propagation velocities of a plane wave in an infinite uniform medium. Substituting them back into Eq. (2.3.10), the corresponding eigenvectors can be obtained. Here, the eigenvector physically represents the vibration amplitude and vibration direction of the particle. Substituting Eq. (2.3.12) into Eq. (2.3.9) leads to

$$\left[(\lambda + \mu)n_j n_i - (\lambda + \mu)\delta_{ij}\right]a_j = 0, \tag{2.3.14}$$

or

$$(\lambda + \mu)\left(n_j n_i a_j - a_i\right) = 0. \tag{2.3.15}$$

For the equation above to be true, it has to satisfy

$$n_m a_m n_i - a_i = 0. \tag{2.3.16}$$

The vector form of the above equation is

$$(\boldsymbol{n} \cdot \boldsymbol{a})\boldsymbol{n} = \boldsymbol{a}. \tag{2.3.17}$$

By performing vector product on both sides of equation with **n**, we obtain

$$\boldsymbol{n} \times \boldsymbol{a} = (\boldsymbol{n} \cdot \boldsymbol{a})\boldsymbol{n} \times \boldsymbol{n} = 0. \tag{2.3.18}$$

It can be seen that the particle vibration vector $\boldsymbol{a}$ is consistent with the wave propagation direction $\boldsymbol{n}$ for a plane wave propagating at a velocity of c_{p}. This kind of wave is the longitudinal waves mentioned earlier. Substituting Eq. (2.3.13) into Eq. (2.3.9), we obtain

$$(\lambda + \mu)n_j n_i a_j = 0, \tag{2.3.19}$$

or

$$n_j a_j = \boldsymbol{n} \cdot \boldsymbol{a} = 0. \tag{2.3.20}$$

It can be seen that the particle vibration vector $\boldsymbol{a}$ is perpendicular to the wave propagation direction $\boldsymbol{n}$ for a plane wave propagating at a speed of c_s, which is the shear wave mentioned above.

In summary, in the infinite isotropic homogeneous elastic medium, there are only two possible modes of plane waves, namely longitudinal wave and transverse wave, whose propagation velocities are c_p and c_s, respectively. Moreover, the propagation velocity of longitudinal wave, i.e. c_p, is greater than that of shear wave, i.e. c_s, and the direction of particle vibration of longitudinal wave is the same as that of wave propagation. The particle vibration direction of the shear wave is perpendicular to the wave propagation direction.

2.3.2 The Stress Distribution on the Wavefront

It has been discussed above that there are only two modes of plane waves in infinite uniform elastic medium. No matter longitudinal wave or transverse wave, their displacement fields can be uniformly expressed as [2]

$$u_i = a_i f\left(n_j x_j - ct\right). \tag{2.3.21}$$

Substituting it into Hooke's law, i.e.

$$\sigma_{ij} = \lambda u_{k,k}\delta_{ij} + \mu\left(u_{i,j} + u_{j,i}\right), \tag{2.3.22}$$

we can obtain the stress on the wavefront (equal phase plane)[3]

$$\begin{aligned}\sigma_{ni} = \sigma_{ij}n_j &= \left[\lambda a_k n_k \delta_{ij} + \mu\left(a_i n_j + a_j n_i\right)\right] f'(\xi) n_j \\ &= \left[\lambda a_k n_k n_i + \mu\left(a_i n_j n_j + a_j n_i n_j\right)\right] f'(\xi) \\ &= [(\lambda + \mu) a_k n_k n_i + \mu a_i] f'(\xi).\end{aligned} \tag{2.3.23}$$

For a plane longitudinal or compressional wave, the particle vibration vector a is in the same direction as the wave propagation vector $\boldsymbol{n}$. So there is

$$a_k n_k = a,$$

or

$$a n_i = a_i.$$

Substituting them into Eq. (2.3.23), we can obtain

$$\sigma_{ni} = (\lambda + 2\mu) a_i f'(\xi). \tag{2.3.24}$$

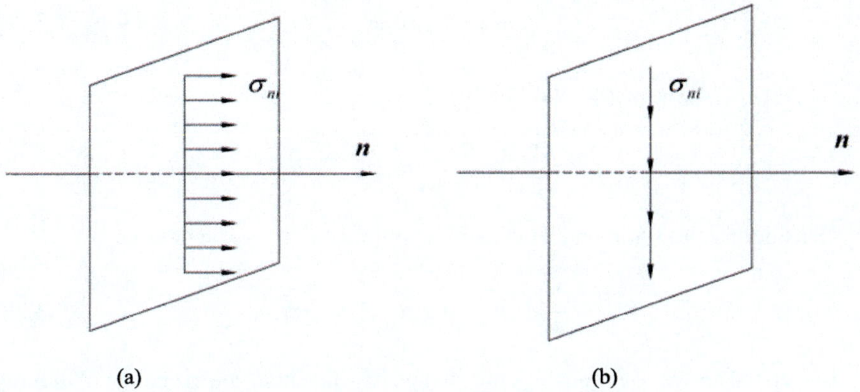

Fig. 2.3 Stress distribution on the plane wavefront. **a** Only normal stress existing on the wavefront of longitudinal wave. **b** Only shear stress existing on the wavefront of transversal wave

It can be seen that the direction of the stress vector on the wavefront is the same as the direction of particle vibration; that is, there is only normal stress on the wavefront without shear stress, as shown in Fig. 2.3a.

For transversal or shear plane waves, the particle vibration vector $\boldsymbol{a}$ is perpendicular to the wave propagation vector $\boldsymbol{n}$. So there is

$$a_k n_k = \boldsymbol{a} \cdot \boldsymbol{n} = 0. \tag{2.3.25}$$

Substituting it into Eq. (2.3.23), we can obtain

$$\sigma_{ni} = \mu a_i f'(\xi). \tag{2.3.26}$$

It can be seen that the stress vector on the wavefront is still consistent with the direction of particle vibration; that is, there is only shear stress on the wavefront without normal stress, as shown in Fig. 2.3b.

2.3.3 The Energy Flow Density of a Plane Wave

When the elastic wave propagates in an elastic medium, the material point vibrates periodically at its own equilibrium position and thus has kinetic energy. The mechanical interaction between the material points causes the neighboring material points to vibrate and thus also has kinetic energy. At the same time, the interaction also causes the deformation of the microelement volume and thus generates potential energy. The propagation of such waves inevitably leads to the propagation of energy. Wherever the wave reaches, the material has both kinetic and potential energy.

The kinetic energy per unit volume can be calculated as follows

$$w_k = \frac{1}{2}\rho\dot{u}_i\dot{u}_i. \tag{2.3.27}$$

The potential energy per unit volume is

$$w_p = \frac{1}{2}\sigma_{ij}\varepsilon_{ij} = \frac{1}{4}\sigma_{ij}(u_{i,j} + u_{j,i}).$$

Considering the symmetry of the stress component, it can be obtained

$$w_p = \frac{1}{4}(\sigma_{ij}u_{i,j} + \sigma_{ji}u_{j,i}) = \frac{1}{2}\sigma_{ij}u_{i,j}. \tag{2.3.28}$$

In fact, the above equation does not change its value after the exchange of dummy index i and j in the second term, namely

$$\begin{aligned} w_p &= \frac{1}{4}(\sigma_{ij}u_{i,j} + \sigma_{ji}u_{j,i}) = \frac{1}{2}\sigma_{ij}u_{i,j} \\ &= \frac{1}{2}[\lambda u_{k,k}\delta_{ij} + \mu(u_{i,j} + u_{j,i})]u_{i,j}. \end{aligned} \tag{2.3.29}$$

The total mechanical energy per unit volume, called the energy density, can be expressed as

$$w = w_k + w_p. \tag{2.3.30}$$

Due to the time dependence of particle motion, the energy density of the microelement volume is different at different moments. The time change rate of the energy density is discussed below.

The time change rate of kinetic energy density is

$$\frac{\partial w_k}{\partial t} = \frac{1}{2}\rho\frac{\partial}{\partial t}(\dot{u}_i\dot{u}_i) = \rho\dot{u}_i\ddot{u}_i = \dot{u}_i\sigma_{ij,j}. \tag{2.3.31}$$

The time change rate of potential energy density is

$$\begin{aligned} \frac{\partial w_p}{\partial t} &= \frac{1}{2}\frac{\partial}{\partial t}[\lambda u_{k,k}^2 + \mu u_{i,j}^2 + \mu u_{i,j}u_{j,i}] \\ &= [\lambda u_{k,k}\delta_{ij} + \mu(u_{i,j} + u_{j,i})]\dot{u}_{i,j}. \\ &= \sigma_{ij}\dot{u}_{i,j}. \end{aligned} \tag{2.3.32}$$

So the time rate of change of the total mechanical energy is

$$\begin{aligned}\frac{\partial w}{\partial t} &= \frac{\partial w_k}{\partial t} + \frac{\partial w_p}{\partial t} \\ &= \dot{u}_i \sigma_{ij,j} + \sigma_{ij} \dot{u}_{i,j} \\ &= \left(\sigma_{ij} \dot{u}_i\right)_{,j}.\end{aligned} \tag{2.3.33}$$

Let's define

$$I_j = -\sigma_{ij} \dot{u}_i. \tag{2.3.34}$$

Therefore, Eq. (2.3.33) can be rewritten as

$$\frac{\partial w}{\partial t} + \nabla \cdot \boldsymbol{I} = 0. \tag{2.3.35}$$

By integrating both sides of the above expression over a finite region V, we can get

$$\frac{\partial}{\partial t} \int_V w \mathrm{d}V = -\int_V \nabla \cdot \boldsymbol{I} \mathrm{d}V = -\int_S \boldsymbol{I} \cdot \boldsymbol{n} \mathrm{d}s. \tag{2.3.36}$$

By comparing Eqs. (2.3.35) and (2.3.36) with the law of conservation of mass, i.e. Eqs. (1.2.4) and (1.2.5), Eqs. (2.3.35) and (2.3.36) can be understood as differential and integral forms of energy conservation. Here $\boldsymbol{I}$ is the counterpart of the mass flow vector $\rho \dot{\boldsymbol{u}}$ and is usually called the energy flow vector. $\nabla \cdot \boldsymbol{I}$ represents the divergence or "source of distribution" of the energy flow field. $\boldsymbol{I}$ represents the energy passing through the unit area of the wavefront per unit time, and its direction represents the energy flow direction, that is, the propagation direction of the wave.

Consider a plane wave whose displacement field is expressed as

$$u_i = a_i f\left(n_j x_j - ct\right) = a_i f(\xi). \tag{2.3.37}$$

Let us derive the specific expression of the energy flow vector for the plane longitudinal wave and plane transverse wave, respectively. For the plane compressional wave, it is

$$n_j a_j = a, \tag{2.3.38}$$

and

$$c = c_p. \tag{2.3.39}$$

So

$$w_k = \frac{1}{2} \rho \dot{u}_i \dot{u}_i = \frac{1}{2} \rho a^2 c_p^2 \left[f'(\xi)\right]^2, \tag{2.3.40}$$

$$w_p = \frac{1}{2}\sigma_{ij}\varepsilon_{ij} = \frac{1}{2}\left(\lambda\theta\delta_{ij} + 2\mu\varepsilon_{ij}\right)\varepsilon_{ij}$$
$$= \frac{1}{2}\left(\lambda\theta^2 + 2\mu\varepsilon_{ij}\varepsilon_{ij}\right). \tag{2.3.41}$$

Considering

$$\theta = u_{i,i} = a_i n_i f'(\xi) = af'(\xi), \tag{2.3.42}$$

$$\varepsilon_{ij} = \frac{1}{2}\left(u_{i,j} + u_{j,i}\right)$$
$$= \frac{1}{2}\left(a_i n_j + a_j n_i\right) f'(\xi)$$
$$= \frac{1}{2}\left(a_i n_j + a_j n_i\right) f'(\xi)$$
$$= a n_i n_j f'(\xi), \tag{2.3.43}$$

and

$$\varepsilon_{ij}\varepsilon_{ij} = a^2\left[f'(\xi)\right]^2, \tag{2.3.44}$$

we can obtain

$$w_p = \frac{1}{2}\lambda a^2\left[f'(\xi)\right]^2 + \mu a^2\left[f'(\xi)\right]^2$$
$$= \frac{1}{2}(\lambda + 2\mu)a^2\left[f'(\xi)\right]^2$$
$$= \frac{1}{2}\rho a^2 c_p^2\left[f'(\xi)\right]^2. \tag{2.3.45}$$

The total mechanical energy density is

$$w = w_k + w_p = \rho a^2 c_p^2\left[f'(\xi)\right]^2. \tag{2.3.46}$$

It can be seen that the magnitudes of kinetic energy density, potential energy density and total mechanical energy density are proportional to the square of the amplitudes. So the amplitude of a wave is usually an index to measure the intensity of a wave.

The energy flow vector is

$$I_i = -\sigma_{ij}\dot{u}_j$$
$$= -\left(\lambda\theta\delta_{ij} + 2\mu\varepsilon_{ij}\right)\dot{u}_j$$
$$= \left(\lambda a a_i c_p + 2\mu a^2 n_i c_p\right)\left[f'(\xi)\right]^2$$

$$
\begin{aligned}
&= (\lambda + 2\mu) a^2 c_p n_i \left[f'(\xi)\right]^2 \\
&= \rho a^2 c_p^2 \left[f'(\xi)\right]^2 c_p n_i \\
&= w c_p n_i,
\end{aligned} \tag{2.3.47}
$$

or, in the vector form,

$$
\boldsymbol{I} = w c_p \boldsymbol{n}. \tag{2.3.48}
$$

That is, **the direction of the energy flow vector is the same as the propagation direction of the wave, and its magnitude is equal to the product of the energy density and the propagation velocity.**

For a transverse wave, it is

$$
a_i n_i = 0, \tag{2.3.49}
$$

and

$$
c = c_s. \tag{2.3.50}
$$

Thus,

$$
w_k = \frac{1}{2} \rho \dot{u}_i \dot{u}_i = \frac{1}{2} \rho a^2 c_s^2 \left[f'(\xi)\right]^2, \tag{2.3.51}
$$

$$
\begin{aligned}
w_p &= \frac{1}{2} \sigma_{ij} \varepsilon_{ij} \\
&= \frac{1}{2} \left(\lambda \theta \delta_{ij} + 2\mu \varepsilon_{ij}\right) \varepsilon_{ij} \\
&= \frac{1}{2} \left(\lambda \theta^2 + 2\mu \varepsilon_{ij} \varepsilon_{ij}\right).
\end{aligned} \tag{2.3.52}
$$

Considering

$$
\theta = u_{i,i} = a_i n_i f'(\xi) = 0, \tag{2.3.53}
$$

$$
\begin{aligned}
\varepsilon_{ij} &= \frac{1}{2} \left(u_{i,j} + u_{j,i}\right) \\
&= \frac{1}{2} \left(a_i n_j + a_j n_i\right) f'(\xi),
\end{aligned} \tag{2.3.54}
$$

and

$$
\varepsilon_{ij} \varepsilon_{ij} = \frac{1}{4} \left(a_i n_j + a_j n_i\right) \left(a_i n_j + a_j n_i\right) \left[f'(\xi)\right]^2
$$

$$
\begin{aligned}
&= \frac{1}{4}\left(a^2 + 2a_i a_j n_i n_j + a^2\right)\left[f'(\xi)\right]^2 \\
&= \frac{1}{2}a^2\left[f'(\xi)\right]^2,
\end{aligned} \tag{2.3.55}
$$

we can obtain

$$
\begin{aligned}
w_p &= \frac{1}{2}\mu a^2\left[f'(\xi)\right]^2 \\
&= \frac{1}{2}\rho c_s^2 a^2\left[f'(\xi)\right]^2.
\end{aligned} \tag{2.3.56}
$$

The total mechanical energy density is

$$
w = w_k + w_p = \rho c_s^2 a^2\left[f'(\xi)\right]^2. \tag{2.3.57}
$$

The energy flow vector is

$$
\begin{aligned}
I_i &= -\sigma_{ij}\dot{u}_j \\
&= -\left(\lambda\theta\delta_{ij} + 2\mu\varepsilon_{ij}\right)\dot{u}_j \\
&= \mu\left(a_i n_j + a_j n_i\right)a_j c_s\left[f'(\xi)\right]^2 \\
&= \mu a^2 c_s n_i\left[f'(\xi)\right]^2 \\
&= \rho a^2 c_s^2\left[f'(\xi)\right]^2 c_s n_i \\
&= w c_s n_i,
\end{aligned} \tag{2.3.58}
$$

or, in vector form,

$$
\boldsymbol{I} = w c_s \boldsymbol{n}. \tag{2.3.59}
$$

That is, **the direction of the energy flow vector is the same as the propagation direction of the wave, and its magnitude is equal to the product of the energy flow density and the propagation velocity.**

For wave propagation problem, the energy density and energy flow vector are all periodic functions of time due to the time dependence feature, and their average value in a period determines the intensity of wave and thus is of great significance [8, 9, 10]. Their mean values in a period T are represented by $\langle w\rangle$ and $\langle I\rangle$, then

$$
\langle w\rangle = \frac{1}{T}\int_0^T w(t)\mathrm{d}t, \tag{2.3.60}
$$

and

$$\langle I\rangle = \frac{1}{T}\int_0^T I(t)\mathrm{d}t. \tag{2.3.61}$$

The ratio of the average energy flow to the average energy density is defined as the energy propagation velocity, i.e.

$$c_E = \frac{\langle I\rangle}{\langle w\rangle}. \tag{2.3.62}$$

From the previous discussion, there is

$$c_E = c_P, \tag{2.3.63}$$

for a plane compressional wave, and

$$c_E = c_S, \tag{2.3.64}$$

for a plane transversal wave. Therefore, **the velocity of energy propagation is equal to the phase velocity**. This is true only for monochromatic waves. For a wave packet formed by the mutual interference of multiple monochromatic waves, the energy propagation velocity is equal to the group velocity of the wave packet, i.e.

$$c_E = c_G. \tag{2.3.65}$$

As real physical quantities, displacement, stress and strain are all real numbers. When they are expressed in complex numbers, it is only for the convenience of mathematical processing. The real or imaginary part of the complex number is the real physical quantity. For calculating the energy flow density, if the displacement and stress are expressed in complex numbers, it should be written as

$$\begin{aligned} I_i &= -\mathrm{Re}\sigma_{ij} \cdot Re\dot{u}_j \\ &= -\frac{1}{2}\left(\sigma_{ij} + \sigma_{ij}^*\right) \cdot \frac{1}{2}\left(\dot{u}_j + \dot{u}_j^*\right) \\ &= -\frac{1}{4}\left(\sigma_{ij}\dot{u}_j^* + \sigma_{ij}^*\dot{u}_j\right) - \frac{1}{4}\left(\sigma_{ij}\dot{u}_j + \sigma_{ij}^*\dot{u}_j^*\right), \end{aligned} \tag{2.3.66a}$$

or

$$\begin{aligned} I_i &= -Im\sigma_{ij} \cdot Im\dot{u}_j \\ &= -\frac{1}{2i}\left(\sigma_{ij} - \sigma_{ij}^*\right) \cdot \frac{1}{2i}\left(\dot{u}_j - \dot{u}_j^*\right) \\ &= -\frac{1}{4}\left(\sigma_{ij}\dot{u}_j^* + \sigma_{ij}^*\dot{u}_j\right) + \frac{1}{4}\left(\sigma_{ij}\dot{u}_j + \sigma_{ij}^*\dot{u}_j^*\right). \end{aligned} \tag{2.3.66b}$$

Considering

$$\frac{1}{T}\int_0^T e^{i\omega t}dt = 0,$$

and

$$\sigma_{ij}^* \dot{u}_j = (\sigma_{ij}\dot{u}_j^*)^*,$$

the calculation formula of the average energy flow density is

$$\begin{aligned}\langle I_i \rangle &= \frac{1}{T}\int_0^T I_i(t)\mathrm{d}t \\ &= -\frac{1}{4T}\int_0^T (\sigma_{ij}\dot{u}_j^* + \sigma_{ij}^*\dot{u}_j)\mathrm{d}t \\ &= -\frac{1}{2T}\mathrm{Re}\int_0^T (\sigma_{ij}\dot{u}_j^*)\mathrm{d}t. \end{aligned} \tag{2.3.67}$$

Example 1 Given the scalar potential $\varphi = A\exp[i(k_1x_1 - \omega t)]$ of a one-dimensional displacement vector field, where A is a real constant, we need to calculate (1) the components of strain tensor, (2) the component of stress tensor, (3) energy flow density vector and (4) wave intensity and propagation velocity.

Solution: Since there is only a scalar potential and no vector potential, the wave is longitudinal wave.

(1) components of strain tensor

The displacement is

$$\boldsymbol{u} = \nabla\varphi = \frac{\partial\varphi}{\partial x_1}\boldsymbol{i} = ik_1\varphi\boldsymbol{i}, \text{ i.e } u_1 = ik_1\varphi,\ u_2 = u_3 = 0.$$

The speed is

$$\dot{u}_1 = \frac{\partial u_1}{\partial t} = -(ik_1)(i\omega)\varphi = k_1\omega\varphi.$$

According to the formula $\varepsilon_{ij} = \frac{1}{2}(u_{i,j} + u_{j,i})$, the nonzero strain component is only ε_{11}, i.e.

$$\varepsilon_{11} = \frac{\partial u_1}{\partial x_1} = (ik_1)^2\varphi = -k_1^2\varphi,$$

and

$$\varepsilon_{22} = \varepsilon_{33} = \varepsilon_{12} = \varepsilon_{13} = \varepsilon_{23} = 0.$$

(2) the components of stress tensor

According to the generalized Hooke's law, $\sigma_{ij} = \lambda\theta\delta_{ij} + 2\mu\varepsilon_{ij}$, we can get

$$\sigma_{11} = \lambda\varepsilon_{11} + 2\mu\varepsilon_{11} = -(\lambda + 2\mu)k_1^2\varphi,$$

$$\sigma_{22} = \lambda\varepsilon_{11} = -\lambda k_1^2\varphi,$$

$$\sigma_{33} = \lambda\varepsilon_{11} = -\lambda k_1^2\varphi,$$

and

$$\sigma_{12} = \sigma_{13} = \sigma_{23} = 0.$$

(3) energy flow density vector

The displacement components, strain components and stress components derived above are all expressed in complex numbers. The real displacement field, strain field and stress field should take the real part or the imaginary part. Here we take the real part to compute the energy flow density vector, i.e.

$$I_1 = -\mathrm{Re}\sigma_{11} \cdot \mathrm{Re}\frac{\partial u_1}{\partial t} = \rho\frac{\omega^4}{c_p}A^2\cos^2(k_1x_1 - \omega t),$$

and

$$I_2 = I_3 = 0.$$

(4) wave intensity and propagation velocity

$$\langle I_1\rangle = -\frac{1}{2T}\mathrm{Re}\int_0^T(\sigma_{11}\cdot\dot{u}_1^*)\mathrm{d}t$$

$$= \frac{1}{2T}\mathrm{Re}\int_0^T\left(\rho\frac{\lambda + 2\mu}{\rho}k_1^3\omega AA^*\right)\mathrm{d}t$$

$$= \rho \frac{\omega^4}{2c_P} A^2,$$

or

$$\begin{aligned}
\langle I \rangle &= \frac{1}{T} \int_0^T I_1 \mathrm{d}t \\
&= \rho \frac{\omega^4}{c_p T} A^2 \int_0^T \cos^2(k_1 x_1 - \omega t)\mathrm{d}t \\
&= \frac{\rho \omega^4}{2c_p} A^2,
\end{aligned}$$

$$w_k = \frac{1}{2}\rho(\dot{u}_i \dot{u}_i) = \frac{1}{2}\rho c_p^2 A^2 k^4 \cos^2(k_1 x_1 - \omega t),$$

$$w_p = \frac{1}{2}\sigma_{ij}\varepsilon_{ij} = \frac{1}{2}\sigma_{11}\varepsilon_{11} = \frac{1}{2}\rho A^2 c_p^2 k^4 \cos^2(k_1 x_1 - \omega t),$$

$$w = w_k + w_p = \rho c_p^2 A^2 k^4 \cos^2(k_1 x_1 - \omega t),$$

$$\begin{aligned}
\langle w \rangle &= \frac{1}{T} \int_0^T w \mathrm{d}t \\
&= \frac{\rho c_p^2 A^2 k^4}{T} \int_0^T \cos^2(k_1 x_1 - \omega t)\mathrm{d}t \\
&= \frac{\rho \omega^4}{2c_p^2} A^2,
\end{aligned}$$

and

$$c = \frac{\langle I \rangle}{\langle w \rangle} = \frac{\rho \omega^4 A^2 / 2c_p}{\rho \omega^4 A^2 / 2c_p^2} = c_p.$$

Example 2 Given the scalar potential $\varphi = A \exp[i(k_1 x_1 + k_2 x_2 - \omega t)]$ of the two-dimensional displacement field, where A is a real constant, we need to calculate (1) the components of strain tensor, (2) the components of stress tensor, (3) energy flow density vector and (4) wave intensity and propagation velocity.

Solution: Since there is only a scalar potential and no vector potential, the wave is longitudinal wave.

(1) the components of strain tensor

The displacement is

$$\boldsymbol{u} = \nabla\varphi = \frac{\partial\varphi}{\partial x_1}\boldsymbol{i} + \frac{\partial\varphi}{\partial x_2}\boldsymbol{j} = i(k_1\boldsymbol{i} + k_2\boldsymbol{j})\varphi.$$

That is, $u_j = ik_j\varphi$. For two-dimensional problems, there is

$$u_1 = ik_1\varphi,\ u_2 = ik_2\varphi \text{ and } u_3 = 0.$$

The vibration speed is

$$\dot{u}_1 = k_1\omega\varphi,\ \dot{u}_2 = k_2\omega\varphi \text{ and } \dot{u}_3 = 0.$$

According to the formula

$$\varepsilon_{ij} = \frac{1}{2}\left(u_{i,j} + u_{j,i}\right) = \frac{1}{2}\left(ik_i \cdot ik_j + ik_j \cdot ik_i\right)\varphi = -k_ik_j\varphi,$$

we can get

$$\varepsilon_{11} = -k_1^2\varphi,\ \varepsilon_{22} = -k_2^2\varphi,\ \varepsilon_{12} = -k_1k_2\varphi,$$

and

$$\varepsilon_{33} = \varepsilon_{23} = \varepsilon_{31} = 0.$$

(2) the components of stress tensor

According to the generalized Hooke's law, i.e.

$$\begin{aligned}\sigma_{ij} &= \lambda\theta\delta_{ij} + 2\mu\varepsilon_{ij} = \lambda u_{l,l}\delta_{ij} + 2\mu\varepsilon_{ij}\\ &= \lambda(ik_l)\cdot(ik_l)\varphi\delta_{ij} + 2\mu\left(-k_ik_j\varphi\right)\\ &= \left(-\lambda k^2\delta_{ij} - 2\mu k_ik_j\right)\varphi,\quad \left(k^2 = k_1^2 + k_2^2\right)\end{aligned}$$

we can get

$$\sigma_{11} = -\left[\lambda k^2 + 2\mu k_1^2\right]\varphi,\ \sigma_{22} = -\left[\lambda k^2 + 2\mu k_2^2\right]\varphi,\ \sigma_{33} = -\lambda k^2\varphi,$$

$$\sigma_{12} = -2\mu k_1k_2\varphi,$$

and

$$\sigma_{13} = \sigma_{23} = 0.$$

(3) energy flow density vector

$$
\begin{aligned}
I_j &= -\mathrm{Re}\sigma_{ij} \cdot \mathrm{Re}\frac{\partial u_i}{\partial t} \\
&= -\left[\left(-\lambda k^2 \delta_{ij} - 2\mu k_i k_j\right)\mathrm{Re}\varphi\right]\left[ik_i \cdot (-i\omega)\mathrm{Re}\varphi\right] \\
&= \left(\lambda k^2 \delta_{ij} + 2\mu k_i k_j\right) k_i \omega A^2 \cos^2(k_1 x_1 + k_2 x_2 - \omega t) \\
&= \left(\lambda k^2 k_j + 2\mu k^2 k_j\right) \omega A^2 \cos^2(k_1 x_1 + k_2 x_2 - \omega t) \\
&= (\lambda + 2\mu) k^2 \omega A^2 k_j \cos^2(k_1 x_1 + k_2 x_2 - \omega t) \\
&= \rho\omega^3 A^2 k_j \cos^2(k_1 x_1 + k_2 x_2 - \omega t),
\end{aligned}
$$

$$
I_1 = \rho\omega^3 A^2 k_1 \cos^2(k_1 x_1 + k_2 x_2 - \omega t),
$$

$$
I_2 = \rho\omega^3 A^2 k_2 \cos^2(k_1 x_1 + k_2 x_2 - \omega t),
$$

$$
I_3 = 0,
$$

and

$$
\begin{aligned}
I &= \sqrt{I_1^2 + I_2^2 + I_3^2} \\
&= \rho\omega^3 A^2 \sqrt{\left(k_1^2 + k_2^2\right)} \cos^2(k_1 x_1 + k_2 x_2 - \omega t) \\
&= \rho\omega^3 A^2 k \cos^2(k_1 x_1 + k_2 x_2 - \omega t) \\
&= \frac{\rho\omega^4 A^2}{c_p} \cos^2(k_1 x_1 + k_2 x_2 - \omega t).
\end{aligned}
$$

(4) wave intensity and propagation velocity

$$
\langle I \rangle = \frac{1}{T}\int_0^T I \mathrm{d}t = \frac{\rho\omega^4 A^2}{T c_p}\int_0^T \cos^2(k_1 x_1 + k_2 x_2 - \omega t)\mathrm{d}t = \frac{\rho\omega^4 A^2}{2c_p},
$$

$$
\begin{aligned}
w_k &= \frac{1}{2}\rho\left(\dot{u}_j \dot{u}_j\right) \\
&= \frac{1}{2}\rho\left[ik_j(-i\omega)A\cos(k_1 x_1 + k_2 x_2 - \omega t)\right]^2 \\
&= \frac{1}{2}\rho k^2 \omega^2 A^2 \cos^2(k_1 x_1 + k_2 x_2 - \omega t) \\
&= \frac{\rho\omega^4 A^2}{2c_p^2} \cos^2(k_1 x_1 + k_2 x_2 - \omega t),
\end{aligned}
$$

$$\begin{aligned} w_p &= \frac{1}{2}\sigma_{ij}\varepsilon_{ij} \\ &= \frac{1}{2}\left[\left(-\lambda k^2\delta_{ij} - 2\mu k_i k_j\right)A\cos(k_1x_1 + k_2x_2 - \omega t)\right] \\ &\quad \times \left[-k_i k_j A\cos(k_1x_1 + k_2x_2 - \omega t)\right] \\ &= \frac{1}{2}\left(\lambda k^2k^2 + 2\mu k^2k^2\right)A^2\cos^2(k_1x_1 + k_2x_2 - \omega t) \\ &= \frac{\rho\omega^4A^2}{2c_p^2}\cos^2(k_1x_1 + k_2x_2 - \omega t), \end{aligned}$$

$$w = w_k + w_p = \frac{\rho\omega^4A^2}{c_p^2}\cos^2(k_1x_1 + k_2x_2 - \omega t),$$

$$\begin{aligned} \langle w\rangle &= \frac{1}{T}\int_0^T w\mathrm{d}t \\ &= \frac{\rho\omega^4A^2}{c_p^2}\frac{1}{T}\int_0^T \cos^2(k_1x_1 + k_2x_2 - \omega t)\mathrm{d}t \\ &= \frac{\rho\omega^4A^2}{2c_p^2}, \end{aligned}$$

$$c = \frac{\langle I\rangle}{\langle w\rangle} = \frac{\frac{\rho\omega^4A^2}{2c_p}}{\frac{\rho\omega^4A^2}{2c_p^2}} = c_p.$$

Example 3 Given that the vector potential of the displacement field is $\boldsymbol{\psi} = (B_1\boldsymbol{i} + B_3\boldsymbol{k})\exp[i(k_1x_1 + k_3x_3 - \omega t)]$, we need to calculate: (1) the components of strain tensor, (2) the components of stress tensor, (3) energy flow density vector and (4) wave intensity and propagation velocity.

Solution: Since there is only a vector potential and no scalar potential, the wave is a shear wave. (Note: $k_2 = 0$, $B_2 = 0$.)

The displacement is

$$\boldsymbol{u} = \nabla \times \boldsymbol{\psi} = \frac{\partial}{\partial x_m}\boldsymbol{e}_m \times \psi_n\boldsymbol{e}_n = \frac{\partial \psi_n}{\partial x_m}(\boldsymbol{e}_m \times \boldsymbol{e}_n) = e_{jmn}\frac{\partial \psi_n}{\partial x_m}\boldsymbol{e}_j,$$

and

$$u_j = e_{jmn}\frac{\partial \psi_n}{\partial x_m} = e_{jmn}(ik_m)B_n\exp[i(k_1x_1 + k_3x_3 - \omega t)].$$

Thus, we can get

$$u_1 = 0, u_2 = i(k_3B_1 - k_1B_3)\exp[i(k_1x_1 + k_3x_3 - \omega t)], \text{ and } u_3 = 0.$$

The vibration speed is

$$\begin{aligned}\dot{u}_j &= e_{jmn}(ik_m)(-i\omega)B_n\exp[i(k_1x_1 + k_3x_3 - \omega t)]\\ &= e_{jmn}k_mB_n\omega\exp[i(k_1x_1 + k_3x_3 - \omega t)].\end{aligned}$$

Thus, we can get

$$\dot{u}_1 = 0, \dot{u}_2 = \omega(k_3B_1 - k_1B_3)\exp[i(k_1x_1 + k_3x_3 - \omega t)], \text{ and } \dot{u}_3 = 0.$$

(1) The components of strain tensor

$$\begin{aligned}\varepsilon_{ij} &= \frac{1}{2}\left(u_{i,j} + u_{j,i}\right)\\ &= \frac{1}{2}\left[i\left(k_je_{imn} + k_ie_{jmn}\right)\right]ik_mB_n\exp[i(k_1x_1 + k_3x_3 - \omega t)]\\ &= -\frac{1}{2}\left(k_je_{imn} + k_ie_{jmn}\right)k_mB_n\exp[i(k_1x_1 + k_3x_3 - \omega t)].\end{aligned}$$

Thus, we can get

$$\varepsilon_{11} = -\frac{1}{2}(k_1k_2B_3 - k_1k_3B_2 + k_1k_2B_3 - k_1k_3B_2)\exp[i(k_1x_1 + k_3x_3 - \omega t)] = 0,$$

$$\varepsilon_{22} = -\frac{1}{2}(k_2k_3B_1 - k_2k_1B_3 + k_2k_3B_1 - k_2k_1B_3)\exp[i(k_1x_1 + k_3x_3 - \omega t)] = 0,$$

$$\varepsilon_{33} = -\frac{1}{2}(k_3k_1B_2 - k_3k_2B_1 + k_3k_1B_2 - k_3k_2B_1)\exp[i(k_1x_1 + k_3x_3 - \omega t)] = 0,$$

$$\begin{aligned}\varepsilon_{12} &= -\frac{1}{2}(k_2k_2B_3 - k_2k_3B_2 + k_1k_3B_1 - k_1k_1B_3)\exp[i(k_1x_1 + k_3x_3 - \omega t)]\\ &= -\frac{1}{2}k_1(k_3B_1 - k_1B_3)\exp[i(k_1x_1 + k_3x_3 - \omega t)],\end{aligned}$$

$$\begin{aligned}\varepsilon_{23} &= -\frac{1}{2}(k_3k_3B_1 - k_3k_1B_3 + k_2k_1B_2 - k_2k_2B_1)\exp[i(k_1x_1 + k_3x_3 - \omega t)]\\ &= -\frac{1}{2}k_3(k_3B_1 - k_1B_3)\exp[i(k_1x_1 + k_3x_3 - \omega t)],\end{aligned}$$

and

$$\varepsilon_{31} = -\frac{1}{2}(+k_1 k_1 B_2 - k_1 k_2 B_1 + k_3 k_2 B_3 - k_3 k_3 B_2)\exp[i(k_1 x_1 + k_3 x_3 - \omega t)]$$
$$= 0.$$

$$(\text{Note}: k_2 = 0, \text{and} B_2 = 0.)$$

(2) The components of stress tensor

According to the generalized Hooke's law, we can obtain

$$\begin{aligned}\sigma_{ij} &= \lambda\theta\delta_{ij} + 2\mu\varepsilon_{ij} \\ &= \lambda u_{l,l}\delta_{ij} + 2\mu\varepsilon_{ij} \\ &= \lambda(ik_l)ie_{lmn}k_m B_n \exp[i(k_1 x_1 + k_3 x_3 - \omega t)]\delta_{ij} + 2\mu\varepsilon_{ij} \\ &= 2\mu\varepsilon_{ij},\end{aligned}$$

$$\sigma_{11} = 2\mu\varepsilon_{11} = 0,\ \sigma_{22} = 2\mu\varepsilon_{22} = 0,\ \sigma_{33} = 2\mu\varepsilon_{33} = 0,$$

$$\sigma_{12} = 2\mu\varepsilon_{12} = \mu k_1(k_1 B_3 - k_3 B_1)\exp[i(k_1 x_1 + k_3 x_3 - \omega t)],$$

$$\sigma_{23} = 2\mu\varepsilon_{23} = \mu k_3(k_1 B_3 - k_3 B_1)\exp[i(k_1 x_1 + k_3 x_3 - \omega t)],$$

and

$$\sigma_{31} = 2\mu\varepsilon_{31} = 0.$$

(3) Energy flow density vector

$$\begin{aligned}I_i &= -\sigma_{ij}\frac{\partial u_j}{\partial t} \\ &= -2\mu\varepsilon_{ij}\frac{\partial u_j}{\partial t} \\ &= -2\mu\left[-\frac{1}{2}(k_j e_{imn} + k_i e_{jmn})k_m B_n \cos(k_1 x_1 + k_3 x_3 - \omega t)\right] \\ &\quad \times \left[(-i\omega)\cdot ie_{jrs}k_r B_s \cos(k_1 x_1 + k_3 x_3 - \omega t)\right] \\ &= \mu\big(k_j e_{imn} + k_i e_{jmn}\big)k_m B_n \cos(k_1 x_1 + k_3 x_3 - \omega t)\omega e_{jrs}k_r B_s \cos(k_1 x_1 + k_3 x_3 - \omega t) \\ &= \mu\omega\cos^2(k_1 x_1 + k_3 x_3 - \omega t)\big[\big(k_j e_{imn} + k_i e_{jmn}\big)k_m B_n\big]e_{jrs}k_r B_s,\end{aligned}$$

$$\begin{aligned}I_1 &= \mu\omega\cos^2(k_1 x_1 + k_3 x_3 - \omega t)[k_1(k_3 B_1 - k_1 B_3)](k_3 B_1 - k_1 B_3) \\ &= \mu\omega\cos^2(k_1 x_1 + k_3 x_3 - \omega t)k_1(k_3 B_1 - k_1 B_3)^2,\end{aligned}$$

$$I_2 = 0,$$

$$I_3 = \mu\omega\cos^2(k_1x_1 + k_3x_3 - \omega t)[k_3(k_3B_1 - k_1B_3)](k_3B_1 - k_1B_3)$$
$$= \mu\omega\cos^2(k_1x_1 + k_3x_3 - \omega t)k_3(k_3B_1 - k_1B_3)^2,$$

and

$$\begin{aligned}
I &= \sqrt{\left(I_1^2 + I_2^2 + I_3^2\right)} \\
&= \mu\omega\cos^2(k_1x_1 + k_3x_3 - \omega t)(k_3B_1 - k_1B_3)^2\sqrt{\left(k_1^2 + k_3^2\right)} \\
&= \mu\omega k\cos^2(k_1x_1 + k_3x_3 - \omega t)(k_3B_1 - k_1B_3)^2 \\
&= \rho c_s\omega^2(k_3B_1 - k_1B_3)^2\cos^2(k_1x_1 + k_3x_3 - \omega t).
\end{aligned}$$

(4) Wave intensity and propagation velocity

$$\begin{aligned}
\langle I\rangle &= \frac{1}{T}\int_0^T I\mathrm{d}t \\
&= \rho c_s\omega^2(k_3B_1 - k_1B_3)^2\frac{1}{T}\int_0^T \cos^2[i(k_1x_1 + k_3x_3 - \omega t)]\mathrm{d}t \\
&= \rho c_s\omega^2(k_3B_1 - k_1B_3)^2\frac{1}{T}\frac{T}{2} \\
&= \frac{1}{2}\rho c_s\omega^2(k_3B_1 - k_1B_3)^2,
\end{aligned}$$

$$\begin{aligned}
w_k &= \frac{1}{2}\rho\left(\dot{u}_j\dot{u}_j\right) \\
&= \frac{1}{2}\rho\left[(-i\omega)\cdot ie_{jmn}k_mB_n\cos(k_1x_1 + k_3x_3 - \omega t)\right]^2 \\
&= \frac{1}{2}\rho\omega^2(k_3B_1 - k_1B_3)^2\cos^2(k_1x_1 + k_3x_3 - \omega t),
\end{aligned}$$

$$\begin{aligned}
w_p &= \frac{1}{2}\sigma_{ij}\varepsilon_{ij} = \frac{1}{2}\left(2\mu\varepsilon_{ij}\right)\varepsilon_{ij} = \mu\varepsilon_{ij}^2 \\
&= \mu\left(\varepsilon_{12}^2 + \varepsilon_{21}^2 + \varepsilon_{23}^2 + \varepsilon_{32}^2\right) \\
&= \frac{1}{2}\mu(k_1B_3 - k_3B_1)^2\cos^2(k_1x_1 + k_3x_3 - \omega t)\left(k_1^2 + k_3^2\right) \\
&= \frac{1}{2}\mu(k_1B_3 - k_3B_1)^2\cos^2(k_1x_1 + k_3x_3 - \omega t)k^2 \\
&= \frac{1}{2}\rho\omega^2(k_1B_3 - k_3B_1)^2\cos^2(k_1x_1 + k_3x_3 - \omega t),
\end{aligned}$$

$$\langle w\rangle = w_k + w_p = \rho\omega^2(k_3B_1 - k_1B_3)^2\cos^2(k_1x_1 + k_3x_3 - \omega t),$$

$$w = \frac{1}{T}\int_0^T w\mathrm{d}t$$

$$= \rho\omega^2(k_3B_1 - k_1B_3)^2\frac{1}{T}\int_0^T \cos^2(k_1x_1 + k_3x_3 - \omega t)\mathrm{d}t$$

$$= \frac{1}{2}\rho\omega^2(k_3B_1 - k_1B_3)^2,$$

and

$$c = \frac{\langle I\rangle}{\langle w\rangle} = \frac{\frac{1}{2}\rho c_s\omega^2(k_3B_1 - k_1B_3)^2}{\frac{1}{2}\rho\omega^2(k_3B_1 - k_1B_3)^2} = c_s.$$

2.4 Inhomogeneous Plane Wave

When a wave propagates in a dissipative medium, the vibration amplitude will gradually decrease until finally disappears due to the dissipation of energy. Because of the different excitation modes of the wave, the direction of attenuation may be inconsistent with the direction of propagation. This will lead to the inhomogeneous distribution of the vibration amplitude on the equiphase plane. The wave having this property is generally called inhomogeneous wave [5, 8, 10].

Usually, the inhomogeneous wave can be expressed as

$$\boldsymbol{u} = \boldsymbol{A}e^{i(\boldsymbol{k}\cdot\boldsymbol{r}-\omega t)}, \tag{2.4.1}$$

where the vibration amplitude and wave vector are complex quantities expressed by

$$\boldsymbol{A} = \boldsymbol{A}_r + i\boldsymbol{A}_m, \tag{2.4.2}$$

and

$$\boldsymbol{k} = \boldsymbol{k}_r + i\boldsymbol{k}_m. \tag{2.4.3}$$

Substituting Eq. (2.4.3) into Eq. (2.4.1), we obtain

$$\boldsymbol{u} = \boldsymbol{A}e^{-\boldsymbol{k}_m\cdot\boldsymbol{r}}e^{i(\boldsymbol{k}_r\cdot\boldsymbol{r}-\omega t)}.$$

It can be shown that the vibration amplitude decay exponentially with the function $e^{-\boldsymbol{k}_m\cdot\boldsymbol{r}}$ in the process of wave propagation. The imaginary part $\boldsymbol{k}_m$ of the wave vector denotes the attenuation of wave. All the points on the plane satisfying

$$\boldsymbol{k}_m \cdot \boldsymbol{r} = \text{const} \tag{2.4.4}$$

have same vibration amplitudes, and this plane is generally called **constant amplitude plane**. The normal direction of the constant amplitude plane is called attenuation direction. The direction of the imaginary part $\boldsymbol{k}_m$ of the wave vector denotes attenuation direction, the magnitude of $\boldsymbol{k}_m$ denotes the magnitude of attenuation and is also called attenuation coefficient.

The direction and magnitude of the real part, i.e. $\boldsymbol{k}_r$, of the wave vector denote the direction of wave propagation and wave number, respectively. The equation of equiphase plane is written as

$$\boldsymbol{k}_r \cdot \boldsymbol{r} - \omega t = \text{const.} \tag{2.4.5}$$

And, thus, the phase velocity is

$$c = \frac{\omega}{k_r}(k_r = |\boldsymbol{k}_r|). \tag{2.4.6}$$

Generally, the directions of $\boldsymbol{k}_r$ and $\boldsymbol{k}_m$ are different. There are several possible cases in the practical problem which need to be discussed:

(1) $\boldsymbol{k}_r$ and $\boldsymbol{k}_m$ have same directions. Then, although the wave is attenuated in the process of wave propagation, the distribution of vibration amplitude in equiphase plane is still homogeneous. This type of waves, which only exists in dissipative medium, is called evanescent wave or homogeneous wave. It is noted that the evanescent wave is also used to denote the wave which has $\boldsymbol{k}_r = \boldsymbol{0}$ in some existing literatures.
(2) $\boldsymbol{k}_r$ and $\boldsymbol{k}_m$ have orthogonal directions. Then, the distribution of vibration amplitude in equiphase plane is inhomogeneous and decays with the exponential law $e^{-\boldsymbol{k}_m \cdot \boldsymbol{r}}$. But the vibration amplitude doesn't decay along the direction of propagation. So, this type of waves may also exist in a non-dissipative medium.
(3) $\boldsymbol{k}_r$ and $\boldsymbol{k}_m$ are neither parallel nor perpendicular. Then, the constant amplitude plane is not consistent with the equiphase plane. The distribution of vibration amplitude in equiphase plane is inhomogeneous. This type of wave should be called inhomogeneous wave. However, it is also improperly called homogeneous wave in some existing literatures. The angle between $\boldsymbol{k}_r$ and $\boldsymbol{k}_m$ cannot be determined by the parameters of materials, but still dependent on the excitation mode of wave. The magnitudes of $\boldsymbol{k}_r$ and $\boldsymbol{k}_m$ are both dependent on the parameters of materials and excitation frequency. In general, the relation k_r~ω is called dispersion relation while k_m~ω is called amplitude-frequency relation.

The wave equation in a dissipative medium can be written as

$$\nabla^2 \boldsymbol{u} - \frac{\boldsymbol{k} \cdot \boldsymbol{k}}{\omega^2}\frac{\partial^2 u}{\partial t^2} = 0, \tag{2.4.7}$$

where the coefficient $\frac{\boldsymbol{k}\cdot\boldsymbol{k}}{\omega^2}$ is determined by the parameters of materials. Since the constitutive parameters of the dissipative medium are complex, the coefficient $\frac{\boldsymbol{k}\cdot\boldsymbol{k}}{\omega^2}$ is also a complex. The angular frequency ω is a real number, and its range of values is $(0, +\infty)$. So $\boldsymbol{k}$ must be a complex vector, namely

$$\boldsymbol{k}\cdot\boldsymbol{k} = a + ib. \tag{2.4.8}$$

Based on the thermodynamic constraints, it demands that $b > 0$, which means that the wave must propagate in a dissipative medium. Due to the attenuation of energy, the wave is attenuated in the process of propagation. Equation (2.4.8) can be written as

$$k_{\mathrm{r}}^2 - k_{\mathrm{m}}^2 = a, \tag{2.4.9a}$$

and

$$2k_r \cdot k_m \cos\gamma = b. \tag{2.4.9b}$$

Equations (2.4.9a) and (2.4.9b) include three parameters, i.e. k_{r}, k_{m} and γ, which is the angle between vectors $\boldsymbol{k}_r$ and $\boldsymbol{k}_m$. Even if all parameters of materials are known, namely a and b are given, k_{r}, k_{m} and γ cannot be obtained by solving Eqs. (2.4.9a) and (2.4.9b). Unless further information is added, the angle γ is usually related to the mode of excitation.

Next, we discuss the complex vector of the vibration amplitude. First, a theorem is given before discussing it in depth.

Theorem 1 *If there are orthogonal vectors* $\boldsymbol{a}$ *and* $\boldsymbol{b}$ *satisfying* $\boldsymbol{a}\cdot\boldsymbol{b} = 0$, *then any complex vector* $\boldsymbol{A} = \boldsymbol{A}_r + i\boldsymbol{A}_m$ *can be expressed as* [5].

$$\boldsymbol{A}_r + i\boldsymbol{A}_m = e^{i\beta}(\boldsymbol{a} + i\boldsymbol{b}). \tag{2.4.10}$$

Prove: If Eq. (2.4.10) is true, we have

$$(\boldsymbol{a} + i\boldsymbol{b})\cdot(\boldsymbol{a} + i\boldsymbol{b}) = e^{-2i\beta}(\boldsymbol{A}_r + i\boldsymbol{A}_m)\cdot(\boldsymbol{A}_r + i\boldsymbol{A}_m), \tag{2.4.11}$$

and

$$\boldsymbol{a}\cdot\boldsymbol{a} - \boldsymbol{b}\cdot\boldsymbol{b} + i2\boldsymbol{a}\cdot\boldsymbol{b} = e^{-2i\beta}(\boldsymbol{A}_r\cdot\boldsymbol{A}_r - \boldsymbol{A}_m\cdot\boldsymbol{A}_m + i2\boldsymbol{A}_r\cdot\boldsymbol{A}_m). \tag{2.4.12}$$

Due to $\boldsymbol{a}\cdot\boldsymbol{b} = 0$, we have

$$\begin{aligned}\boldsymbol{a}\cdot\boldsymbol{a} - \boldsymbol{b}\cdot\boldsymbol{b} &= \cos 2\beta(\boldsymbol{A}_r\cdot\boldsymbol{A}_r - \boldsymbol{A}_m\cdot\boldsymbol{A}_m) + 2\sin 2\beta\boldsymbol{A}_r\cdot\boldsymbol{A}_m \\ &-i[\sin 2\beta(\boldsymbol{A}_r\cdot\boldsymbol{A}_r - \boldsymbol{A}_m\cdot\boldsymbol{A}_m) - 2\cos 2\beta\boldsymbol{A}_r\cdot\boldsymbol{A}_m],\end{aligned} \tag{2.4.13}$$

so

$$\boldsymbol{a}\cdot\boldsymbol{a}-\boldsymbol{b}\cdot\boldsymbol{b}=\cos 2\beta(\boldsymbol{A}_r\cdot\boldsymbol{A}_r-\boldsymbol{A}_m\cdot\boldsymbol{A}_m)+2\sin 2\beta\boldsymbol{A}_r\cdot\boldsymbol{A}_m, \tag{2.4.14a}$$

and

$$0=\sin 2\beta(\boldsymbol{A}_r\cdot\boldsymbol{A}_r-\boldsymbol{A}_m\cdot\boldsymbol{A}_m)-2\cos 2\beta\boldsymbol{A}_r\cdot\boldsymbol{A}_m. \tag{2.4.14b}$$

When $\boldsymbol{A}_r\cdot\boldsymbol{A}_r\neq\boldsymbol{A}_m\cdot\boldsymbol{A}_m$, we get from above equation

$$\tan 2\beta=\frac{2\boldsymbol{A}_r\cdot\boldsymbol{A}_m}{\boldsymbol{A}_r\cdot\boldsymbol{A}_r-\boldsymbol{A}_m\cdot\boldsymbol{A}_m}. \tag{2.4.15}$$

If β is known, using the values of $\boldsymbol{A}_r$ and $\boldsymbol{A}_m$, $\boldsymbol{a}$ and $\boldsymbol{b}$ can be determined by

$$\boldsymbol{a}=\cos\beta\boldsymbol{A}_r+\sin\beta\boldsymbol{A}_m, \tag{2.4.16a}$$

and

$$\boldsymbol{b}=\cos\beta\boldsymbol{A}_m-\sin\beta\boldsymbol{A}_r. \tag{2.4.16b}$$

When $\boldsymbol{A}_r\cdot\boldsymbol{A}_r=\boldsymbol{A}_m\cdot\boldsymbol{A}_m$, we have $(\boldsymbol{A}_r-\boldsymbol{A}_m)\cdot(\boldsymbol{A}_r+\boldsymbol{A}_m)=0$.
Let

$$\boldsymbol{a}=\frac{1}{\sqrt{2}}(\boldsymbol{A}_r-\boldsymbol{A}_m),\ \boldsymbol{b}=\frac{1}{\sqrt{2}}(\boldsymbol{A}_r+\boldsymbol{A}_m), \tag{2.4.17}$$

so

$$\boldsymbol{A}_r+i\boldsymbol{A}_m=e^{-i\frac{\pi}{4}}(\boldsymbol{a}+i\boldsymbol{b}). \tag{2.4.18}$$

Now $\boldsymbol{a}\perp\boldsymbol{b}$ and $\beta=-\frac{\pi}{4}$. Qed.
Using the above theorem, the displacement with physical meaning is

$$\boldsymbol{u}=\mathrm{Re}\big[e^{-\boldsymbol{k}_m\cdot\boldsymbol{r}}(\boldsymbol{a}+i\boldsymbol{b})e^{i(\boldsymbol{k}_r\cdot\boldsymbol{r}-\omega t+\beta)}\big]. \tag{2.4.19}$$

Let $\xi=\boldsymbol{k}_r\cdot\boldsymbol{r}-\omega t+\beta$, we obtain

$$\boldsymbol{u}=e^{-\boldsymbol{k}_m\cdot\boldsymbol{r}}(\boldsymbol{a}\cos\xi-\boldsymbol{b}\sin\xi). \tag{2.4.20}$$

Due to $\boldsymbol{a}\cdot\boldsymbol{b}=0$, let the coordinate axis be along the directions of $\boldsymbol{a}$ and $\boldsymbol{b}$, we obtain

$$\begin{aligned}u_x&=e^{-\boldsymbol{k}_m\cdot\boldsymbol{r}}\cdot|\boldsymbol{a}|\cos\xi\\ u_y&=e^{-\boldsymbol{k}_m\cdot\boldsymbol{r}}\cdot|\boldsymbol{b}|\sin\xi\end{aligned}, \tag{2.4.21}$$

and

$$\frac{u_x^2}{\left(|\boldsymbol{a}|e^{-\boldsymbol{k}_m\cdot\boldsymbol{r}}\right)^2}+\frac{u_y^2}{\left(|\boldsymbol{b}|e^{-\boldsymbol{k}_m\cdot\boldsymbol{r}}\right)^2}=1. \tag{2.4.22}$$

The polarization plane of the mass point is elliptical, whose semi-major and semi-minor axis are $|\boldsymbol{a}|e^{-\boldsymbol{k}_m\cdot\boldsymbol{r}}$ and $|\boldsymbol{b}|e^{-\boldsymbol{k}_m\cdot\boldsymbol{r}}$, respectively. The polarization plane of the mass point is determined by $\boldsymbol{A}_r$ and $\boldsymbol{A}_m$. If $|\boldsymbol{a}|=|\boldsymbol{b}|$, the polarization of the mass point is a circle. The above discussion is suitable to the case with $\boldsymbol{A}_r\times\boldsymbol{A}_m\neq\boldsymbol{0}$. If $\boldsymbol{A}_r\times\boldsymbol{A}_m=\boldsymbol{0}$, then $\boldsymbol{A}_r$ and $\boldsymbol{A}_m$ have the same directions, we get

$$\boldsymbol{A}_r+i\boldsymbol{A}_m=Ae^{i\beta}\boldsymbol{e}_A, \tag{2.4.23}$$

where $\boldsymbol{e}_A$ is unit vector of $\boldsymbol{A}_r$ or $\boldsymbol{A}_m$. $A^2=|\boldsymbol{A}_r|^2+|\boldsymbol{A}_m|^2$. Thus, the displacement having physical meaning is

$$\begin{aligned}\boldsymbol{u}&=\mathrm{Re}\left\{(\boldsymbol{A}_r+i\boldsymbol{A}_m)e^{i(\boldsymbol{k}_r\cdot\boldsymbol{r}-\omega t)}e^{-\boldsymbol{k}_m\cdot\boldsymbol{r}}\right\}\\&=\mathrm{Re}\left\{A\boldsymbol{e}_Ae^{-\boldsymbol{k}_m\cdot\boldsymbol{r}}\cdot e^{i(\boldsymbol{k}_r\cdot\boldsymbol{r}-\omega t+\beta)}\right\}\\&=A\boldsymbol{e}_Ae^{-\boldsymbol{k}_m\cdot\boldsymbol{r}}\cdot\cos\xi\,(\xi=\boldsymbol{k}_r\cdot\boldsymbol{r}-\omega t+\beta).\end{aligned} \tag{2.4.24}$$

The above equation shows that if $\boldsymbol{A}_r\times\boldsymbol{A}_m=\boldsymbol{0}$, the polarization of the mass point is a straight line.

Next, we define the longitudinal and transverse waves based on the polarization characteristics of the point. Similar to the definition of the longitudinal and transverse waves in classical elastic media, let the inhomogeneous longitudinal wave satisfies

$$\boldsymbol{A}=\varphi\boldsymbol{k}, \tag{2.4.25}$$

where φ is a scalar complex. From Eq. (2.4.25), one can find

$$\boldsymbol{A}\times\boldsymbol{k}=\boldsymbol{0}. \tag{2.4.26}$$

If $\boldsymbol{A}$ and $\boldsymbol{k}$ are all real vectors, the above equation means that the polarization direction of the point is consistent with the propagation direction of the wave. But in the case of complex vectors, it does not mean the polarization direction is consistent with the propagation direction. But it can indicate

$$\nabla\times\boldsymbol{u}=\boldsymbol{0}. \tag{2.4.27}$$

Let the inhomogeneous transverse wave satisfies

$$\boldsymbol{A}=\boldsymbol{k}\times\boldsymbol{\varphi}, \tag{2.4.28}$$

where $\boldsymbol{\varphi}$ is a complex vector. Recalling the relation of the vector cross product

$$\boldsymbol{k} \times (\boldsymbol{A} \times \boldsymbol{k}) = (\boldsymbol{k} \cdot \boldsymbol{k})\boldsymbol{A} - (\boldsymbol{k} \cdot \boldsymbol{A})\boldsymbol{k},$$

we obtain from Eq. (2.4.28)

$$\boldsymbol{A} \cdot \boldsymbol{k} = 0. \tag{2.4.29}$$

If $\boldsymbol{A}$ and $\boldsymbol{k}$ are both real vectors, the above equation indicates that the polarization direction of point is perpendicular to the direction of wave propagation, but it does not hold in the case of complex vector. But it can indicate

$$\nabla \cdot \boldsymbol{u} = 0. \tag{2.4.30}$$

Next, let's discuss the energy carried by the inhomogeneous waves. Recalling the relation

$$\nabla \cdot (\boldsymbol{T} \cdot \boldsymbol{a}) = \boldsymbol{T} : \nabla \boldsymbol{a} + (\nabla \cdot \boldsymbol{T}) \cdot \boldsymbol{a}, \tag{2.4.31}$$

we have

$$(\nabla \cdot \boldsymbol{\sigma}) \cdot \dot{\boldsymbol{u}} + \boldsymbol{\sigma} : \dot{\boldsymbol{\varepsilon}} = \nabla \cdot (\boldsymbol{\sigma} \cdot \dot{\boldsymbol{u}}). \tag{2.4.32}$$

Considering $\nabla \cdot \boldsymbol{\sigma} = \rho \ddot{\boldsymbol{u}}$, we get

$$\rho \ddot{\boldsymbol{u}} \cdot \dot{\boldsymbol{u}} + \boldsymbol{\sigma} : \dot{\boldsymbol{\varepsilon}} = \nabla \cdot (\boldsymbol{\sigma} \cdot \dot{\boldsymbol{u}}). \tag{2.4.33}$$

Let $\dot{K} = \rho \ddot{\boldsymbol{u}} \cdot \dot{\boldsymbol{u}}$, $\dot{E} = \boldsymbol{\sigma} : \dot{\boldsymbol{\varepsilon}}$ and $\boldsymbol{J} = -\boldsymbol{\sigma} \cdot \dot{\boldsymbol{u}}$, we have

$$\dot{K} + \dot{E} = -\nabla \cdot \boldsymbol{J}. \tag{2.4.34}$$

The integral form of Eq. (2.4.34) is

$$\frac{\mathrm{d}}{\mathrm{d}t} \int_v (K + E)\mathrm{d}v = -\int_v (\nabla \cdot \boldsymbol{J})\mathrm{d}v = -\oint_s \boldsymbol{J} \cdot \boldsymbol{n}\mathrm{d}s. \tag{2.4.35}$$

The above relation is the energy balance equation; i.e. the time change rate of the sum of the kinetic energy and deformation energy equals to the flux of the surface energy flow vector through the surface surrounding the volume element. The more general form of the energy balance equation can be expressed as

$$\frac{\mathrm{d}}{\mathrm{d}t}(K + E) = -\nabla \cdot \boldsymbol{J} - \nabla \cdot \boldsymbol{q} + \rho \boldsymbol{b} \cdot \dot{\boldsymbol{u}} + \rho r, \tag{2.4.36}$$

where $\boldsymbol{q}$ is the heat-flow vector (along the outer normal direction), r is the heat flow per unit mass, $\boldsymbol{b}$ is the force applied on unit mass. Considering the energy dissipation in a dissipative medium, the energy balance Eq. (2.4.34) can be modified as

$$\frac{\mathrm{d}}{\mathrm{d}t}(K+E)+D=-\nabla\cdot\boldsymbol{J}, \tag{2.4.37}$$

where D denotes the energy dissipation rate per unit mass.

Finally, let's derive the specific expression of surface energy flow

$$\begin{aligned}\boldsymbol{J}&=-\mathrm{Re}(\boldsymbol{\sigma})\cdot\mathrm{Re}(\dot{\boldsymbol{u}})\\&=-\frac{1}{2}(\boldsymbol{\sigma}+\boldsymbol{\sigma}^*)\cdot\frac{1}{2}(\dot{\boldsymbol{u}}+\dot{\boldsymbol{u}}^*)\\&=-\frac{1}{4}(\boldsymbol{\sigma}\cdot\dot{\boldsymbol{u}}^*+\boldsymbol{\sigma}^*\cdot\dot{\boldsymbol{u}})-\frac{1}{4}(\boldsymbol{\sigma}\cdot\dot{\boldsymbol{u}}+\boldsymbol{\sigma}^*\cdot\dot{\boldsymbol{u}}^*).\end{aligned} \tag{2.4.38}$$

Recalling the relation

$$\frac{1}{T}\int_0^T \mathrm{e}^{\pm\mathrm{i}\omega t}\mathrm{d}t=0, \tag{2.4.39}$$

the average energy flow in a period is

$$\begin{aligned}\langle\boldsymbol{J}\rangle&=\frac{1}{T}\int_0^T \boldsymbol{J}(t)\mathrm{d}t\\&=-\frac{1}{4T}\int_0^T(\boldsymbol{\sigma}\cdot\dot{\boldsymbol{u}}^*+\boldsymbol{\sigma}^*\cdot\dot{\boldsymbol{u}})\mathrm{d}t\\&=-\frac{1}{2T}\int_0^T\mathrm{Re}(\boldsymbol{\sigma}\cdot\dot{\boldsymbol{u}}^*)\mathrm{d}t.\end{aligned} \tag{2.4.40}$$

The energy flow along the direction of wave propagation $\boldsymbol{n}$ is expressed as

$$\langle J_n\rangle=\langle\boldsymbol{J}\rangle\cdot\boldsymbol{n}. \tag{2.4.41}$$

Let

$$\boldsymbol{u}=\boldsymbol{U}_0(\omega)e^{i(\boldsymbol{k}\cdot\boldsymbol{r}-\omega t)},\ \boldsymbol{k}=\boldsymbol{k}_r+i\boldsymbol{k}_m,$$

we obtain

$$\boldsymbol{\sigma} = \lambda tr(\boldsymbol{\varepsilon})\boldsymbol{I} + 2\mu\boldsymbol{\varepsilon}, \tag{2.4.42a}$$

$$\boldsymbol{\sigma} \cdot \boldsymbol{n} = \lambda tr(\boldsymbol{\varepsilon})\boldsymbol{n} + 2\mu\boldsymbol{\varepsilon} \cdot \boldsymbol{n}, \tag{2.4.42b}$$

$$\varepsilon = \frac{1}{2}(\nabla\boldsymbol{u} + \boldsymbol{u}\nabla) = \frac{1}{2}(i\boldsymbol{k} \otimes \boldsymbol{U}_0 + \boldsymbol{U}_0 \otimes i\boldsymbol{k})e^{i(\boldsymbol{k}\cdot\boldsymbol{r}-\omega t)}, \tag{2.4.42c}$$

$$tr(\boldsymbol{\varepsilon}) = \boldsymbol{\varepsilon} : \boldsymbol{I} = i\boldsymbol{k} \cdot \boldsymbol{u}, \tag{2.4.42d}$$

$$\langle \boldsymbol{J} \rangle = \frac{1}{2}\omega\mathrm{Re}\left\{e^{-2\boldsymbol{k}_m\cdot\boldsymbol{r}}\left[\mu\boldsymbol{k}\left(\boldsymbol{U}_0 \cdot \boldsymbol{U}_0^*\right) + \mu\boldsymbol{U}_0\left(\boldsymbol{k} \cdot \boldsymbol{U}_0^*\right) + \lambda(\boldsymbol{k} \cdot \boldsymbol{U}_0)\boldsymbol{U}_0^*\right]\right\}, \tag{2.4.42e}$$

and

$$\begin{aligned} \langle J_n \rangle = \frac{1}{2}\omega e^{-2\boldsymbol{k}_m\cdot\boldsymbol{r}} \cdot \mathrm{Re}[&\mu\left(\boldsymbol{k} \cdot \boldsymbol{U}_0^*\right)(\boldsymbol{U}_0 \cdot \boldsymbol{n}) + \mu\left(\boldsymbol{U}_0 \cdot v_0^*\right)(\boldsymbol{k} \cdot \boldsymbol{n}) \\ &+ \lambda(\boldsymbol{k} \cdot \boldsymbol{U}_0)\left(\boldsymbol{U}_0^* \cdot \boldsymbol{n}\right)]. \end{aligned} \tag{2.4.42f}$$

For the inhomogeneous longitudinal wave, we have [5]

$$\boldsymbol{k} \cdot \boldsymbol{k} = \frac{\rho\omega^2}{\lambda + 2\mu}, \ \boldsymbol{U}_0 = i\boldsymbol{k}\Phi, \tag{2.4.43}$$

where Φ is the complex function of the scalar potentials, i.e. $\boldsymbol{u} = \nabla\Phi$. Then,

$$\langle \boldsymbol{J} \rangle = \frac{1}{2}|\Phi|^2\omega e^{-2\boldsymbol{k}_m\cdot\boldsymbol{r}}\left[\rho\omega^2\boldsymbol{k}_r - 4(\boldsymbol{k}_r \times \boldsymbol{k}_m) \times (u_r\boldsymbol{k}_m + u_m\boldsymbol{k}_r)\right], \tag{2.4.44}$$

and

$$\langle J_n \rangle = \frac{1}{2}|\Phi|^2\omega e^{-2\boldsymbol{k}_m\cdot\boldsymbol{r}}\left[\rho\omega^2|\boldsymbol{k}_r| - 4u_r|\boldsymbol{k}_r||\boldsymbol{k}_m|^2\sin^2\theta\right], \tag{2.4.45}$$

where $\mu = \mu_r + \mathrm{i}\mu_m$. θ is the angle between $\boldsymbol{k}_r$ and $\boldsymbol{k}_m$.

For the inhomogeneous transverse wave, we have

$$\boldsymbol{k} \cdot \boldsymbol{k} = \frac{\rho\omega^2}{\mu}, \ \boldsymbol{U}_0 = i\boldsymbol{k} \times \boldsymbol{\Psi}, \tag{2.4.46}$$

where $\boldsymbol{\Psi}$ is the complex function of the vector potentials, i.e. $\boldsymbol{u} = \nabla \times \boldsymbol{\Psi}$, and

$$\begin{aligned} \left(\boldsymbol{U}_0^* \cdot \boldsymbol{k}\right)\boldsymbol{U}_0 &= -\left[(\boldsymbol{k} \times \boldsymbol{\Psi}^*) \cdot \boldsymbol{k}^*(\boldsymbol{k} \times \boldsymbol{\Psi})\right] \\ &= -\left[\boldsymbol{k}^* \times (\boldsymbol{k} \times \boldsymbol{\Psi}) \times (\boldsymbol{k} \times \boldsymbol{\Psi}^*) + (\boldsymbol{k} \times \boldsymbol{\Psi}^*) \cdot (\boldsymbol{k} \times \boldsymbol{\Psi})\boldsymbol{k}^*\right]. \end{aligned} \tag{2.4.47}$$

Due to $\boldsymbol{k} \cdot \boldsymbol{\Psi} = 0$,

$$(\boldsymbol{k} \times \boldsymbol{\Psi}^*) \cdot (\boldsymbol{k} \times \boldsymbol{\Psi}) = [\boldsymbol{\Psi}^* \times (\boldsymbol{k} \times \boldsymbol{\Psi})] \cdot \boldsymbol{k} = (\boldsymbol{\Psi}^* \cdot \boldsymbol{\Psi})(\boldsymbol{k} \cdot \boldsymbol{k}), \tag{2.4.48}$$

then

$$\begin{aligned}(\boldsymbol{U}_0^* \cdot \boldsymbol{k})\boldsymbol{U}_0 &= -\{[(\boldsymbol{k} \cdot \boldsymbol{\Psi})\boldsymbol{k} - (\boldsymbol{k}^* \cdot \boldsymbol{k})\boldsymbol{\Psi}] \times (\boldsymbol{k} \times \boldsymbol{\Psi}^*) + (\boldsymbol{k} \cdot \boldsymbol{k})(\boldsymbol{\Psi}^* \cdot \boldsymbol{\Psi})\boldsymbol{k}^*\} \\ &= (\boldsymbol{k}^* \cdot \boldsymbol{\Psi})(\boldsymbol{k} \cdot \boldsymbol{k})\boldsymbol{\Psi}^* - (\boldsymbol{k}^* \cdot \boldsymbol{\Psi})(\boldsymbol{k} \cdot \boldsymbol{\Psi}^*)\boldsymbol{k} \\ &\quad + (\boldsymbol{k}^* \cdot \boldsymbol{k})(\boldsymbol{\Psi}^* \cdot \boldsymbol{\Psi})\boldsymbol{k} - (\boldsymbol{k} \cdot \boldsymbol{k})(\boldsymbol{\Psi}^* \cdot \boldsymbol{\Psi})\boldsymbol{k}^* \\ &= (\boldsymbol{k}^* \cdot \boldsymbol{\Psi})(\boldsymbol{k} \cdot \boldsymbol{k})\boldsymbol{\Psi}^* - (\boldsymbol{k}^* \cdot \boldsymbol{\Psi})(\boldsymbol{k} \cdot \boldsymbol{\Psi}^*)\boldsymbol{k} \\ &\quad + (\boldsymbol{\Psi}^* \cdot \boldsymbol{\Psi})[(\boldsymbol{k}^* \cdot \boldsymbol{k})\boldsymbol{k} - (\boldsymbol{k} \cdot \boldsymbol{k})\boldsymbol{k}^*],\end{aligned}$$

and

$$\begin{aligned}\boldsymbol{U}_0 \cdot \boldsymbol{U}_0^* &= (\boldsymbol{\Psi}^* \times \boldsymbol{\Psi}) \cdot (\boldsymbol{\Psi} \times \boldsymbol{k}) = \boldsymbol{\Psi} \cdot [\boldsymbol{k} \times (\boldsymbol{\Psi}^* \times \boldsymbol{k}^*)] \\ &= (\boldsymbol{k} \cdot \boldsymbol{k}^*)(\boldsymbol{\Psi} \cdot \boldsymbol{\Psi}^*) - (\boldsymbol{k} \cdot \boldsymbol{\Psi}^*)(\boldsymbol{k}^* \cdot \boldsymbol{\Psi}).\end{aligned} \tag{2.4.49}$$

In summary, we obtain

$$\begin{aligned}(\boldsymbol{U}_0 \cdot \boldsymbol{U}_0^*)\boldsymbol{k} + (\boldsymbol{k} \cdot \boldsymbol{U}_0^*)\boldsymbol{U}_0 &= (\boldsymbol{\Psi} \times \boldsymbol{\Psi}^*)[2(\boldsymbol{k} \cdot \boldsymbol{k}^*) - (\boldsymbol{k} \cdot \boldsymbol{k})\boldsymbol{k}^*] \\ &\quad - 2(\boldsymbol{k} \cdot \boldsymbol{\Psi}^*)(\boldsymbol{k}^* \cdot \boldsymbol{\Psi})\boldsymbol{k} + (\boldsymbol{k}^* \cdot \boldsymbol{\Psi})(\boldsymbol{k} \cdot \boldsymbol{k})\boldsymbol{\Psi}^*.\end{aligned} \tag{2.4.50}$$

Substituting Eq. (2.4.50) into Eq. (2.4.42e), we obtain

$$\langle J \rangle = \frac{1}{2}\omega e^{-2\boldsymbol{k}_m \cdot \boldsymbol{r}} \mathrm{Re}\begin{Bmatrix} \mu(\boldsymbol{\Psi} \cdot \boldsymbol{\Psi}^*)[\boldsymbol{k} \times (\boldsymbol{k} \cdot \boldsymbol{k}^*) + (\boldsymbol{k}^* \cdot \boldsymbol{k})\boldsymbol{k}] \\ -2(\boldsymbol{k} \cdot \boldsymbol{\Psi}^*)(\boldsymbol{k}^* \cdot \boldsymbol{\Psi})\boldsymbol{k} + (\boldsymbol{k}^* \cdot \boldsymbol{\Psi})(\boldsymbol{k} \cdot \boldsymbol{k})\boldsymbol{\Psi}^* \end{Bmatrix}. \tag{2.4.51}$$

2.5 Spectrum Analysis of Plane Wave

Suppose two plane waves with different frequencies propagating along the x direction

$$u_1(x, t) = A\cos(k_1 x - \omega_1 t), \tag{2.5.1}$$

$$u_2(x, t) = A\cos(k_2 x - \omega_2 t). \tag{2.5.2}$$

The result of superposition of these two waves in the process of the wave propagation is

$$u(x, t) = u_1(x, t) + u_2(x, t) = 2A\cos(\overline{k}x - \overline{\omega}t)\cos(kx - \omega t), \tag{2.5.3}$$

where

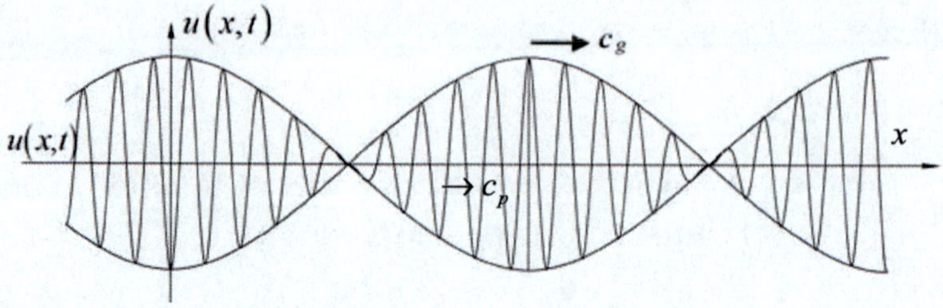

Fig. 2.4 Schematic diagram of the group velocity and phase velocity

$$\overline{k} = \frac{1}{2}(k_1 - k_2), \tag{2.5.4}$$

$$\overline{\omega} = \frac{1}{2}(\omega_1 - \omega_2), \tag{2.5.5}$$

$$k = \frac{1}{2}(k_1 + k_2), \tag{2.5.6}$$

$$\omega = \frac{1}{2}(\omega_1 + \omega_2). \tag{2.5.7}$$

The profile is shown in Fig. 2.4.
We usually call

$$u'(x, t) = 2A\cos(\overline{k}x - \overline{\omega}t) \tag{2.5.8}$$

amplitude-modulated wave (AM wave) and call

$$u''(x, t) = \cos(kx - \omega t) \tag{2.5.9}$$

carrier wave and call $u(x, t)$ modulated wave. When the frequencies of two waves are close, we have

$$\overline{\omega} \approx 0, \tag{2.5.10}$$

$$\omega_1 \approx \omega_2 \approx \omega. \tag{2.5.11}$$

Thus, the AM wave is low-frequency wave, and the carrier wave is high-frequency wave (compared with AM wave). Two waves have not only different frequencies but also different propagation speeds. The propagation velocity of the high-frequency carrier wave is known as phase velocity, and the propagation velocity of the low-frequency AM wave (also known as wave packet) is defined as group velocity. The phase velocity and group velocity are denoted by c_p and c_g, respectively. If let

$$\overline{k}x - \overline{\omega}t = \text{const}, \tag{2.5.12}$$

we can obtain

$$c_g = \frac{dx}{dt} = \frac{\overline{\omega}}{\overline{k}} \approx \frac{d\omega}{dk}. \tag{2.5.13}$$

If let

$$kx - \omega t = \text{const}, \tag{2.5.14}$$

we have

$$c_p = \frac{dx}{dt} = \frac{\omega}{k} \approx \frac{\omega_1}{k_1} \approx \frac{\omega_2}{k_2}. \tag{2.5.15}$$

For the different media, the dispersion curves

$$\omega = \omega(k) \tag{2.5.16}$$

have different forms. Therefore, the group velocity c_g is generally not the same as the phase velocity c_p. But in a homogeneous medium, we have

$$\omega = ck, \tag{2.5.17}$$

where velocity c is a constant determined by the elastic modulus and density of the medium. Thus,

$$c_g = c_p = c, \tag{2.5.18}$$

such a medium is usually called **non-dispersive medium.** A **dispersive medium** satisfies $c_g \neq c_p$. If

$$c_g < c_p, \tag{2.5.19}$$

it is called **normal dispersion.** If

$$c_g > c_p, \tag{2.5.20}$$

it is called **abnormal dispersion.**

The above discussion is suited to the superposition of two monochromatic plane waves of different frequencies. The waves encountered in practice are often the superposition of many plane waves of different frequencies. Assume one-dimensional plane wave

$$u(x, t) = \int_{-\infty}^{\infty} A(\omega) e^{i(k(\omega)x - \omega t)} d\omega, \tag{2.5.21}$$

where the relationship curve $A(\omega)$~ω is called **amplitude-frequency curve**. A continuous amplitude-frequency curve means that the wave packet contains monochromatic waves of any frequency. Such wave packet is called **transient wave**, whose displacement field $u(x, t)$ is aperiodic function of coordinates and time. Assume another one-dimensional plane wave given by

$$u(x, t) = \sum_{n=-\infty}^{+\infty} A_n(\omega_n)\, e^{i[k_n(\omega_n)x - \omega_n t]}, \tag{2.5.22}$$

where the relationship $A_n(\omega_n)$~ω_n is discrete. Such wave packet and its displacement field $u(x, t)$ are periodic functions of coordinates and time, which is usually called **steady-state wave**. If we call $A_n(\omega_n)$ the **frequency spectrum** corresponding to frequency ω_n, then it can be said that the transient wave has a continuous frequency spectrum while the steady-state wave has a discrete frequency spectrum.

Chapter 3
Reflection and Transmission of Elastic Waves at Interfaces

The solution of the wave equation in an infinite elastic medium and the problem of elastic wave propagation do not need to consider the boundary conditions. However, the infinite homogeneous medium is an ideal geometric model, and the actual medium carrying elastic waves is always finite-sized. Even a medium as huge as the Earth is finite and inhomogeneous. The crust and the interior of the Earth can be roughly thought of as a layered structure while the ground surface is the boundary. For a finite medium, due to the existence of boundary, the interaction between the wave and the boundary occurs in the process of elastic wave propagation and eventually causes the change of the direction of elastic wave propagation. This chapter focuses on the reflection and transmission phenomena of plane waves at free surfaces and at the interfaces of elastic media.

The problem of plane wave propagation is fundamental to the theory of wave motion. The main causes includes: (1) although plane waves propagation is relative simple in mathematical treatment, the physical essences of wave propagation problem is included; (2) an elastic wave with a curved wavefront can be expressed as an integral form of plane wave; (3) certain propagation laws of plane waves can be approximately used to study the propagation of non-plane waves; especially, the spherical waves at enough long distances are essentially similar to plane waves. The propagation of seismic waves in the stratum is an example of the elastic wave propagation in a stratified medium. The wavefront of seismic waves generated by an earthquake source in the isotropic homogeneous medium is a spherical surface, and the wavefront will change when the spherical wave encounters the interfaces. When the earthquake source is far enough away, which means that the distance between the source and the receiver is much larger than the wavelength ($r \gg \lambda$), the spherical wave can be approximated as a plane wave. Also, when the wavelength is much smaller than the curvature radius of the interface, i.e. $\lambda \ll \rho$, the curve interface can be approximated as a planse. In this way, the discussion can be greatly simplified and does not affect the revelation of the essence of many phenomena.

© Science Press 2022
P. Wei, *Theory of Elastic Waves*,
https://doi.org/10.1007/978-981-19-5662-1_3

3.1 Classification of Interfaces and Plane Waves

3.1.1 Perfect Interface and Imperfect Interface

In the previous discussion, we know that there are two forms of wave propagation in an infinite homogeneous elastic medium; one is an elastic longitudinal wave propagating with velocity c_p, and the other is an elastic transverse wave propagating with velocity c_s. When the boundary of the medium is far from the source of the disturbance, or only consider this phase of the propagation process that the wave has not yet reached the boundary, we can ignore the influence of the boundary. However, in practical problems, two or multiple different mediums are always jointed together and jointed with surrounding mediums through boundaries. Waves at the interface where the discontinuity of the material properties occurs will produce complex reflections and transmissions, and mode conversions, namely waves of a different type from the original incident wave, are usually produced. In this case, it becomes necessary to consider the influences of the boundary on wave propagation.

The so-called interface is the curve or plane surface where different media contact each other. On both sides of this surface, the properties of the medium are significantly different. For example, Fig. 3.1 represents two media in contact with each other. The densities ρ, Lamé coefficients λ, μ of the two media are different, thus the wave velocities

$$c_P = \left(\frac{\lambda + 2\mu}{\rho}\right)^{\frac{1}{2}}, c_S = \left(\frac{\mu}{\rho}\right)^{\frac{1}{2}} \tag{3.1.1}$$

of the media may differ as well.

The **perfect interface** assumes that the two adjacent media are closely connected or welded at the interface, such that the displacement and stress components are continuous across the interface, the interface condition can be expressed as:

$$u_1^+ = u_1^-, \quad u_2^+ = u_2^-, \quad u_3^+ = u_3^-, \tag{3.1.2a}$$

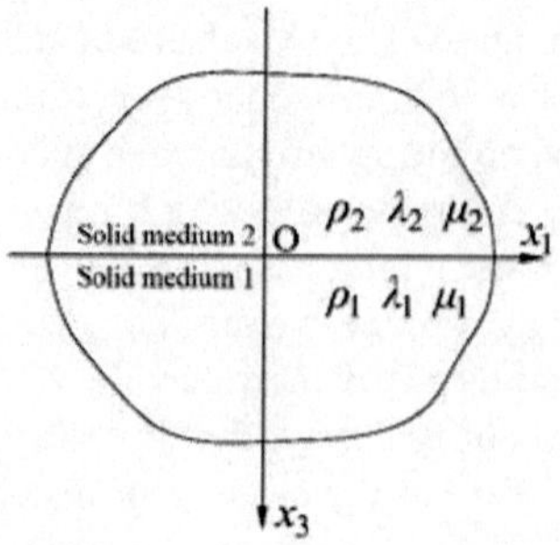

Fig. 3.1 Interface of two solid mediums

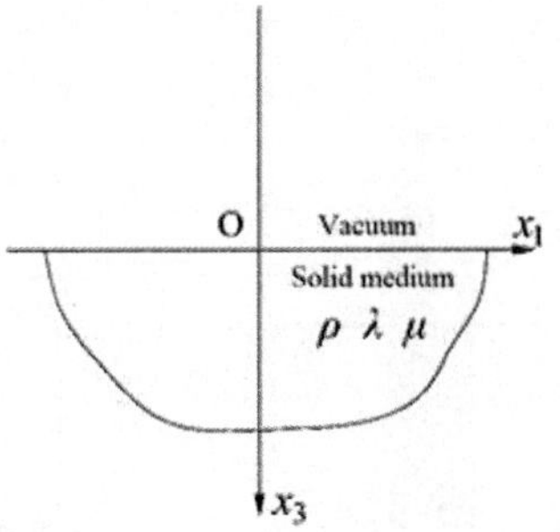

Fig. 3.2 Free surface of solid medium

$$\sigma_{33}^{+} = \sigma_{33}^{-}, \quad \sigma_{32}^{+} = \sigma_{32}^{-}, \quad \sigma_{31}^{+} = \sigma_{31}^{-}. \tag{3.1.2b}$$

The "+" and "−" in the above equation represents the mechanical quantities on both sides of the interface. If the second medium does not exist at all, i.e. that part of the space is the vacuum, then the interface between first medium and the vacuum is the **free surface** of first medium, as shown in Fig. 3.2. Stress is zero while the displacement is unrestricted on the free surface. The free surface condition can be expressed as

$$\sigma_{33}^{+} = \sigma_{33}^{-} = 0, \quad \sigma_{32}^{+} = \sigma_{32}^{-} = 0, \quad \sigma_{31}^{+} = \sigma_{31}^{-} = 0. \tag{3.1.3}$$

For the interface between solid and liquid (or gas), considering that the liquid (or gas) cannot be subjected to shear stress, as well as the liquid (or gas) can flow relative to the solid, the interface conditions can be expressed as

$$\sigma_{33}^{+} = \sigma_{33}^{-} = p, \quad \sigma_{32}^{+} = \sigma_{32}^{-} = 0, \quad \sigma_{31}^{+} = \sigma_{31}^{-} = 0, \tag{3.1.4a}$$

$$u_3^{+} = u_3^{-}. \tag{3.1.4b}$$

where p is the pressure of the liquid or gas.

The interface between two different media, typically due to the diffusion of substances or the use of an adhesive, forms a thin layer called the interfacial layer or interfacial phase. Due to interfacial damage, interfacial cracking and debonding, the bonding state of the interfacial layer is always intermediate state between a complete bond and complete debonding. The displacement or surface force is discontinuous on both sides of the interface. Such interface is usually referred to as **imperfect interface**. In addition, when the thickness of the interface is very small relative to the wavelength $\left(\frac{a}{\lambda} \ll 1\right)$, interfacial layer is commonly viewed mathematically as an interface without thickness while the mechanical properties of an interfacial layer remain. There are many models for imperfect interfaces. Three of the most commonly used models are introduced here: (1) **spring model** (neglecting the inertia of the

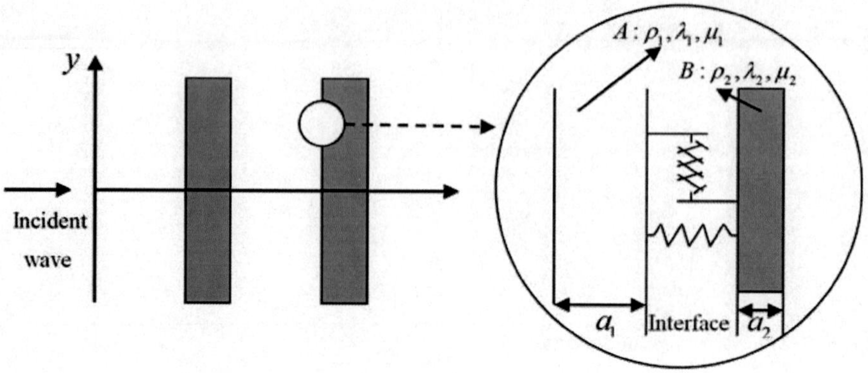

Fig. 3.3 Spring model of imperfect interface

interfacial layer); (2) **mass model** (neglecting the elasticity of the interfacial layer); and (3) **mass-spring model** (considering the intertia and elasticity of the interfacial layer). In the spring model of the imperfect interface, the displacement components in the medium on both sides of the interface are discontinuous, but the surface force components are still continuous. The interface connecting the two mediums is like a distributed spring along the interface, as shown in Fig. 3.3. The interface condition can be expressed as

$$\sigma_{33}^{+} = \sigma_{33}^{-}, \quad \sigma_{32}^{+} = \sigma_{32}^{-}, \quad \sigma_{31}^{+} = \sigma_{31}^{-}, \tag{3.1.5a}$$

$$u_{3}^{+} - u_{3}^{-} = f_{3} \cdot \sigma_{33}^{+}, \quad u_{2}^{+} - u_{2}^{-} = f_{2} \cdot \sigma_{32}^{+}, \quad u_{1}^{+} - u_{1}^{-} = f_{1} \cdot \sigma_{31}^{+}. \tag{3.1.5b}$$

where f_1, f_2 and f_3 are the normal and tangential spring flexibility coefficients.

In the mass model of the imperfect interface, the displacement components in the medium on both sides of the interface are continuous, but the surface force components are discontinuous. The interface connecting the two media is like a distributed mass along the interface. Due to their inertia, a discontinuity of forces on both sides of the interface generates, as shown in Fig. 3.4 the interface condition can be expressed as:

$$\sigma_{33}^{+} - \sigma_{33}^{-} = g_{3}u_{3}^{+}, \quad \sigma_{32}^{+} - \sigma_{32}^{-} = g_{2}u_{2}^{+}, \quad \sigma_{31}^{+} - \sigma_{31}^{-} = g_{1}u_{1}^{+}, \tag{3.1.6a}$$

$$u_{3}^{+} = u_{3}^{-}, \quad u_{2}^{+} = u_{2}^{-}, \quad u_{1}^{+} = u_{1}^{-}. \tag{3.1.6b}$$

where g_1, g_2 and g_3 are mass factors.

In the spring-mass model of imperfect interfaces, as shown in Fig. 3.5, the displacements and surface forces are both discontinuous. The interface conditions can be expressed as

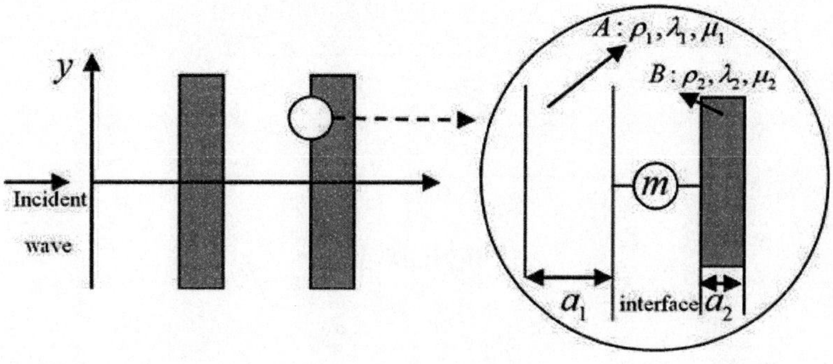

Fig. 3.4 Mass model of imperfect interface

$$\sigma_{33}^{+} - \sigma_{33}^{-} = g_3 \cdot \frac{1}{2}\left(u_3^{+} + u_3^{-}\right), \tag{3.1.7a}$$

$$\sigma_{32}^{+} - \sigma_{32}^{-} = g_2 \cdot \frac{1}{2}\left(u_2^{+} + u_2^{-}\right), \tag{3.1.7b}$$

$$\sigma_{31}^{+} - \sigma_{31}^{-} = g_1 \cdot \frac{1}{2}\left(u_1^{+} + u_1^{-}\right), \tag{3.1.7c}$$

$$u_3^{+} - u_3^{-} = f_3 \cdot \frac{1}{2}\left(\sigma_{33}^{+} + \sigma_{33}^{-}\right), \tag{3.1.7d}$$

$$u_2^{+} - u_2^{-} = f_2 \cdot \frac{1}{2}\left(\sigma_{32}^{+} + \sigma_{32}^{-}\right), \tag{3.1.7e}$$

$$u_1^{+} - u_1^{-} = f_1 \cdot \frac{1}{2}\left(\sigma_{31}^{+} + \sigma_{31}^{-}\right). \tag{3.1.7f}$$

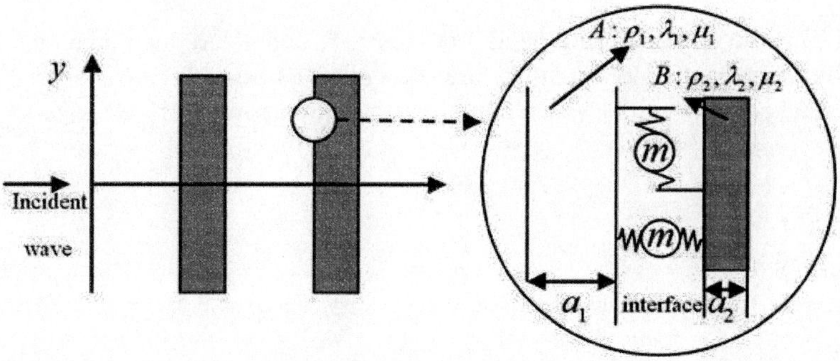

Fig. 3.5 Spring-mass model for imperfect interfaces

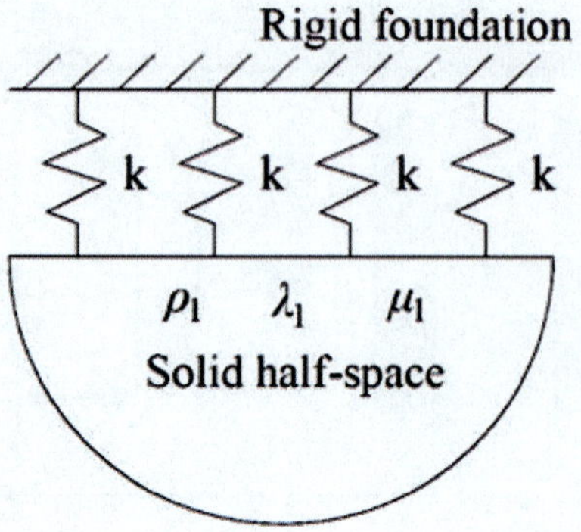

Fig. 3.6 Elastic support boundary

When the mass factor $g_1 = g_2 = g_3 = 0$, the spring-mass model of the imperfect interface degenerates to the spring model. When the spring flexibility factors $f_1 = f_2 = f_3 = 0$, the spring-mass model of the imperfect interface degenerates to the mass model. The imperfect interface degenerates to the perfect interface when both the mass factor and the spring factor are equal to zero.

When one of the two mediums is enough hard compared with the other, it can also be approximated as rigid; namely, the deformation is very small and negligible. At this time, for the softer medium, the boundary conditions can be expressed as

$$u_3^+ = u_3^- = 0, \quad u_2^+ = u_2^- = 0, \quad u_1^+ = u_1^- = 0. \tag{3.1.8}$$

That is, the surface displacement is zero while the surface force is not zero. Such a boundary is referred to as the **rigid support boundary**. If the surface displacement and the surface force are both not zero, such a boundary is referred to as **elastic support boundary,** as shown in Fig. 3.6. Its boundary conditions are expressed as

$$u_3^+ = f_3 \cdot \sigma_{33}^+, \quad u_2^+ = f_2 \cdot \sigma_{32}^+, \quad u_1^+ = f_1 \cdot \sigma_{31}^+. \tag{3.1.9}$$

where f_3 is the normal spring flexibility factor; f_1 and f_2 are the tangential spring flexibility factors. In addition, there are **slide support boundary** (tangential stress component equal to zero), whose boundary conditions can be expressed as

$$u_3^+ = 0, \tag{3.1.10a}$$

$$\sigma_{32}^+ = 0, \tag{3.1.10b}$$

$$\sigma_{31}^+ = 0. \tag{3.1.10c}$$

Considering the viscosity of the clay layer, the stratum on the upper part of the clay layer can be regarded as a solid medium with **viscoelastic support**, as shown in

Fig. 3.7. Viscoelastic support is more extensive than elastic support. It is the support model where viscous effect has further been taken into consideration based on the elastic effect. There are several models of viscoelastic support. The most commonly used models are Maxwell model, Kelvin model and standard linear solid model (also called Zenner model). The Maxwell model can be viewed as a linear spring and a dashpot in series, as shown in Fig. 3.8a, where the spring satisfies the elastic Hooke's law ($F = ku$) and the dashpot satisfies Newton law of viscosity ($F = \eta\dot{u}$). The force acting on it and the displacement satisfied the following relationship

$$u_z = \sigma_{zz}\left(\frac{1}{k} + \frac{1}{i\eta\omega}\right). \tag{3.1.11}$$

The Kelvin model can be viewed as consisting of a linear spring and a dashpot in parallel, as shown in Fig. 3.8b. The force acting on it and the displacement satisfied the following relationship

$$\sigma_{zz} = (k + i\eta\omega)u_z. \tag{3.1.12}$$

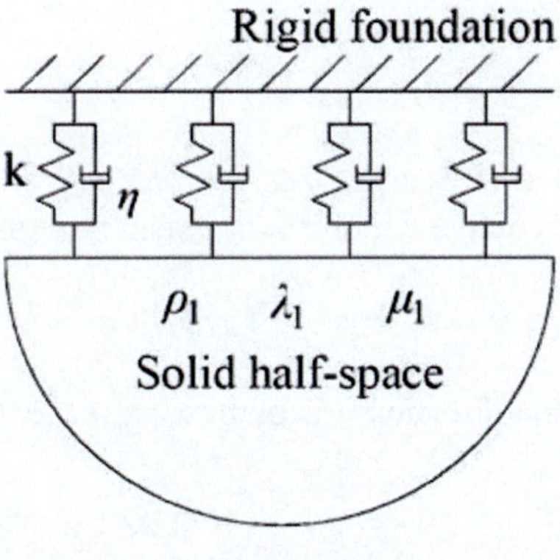

Fig. 3.7 Half-space with viscoelastic support boundary

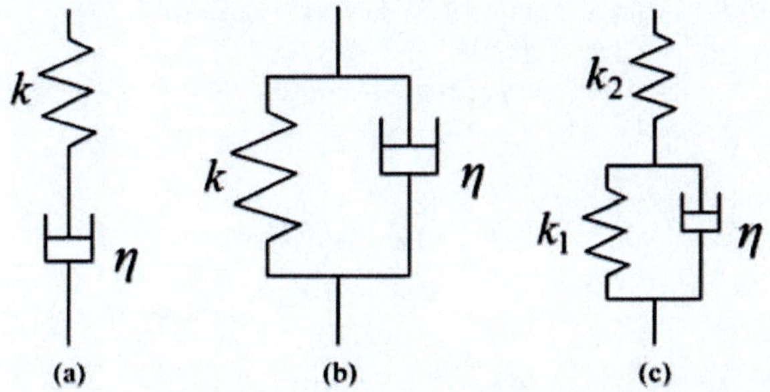

Fig. 3.8 Commonly used viscoelastic models. **a** Maxwell model; **b** Kelvin model; **c** standard linear solid model

The standard linear solid model consists of two linear springs and a dashpot, as shown in Fig. 3.8c. The constitutive relationship can be expressed as

$$u_z = \sigma_{zz}\left(\frac{1}{k_2} + \frac{1}{k_1 + i\eta\omega}\right). \tag{3.1.13}$$

3.1.2 P Wave, S Wave and SH Wave

For a general plane wave, the wave function has the following form

$$\varphi = \varphi(\mathbf{x}\cdot\mathbf{n} - c_P t) = \varphi(x_j n_j - c_P t), \tag{3.1.14}$$

$$\psi_k = \psi_k(\mathbf{x}\cdot\mathbf{n} - c_S t) = \psi_k(x_j n_j - c_S t), \tag{3.1.15}$$

where $\mathbf{n}$ is the unit vector in the direction of wave propagation. The displacement field can be expressed by

$$\mathbf{u} = \mathbf{u}_P + \mathbf{u}_S = \nabla\varphi + \nabla\times\psi, \tag{3.1.16}$$

where $\Psi = \psi_k\mathbf{e}_k$, and $\mathbf{e}_k (k = 1, 2, 3)$ is the unit vector in the direction of the coordinate x_k. Thus, the component of $\mathbf{u} = u_i\mathbf{e}_i$ can be expressed as

$$u_i = \varphi_{,i} + e_{jki}\psi_{k,j}. \tag{3.1.17}$$

If the wave propagation direction $\mathbf{n}$ is in the x_1x_3 plane, then

$$n_2 = 0, \quad \frac{\partial}{\partial x_2} = 0. \tag{3.1.18}$$

Thus, Eqs. 3.1.14 and 3.1.15 are rewritten as

$$\begin{aligned} \varphi &= \varphi(x_1 n_1 + x_3 n_3 - c_P t), \\ \psi_k &= \psi_k(x_1 n_1 + x_3 n_3 - c_S t), \end{aligned} \tag{3.1.19}$$

and

$$u_1 = \frac{\partial\varphi}{\partial x_1} - \frac{\partial\psi_2}{\partial x_3}, \tag{3.1.20a}$$

$$u_2 = \frac{\partial\psi_1}{\partial x_3} - \frac{\partial\psi_3}{\partial x_1}, \tag{3.1.20b}$$

$$u_3 = \frac{\partial \varphi}{\partial x_3} + \frac{\partial \psi_2}{\partial x_1}. \tag{3.1.20c}$$

From the above equations, it can be seen that φ and ψ_2 are related to the displacement components u_1 and u_3 in the x_1x_3 plane, while ψ_1 and ψ_3 are only related to the displacement component u_2 perpendicular to the x_1x_3 plane.

In general, the direction of displacement of a mass point can be a random angle respect to the direction of wave propagation. In order to simplify the problem, the displacement can be decomposed as follows, as shown in Fig. 3.9. Suppose the plane wave propagates in the incident plane x_1x_3 along the direction $\boldsymbol{n}$, and the displacement $\boldsymbol{u}$ (not in the x_1x_3 plane) has a random angle α from $\boldsymbol{n}$, and $\boldsymbol{n} \perp \boldsymbol{n}_1$, where $\boldsymbol{n}_1$ is also in the x_1x_3 plane. The component of the displacement $\boldsymbol{u}$ along the $\mathbf{n}$ direction is denoted by $\boldsymbol{u}_P$, and the wave corresponding to it is the longitudinal wave, also called the P wave. The component of the displacement $\boldsymbol{u}$ in the plane normal to $\boldsymbol{n}$ is denoted by $\boldsymbol{u}_S$, and the corresponding wave is transverse wave, also called S wave. The displacement $\boldsymbol{u}_S$ of the S wave can be further decomposed into the component $\boldsymbol{u}_{\text{SV}}$ in the x_1x_3 plane and the component $\boldsymbol{u}_{\text{SH}}$ perpendicular to the x_1x_3 plane (along the x_2 direction). The wave corresponding with the former is SV wave (shear vertical wave), and the wave corresponding with the latter is SH wave (shear horizontal wave). If the component of the displacement vector in the x_1x_2 plane is $\boldsymbol{u}'_P$ and the component in the x_3 direction is $\boldsymbol{u}_Q$, then

$$\begin{aligned}
\boldsymbol{u} &= \boldsymbol{u}'_P + \boldsymbol{u}_Q \\
&= (\boldsymbol{u}_{\text{SH}} + \boldsymbol{u}_{P1}) + \boldsymbol{u}_Q \\
&= [\boldsymbol{u}_{\text{SH}} + (\boldsymbol{u}_{P11} + \boldsymbol{u}_{P12})] + (\boldsymbol{u}_{Q1} + \boldsymbol{u}_{Q2}) \\
&= \left[\boldsymbol{u}_{\text{SH}} + \left(\boldsymbol{u}_{P12} + \boldsymbol{u}_{Q2}\right)\right] + \left(\boldsymbol{u}_{P11} + \boldsymbol{u}_{Q1}\right) \\
&= \boldsymbol{u}_{\text{SH}} + \boldsymbol{u}_{\text{SV}} + \boldsymbol{u}_P \\
&= \boldsymbol{u}_S + \boldsymbol{u}_P,
\end{aligned} \tag{3.1.21}$$

$$\left.\begin{matrix} \boldsymbol{n} \perp \boldsymbol{n}_1 \\ \boldsymbol{n} \perp \boldsymbol{u}_{\text{SH}} \end{matrix}\right\} \Rightarrow \boldsymbol{n} \perp \boldsymbol{u}_S \Rightarrow \boldsymbol{u}_P \perp \boldsymbol{u}_S.$$

When an elastic wave is reflected and transmitted on the interface of two media, the wave in each medium can be decomposed into three types of waves: P wave, SV wave and SH wave. From the previous analysis, it can be seen that the motions u_1 and u_3 derived from φ and ψ_2 are in the x_1x_3 plane which are plane P and SV waves, respectively, while the motions u_2 derived from ψ_1 and ψ_3 are in the direction perpendicular to the x_1x_3 plane and thus are SH waves.

The stress fields induced by P, SV and SH waves propagating in the x_1x_3 plane are

$$\sigma_{33} = \lambda\theta + 2\mu\varepsilon_{33}$$

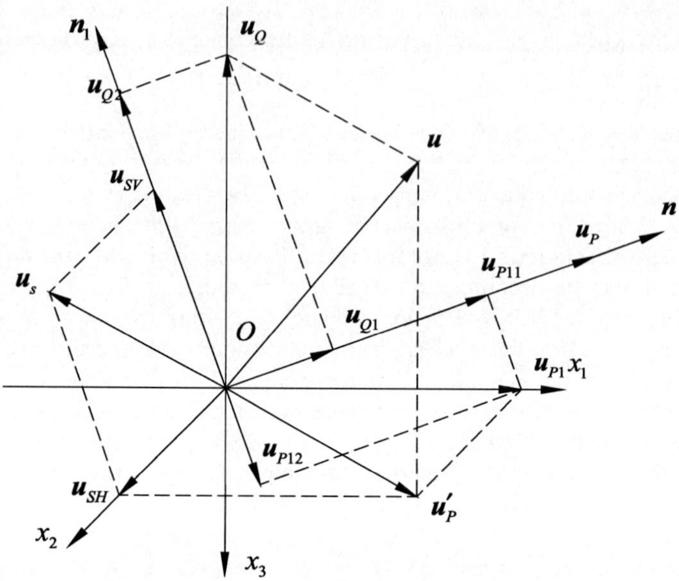

Fig. 3.9 Decomposition of arbitrary plane waves into P wave, SV wave and SH wave

$$
\begin{aligned}
&= \lambda\left(\frac{\partial u_1}{\partial x_1} + \frac{\partial u_3}{\partial x_3}\right) + 2\mu\left(\frac{\partial u_3}{\partial x_3}\right) \\
&= \lambda\left(\frac{\partial^2 \varphi}{\partial x_1^2} + \frac{\partial^2 \varphi}{\partial x_3^2}\right) + 2\mu\left(\frac{\partial^2 \varphi}{\partial x_3^2} + \frac{\partial^2 \psi_2}{\partial x_1 \partial x_3}\right) \\
&= \lambda \nabla^2 \varphi + 2\mu\left(\frac{\partial^2 \varphi}{\partial x_3^2} + \frac{\partial^2 \psi_2}{\partial x_1 \partial x_3}\right) \\
&= \frac{\lambda}{c_P^2}\frac{\partial^2 \varphi}{\partial t^2} + 2\mu\left(\frac{\partial^2 \varphi}{\partial x_3^2} + \frac{\partial^2 \psi_2}{\partial x_1 \partial x_3}\right) \\
&= \left(\rho - \frac{2\mu}{c_P^2}\right)\frac{\partial^2 \varphi}{\partial t^2} + 2\mu\left(\frac{\partial^2 \varphi}{\partial x_3^2} + \frac{\partial^2 \psi_2}{\partial x_1 \partial x_3}\right) \\
&= \rho\frac{\partial^2 \varphi}{\partial t^2} + 2\mu\left(-\frac{1}{c_P^2}\ddot{\varphi} + \frac{\partial^2 \varphi}{\partial x_3^2} + \frac{\partial^2 \psi_2}{\partial x_1 \partial x_3}\right) \\
&= \rho\frac{\partial^2 \varphi}{\partial t^2} + 2\mu\left(\frac{\partial^2 \psi_2}{\partial x_1 \partial x_3} - \frac{\partial^2 \varphi}{\partial x_1^2}\right) \\
&= \rho\left[c_P^2 \nabla^2 \varphi + 2c_S^2\left(\frac{\partial^2 \psi_2}{\partial x_1 \partial x_3} - \frac{\partial^2 \varphi}{\partial x_1^2}\right)\right],
\end{aligned}
\tag{3.1.22}
$$

$$\sigma_{31} = 2\mu\varepsilon_{31}$$

$$
\begin{aligned}
&= \mu\left(\frac{\partial u_1}{\partial x_3} + \frac{\partial u_3}{\partial x_1}\right) \\
&= \mu\left(2\frac{\partial^2 \varphi}{\partial x_1 \partial x_3} + \frac{\partial^2 \psi_2}{\partial x_1^2} - \frac{\partial^2 \psi_2}{\partial x_3^2}\right) \\
&= \rho c_S^2\left(2\frac{\partial^2 \varphi}{\partial x_1 \partial x_3} + \frac{\partial^2 \psi_2}{\partial x_1^2} - \frac{\partial^2 \psi_2}{\partial x_3^2}\right),
\end{aligned}
\tag{3.1.23}
$$

$$
\begin{aligned}
\sigma_{32} = 2\mu\varepsilon_{32} &= \mu\left(\frac{\partial u_2}{\partial x_3} + \frac{\partial u_3}{\partial x_2}\right) = \mu\left(\frac{\partial u_2}{\partial x_3}\right) \\
&= \mu\left(\frac{\partial^2 \psi_1}{\partial x_3^2} - \frac{\partial^2 \psi_3}{\partial x_1 \partial x_3}\right) \\
&= \rho c_S^2\left(\frac{\partial^2 \psi_1}{\partial x_3^2} - \frac{\partial^2 \psi_3}{\partial x_1 \partial x_3}\right).
\end{aligned}
\tag{3.1.24}
$$

From Eqs. (3.1.22)–(3.1.24), it can be seen that u_2 appears in the expression σ_{32}, while u_1 and u_3 appear in the expressions σ_{33} and σ_{31}. In the plane strain problem, there are only displacement components u_1 and u_3, that is, only P and SV waves. In the out-of-plane problem (also called inverse plane problem), there are only displacement components u_3; that is, only SH waves exist. In the interface reflection and transmission problem, the P and SV waves are coupled to each other, independent of the SH waves. Therefore, only P and SV waves need to be discussed together, and SH waves can be discussed separately. This chapter will follow this way to discuss the reflection and transmission problem.

3.2 Reflection of Elastic Waves on Free Surface

When a plane wave propagates in a homogeneous and isotropic medium, its waveform, velocity and propagation direction will not change. However, when the density and elastic constants of the medium changed, reflection and transmission will occur. This section deals with the reflection of plane waves on a free surface of a homogeneous isotropic half-space (elastic half-space for short).

3.2.1 Reflection of P Wave on Free Surface

As shown in Fig. 3.10, the x_1x_2 plane is the free surface of the medium, the x_1x_3 plane is the incident plane of the wave, or, the plane of polarization plane of mass point. The space where $x_3 < 0$ is vacuum, and the region where $x_3 \geq 0$ is the solid medium. The density of the solid medium is ρ, and the Lame coefficients are λ and μ. Consider a train of P wave propagating in the plane x_1x_3. When they reach the free surface, due to the coupling of displacement u_1 and u_3, reflected P waves and reflected SV wave will be excited.

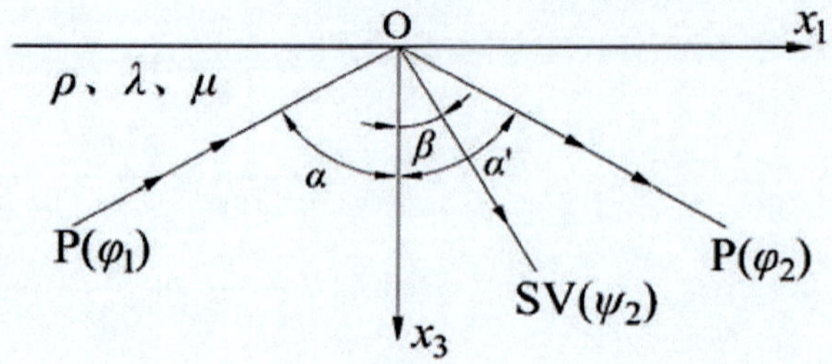

Fig. 3.10 Reflection of P waves on a free surface

Let the wave functions of incident P wave, reflected P wave and SV wave be, respectively,

$$\varphi_1 = A_1 \exp[i(k_{P1}x_1 - k_{P3}x_3 - \omega_P t)], \tag{3.2.1}$$

$$\varphi_2 = A_2 \exp\left[i\left(k'_{P1}x_1 + k'_{P3}x_3 - \omega'_P t\right)\right], \tag{3.2.2}$$

$$\psi_1 \mathbf{e}_2 = B_2 \exp[i(k_{S1}x_1 + k_{P3}x_3 - \omega_S t)]\mathbf{e}_2, \tag{3.2.3}$$

where A_1, A_2, B_2 and ω_P, ω'_P, ω_S are the amplitudes and angular frequencies of the corresponding waves, and

$$k_{P1} = \frac{\omega_P}{c_P}\sin\alpha, \quad k_{P3} = \frac{\omega_P}{c_P}\cos\alpha, \tag{3.2.4}$$

$$k'_{P1} = \frac{\omega'_P}{c_P}\sin\alpha', \quad k'_{P3} = \frac{\omega'_P}{c_P}\cos\alpha', \tag{3.2.5}$$

$$k_{S1} = \frac{\omega_S}{c_S}\sin\beta, \quad k_{S3} = \frac{\omega_S}{c_S}\cos\beta, \tag{3.2.6}$$

are the wave vectors of the incident P wave, reflected P wave and reflected SV wave, respectively.

Since there is no medium in the region $x_3 < 0$, that is, the solid medium surface is a free surface, the boundary conditions can be expressed as

$$\sigma_{33}|_{x_3=0} = 0, \quad \sigma_{32}|_{x_3=0} = 0, \quad \sigma_{31}|_{x_3=0} = 0. \tag{3.2.7}$$

Also, u_1 and u_3 only appear in σ_{33} and σ_{31}, σ_{32} is independent of them, so only boundary conditions

$$\sigma_{33}|_{x_3=0} = 0, \quad \sigma_{31}|_{x_3=0} = 0, \tag{3.2.8}$$

is validated. According to Eqs. (3.1.22) and (3.1.23), the stress boundary conditions of the free surface can be rewritten as

$$\left\{\lambda\nabla^2(\varphi_1+\varphi_2)+2\mu\left[\frac{\partial^2(\varphi_1+\varphi_2)}{\partial x_3^2}+\frac{\partial^2\psi_2}{\partial x_1\partial x_3}\right]\right\}\bigg|_{x_3=0}=0, \tag{3.2.9}$$

$$\left[2\frac{\partial^2(\varphi_1+\varphi_2)}{\partial x_1\partial x_3}+\frac{\partial^2\psi_2}{\partial x_1^2}-\frac{\partial^2\psi_2}{\partial x_3^2}\right]\bigg|_{x_3=0}=0. \tag{3.2.10}$$

Substitute Eqs. (3.2.1)–(3.2.3) into the above equations, and let $x_3=0$, then

$$\begin{aligned}&\lambda\{k_P^2 A_1\exp[i(k_{P1}x_1-\omega_P t)]+k'_P A_2\exp[i(k'_{P1}x_1-\omega'_P t)]\}\\&+2\mu\{k_{P3}^2 A_1\exp[i(k_{P1}x_1-\omega_P t)]+k'^2_{P3}A_2\exp[i(k'_{P1}x_1-\omega'_P t)]\\&+k_S^2\sin\beta\cos\beta\cdot B_2\exp[i(k_{S1}x_1-\omega_S t)]\}=0,\end{aligned} \tag{3.2.11}$$

$$\begin{aligned}&2\{k_{P1}k_{P3}A_1\exp[i(k_{P1}x_1-\omega_P t)]-k'_{P1}k'_{P3}A_2\exp[i(k'_{P1}x_1-\omega'_P t)]\}\\&+k_S^2\cos 2\beta\cdot B_2\exp[i(k_{S1}x_1-\omega_S t)]\}=0.\end{aligned} \tag{3.2.12}$$

In order to make Eqs. (3.2.11) and (3.2.12) hold for any x_1 and t, we must let

$$\omega_P=\omega'_P=\omega_S, \tag{3.2.13}$$

$$k_{P1}=k'_{P1}=k_{S1}=k_c. \tag{3.2.14}$$

It can be seen that the boundary conditions require that the frequencies of the reflected wave and the incident wave must be the same, and the projections of the wave vector of the reflected wave and the incident wave on the surface must be the same, which is generally called the **apparent wave number**. The same apparent wave number of each wave can also be expressed as

$$\frac{\sin\alpha}{c_P}=\frac{\sin\alpha'}{c_P}=\frac{\sin\beta}{c_S}, \tag{3.2.15}$$

so

$$\alpha=\alpha', \tag{3.2.16}$$

that is, the reflection angle of the reflected P wave is equal to the incidence angle. The incident angle of the incident P wave and the reflection angle of the reflected SV wave satisfy the relationship

$$\frac{\sin\alpha}{\sin\beta}=\frac{c_P}{c_S}. \tag{3.2.17}$$

Equation (3.2.17) is called Snell's law. It can be seen that the reflection angle of the reflected SV wave is smaller than that of the reflected P wave, i.e. $\beta<\alpha$.

Applying the trigonometric function formula to Eqs. (3.2.11) and (3.2.12) leads to

$$(A_1 + A_2)\cos 2\beta + B_2 \sin 2\beta = 0, \tag{3.2.18}$$

$$(A_1 - A_2)c_S^2 \sin 2\alpha + B_2 c_P^2 \cos 2\beta = 0, \tag{3.2.19}$$

or in matrix form

$$\begin{pmatrix} \cos 2\beta & \sin 2\beta \\ -c_S^2 \sin 2\alpha & c_P^2 \cos 2\beta \end{pmatrix} \begin{Bmatrix} A_2 \\ B_2 \end{Bmatrix} = A_1 \begin{Bmatrix} -\cos 2\beta \\ -c_S^2 \sin 2\alpha \end{Bmatrix}.$$

When the amplitude A_1 of the incident wave is given, the amplitudes A_2 and B_2 of the reflected wave can be obtained from the above equation. The reflection coefficient of reflected P wave is defined as the ratio of the amplitude of the potential function of reflected P wave to the incident P wave, i.e.

$$F_{\mathrm{PP}} = \frac{A_2}{A_1} = \frac{c_S^2 \sin 2\alpha \sin 2\beta - c_P^2 \cos^2 2\beta}{c_S^2 \sin 2\alpha \sin 2\beta + c_P^2 \cos^2 2\beta}, \tag{3.2.20}$$

The reflection coefficient of the reflected SV wave is defined as the ratio of the amplitude of the reflected SV wave to the incident P wave, i.e.

$$F_{\mathrm{PV}} = \frac{B_2}{A_1} = -\frac{2c_S^2 \sin 2\alpha \cos 2\beta}{c_S^2 \sin 2\alpha \sin 2\beta + c_P^2 \cos^2 2\beta}. \tag{3.2.21}$$

From Eqs. (3.2.20) and (3.2.21), we can obtain the following conclusions:

(1) Vertical incidence

When P wave is incident vertically, i.e. $\alpha = 0$, there is

$$F_{\mathrm{PP}} = \frac{A_2}{A_1} = -1, \quad F_{\mathrm{PV}} = \frac{B_2}{A_1} = 0. \tag{3.2.22}$$

It shows that there are only reflected P wave and no reflected SV wave in this case. If A_2 and A_1 are treated as complex numbers, then $F_{\mathrm{PP}} = A_2/A_1 = -1$ indicates that there is a phase shift of 180° between the reflected wave and the incident wave.

(2) Grazing incidence

When the incidence angle $\alpha = \frac{\pi}{2}$, the direction of wave propagation is parallel to the boundary, and we call this situation as grazing incidence. By substituting $\alpha = \frac{\pi}{2}$ into Eqs. (3.2.20) and (3.2.21), we can obtain

$$F_{\mathrm{PP}} = \frac{A_2}{A_1} = -1, \quad F_{\mathrm{PV}} = \frac{B_2}{A_1} = 0. \tag{3.2.23}$$

The above equation shows that there is still no reflected SV wave. The reflected P wave maintains the same amplitude as the incident P wave, but with a phase shift of 180°. This 180° phase shift allows the superposition of the incident and reflected waves to guarantee the stress-free condition of the free surface. Without the phase shift of 180°, the superposition of two waves propagating in the same direction cannot satisfy the stress-free condition of surface.

(3) Mode conversion of waves

If $\frac{A_2}{A_1} = 0$, it indicates that the reflected wave only has SV wave; that is, the incident P wave completely changes into SV wave after reflection, and this phenomenon is called wave pattern conversion. According to Eq. (3.2.20), the incident angle corresponding to this special case can be obtained as

$$c_S^2 \sin 2\alpha \sin 2\beta - c_P^2 \cos^2 2\beta = 0. \tag{3.2.24}$$

Sometimes, the reflection coefficient is also defined as the amplitude ratio of the displacement. Now let us consider the relation between the amplitude ratio of displacement and the amplitude ratio of potential function.

Suppose the wave function of the incident P wave is

$$\varphi_1 = A_1 \exp[i(k_{P1}x_1 - k_{P3}x_3 - \omega_P t)]. \tag{3.2.25}$$

From Eq. (3.1.17), we can see

$$u_1 = ik_{P1}A_1 \exp[i(k_{P1}x_1 - k_{P3}x_3 - \omega_P t)], \tag{3.2.26}$$

$$u_3 = -ik_{P3}A_1 \exp[i(k_{P1}x_1 - k_{P3}x_3 - \omega_P t)]. \tag{3.2.27}$$

Let the displacement amplitude of the incident P wave be C, then

$$C = \sqrt{|u_1|^2 + |u_3|^2} = ik_P A_1 = \frac{i\omega A_1}{c_P}. \tag{3.2.28}$$

Define the wave function of reflected P wave as

$$\varphi_2 = A_2 \exp[i(k_{P1}x_1 + k_{P3}x_3 - \omega_P t)], \tag{3.2.29}$$

where A_2 is the amplitude of the potential function. Similarly, the displacement amplitude C_1 of the reflected P wave can be obtained, i.e.

$$C_1 = \frac{i\omega A_2}{c_P}, \tag{3.2.30}$$

therefore

$$\frac{C_1}{C} = \frac{i\omega A_2}{i\omega A_1}\frac{c_P}{c_P} = \frac{A_2}{A_1}. \tag{3.2.30}$$

As above, the displacement amplitude C_2 of the reflected SV wave can be obtained as

$$C_2 = \frac{i\omega B_2}{c_S}, \tag{3.2.31}$$

thus

$$\frac{C_2}{C} = \frac{B_2}{A_1}\frac{c_P}{c_S}. \tag{3.2.32}$$

It can be seen that **the displacement amplitude ratio is equal to the amplitude ratio of potential function multiplied by the inverse of the corresponding wave velocity ratio.**

The displacement reflection coefficient of P wave is defined as the ratio of the displacement amplitude of the reflected P wave to the incident P wave, i.e.

$$R_{\text{PP}} = \frac{A_2}{A_1} \cdot \frac{c_P}{c_P} = \frac{A_2}{A_1} = \frac{c_S^2 \sin 2\alpha \sin 2\beta - c_P^2 \cos^2 2\beta}{c_S^2 \sin 2\alpha \sin 2\beta + c_P^2 \cos^2 2\beta}. \tag{3.2.33}$$

The displacement reflection coefficient of the reflected SV wave is defined as the ratio of the displacement amplitude of the reflected SV wave to the incident P wave, i.e.

$$R_{\text{PV}} = \frac{B_2}{A_1} \cdot \frac{c_P}{c_S} = -\frac{2c_S c_P \sin 2\alpha \cos 2\beta}{c_S^2 \sin 2\alpha \sin 2\beta + c_P^2 \cos^2 2\beta}. \tag{3.2.34}$$

The reflection coefficient can also be defined in terms of the average energy flow density ratio. Let us first calculate the average energy flow of the incident and reflected waves.

(1) The average energy flow of the incident P wave

$$\boldsymbol{u} = \nabla\varphi_1 = \frac{\partial\varphi_1}{\partial x_1}\boldsymbol{i} + \frac{\partial\varphi_1}{\partial x_3}\boldsymbol{k} = i(k_{P1}\boldsymbol{i} - k_{P3}\boldsymbol{k})\varphi_1, \tag{3.2.35}$$

$$u_1 = ik_{P1}\varphi_1, \quad u_2 = 0, \quad u_3 = -ik_{P3}\varphi_1, \tag{3.2.36}$$

$$\begin{aligned} &\varepsilon_{11} = -k_{P1}^2\varphi_1, \quad \varepsilon_{22} = 0, \quad \varepsilon_{33} = -k_{P3}^2\varphi_1, \\ &\varepsilon_{12} = \varepsilon_{23} = 0, \quad \varepsilon_{31} = k_{P1}k_{P3}\varphi_1, \quad \theta = -k_P^2\varphi_1. \end{aligned}$$

Since $\sigma_{ij} = \lambda\theta\delta_{ij} + 2\mu\varepsilon_{ij}$, we get

$$\sigma_{11} = -\left(\lambda k_P^2 + 2\mu k_{P1}^2\right)\varphi_1, \quad \sigma_{22} = -\lambda k_P^2\varphi_1, \quad \sigma_{33} = -\left(\lambda k_P^2 + 2\mu k_{P3}^2\right)\varphi_1,$$
$$\sigma_{13} = 2\mu k_{P1}k_{P3}\varphi_1, \quad \sigma_{12} = \sigma_{23} = 0. \tag{3.2.37}$$

Because the energy flow density $I_j = -\mathrm{Re}\,\sigma_{ij}\cdot\mathrm{Re}\,\frac{\partial u_i}{\partial t}$, we can obtain

$$\begin{aligned} I_1 &= -\left(\mathrm{Re}\,\sigma_{11}\cdot\mathrm{Re}\,\frac{\partial u_1}{\partial t} + \mathrm{Re}\,\sigma_{31}\cdot\mathrm{Re}\,\frac{\partial u_3}{\partial t}\right) \\ &= \rho\omega_P^3 A_1^2 k_{P1}\cos^2(k_{P1}x_1 - k_{P3}x_3 - \omega_P t), \end{aligned} \tag{3.2.38a}$$

$$I_2 = 0, \tag{3.2.38b}$$

$$I_3 = \rho\omega_P^3 A_1^2 k_{P3}\cos^2(k_{P1}x_1 - k_{P3}x_3 - \omega_P t), \tag{3.2.38c}$$

$$I_{P1} = \sqrt{\left(I_1^2 + I_2^2 + I_3^2\right)} = \frac{\rho\omega_P^4 A_1^2}{c_P}\cos^2(k_{P1}x_1 - k_{P3}x_3 - \omega_P t). \tag{3.2.39}$$

The average energy flow is

$$\langle I_{P1}\rangle = \frac{1}{T}\int_0^T I\mathrm{d}t = \frac{\rho\omega_P^4 A_1^2}{Tc_P}\int_0^T \cos^2(k_{P1}x_1 - k_{P3}x_3 - \omega_P t)\mathrm{d}t = \frac{\rho\omega_P^4 A_1^2}{2c_P}. \tag{3.2.40}$$

(2) The average energy flow of reflected P wave

$$\boldsymbol{u} = \nabla\varphi = \frac{\partial\varphi}{\partial x_1}\boldsymbol{i} + \frac{\partial\varphi}{\partial x_3}\boldsymbol{k} = i\left(k'_{P1}\boldsymbol{i} + k'_{P3}\boldsymbol{k}\right)\varphi_2, \tag{3.2.41}$$

$$u_1 = ik'_{P1}\varphi_2, \quad u_2 = 0, \quad u_3 = ik'_{P3}\varphi_2, \tag{3.2.42}$$

$$\varepsilon_{11} = -k'^2_{P1}\varphi_2, \quad \varepsilon_{22} = 0, \quad \varepsilon_{33} = -k'^2_{P3}\varphi_2,$$
$$\varepsilon_{12} = \varepsilon_{23} = 0, \quad \varepsilon_{31} = -k'_{P1}k'_{P3}\varphi_2, \quad \theta = -k'^2_P\varphi_2, \tag{3.2.43}$$

$$\sigma_{11} = -\left(\lambda k'^2_P + 2\mu k'^2_{P1}\right)\varphi_2, \quad \sigma_{22} = -\lambda k'^2_P\varphi_2, \quad \sigma_{33} = -\left(\lambda k'^2_P + 2\mu k'^2_{P3}\right)\varphi_2,$$
$$\sigma_{13} = -2\mu k'_{P1}k'_{P3}\varphi_2, \quad \sigma_{12} = \sigma_{23} = 0, \tag{3.2.44}$$

$$\begin{aligned} I_1 &= -\left(\operatorname{Re}\sigma_{11}\cdot\operatorname{Re}\frac{\partial u_1}{\partial t}+\operatorname{Re}\sigma_{31}\cdot\operatorname{Re}\frac{\partial u_3}{\partial t}\right) \\ &= \rho\omega_P'^3A_2^2k_{P1}'\cos^2\left(k_{P1}'x_1+k_{P3}'x_3-\omega_Pt\right), \end{aligned} \tag{3.2.45a}$$

$$I_2 = 0, \tag{3.2.45b}$$

$$I_3 = \rho\omega_P'^3A_2^2k_{P3}'\cos^2\left(k_{P1}'x_1+k_{P3}'x_3-\omega_Pt\right). \tag{3.2.45c}$$

$$I_{P2} = \sqrt{\left(I_1^2+I_2^2+I_3^2\right)} = \frac{\rho\omega_P'^4A_2^2}{c_P}\cos^2\left(k_{P1}'x_1+k_{P3}'x_3-\omega_Pt\right). \tag{3.2.46}$$

The average energy flow is

$$\langle I_{P2}\rangle = \frac{1}{T}\int_0^T I\mathrm{d}t = \frac{\rho\omega_P'^4A_2^2}{Tc_P}\int_0^T\cos^2\left(k_{P1}'x_1+k_{P3}'x_3-\omega_Pt\right)\mathrm{d}t = \frac{\rho\omega_P'^4A_2^2}{2c_P}. \tag{3.2.47}$$

(3) The average energy flow of the reflected SV wave

$$\boldsymbol{u} = \nabla\times\boldsymbol{\psi} = \frac{\partial}{\partial x_m}\boldsymbol{e}_m\times\psi_n\boldsymbol{e}_n = \frac{\partial\psi_n}{\partial x_m}(\boldsymbol{e}_m\times\boldsymbol{e}_n) = e_{jmn}\frac{\partial\psi_n}{\partial x_m}\boldsymbol{e}_j, \tag{3.2.48}$$

$$u_1 = -\frac{\partial\psi_2}{\partial x_3} = -ik_{S3}\psi_2,\quad u_2 = 0,\quad u_3 = \frac{\partial\psi_2}{\partial x_1} = ik_{S1}\psi_2, \tag{3.2.49}$$

$$\begin{aligned} &\varepsilon_{11} = k_{S1}k_{S3}\psi_2,\quad \varepsilon_{22} = 0,\quad \varepsilon_{33} = -k_{S1}k_{S3}\psi_2, \\ &\varepsilon_{12} = \varepsilon_{23} = 0,\quad \varepsilon_{31} = \frac{1}{2}\left(k_{S3}^2-k_{S1}^2\right)\psi_2,\quad \theta = 0, \end{aligned} \tag{3.2.50}$$

$$\begin{aligned} &\sigma_{11} = 2\mu k_{S1}k_{S3}\psi_2,\quad \sigma_{22} = 0,\quad \sigma_{33} = -2\mu k_{S1}k_{S3}\psi_2, \\ &\sigma_{13} = \mu\left(k_{S3}^2-k_{S1}^2\right)\psi_2,\quad \sigma_{12} = \sigma_{23} = 0, \end{aligned} \tag{3.2.51}$$

$$\begin{aligned} I_1 &= -\left(\operatorname{Re}\sigma_{11}\cdot\operatorname{Re}\frac{\partial u_1}{\partial t}+\operatorname{Re}\sigma_{31}\cdot\operatorname{Re}\frac{\partial u_3}{\partial t}\right) \\ &= \rho\omega_S^3B_2^2k_{S1}\cos^2(k_{S1}x_1+k_{S3}x_3-\omega_St), \end{aligned} \tag{3.2.52a}$$

$$I_2 = 0, \tag{3.2.52b}$$

$$I_3 = \rho\omega_S^3 B_2^2 k_{S3}\cos^2(k_{S1}x_1 + k_{S3}x_3 - \omega_S t), \tag{3.2.52c}$$

$$I_S = \sqrt{\left(I_1^2 + I_2^2 + I_3^2\right)} = \frac{\rho\omega_S^4 B_2^2}{c_S}\cos^2(k_{S1}x_1 + k_{S3}x_3 - \omega_S t), \tag{3.2.53}$$

The average energy flow is

$$\langle I_S\rangle = \frac{1}{T}\int\limits_0^T I\mathrm{d}t = \frac{\rho\omega_S^4 B_2^2}{Tc_S}\int\limits_0^T \cos^2(k_{S1}x_1 + k_{S3}x_3 - \omega_S t)\mathrm{d}t = \frac{\rho\omega_S^4 B_2^2}{2c_S}. \tag{3.2.54}$$

In summary, the average energy flow density ratio of the reflected wave to the incident wave is

$$\frac{\langle I_{P2}\rangle}{\langle I_{P1}\rangle} = \frac{\frac{\rho\omega_P'^4 A_2^2}{2c_P}}{\frac{\rho\omega_P^4 A_1^2}{2c_P}} = \frac{A_2^2}{A_1^2}, \tag{3.2.55}$$

$$\frac{\langle I_S\rangle}{\langle I_{P1}\rangle} = \frac{\frac{\rho\omega_S^4 B_2^2}{2c_S}}{\frac{\rho\omega_P^4 A_1^2}{2c_P}} = \frac{c_P B_2^2}{c_S A_1^2}. \tag{3.2.56}$$

It can be seen that **the energy reflection coefficient of P wave is the square of the amplitude ratio of the potential function of reflected P wave to incident P wave**, i.e.

$$D_{\mathrm{PP}} = \frac{A_2^2}{A_1^2}. \tag{3.2.57}$$

The energy reflection coefficient of SV wave is the square of the amplitude ratio of the potential function of the reflected SV wave to the incident P wave multiplied by the reciprocal of the velocity ratio of the corresponding wave, i.e.

$$D_{\mathrm{PV}} = \frac{c_P B_2^2}{c_S A_1^2}. \tag{3.2.58}$$

From the point of view of energy, the energy carried by the incident wave, after the reflection at the surface, is all transferred to the reflected wave. Then, the physical principle of energy conservation be expressed by the following formula

$$D_{\mathrm{PP}} + D_{\mathrm{PV}} = 1, \tag{3.2.59}$$

or

$$\langle I_{P2}\rangle + \langle I_S\rangle = \langle I_{P1}\rangle. \tag{3.2.60}$$

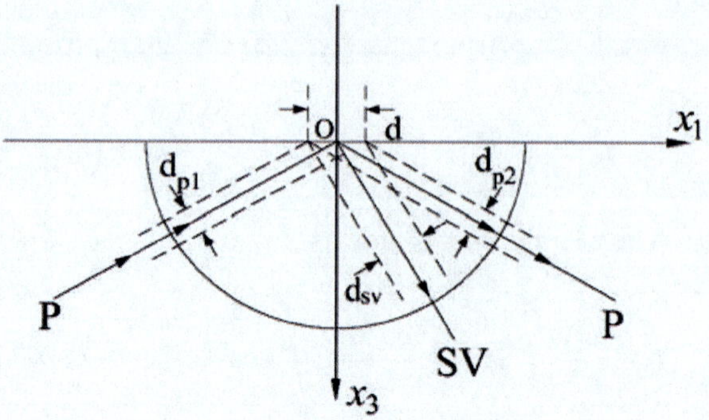

Fig. 3.11 Schematic diagram of energy conservation of incident and reflected waves

Substitute Eqs. (3.2.20) and (3.2.21) into Eqs. (3.2.57) and (3.2.58) and further into Eq. (3.2.59), we find that Eq. (3.2.59) cannot be satisfied. Then, how the principle of conservation of energy should be expressed? Referring to Fig. 3.11, we let the width of the incident wave beam be $d_{P1} = 1$, and the width becomes $d = 1/\cos\alpha$ when the incident wave is projected onto the surface. The width of the P wave and the SV wave reflected from surface of the width d are $d_{P2} = 1$ and $d_{\mathrm{SV}} = d\cos\beta$, respectively. Considering that the energy flow is the energy flowing through the wavefront per unit area and per unit time, the law of energy conservation should be expressed as

$$\langle I_{P2}\rangle \times d_{P2} + \langle I_S\rangle \times d_{\mathrm{SV}} = \langle I_{P1}\rangle \times d_{P1}. \tag{3.2.61}$$

It implies that **the energy flowing into the surface of a given length is equal to the energy flowing out of the surface of the same length.** The above equation can also be written as

$$\langle I_{P2}\rangle + \langle I_S\rangle \frac{\cos\beta}{\cos\alpha} = \langle I_{P1}\rangle, \tag{3.2.62}$$

$$D_{\mathrm{PP}} + D_{\mathrm{PV}} \frac{\cos\beta}{\cos\alpha} = 1. \tag{3.2.63}$$

If Eq. (3.2.62) is rewritten as

$$\langle I_{P2}\rangle \cos\alpha + \langle I_S\rangle \cos\beta = \langle I_{P1}\rangle \cos\alpha,$$

or

$$\langle I_{P2}\cos\alpha\rangle + \langle I_S\cos\beta\rangle = \langle I_{P1}\cos\alpha\rangle.$$

The above equation can also be understood as

$$\langle I_3(P2)\rangle + \langle I_3(\mathrm{SV})\rangle - \langle I_3(P1)\rangle = 0. \tag{3.2.64}$$

The above equation indicates that **the average energy flow of incident wave along the normal of surface is equal to the average energy flow of reflected wave flowing out along the normal.**

3.2.2 Reflection of SH Waves on Free Surface

Since SH wave only has displacement in the direction of $\boldsymbol{e}_2$, on the free interface, the incident SH wave will only generate reflected SH wave, the reflected P and SV wave cannot generate, as shown in Fig. 3.12. Considering that the potential function of SH wave is a vector function with two components, it is more convenient to directly assume the displacement field. Therefore, when discussing the reflection and transmission of SH wave, the displacement field is directly assumed. Suppose the displacement fields of incident SH wave and reflected SH wave are

$$u_2 = C_1 \exp[i(k_{S1}x_1 - k_{S3}x_3 - \omega_S t)], \tag{3.2.65}$$

$$u_2' = C_2 \exp\left[i\left(k_{S1}' x_1 + k_{S3}' x_3 - \omega_S' t\right)\right]. \tag{3.2.66}$$

where

$$k_{S1} = \frac{\omega_S}{c_S}\sin\beta, \quad k_{S3} = \frac{\omega_S}{c_S}\cos\beta,$$

$$k_{S1}' = \frac{\omega_S'}{c_S}\sin\beta', \quad k_{S3}' = \frac{\omega_S'}{c_S}\cos\beta'.$$

Since u_2 appears only in σ_{32}, the boundary condition can be expressed as

$$\sigma_{32}|_{x_3=0} = 0. \tag{3.2.67}$$

From Eq. (3.1.24), we can get

$$C_1 k_{S3} \exp[i(k_{S1}x_1 - \omega_S t)] - C_2 k_{S3}' \exp\left[i\left(k_{S1}' x_1 - \omega_S' t\right)\right] = 0. \tag{3.2.68}$$

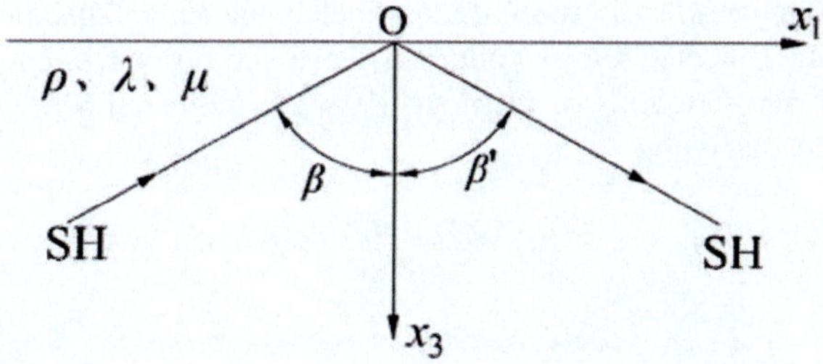

Fig. 3.12 Reflection of SH wave on a free surface

In order for Eq. (3.2.68) to hold for any x_1 and t, the frequency and wavenumber must satisfy

$$\omega_S = \omega'_S, \quad k_{S1} = k'_{S1}. \tag{3.2.69}$$

then

$$\beta = \beta', \quad k_{S3} = k'_{S3}. \tag{3.2.70}$$

According to Eq. (3.2.68), we can obtain

$$C_1 = C_2. \tag{3.2.71}$$

The displacement reflection coefficient of the reflected SH wave is defined as the ratio of the displacement amplitude of the reflected SH wave to the incident SH wave, i.e.

$$R_{\mathrm{HH}} = \frac{C_2}{C_1} = 1. \tag{3.2.72}$$

It shows that the reflected angle is equal to the incident angle when SH wave is reflected from a free surface, the frequency does not change, and the amplitude and phase also remain unchanged. From the point of view of conservation of energy, there should be

$$D_{\mathrm{HH}} = \frac{\langle I_{\mathrm{SH}}^{\mathrm{re}} \rangle}{\langle I_{\mathrm{SH}}^{\mathrm{in}} \rangle} = 1. \tag{3.2.73}$$

3.2.3 *Reflection of SV Waves on Free Surface*

As shown in Fig. 3.13, plane x_1x_2 is the free surface of the medium, plane x_1x_3 is the incident plane of the wave, space $x_3 < 0$ is the vacuum, and the region of $x_3 \geq 0$ is the elastic medium. The density and Lame coefficients of the medium are ρ, λ, μ, respectively. Since the particle vibrates in the plane x_1x_3 for the incident SV wave, the particle vibration of its reflected wave must also be in the plane x_1x_3. Therefore, the reflected waves contain P wave and SV wave, but there is no SH wave. Let the wave functions of incident SV wave, reflected SV wave and reflected P wave be, respectively:

$$\psi_2 = A_1 \exp[i(k_{S1}x_1 - k_{S3}x_3 - \omega_S t)], \tag{3.2.74a}$$

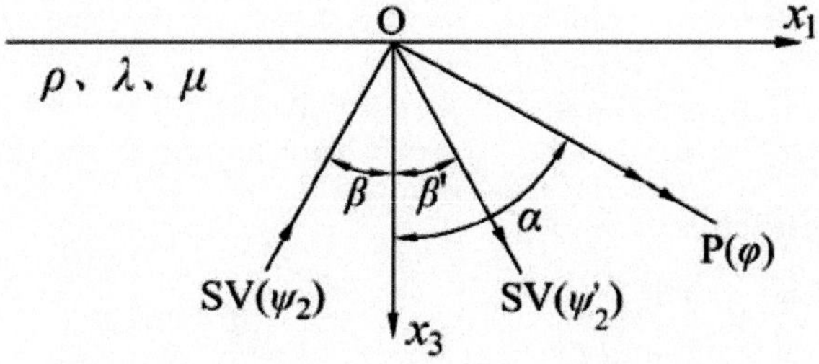

Fig. 3.13 Reflection of SV wave on a free surface

$$\psi_2' = A_2 \exp\left[i\left(k_{S1}'x_1 + k_{S3}'x_3 - \omega_S' t\right)\right], \tag{3.2.74b}$$

$$\varphi = B \exp[i(k_{P1}x_1 + k_{P3}x_3 - \omega_P t)]. \tag{3.2.74c}$$

where

$$k_{S1} = \frac{\omega_S}{c_S}\sin\beta,\quad k_{S3} = \frac{\omega_S}{c_S}\cos\beta,$$

$$k_{S1}' = \frac{\omega_S'}{c_S}\sin\beta',\quad k_{S3}' = \frac{\omega_S'}{c_S}\cos\beta',$$

$$k_{P1} = \frac{\omega_P}{c_P}\sin\alpha,\quad k_{P3} = \frac{\omega_P}{c_P}\cos\alpha.$$

Since u_1 and u_3 only appear in σ_{33} and σ_{31} and are irrelevant to σ_{32}, the boundary conditions of the free surface can be expressed as

$$\sigma_{33}|_{x_3=0} = 0,\ \sigma_{31}|_{x_3=0} = 0. \tag{3.2.75}$$

According to Eqs. (3.1.22) and (3.1.23), the stress boundary equations of the free surface can be written as

$$\begin{aligned} &-\lambda k_P^2 B \exp[i(k_{P1}x_1 - \omega_P t)] + 2\mu\left\{-Bk_{P3}^2 \exp[i(k_{P1}x_1 - \omega_P t)]\right. \\ &\left.+A_1 k_{S1}k_{S3}\exp[i(k_{S1}x_1 - \omega_S t)] - A_2 k_{S1}'k_{S3}'\exp\left[i\left(k_{S1}'x_1 - \omega_S' t\right)\right]\right\} = 0, \end{aligned} \tag{3.2.76a}$$

$$\begin{aligned} &-2Bk_{P1}k_{P3}\exp[i(k_{P1}x_1 - \omega_P t)] - \left\{A_1 k_{S1}^2 \exp[i(k_{S1}x_1 - \omega_S t)]\right. \\ &\left.+A_2 k_{S1}'^2 \exp\left[i\left(k_{S1}'x_1 - \omega_s' t\right)\right]\right\} + \left\{A_1 k_{S3}^2 \exp[i(k_{S1}x_1 - \omega_S t)]\right. \\ &\left.+A_2 k_{S3}'^2 \exp\left[i\left(k_{S1}'x_1 - \omega_S' t\right)\right]\right\} = 0. \end{aligned} \tag{3.2.76b}$$

If the above formula is valid for any x_1 and t, there must be

$$\omega_S = \omega_S' = \omega_P,\quad k_{S1} = k_{S1}' = k_{P1}, \tag{3.2.77}$$

namely

$$\frac{\sin\beta}{c_S} = \frac{\sin\beta'}{c_S} = \frac{\sin\alpha}{c_P}. \tag{3.2.78}$$

Hence,

$$\beta = \beta'.$$

It shows that the frequency of the reflected wave is the same as the incident wave, the reflected angle of the reflected SV wave is equal to the incident angle, and the incident angle of the incident SV wave satisfies the relationship

$$\frac{\sin\beta}{\sin\alpha} = \frac{c_S}{c_P} \tag{3.2.79}$$

with the reflected angle of the reflected P wave. In other words, the SV wave also follows Snell's law at the free surface, and the reflected angle of the reflected SV wave is smaller than that of the reflected P wave.

Inserting Eq. (3.2.77) into Eqs. (3.2.76a, 3.2.76b) leads to

$$B\cos 2\beta - (A_1 - A_2)\sin 2\beta = 0, \tag{3.2.80a}$$

$$-Bc_S^2\sin 2\alpha + c_P^2(A_1 + A_2)\cos 2\beta = 0. \tag{3.2.80b}$$

The reflection coefficient of the reflected SV wave is

$$F_{\mathrm{VV}} = \frac{A_2}{A_1} = \frac{c_S^2\sin 2\alpha\sin 2\beta - c_P^2\cos^2 2\beta}{c_S^2\sin 2\alpha\sin 2\beta + c_P^2\cos^2 2\beta}. \tag{3.2.81}$$

The reflection coefficient of the reflected P wave is

$$F_{\mathrm{VP}} = \frac{B}{A_1} = \frac{2c_P^2\sin 2\beta\cos 2\beta}{c_S^2\sin 2\alpha\sin 2\beta + c_P^2\cos^2 2\beta}. \tag{3.2.82}$$

According to Eqs. (3.2.81) and (3.2.82), we can get the following conclusions:

(1) Vertical incidence

When SV wave is incident vertically, i.e. $\beta = 0$, there are

$$F_{\mathrm{VV}} = \frac{A_2}{A_1} = -1, \quad F_{\mathrm{VP}} = \frac{\mathrm{B}}{A_1} = 0. \tag{3.2.83}$$

It shows that there is only the reflected SV wave in this case, no reflected P wave, and the wave pattern does not change.

(2) Mode conversion of wave motion

If $F_{VV} = \frac{A_2}{A_1} = 0$, namely

$$c_S^2 \sin 2\alpha \sin 2\beta - c_P^2 \cos^2 2\beta = 0. \tag{3.2.84}$$

Then, the reflected SV wave disappears and all the reflected waves are P waves. In this case, the reflection coefficient of P wave is

$$F_{VP} = \frac{B}{A_1} = \tan 2\beta. \tag{3.2.85}$$

(3) Critical reflection and total reflection of SV wave

In case of SV wave incident, if the reflection angle of P wave satisfies $\alpha = \pi/2$, the following equation can be obtained

$$\frac{\sin\beta}{\sin\alpha} = \frac{c_S}{c_P} = \left(\frac{\mu}{\lambda + 2\mu}\right)^{\frac{1}{2}}. \tag{3.2.86}$$

The incident angle β determined by Eq. (3.2.86) is called **the critical angle** of SV wave incidence and can be denoted by β_{cr}. The corresponding reflection process is called **critical reflection** (as shown in Fig. 3.14), obviously,

$$\beta_{cr} = \arcsin\left(\left(\frac{\mu}{\lambda + 2\mu}\right)^{\frac{1}{2}}\right). \tag{3.2.87}$$

It shows that the critical angle depends on the material constant. In this case, the reflection coefficient of SV wave is

$$F_{VV} = \frac{A_2}{A_1} = -1. \tag{3.2.88}$$

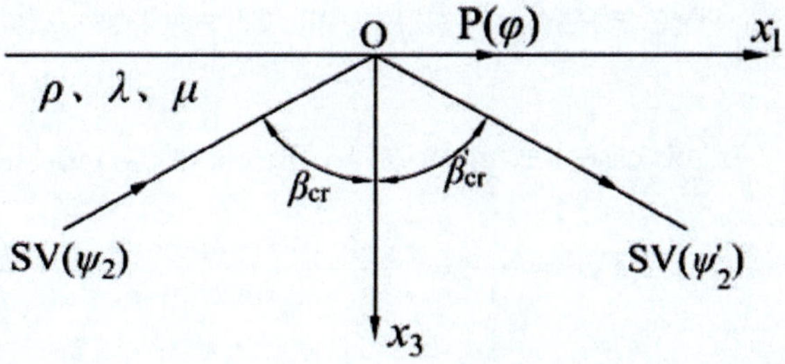

Fig. 3.14 Critical reflection of SV wave

When incidence angle $\beta > \beta_{cr} = \arcsin\left(\frac{\mu}{\lambda+2\mu}\right)^{\frac{1}{2}}$, $\sin\alpha = \frac{c_P}{c_S}\sin\beta > 1$, we can know that α can no longer be a real number (although β is a real number); then, we can think that $\alpha = \frac{\pi}{2} - i\gamma$, and thereby

$$\sin\alpha = \sin\left(\frac{\pi}{2} - i\gamma\right) = \cos(i\gamma) = \mathrm{ch}\gamma > 1. \tag{3.2.89}$$

where $\gamma = \mathrm{arcch}\left[\left(\frac{\lambda+2\mu}{\mu}\right)^{\frac{1}{2}}\sin\beta\right]$, and

$$\cos\alpha = \cos\left(\frac{\pi}{2} - i\gamma\right) = \sin(i\gamma) = i\,\mathrm{sh}\gamma. \tag{3.2.90}$$

In this case, the potential function of reflected P wave can be expressed as

$$\begin{aligned}\phi &= B\exp[i(k_{P1}x_1 + k_{P3}x_3 - \omega_P t)] \\ &= B\exp(-k_P\mathrm{sh}\gamma\cdot x_3)\exp[i(k_P\mathrm{ch}\gamma\cdot x_1 - \omega_P t)].\end{aligned} \tag{3.2.91}$$

It can be seen that the reflected P wave has become a surface wave that propagates along the surface and decays along the positive direction of x_3. It is an **inhomogeneous wave** because the direction of attenuation is not consistent with the direction of propagation. The velocity of the inhomogeneous wave along the surface is

$$c'_P = \frac{\omega_P}{k_P\mathrm{ch}\gamma} < c_P. \tag{3.2.92}$$

Considering that

$$\mathrm{ch}^2\gamma - \mathrm{sh}^2\gamma = 1,$$

$$\sin 2\alpha = 2\sin\alpha\cos\alpha = 2i\,\mathrm{ch}\gamma\,\mathrm{sh}\gamma = 2i\,\mathrm{sh}2\gamma.$$

We can also calculate like this

$$\cos\alpha = \cos\left(\frac{\pi}{2} - i\gamma\right) = \sin(i\gamma) = i\,\mathrm{sh}\gamma = i\sqrt{\mathrm{ch}^2\gamma - 1} = i\sqrt{\sin^2\alpha - 1}. \tag{3.2.93}$$

In this case, the reflection coefficient of the reflected SV wave

$$\begin{aligned}F_{VV} &= \frac{A_2}{A_1} = \frac{c_S^2\sin 2\alpha\sin 2\beta - c_P^2\cos^2 2\beta}{c_S^2\sin 2\alpha\sin 2\beta + c_P^2\cos^2 2\beta} \\ &= \frac{-c_P^2\cos^2 2\beta + i2c_S^2\mathrm{sh}2\gamma\sin 2\beta}{c_P^2\cos^2 2\beta + i2c_S^2\mathrm{sh}2\gamma\sin 2\beta} = \frac{-a+ib}{a+ib}.\end{aligned} \tag{3.2.94}$$

where

$$a = c_P^2 \cos^2 2\beta, \quad b = c_S^2 \mathrm{sh} 2\gamma \sin 2\beta. \tag{3.2.95}$$

Therefore, the amplitude of the reflection coefficient equal to 1, i.e. $|F_{\mathrm{VV}}| = 1$, while the phase shift $\delta = \arctan \frac{2ab}{b^2 - a^2}$. This situation is called **total reflection of SV wave**.

(4) Grazing incident

When the incident angle $\beta = \frac{\pi}{2}$, there are

$$F_{\mathrm{VV}} = \frac{A_2}{A_1} = -1, \quad F_{\mathrm{VP}} = \frac{B}{A_1} = 0. \tag{3.2.96}$$

It can be found that the reflected P wave will disappear and only the reflected SV wave will exist. Reflected SV wave is a wave with the same amplitude in the same direction as the incident wave but with a phase shift of 180°. The superposition of the incident wave and the reflected wave ensures the free surface to be stress-free.

The displacement reflection coefficient of SV wave is defined as the ratio of the displacement amplitude of the reflected SV wave to the incident SV wave, i.e.

$$R_{\mathrm{VV}} = \frac{A_2}{A_1} \cdot \frac{c_S}{c_S} = \frac{c_S^2 \sin 2\alpha \sin 2\beta - c_P^2 \cos^2 2\beta}{c_S^2 \sin 2\alpha \sin 2\beta + c_P^2 \cos^2 2\beta}. \tag{3.2.97a}$$

The displacement reflection coefficient of P wave is defined as the ratio of the displacement amplitude of the reflected P wave to the incident SV wave, i.e.

$$R_{\mathrm{VP}} = \frac{B}{A_1} \cdot \frac{c_S}{c_P} = \frac{2c_S c_P \cos 2\beta \sin 2\beta}{c_S^2 \sin 2\alpha \sin 2\beta + c_P^2 \cos^2 2\beta}. \tag{3.2.97b}$$

(5) The law of conservation of energy

In case of SV wave incident, the energy reflection coefficient of reflected P wave and reflected SV wave are

$$D_{\mathrm{VV}} = \frac{A_2^2}{A_1^2}, \tag{3.2.98a}$$

$$D_{\mathrm{VP}} = \frac{B^2}{A_1^2} \frac{c_S}{c_P}. \tag{3.2.98b}$$

The law of conservation of energy should be expressed as

$$\langle I_{P2} \rangle \cos\alpha + \langle I_{\mathrm{SV}2} \rangle \cos\beta = \langle I_{\mathrm{SV}1} \rangle \cos\beta. \tag{3.2.99a}$$

or

$$D_{VP}\frac{\cos\alpha}{\cos\beta}+D_{VV}=1. \tag{3.2.99b}$$

3.2.4 Incident P Wave and SV Wave Simultaneously

In the previous sections, we have discussed the case where P wave, SV wave and SH wave incidence separately. In this section, we will discuss the reflection problem on the free surface when P wave and SV wave incidence appear simultaneously. As discussed above, when P wave and SV wave incidence appears separately, there are reflected P wave and reflected SV wave corresponding to them, respectively. When the P wave and the SV wave incidence occur at the same time, should the reflected waves contain two P waves and two SV waves? Let us assume that there are two reflected SV waves and two reflected P waves (as shown in Fig. 3.15). We will first proved that the actual reflected wave has only one SV wave and one P wave.

It is assumed that the wave functions of incident P wave, SV wave and reflected P wave and SV wave are, respectively,

$$\varphi=A\exp[i(k_{P1}x_1-k_{P3}x_3-\omega_P t)], \tag{3.2.100a}$$

$$\psi_2=B\exp[i(k_{S1}x_1-k_{S3}x_3-\omega_S t)], \tag{3.2.100b}$$

$$\varphi'=A'\exp\left[i\left(k'_{P1}x_1+k'_{P3}x_3-\omega'_P t\right)\right], \tag{3.2.100c}$$

$$\varphi''=A''\exp\left[i\left(k''_{P1}x_1+k''_{P3}x_3-\omega''_P t\right)\right], \tag{3.2.100d}$$

$$\psi'_2=B'\exp\left[i\left(k'_{S1}x_1+k'_{S3}x_3-\omega'_S t\right)\right], \tag{3.2.100e}$$

$$\psi''_2=B''\exp\left[i\left(k''_{S1}x_1+k''_{S3}x_3-\omega''_S t\right)\right]. \tag{3.2.100f}$$

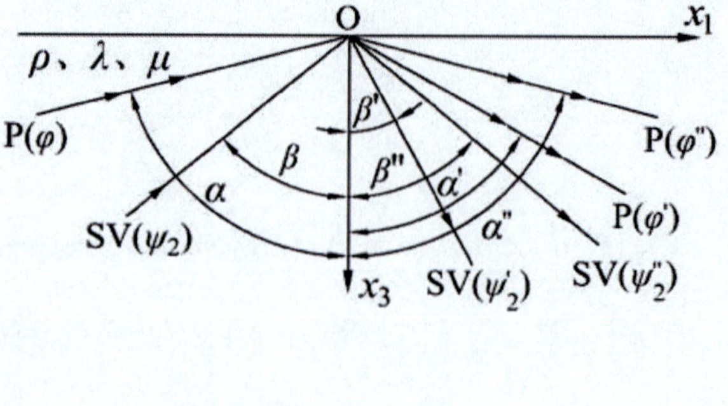

Fig. 3.15 Assumptive reflection waves when P wave and SV wave incidence occur simultaneously

where A, B, A', A'', B', B'' and ω_P, ω_S, ω'_P, ω''_P, ω'_S, ω''_S are the amplitude and angular frequency of each wave, respectively, and

$$k_{P1} = \frac{\omega_P}{c_P}\sin\alpha, \quad k_{P3} = \frac{\omega_P}{c_P}\cos\alpha,$$

$$k'_{P1} = \frac{\omega'_P}{c_P}\sin\alpha', \quad k'_{P3} = \frac{\omega'_P}{c_P}\cos\alpha',$$

$$k''_{P1} = \frac{\omega''_P}{c_P}\sin\alpha'', \quad k''_{P3} = \frac{\omega''_P}{c_P}\cos\alpha'',$$

$$k_{S1} = \frac{\omega_S}{c_S}\sin\beta, \quad k_{S3} = \frac{\omega_S}{c_S}\cos\beta,$$

$$k'_{S1} = \frac{\omega'_S}{c_S}\sin\beta', \quad k'_{S3} = \frac{\omega'_S}{c_S}\cos\beta',$$

$$k''_{S1} = \frac{\omega''_S}{c_S}\sin\beta'', \quad k''_{S3} = \frac{\omega''_S}{c_S}\cos\beta''.$$

The boundary conditions are as follows

$$\sigma_{33}|_{x_3=0} = 0, \quad \sigma_{31}|_{x_3=0} = 0. \tag{3.2.101}$$

According to Eq. (3.1.22), we get

$$\begin{aligned}
\sigma_{33}|_{x_3=0} = {} & \lambda\left[\frac{\partial^2(\varphi+\varphi'+\varphi'')}{\partial x_1^2} + \frac{\partial^2(\varphi+\varphi'+\varphi'')}{\partial x_3^2}\right] \\
& + 2\mu\left[\frac{\partial^2(\varphi+\varphi'+\varphi'')}{\partial x_3^2} + \frac{\partial^2(\psi_2+\psi'_2+\psi''_2)}{\partial x_1 \partial x_3}\right]\bigg|_{x_3=0} \\
= {} & \lambda\left\{-k_{P1}^2 A\exp[i(k_{P1}x_1-\omega_P t)] - (k'_{P1})^2 A'\exp\left[i\left(k'_{P1}x_1-\omega'_P t\right)\right]\right. \\
& \left.-\left(k''_{P1}\right)^2 A''\exp\left[i\left(k''_{P1}x_1-\omega'_P t\right)\right]\right\} \\
& + (\lambda+2\mu)\left\{-k_{P3}^2 A\exp[i(k_{P1}x_1-\omega_P t)]\right. \\
& \left.-\left(k'_{P3}\right)^2 A'\exp\left[i\left(k'_{P1}x_1-\omega'_P t\right)\right] - \left(k''_{P3}\right)^2 A''\exp\left[i\left(k''_{P1}x_1-\omega''_P t\right)\right]\right\} \\
& + 2\mu\left\{k_{S3}k_{S1}B\exp[i(k_{S1}x_1-\omega_S t)] - k'_{S1}k'_{S3}B'\exp\left[i\left(k'_{S1}x_1-\omega'_S t\right)\right]\right. \\
& \left.-k''_{S1}k''_{S3}B''\exp\left[i\left(k''_{S1}x_1-\omega''_S t\right)\right]\right\} = 0.
\end{aligned} \tag{3.2.102}$$

Then, the formula is valid for any x and t requires

$$k_{P1} = k'_{P1} = k''_{P1} = k_{S1} = k'_{S1} = k''_{S1}, \quad \omega_P = \omega'_P = \omega'_P = \omega_S = \omega'_S = \omega''_S. \tag{3.2.103}$$

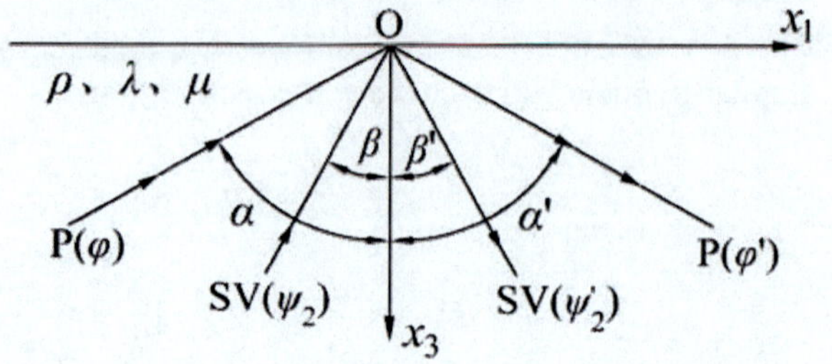

Fig. 3.16 Reflected waves when P wave and SV wave incidence occur simultaneously

Thereby

$$\alpha = \alpha' = \alpha'', \tag{3.2.104a}$$

$$\beta = \beta' = \beta''. \tag{3.2.104b}$$

The above equation indicates that the reflected wave has only one SV wave and one P wave, as shown in Fig. 3.16. And, there are

$$\frac{\sin\alpha}{c_P} = \frac{\sin\beta}{c_S}. \tag{3.2.105}$$

Snell's law is also satisfied when P wave and SV wave incidence occur at the same time.

The wave functions of incident P wave, SV wave and reflected P wave and SV wave are defined by

$$\varphi = A\exp[i(k_{P1}x_1 - k_{P3}x_3 - \omega t)], \tag{3.2.106a}$$

$$\varphi' = A'\exp[i(k_{P1}x_1 + k_{P3}x_3 - \omega t)], \tag{3.2.106b}$$

$$\psi_2 = B\exp[i(k_{S1}x_1 - k_{S3}x_3 - \omega t)], \tag{3.2.106c}$$

$$\psi_2' = B'\exp[i(k_{S1}x_1 + k_{S3}x_3 - \omega t)]. \tag{3.2.106d}$$

According to the stress boundary condition

$$\sigma_{33}|_{x_3=0} = 0, \quad \sigma_{31}|_{x_3=0} = 0. \tag{3.2.107}$$

The stress boundary equations of the free surface can be obtained as

$$\left[\lambda k_{P1}^2 + (\lambda + 2\mu)k_{P3}^2\right](A + A') + 2\mu k_{S1}k_{S3}(B' - B) = 0, \tag{3.2.108a}$$

$$2k_{P1}k_{P3}(A - A') + (k_{S3}^2 - k_{S1}^2)(B + B') = 0. \tag{3.2.108b}$$

Applying the trigonometric formula to the above equation

$$(A + A')\cos 2\beta + (B' - B)\sin 2\beta = 0, \tag{3.2.109a}$$

$$(A - A')\sin 2\alpha c_S^2 + \cos 2\beta c_P^2 (B + B') = 0. \tag{3.2.109b}$$

The following equations are obtained

$$A' = \frac{(\sin 2\alpha \sin 2\beta c_S^2 - \cos^2 2\beta c_P^2)A + 2\cos 2\beta \sin 2\beta c_P^2 B}{\sin 2\alpha \sin 2\beta c_S^2 + \cos^2 2\beta c_P^2}, \tag{3.2.110a}$$

$$B' = \frac{-2\sin 2\alpha \cos 2\beta c_S^2 A + (\sin 2\alpha \sin 2\beta c_S^2 - \cos^2 2\beta c_P^2)B}{\sin 2\alpha \sin 2\beta c_S^2 + \cos^2 2\beta c_P^2}. \tag{3.2.110b}$$

It can also be written in matrix form

$$\begin{pmatrix} A' \\ B' \end{pmatrix} = F\begin{pmatrix} A \\ B \end{pmatrix} = \begin{pmatrix} F_{11} & F_{12} \\ F_{21} & F_{22} \end{pmatrix}\begin{pmatrix} A \\ B \end{pmatrix}. \tag{3.2.111}$$

where

$$F_{11} = \frac{\sin 2\alpha \sin 2\beta c_S^2 - \cos^2 2\beta c_P^2}{\sin 2\alpha \sin 2\beta c_S^2 + \cos^2 2\beta c_P^2}, \tag{3.2.112a}$$

$$F_{12} = \frac{2\cos 2\alpha \sin 2\beta c_P^2}{\sin 2\alpha \sin 2\beta c_S^2 + \cos^2 2\beta c_P^2}, \tag{3.2.112b}$$

$$F_{21} = \frac{-2\sin 2\alpha \cos 2\beta c_S^2}{\sin 2\alpha \sin 2\beta c_S^2 + \cos^2 2\beta c_P^2}, \tag{3.2.112c}$$

$$F_{22} = \frac{\sin 2\alpha \sin 2\beta c_S^2 - \cos^2 2\beta c_P^2}{\sin 2\alpha \sin 2\beta c_S^2 + \cos^2 2\beta c_P^2}. \tag{3.2.112d}$$

The matrix $\boldsymbol{F}$ is defined as the reflection matrix of reflected P wave and reflected SV wave when P wave and SV wave incidences occur simultaneously.

Let us make a discussion on Eq. (3.2.111). When $B = 0$, it degenerates into the case of P wave incidence alone, and F_{11} and F_{21} are respectively the reflection coefficients of reflected P wave and reflected SV wave when P wave incidence alone. When $A = 0$, it degenerates into the case of SV wave incidence alone, and F_{12} and F_{22} are, respectively, the reflection coefficients of reflected P wave and reflected SV wave when SV wave incidence alone.

When P wave and SV wave incidences occur at the same time, we can get the following conclusions:

(1) The incident angle of P wave is equal to the reflected angle of P wave, and the incident angle of SV wave is equal to the reflected angle of SV wave.
(2) Since $\frac{c_P}{c_S} = \frac{\sin\alpha}{\sin\beta}$, and $c_P > c_S$, there is $\alpha > \beta$. If the P wave and SV wave incidence occur at the same time, their incident angles α and β cannot be arbitrary, but must satisfy a certain relationship, and the incident angle β of SV wave is always smaller than the incident angle α of P wave.
(3) The reflection coefficient of P wave in the case of P wave incidence alone is equal to that of SV wave in the case of SV wave incidence alone, i.e. $F_{11} = F_{22}$.

Let the displacement amplitudes of the incident P wave and the SV wave be C_1 and C_2, respectively, and the displacement amplitudes of the reflected P wave and the reflected SV wave be C_1' and C_2', respectively. When P wave incidence occurs separately, the displacement reflection coefficients of P wave and SV wave are denoted by R_{11} and R_{21}, respectively, then

$$R_{11} = F_{11}, \quad R_{21} = F_{21} \cdot \frac{c_P}{c_S}. \tag{3.2.113}$$

When SV wave incidence occurs separately, the displacement reflection coefficients of P wave and SV wave are denoted by R_{12} and R_{22}, respectively, then

$$R_{12} = F_{12} \cdot \frac{c_S}{c_P}, \quad R_{22} = F_{22}. \tag{3.2.114}$$

Therefore,

$$\begin{pmatrix} C_1' \\ C_2' \end{pmatrix} = \begin{pmatrix} R_{11} & R_{12} \\ R_{21} & R_{22} \end{pmatrix} \begin{pmatrix} C_1 \\ C_2 \end{pmatrix}. \tag{3.2.115}$$

Defining the matrix $\boldsymbol{R} = \begin{pmatrix} R_{11} & R_{12} \\ R_{21} & R_{22} \end{pmatrix}$ as the displacement reflection matrix of reflected P wave and reflected SV wave when P wave and SV wave incidences occur at the same time, then

$$R_{11} = \frac{\sin 2\alpha \sin 2\beta c_S^2 - \cos^2 2\beta c_P^2}{\sin 2\alpha \sin 2\beta c_S^2 + \cos^2 2\beta c_P^2}, \tag{3.2.116a}$$

$$R_{12} = \frac{2\cos 2\alpha \sin 2\beta c_p c_s}{\sin 2\alpha \sin 2\beta c_S^2 + \cos^2 2\beta c_P^2}, \tag{3.2.116b}$$

$$R_{21} = \frac{-2\sin 2\alpha \cos 2\beta c_S c_P}{\sin 2\alpha \sin 2\beta c_S^2 + \cos^2 2\beta c_P^2}, \tag{3.2.116c}$$

$$R_{22} = \frac{\sin 2\alpha \sin 2\beta c_S^2 - \cos^2 2\beta c_P^2}{\sin 2\alpha \sin 2\beta c_S^2 + \cos^2 2\beta c_P^2}. \tag{3.2.116d}$$

3.3 Reflection and Transmission of Elastic Waves at the Interface

Discussing the reflection and transmission of plane waves at the plane interface between two semi-infinite elastic media is of great practical significance for studying the propagation law of seismic waves in the earth medium and using seismic waves for seismic exploration. By studying the motion law of seismic wave in layered medium, we can deduce the stratigraphic structure and physical property parameters, which is useful to search for oil, gas and other mineral resources.

In this section, the incidences of P, SV and SH waves at the interfaces of two elastic half-spaces are discussed. Two elastic half-spaces are both homogeneous isotropic elastic media, but the material constants of the two half-spaces are different. The interface of the two half-spaces is assumed to be perfect.

3.3.1 Reflection and Transmission of P Waves at the Interface

As shown in Fig. 3.17, the density and Lame coefficients of medium 1 and medium 2 are denoted by $(\rho, \lambda, \mu)$and (ρ', λ', μ'), respectively. Reflection and transmission waves are formed when incident P wave propagates in plane x_1x_3 of the medium 1 and encounter the boundary plane x_1x_2.

Fig. 3.17 Reflection and transmission of P waves at the interface

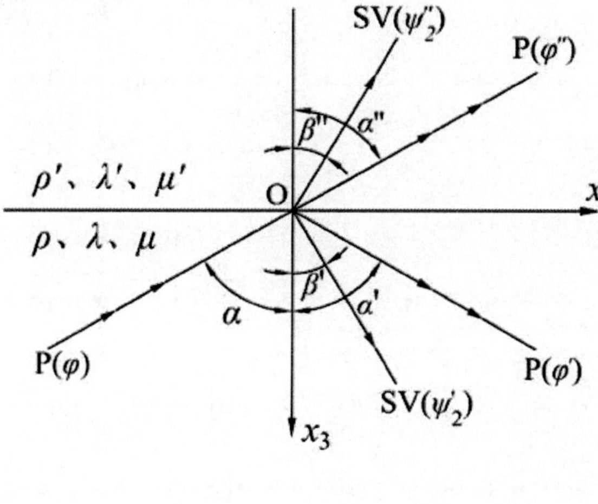

Suppose the wave functions of incident P wave, reflected P wave and SV wave, transmitted P wave and SV wave are, respectively

$$\varphi = A\exp[i(k_{P1}x_1 - k_{P3}x_3 - \omega_P t)], \tag{3.3.1a}$$

$$\varphi' = A'\exp\left[i\left(k'_{P1}x_1 + k'_{P3}x_3 - \omega'_P t\right)\right], \tag{3.3.1b}$$

$$\psi'_2 = B'\exp\left[i\left(k'_{S1}x_1 + k'_{S3}x_3 - \omega'_S t\right)\right], \tag{3.3.1c}$$

$$\varphi'' = A''\exp\left[i\left(k''_{P1}x_1 - k''_{P3}x_3 - \omega''_P t\right)\right], \tag{3.3.1d}$$

$$\psi''_2 = B''\exp\left[i\left(k''_{S1}x_1 - k''_{S3}x_3 - \omega''_S t\right)\right], \tag{3.3.1e}$$

where A, A', A'', B', B'', ω_P, ω'_P, ω''_P, ω'_S and ω''_S are the amplitudes and angular frequencies of corresponding waves, and

$$k_{P1} = \frac{\omega_P}{c_P}\sin\alpha, \quad k_{P3} = \frac{\omega_P}{c_P}\cos\alpha, \tag{3.3.2a}$$

$$k'_{P1} = \frac{\omega'_P}{c_P}\sin\alpha', \quad k'_{P3} = \frac{\omega'_P}{c_P}\cos\alpha', \tag{3.3.2b}$$

$$k''_{P1} = \frac{\omega''_P}{c'_P}\sin\alpha'', \quad k''_{P3} = \frac{\omega''_P}{c'_P}\cos\alpha'', \tag{3.3.2c}$$

$$k'_{S1} = \frac{\omega'_S}{c_S}\sin\beta', \quad k'_{S3} = \frac{\omega'_S}{c_S}\cos\beta', \tag{3.3.2d}$$

$$k''_{S1} = \frac{\omega''_S}{c'_S}\sin\beta'', \quad k''_{S3} = \frac{\omega''_S}{c'_S}\cos\beta''. \tag{3.3.2e}$$

In this case, the boundary conditions at $x_3 = 0$ are

$$u_1 = u'_1, \quad u_2 = u'_2, \quad u_3 = u'_3, \tag{3.3.3a}$$

$$\sigma_{33} = \sigma'_{33}, \quad \sigma_{31} = \sigma'_{31}, \quad \sigma_{32} = \sigma'_{32}. \tag{3.3.3b}$$

Since $u_2 = u'_2 = 0$ and u_1 and u_3 only appear in σ_{33} and σ_{31}, the resultful boundary conditions are

$$u_1|_{x_3=0} = u'_1|_{x_3=0}, \quad u_3|_{x_3=0} = u'_3|_{x_3=0}, \tag{3.3.4a}$$

$$\sigma_{33}|_{x_3=0} = \sigma'_{33}|_{x_3=0}, \quad \sigma_{31}|_{x_3=0} = \sigma'_{31}|_{x_3=0}. \tag{3.3.4b}$$

By using the relations between displacement and potential function, i.e. Eqs. (3.1.20a, 3.1.20b, 3.1.20c), (3.1.22) and (3.1.23), the boundary conditions can be obtained as follows:

$$\left[\frac{\partial(\varphi+\varphi')}{\partial x_1}-\frac{\partial\psi_2'}{\partial x_3}\right]|_{x_3=0}=\left[\frac{\partial\varphi''}{\partial x_1}-\frac{\partial\psi_2''}{\partial x_3}\right]|_{x_3=0}, \tag{3.3.5a}$$

$$\left[\frac{\partial(\varphi+\varphi')}{\partial x_3}+\frac{\partial\psi_2'}{\partial x_1}\right]|_{x_3=0}=\left[\frac{\partial\varphi''}{\partial x_3}+\frac{\partial\psi_2''}{\partial x_1}\right]|_{x_3=0}, \tag{3.3.5b}$$

$$\begin{aligned}&\left\{\lambda\nabla^2(\varphi+\varphi')+2\mu\left[\frac{\partial^2(\varphi+\varphi')}{\partial x_3^2}+\frac{\partial^2\psi_2'}{\partial x_1\partial x_3}\right]\right\}|_{x_3=0}\\&=\left\{\lambda'\nabla^2\varphi''+2\mu'\left[\frac{\partial^2\varphi''}{\partial x_3^2}+\frac{\partial^2\psi_2''}{\partial x_1\partial x_3}\right]\right\}|_{x_3=0},\end{aligned} \tag{3.3.5c}$$

$$\mu\left[2\frac{\partial^2(\varphi+\varphi')}{\partial x_1\partial x_3}+\frac{\partial^2\psi_2'}{\partial x_1^2}-\frac{\partial^2\psi_2'}{\partial x_3^2}\right]|_{x_3=0}=\mu'\left[2\frac{\partial^2\varphi''}{\partial x_1\partial x_3}+\frac{\partial^2\psi_2''}{\partial x_1^2}-\frac{\partial^2\psi_2''}{\partial x_3^2}\right]|_{x_3=0}. \tag{3.3.5d}$$

Substituting Eqs. (3.3.1a, 3.3.1b, 3.3.1c, 3.3.1d, 3.3.1e) into the above equation and set $x_3=0$, we have

$$\begin{aligned}&k_{P1}A\exp[i(k_{P1}x_1-\omega_Pt)]+k_{P1}'A'\exp\left[i\left(k_{P1}'x_1-\omega_P't\right)\right]\\&-k_{S3}'B'\exp\left[i\left(k_{S1}'x_1-\omega_S't\right)\right]\\&=k_{P1}''A''\exp\left[i\left(k_{P1}''x_1-\omega_P''t\right)\right]+k_{S3}''B''\exp\left[i\left(k_{S1}''x_1-\omega_S''t\right)\right],\end{aligned} \tag{3.3.6a}$$

$$\begin{aligned}&-k_{P3}A\exp[i(k_{P1}x_1-\omega_Pt)]+k_{P3}'A'\exp\left[i\left(k_{P1}'x_1-\omega_P't\right)\right]\\&+k_{S1}'B'\exp\left[i\left(k_{S1}'x_1-\omega_S't\right)\right]\\&=-k_{P3}''A''\exp\left[i\left(k_{P1}''x_1-\omega_P''t\right)\right]+k_{S1}''B''\exp\left[i\left(k_{S1}''x_1-\omega_S''t\right)\right],\end{aligned} \tag{3.3.6b}$$

$$\begin{aligned}&\lambda\left\{-k_{P1}^2A\exp[i(k_{P1}x_1-\omega_Pt)]-k_{P1}'^2A'\exp\left[i\left(k_{P1}'x_1-\omega_P't\right)\right]\right\}\\&+(\lambda+2\mu)\left\{-k_{P3}^2A\exp[i(k_{P1}x_1-\omega_Pt)]-k_{P3}'^2A'\exp\left[i\left(k_{P1}'x_1-\omega_P't\right)\right]\right\}\\&-2\mu k_{S1}'k_{S3}'B'\exp\left[i\left(k_{S1}'x_1-\omega_S't\right)\right]\\&=-\lambda'k_{P1}''^2A''\exp\left[i\left(k_{P1}''x_1-\omega_P''t\right)\right]-(\lambda'+2\mu')k_{P3}''^2A''\exp\left[i\left(k_{P1}''x_1-\omega_P't\right)\right]\\&+2\mu'k_{S1}''k_{S3}''B''\exp\left[i\left(k_{S1}''x_1-\omega_S''t\right)\right],\end{aligned} \tag{3.3.6c}$$

$$\begin{aligned}&\mu\{2k_{P1}k_{P3}A\exp[i(k_{P1}x_1-\omega_Pt)]-2k_{P1}'k_{P3}'A'\exp\left[i\left(k_{P1}'x_1-\omega_P't\right)\right]\\&-k_{S1}'^2B'\exp\left[i\left(k_{S1}'x_1-\omega_S't\right)\right]+k_{S3}'^2B'\exp\left[i\left(k_{S1}'x_1-\omega_S't\right)\right]\end{aligned}$$

$$= \mu'\{2k''_{P1}k''_{P3}A''\exp[i(k''_{P1}x_1 - \omega''_P t)] - k''^2_{S1}B''\exp[i(k''_{S1}x_1 - \omega''_S t)]$$
$$+ k''^2_{S3}B''\exp[i(k''_{S1}x_1 - \omega''_S t)]. \quad (3.3.6d)$$

In order to make the above formula hold for any x_1 and t, there must be

$$\omega_P = \omega'_P = \omega''_P = \omega'_S = \omega''_S, \quad (3.3.7a)$$

$$k_{P1} = k'_{P1} = k''_{P1} = k'_{S1} = k''_{S1}. \quad (3.3.7b)$$

It shows that the frequency of the incident, reflected and transmitted waves must be the same, and the apparent wave number must be the same. If we defined

$$\frac{\sin\alpha}{c_P} = \frac{\sin\alpha'}{c_P} = \frac{\sin\beta'}{c_S} = \frac{\sin\alpha''}{c'_P} = \frac{\sin\beta''}{c'_S} = \frac{1}{\overline{c}}, \quad (3.3.8)$$

then $\overline{c}$ can be called the apparent wave velocity. Therefore, the same apparent wave number means the same apparent wave speed, which is often called Snell's law of reflection. From Snell's law, we get

$$\alpha = \alpha'. \quad (3.3.9)$$

Namely, the reflected angle of reflected P wave is equal to the incident angle, and the reflected angle of reflected SV wave is smaller than the reflected angle of reflected P wave, while the transmitted angle of transmitted SV wave is smaller than the transmitted angle of transmitted P wave.

The simplification of formula (3.3.6a, 3.3.6b, 3.3.6c, 3.3.6d) leads to

$$A + A' - p_2B' - A'' - p_4B'' = 0, \quad (3.3.10a)$$

$$p_1(A - A') - B' - p_3A'' + B'' = 0, \quad (3.3.10b)$$

$$[\lambda(1 + p_1^2) + 2\mu p_1^2](A + A') + 2\mu p_2B'$$
$$- [\lambda'(1 + p_3^2) + 2\mu' p_3^2]A'' + 2\mu' p_4B'' = 0, \quad (3.3.10c)$$

$$\mu[2p_1(A - A') - (1 - p_2^2)B'] - \mu'[2p_3A'' - (1 - p_4^2)B''] = 0. \quad (3.3.10d)$$

where

$$p_1 = \frac{k_{P3}}{k_{P1}} = \cot\alpha, \quad p_2 = \frac{k'_{S3}}{k'_{S1}} = \cot\beta',$$

$$p_3 = \frac{k''_{P3}}{k''_{P1}} = \cot\alpha'', \quad p_4 = \frac{k''_{S3}}{k''_{S1}} = \cot\beta''.$$

Assume

$$P = A + A', \quad Q = A - A'. \tag{3.3.11}$$

Then, from Eqs. (3.3.10a) and (3.3.10c), we get

$$P = l_1 A'' + h_1 B'', \quad B' = l_2 A'' + h_2 B''. \tag{3.3.12}$$

From Eqs. (3.3.10b) and (3.3.10d), we get

$$Q = l_3 A'' + h_3 B'', \quad B' = -l_4 A'' - h_4 B''. \tag{3.3.13}$$

where

$$l_1 = \frac{2\mu + \lambda'(1+p_3^2) + 2u' p_3^2}{(\lambda + 2\mu)(1+p_1^2)}, \quad l_2 = \frac{\lambda'(1+p_3^2) + 2\mu' p_3^2 - [\lambda(1+p_1^2) + 2\mu p_1^2]}{(\lambda + 2\mu)(1+p_1^2)p_2},$$

$$l_3 = \frac{2\mu' p_3 - \mu(1-p_2^2)p_3}{\mu p_1(1+p_2^2)}, \quad l_4 = \frac{2(\mu - \mu')p_3}{\mu(1+p_2^2)},$$

$$h_1 = \frac{2(\mu - \mu')p_4}{(\lambda + 2\mu)(1+p_1^2)}, \quad h_2 = -\frac{[2\mu' + \lambda(1+p_1^2) + 2\mu p_1^2]p_4}{(\lambda + 2\mu)(1+p_1^2)p_2},$$

$$h_3 = \frac{\mu(1-p_2^2) - \mu'(1-p_4^2)}{\mu p_1(1+p_2^2)}, \quad h_4 = \frac{\mu'(1-p_4^2) - 2\mu}{\mu(1+p_2^2)}.$$

Inserting Eq. (3.3.11) into Eqs. (3.3.12) and (3.3.13) leads to the following equations

$$-\frac{A'}{A} + l_1\frac{A''}{A} + h_1\frac{B''}{A} = 1, \tag{3.3.14a}$$

$$-\frac{B'}{A} + l_2\frac{A''}{A} + h_2\frac{B''}{A} = 0, \tag{3.3.14b}$$

$$\frac{A'}{A} + l_3\frac{A''}{A} + h_3\frac{B''}{A} = 1, \tag{3.3.14c}$$

$$\frac{B'}{A} + l_4\frac{A''}{A} + h_4\frac{B''}{A} = 0. \tag{3.3.14d}$$

The reflection coefficient of reflected P wave and SV wave is defined as the ratio of the potential function amplitude of reflected P wave and SV wave to the potential function amplitude of incident P wave, namely

$$F_{PP} = \frac{A'}{A} = \frac{(l_1 - l_3)(h_2 + h_4) - (l_2 + l_4)(h_1 - h_3)}{(l_1 + l_3)(h_2 + h_4) - (l_2 + l_4)(h_1 + h_3)}, \tag{3.3.15}$$

$$F_{PV} = \frac{B'}{A} = \frac{2(l_2 h_4 - l_4 h_2)}{(l_1 + l_3)(h_2 + h_4) - (l_2 + l_4)(h_1 + h_3)}. \tag{3.3.16}$$

The transmission coefficient of the potential function of transmitted P wave and SV wave is defined as the ratio of the potential function amplitude of transmitted P wave and SV wave to the potential function amplitude of incident P wave, namely

$$E_{PP'} = \frac{A''}{A} = \frac{2(h_2 + h_4)}{(l_1 + l_3)(h_2 + h_4) - (l_2 + l_4)(h_1 + h_3)}, \tag{3.3.17}$$

$$E_{PV'} = \frac{B''}{A} = \frac{2(l_2 + l_4)}{(l_1 + l_3)(h_2 + h_4) - (l_2 + l_4)(h_1 + h_3)}. \tag{3.3.18}$$

From Eqs. (3.3.15) to (3.3.18), we can get the following conclusions:

(1) When the P wave is incident vertically, then $\alpha' = \beta' = \alpha'' = \beta'' = 0$, while p_1, p_2, p_3 and p_4 tend to be infinite. From Eqs. (3.3.16) and (3.3.18), we get

$$F_{PV} = 0, \quad E_{PV'} = 0. \tag{3.3.19}$$

It indicates that there is no mode conversion when the plane P wave is incident vertically; namely, there is no transverse wave component. The reason is that the longitudinal wave will only cause displacement along the normal direction of the interface when it is incident vertically. In other words, it only causes the medium to expand and contract.

(2) Total reflection

When the two wave velocities violate the following relation, i.e.

$$c'_S < c'_P < c_S < c_P, \tag{3.3.20}$$

total reflection may occur. Assume that

$$c_S < c_P < c'_S < c'_P. \tag{3.3.21}$$

According to Snell's law, we obtain

$$\beta' < \alpha' < \beta'' < \alpha''. \tag{3.3.22}$$

thus

$$\sin\alpha'' = \frac{c'_P \sin\alpha}{c_P}, \quad \sin\beta'' = \frac{c'_S \sin\alpha}{c_P}. \tag{3.3.23}$$

There are two critical angles

$$\alpha_{\mathrm{cr1}} = \arcsin\left(\frac{c_P}{c'_P}\right), \quad \alpha_{\mathrm{cr2}} = \arcsin\left(\frac{c_P}{c'_S}\right). \tag{3.3.24}$$

When $\alpha = \alpha_{\mathrm{cr1}}$, there is $\alpha'' = \frac{\pi}{2}$; When $\alpha = \alpha_{\mathrm{cr2}}$, there is $\beta'' = \frac{\pi}{2}$. If the incident angle $\alpha > \alpha_{\mathrm{cr2}}$, β'' and α'' must be complex, and the incident P wave will be totally reflected.

The displacement reflection coefficient of reflected P wave and SV wave is defined as the ratio of the displacement amplitude of reflected P wave and SV wave to the displacement amplitude of incident P wave, namely

$$R_{\mathrm{PP}} = \frac{A'}{A} \times \frac{c_P}{c_P} = \frac{(l_1 - l_3)(h_2 + h_4) - (l_2 + l_4)(h_1 - h_3)}{(l_1 + l_3)(h_2 + h_4) - (l_2 + l_4)(h_1 + h_3)}. \tag{3.3.25}$$

$$R_{\mathrm{PV}} = \frac{B'}{A} \times \frac{c_P}{c_S} = \frac{2c_P(l_2 h_4 - l_4 h_2)}{c_S[(l_1 + l_3)(h_2 + h_4) - (l_2 + l_4)(h_1 + h_3)]}. \tag{3.3.26}$$

The displacement transmission coefficient of transmitted P wave and SV wave is defined as the ratio of the displacement amplitude of transmitted P wave and SV wave to the displacement amplitude of incident P wave, namely

$$T_{\mathrm{PP'}} = \frac{A''}{A} \times \frac{c_P}{c'_P} = \frac{2c_P(h_2 + h_4)}{c'_P[(l_1 + l_3)(h_2 + h_4) - (l_2 + l_4)(h_1 + h_3)]}, \tag{3.3.27}$$

$$T_{\mathrm{PV'}} = \frac{B''}{A} \times \frac{c_P}{c'_S} = \frac{2c_P(l_2 + l_4)}{c'_S[(l_1 + l_3)(h_2 + h_4) - (l_2 + l_4)(h_1 + h_3)]}. \tag{3.3.28}$$

If the reflection and transmission coefficients are defined by the ratio of the average energy carried by the reflected wave and the transmitted wave in a period along the direction of propagation, then

$$R_{\mathrm{PP}} = \left(\frac{A'}{A}\right)^2, \tag{3.3.29a}$$

$$R_{\mathrm{PV}} = \left(\frac{B'}{A}\right)^2 \cdot \frac{c_P}{c_S}, \tag{3.3.29b}$$

$$T_{\mathrm{PP'}} = \left(\frac{A''}{A}\right)^2 \frac{c_P}{c'_P}, \tag{3.3.29c}$$

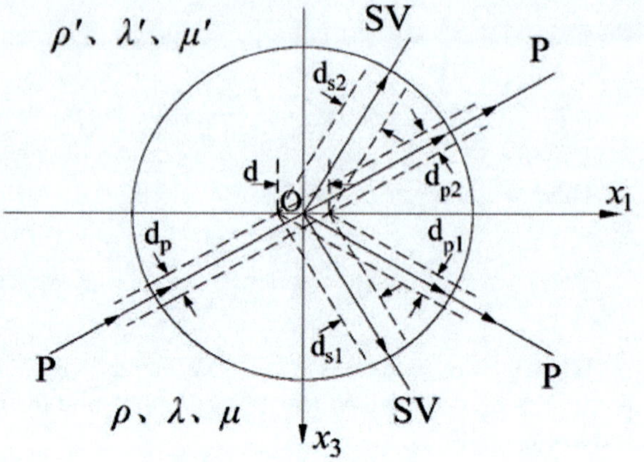

Fig. 3.18 Energy relationship diagram of incident, reflected and transmitted waves

$$R_{PV'} = \left(\frac{B''}{A}\right)^2 \cdot \frac{c_P}{c'_S}. \tag{3.3.29d}$$

Similar to the discussion on the energy conservation relationship between incident wave and reflected wave in Sect. 3.2.1, in this section, with reference to Fig. 3.18, the conservation relationship among reflected wave, transmitted wave and incident wave can be expressed as

$$\langle I_{P'}\rangle \times d_{P'} + \langle I_{S'}\rangle \times d_{S'} + \langle I_{P''}\rangle \times d_{P''} + \langle I_{S''}\rangle \times d_{S''} = \langle I_P\rangle \times d_P. \tag{3.3.30}$$

namely

$$\langle I_{P'}\rangle \times \cos\alpha' + \langle I_{S'}\rangle \times \cos\beta' + \langle I_{P''}\rangle \times \cos\alpha'' + \langle I_{S''}\rangle \times \cos\beta'' = \langle I_P\rangle \times \cos\alpha. \tag{3.3.31}$$

The above equation can also be rewritten as

$$\langle I_3(P')\rangle + \langle I_3(S')\rangle + \langle I_3(P'')\rangle + \langle I_3(S'')\rangle = \langle I_3(P)\rangle. \tag{3.3.32}$$

It indicates that **the average value of the energy flow of the incident wave flowing in along the normal direction of the interface is equal to the sum of the average value of the energy flow of the reflected wave and transmitted wave flowing out along the normal direction of the interface.**

3.3.2 Reflection and Transmission of SH Waves at the Interface

As shown in Fig. 3.19, the density and Lame coefficients of medium 1 and medium 2 are denoted by (ρ, λ, μ) and (ρ', λ', μ'), respectively. When the incident SH wave propagates along the plane x_1x_3 in the medium 1 and encounter the boundary plane x_1x_2, the reflected and transmitted SH waves are formed. Considering that the potential function of SH wave is a vector function with two components, it is more convenient to directly assume the displacement field. Therefore, when discussing the reflection and transmission of SH wave, the displacement field is directly assumed.

Suppose the displacement fields of incident SH wave, reflected SH wave and transmitted SH wave are

$$u_2 = C\exp[i(k_{S1}x_1 - k_{S3}x_3 - \omega_S t)], \tag{3.3.33a}$$

$$u_2' = C'\exp\left[i\left(k_{S1}'x_1 + k_{S3}'x_3 - \omega_S' t\right)\right], \tag{3.3.33b}$$

$$u_2'' = C''\exp\left[i\left(k_{S1}''x_1 - k_{S3}''x_3 - \omega_S'' t\right)\right]. \tag{3.3.33c}$$

where

$$k_{S1} = \frac{\omega_S}{c_S}\sin\beta, \quad k_{S3} = \frac{\omega_S}{c_S}\cos\beta,$$

$$k_{S1}' = \frac{\omega_S'}{c_S}\sin\beta', \quad k_{S3}' = \frac{\omega_S'}{c_S}\cos\beta',$$

$$k_{S1}'' = \frac{\omega_S''}{c_S'}\sin\beta'', \quad k_{S3}' = \frac{\omega_S''}{c_S'}\cos\beta''.$$

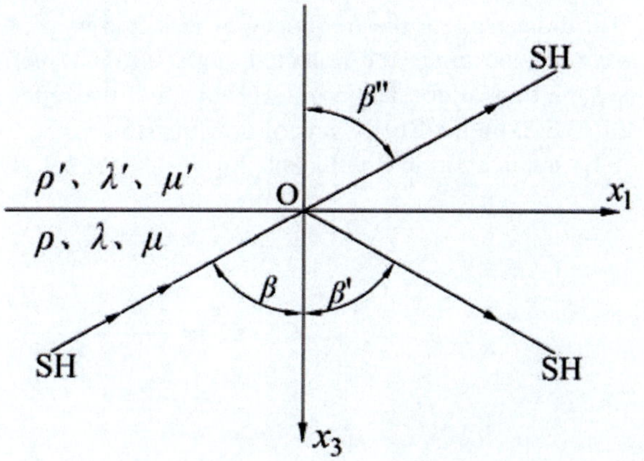

Fig. 3.19 Reflection and transmission of SH wave at the interface

At the interface ($x_3 = 0$), the displacement and stress must satisfy the boundary conditions

$$u_2 = u_2', \tag{3.3.34a}$$

$$\sigma_{32} = \sigma_{32}'. \tag{3.3.34b}$$

According to Eq. (3.1.24), the interface conditions can be written as

$$C\exp[i(k_{S1}x_1 - \omega_S t)] + C'\exp\left[i\left(k_{S1}'x_1 - \omega_S' t\right)\right] = C''\exp\left[i\left(k_{S1}''x_1 - \omega_S'' t\right)\right], \tag{3.3.35a}$$

$$\begin{aligned}&\mu\left\{Ck_{S3}\exp[i(k_{S1}x_1 - \omega_S t)] - C'k_{S3}'\exp\left[i\left(k_{S1}'x_1 - \omega_S' t\right)\right]\right\}\\ &= \mu' C'' k_{S3}''\exp\left[i\left(k_{S1}''x_1 - \omega_S'' t\right)\right].\end{aligned} \tag{3.3.35b}$$

In order for Eqs. (3.3.35a, 3.3.35b) to hold for any x_1 and t, there must be

$$\omega_S = \omega_S' = \omega_S'', \tag{3.3.36}$$

$$k_{S1} = k_{S1}' = k_{S1}''. \tag{3.3.37}$$

Equation (3.3.37) can also be written as

$$\frac{\sin\beta}{c_S} = \frac{\sin\beta'}{c_S} = \frac{\sin\beta''}{c_S'}, \tag{3.3.38a}$$

namely

$$\beta = \beta', \quad \frac{\sin\beta}{\sin\beta''} = \frac{c_S}{c_S'}. \tag{3.3.38b}$$

It indicates that the frequency of reflected wave, incident wave and transmitted wave are the same, the reflected angle of SH wave is equal to the incident angle and the transmitted angle of SH wave and the reflected angle of SH wave satisfy Eq. (3.3.38b), i.e. Snell's law of transmission.

By application of Snell's law, Eqs. (3.3.35a, 3.3.35b) reduces to

$$C + C' = C'', \tag{3.3.39a}$$

$$C - C' = \frac{\rho' c_S' \cos\beta'' C''}{\rho c_S \cos\beta}. \tag{3.3.39b}$$

The displacement reflection coefficient of reflected SH wave is defined as the ratio of the displacement amplitude of reflected SH wave to the displacement amplitude of incident SH wave, i.e.

$$R_{\mathrm{HH}} = \frac{C'}{C} = \frac{\rho c_S \cos\beta - \rho' c'_S \cos\beta''}{\rho c_S \cos\beta + \rho' c'_S \cos\beta''}. \tag{3.3.40a}$$

The displacement transmission coefficient of transmitted SH wave is defined as the ratio of the displacement amplitude of transmitted SH wave to the displacement amplitude of incident SH wave, i.e.

$$T_{\mathrm{HH'}} = \frac{C''}{C} = \frac{2\rho c_S \cos\beta}{\rho c_S \cos\beta + \rho' c'_S \cos\beta''}. \tag{3.3.40b}$$

According to Eqs. (3.3.40a) and (3.3.40b), we can obtain the conclusions as follows

(1) Total transmission

When the incident angle β satisfies

$$\rho c_S \cos\beta - \rho' c'_S \cos\beta'' = 0, \tag{3.3.41}$$

the amplitude of the reflected wave becomes zero. Thus, the reflection coefficient $R_{HH} = 0$, namely, total transmission occurs,

(2) Critical reflection and total reflection of SH wave

According to Snell's law, as long as $c'_s \le c_S$, β' must be a real number, and the transmitted waves always exist. In the case of incident SH wave, if $c_S < c'_s$ and the transmission angle of SH wave $\beta'' = \frac{\pi}{2}$, then the incident angle β determined by

$$\frac{\sin\beta''}{\sin\beta} = \frac{c'_S}{c_S}. \tag{3.3.42a}$$

It is called the critical incident angle of SH wave and denoted as β_{cr}. At this time, the transmitted wave propagates along the interface, and this reflection process is called critical reflection.

If $\beta > \beta_{\mathrm{cr}}$, then β'' becomes a complex number, and then the transmitted wave becomes SH-type surface wave propagating along the interface, and total reflection occurs. At this time, $\beta'' = \frac{\pi}{2} + i\varphi$, where $\varphi = \mathrm{arcch}\left(\frac{c'_S}{c_S}\sin\beta\right)$, and the displacement of transmitted SH-type surface wave is

$$u''_2 = C'' \exp\left(\frac{\omega \mathrm{sh}\phi}{c'_s} \cdot x_3\right) \exp\left[i\omega\left(\frac{\mathrm{ch}\phi}{c'_S} x_1 - t\right)\right]. \tag{3.3.42b}$$

It can be seen from Eq. (3.3.42b), the amplitude of SH-type surface wave decreases exponentially with the increase of depth into the medium 2, in other word, it can only propagate in the surface of the medium 2. At this time, the reflection coefficient of SH wave has become to 1. It indicates that the reflected SH wave has the same amplitude as the incident wave.

3.3.3 *Reflection and Transmission of SV Waves at the Interface*

As shown in Fig. 3.20, the density and Lame coefficients of medium 1 and medium 2 are (ρ, λ, μ) and (ρ', λ', μ'), respectively. When the incident SV wave propagates along the plane x_1x_3 and encounters the interface plane x_1x_2 from the medium 1, the reflected P wave and SV wave, and the transmitted P wave and SV wave are formed in the medium 2.

The wave functions of incident SV wave, reflected P wave and SV wave, transmitted P wave and SV wave are defined as, respectively,

$$\psi_2 = B\exp[i(k_{S1}x_1 - k_{S3}x_3 - \omega_S t)], \tag{3.3.43a}$$

$$\varphi' = A'\exp\left[i\left(k'_{P1}x_1 + k'_{P3}x_3 - \omega'_P t\right)\right], \tag{3.3.43b}$$

$$\psi'_2 = B'\exp\left[i\left(k'_{S1}x_1 + k'_{S3}x_3 - \omega'_S t\right)\right], \tag{3.3.43c}$$

$$\varphi'' = A''\exp\left[i\left(k''_{P1}x_1 - k''_{P3}x_3 - \omega''_P t\right)\right], \tag{3.3.43d}$$

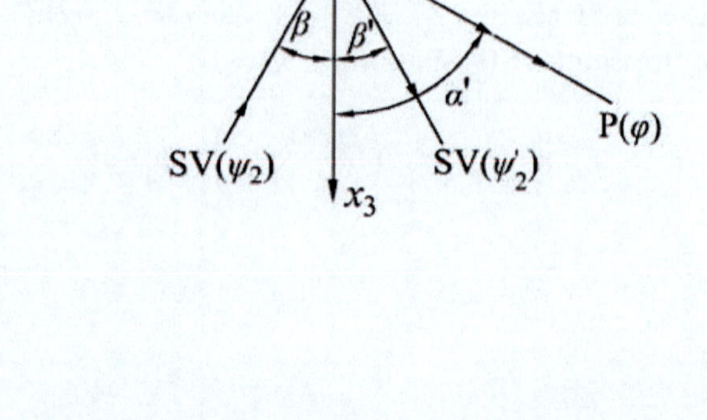

Fig. 3.20 Reflection and transmission of incident SV waves at the interface

$$\psi_2'' = B'' \exp\left[i\left(k_{S1}''x_1 - k_{S3}''x_3 - \omega_S''t\right)\right]. \tag{3.3.43e}$$

where B, B', B'', A', A'' and ω_S, ω_S', ω_S'', ω_P', ω_P'' are the amplitudes and angular frequencies of corresponding waves, and

$$k_{S1} = \frac{\omega_S}{c_S}\sin\beta, \quad k_{S3} = \frac{\omega_S}{c_S}\cos\beta,$$

$$k_{S1}' = \frac{\omega_S'}{c_S}\sin\beta', \quad k_{S3}' = \frac{\omega_S'}{c_S}\cos\beta',$$

$$k_{S1}'' = \frac{\omega_S''}{c_S'}\sin\beta'', \quad k_{S3}'' = \frac{\omega_S''}{c_S'}\cos\beta'',$$

$$k_{P1}' = \frac{\omega_P'}{c_P}\sin\alpha', \quad k_{P3}' = \frac{\omega_P'}{c_P}\cos\alpha',$$

$$k_{P1}'' = \frac{\omega_P''}{c_P'}\sin\alpha'', \quad k_{P3}'' = \frac{\omega_P''}{c_P'}\cos\alpha''.$$

The boundary conditions for $x_3 = 0$ are

$$u_1 = u_1', \quad u_3 = u_3', \tag{3.3.44a}$$

$$\sigma_{33} = \sigma_{33}', \quad \sigma_{31} = \sigma_{31}'. \tag{3.3.44b}$$

The boundary condition equations can be obtained from the relationship between displacement, stress and potential function as follows

$$\left[\frac{\partial\varphi'}{\partial x_1} - \frac{\partial\left(\psi_2 + \psi_2'\right)}{\partial x_3}\right]\bigg|_{x_3=0} = \left[\frac{\partial\varphi''}{\partial x_1} - \frac{\partial\psi_2''}{\partial x_3}\right]\bigg|_{x_3=0} \tag{3.3.45a}$$

$$\left[\frac{\partial\varphi'}{\partial x_3} + \frac{\partial\left(\psi_2 + \psi_2'\right)}{\partial x_1}\right]\bigg|_{x_3=0} = \left[\frac{\partial\varphi''}{\partial x_3} + \frac{\partial\psi_2''}{\partial x_1}\right]\bigg|_{x_3=0}, \tag{3.3.45b}$$

$$\begin{aligned}&\left\{\lambda\nabla^2\varphi' + 2\mu\left[\frac{\partial^2\varphi'}{\partial x_3^2} + \frac{\partial^2\left(\psi_2 + \psi_2'\right)}{\partial x_1\partial x_3}\right]\right\}\bigg|_{x_3=0}\\ &= \left\{\lambda'\nabla^2\varphi'' + 2\mu'\left[\frac{\partial^2\varphi''}{\partial x_3^2} + \frac{\partial^2\psi_2''}{\partial x_1\partial x_3}\right]\right\}\bigg|_{x_3=0},\end{aligned} \tag{3.3.45c}$$

$$\mu\left[2\frac{\partial^2\varphi'}{\partial x_1\partial x_3} + \frac{\partial^2\left(\psi_2 + \psi_2'\right)}{\partial x_1^2} - \frac{\partial^2\left(\psi_2 + \psi_2'\right)}{\partial x_3^2}\right]\bigg|_{x_3=0}$$

$$= \mu'\left[2\frac{\partial^2\varphi''}{\partial x_1\partial x_3} + \frac{\partial^2\psi_2''}{\partial x_1^2} - \frac{\partial^2\psi_2''}{\partial x_3^2}\right]|_{x_3=0}. \tag{3.3.45d}$$

Inserting Eqs. (3.3.43a, 3.3.43b, 3.3.43c, 3.3.43d, 3.3.43e) into the above equation, and let $x_3 = 0$, then

$$\begin{aligned} &k'_{P1}A'\exp\left[i\left(k'_{P1}x_1 - \omega'_P t\right)\right] + k_{S3}B\exp[i(k_{S1}x_1 - \omega_S t)] \\ &\quad - k'_{s3}B'\exp\left[i\left(k'_{S1}x_1 - \omega'_S t\right)\right] \\ &\quad = k''_{P1}A''\exp\left[i\left(k''_{P1}x_1 - \omega''_P t\right)\right] + k''_{S3}B''\exp\left[i\left(k''_{S1}x_1 - \omega''_S t\right)\right] \end{aligned} \tag{3.3.46a}$$

$$\begin{aligned} &k'_{P3}A'\exp\left[i\left(k'_{P1}x_1 - \omega'_P t\right)\right] + k_{S1}B\exp[i(k_{S1}x_1 - \omega_S t)] \\ &\quad + k'_{s1}B'\exp\left[i\left(k'_{S1}x_1 - \omega'_S t\right)\right] \\ &\quad = -k''_{P3}A''\exp\left[i\left(k''_{P1}x_1 - \omega''_P t\right)\right] + k''_{S1}B''\exp\left[i\left(k''_{S1}x_1 - \omega''_S t\right)\right], \end{aligned} \tag{3.3.46b}$$

$$\begin{aligned} &\lambda k'^2_{P1}A'\exp\left[i\left(k'_{P1}x_1 - \omega'_P t\right)\right] + (\lambda + 2\mu)k'^2_{P3}A'\exp\left[i\left(k'_{P1}x_1 - \omega'_P t\right)\right] \\ &\quad + 2\mu\left\{-k_{S1}k_{S3}B\exp[i(k_{S1}x_1 - \omega_S t)] + k'_{S1}k'_{S3}B'\exp\left[i\left(k'_{S1}x_1 - \omega'_S t\right)\right]\right\} \\ &\quad = \lambda' k''^2_{P1}A''\exp\left[i\left(k''_{P1}x_1 - \omega''_P t\right)\right] + \left(\lambda' + 2\mu'\right)k''^2_{P3}A''\exp\left[i\left(k''_{P1}x_1 - \omega''_P t\right)\right] \\ &\quad - 2\mu' k''_{S1}k''_{S3}B''\exp\left[i\left(k''_{S1}x_1 - \omega''_S t\right)\right], \end{aligned} \tag{3.3.46c}$$

$$\begin{aligned} &\mu\{2k'_{P1}k'_{P3}A'\exp\left[i\left(k'_{P1}x_1 - \omega'_P t\right)\right] + k^2_{S1}B\exp[i(k_{S1}x_1 - \omega_S t)] \\ &\quad + k'^2_{S1}B'\exp\left[i\left(k'_{S1}x_1 - \omega'_S t\right)\right] - k^2_{S3}B\exp[i(k_{S1}x_1 - \omega_S t)] \\ &\quad - k'^2_{S3}B'\exp\left[i\left(k'_{S1}x_1 - \omega'_S t\right)\right] = \mu'\{-2k''_{P1}k''_{P3}A''\exp\left[i\left(k''_{P1}x_1 - \omega''_P t\right)\right] \\ &\quad + k''^2_{S1}B''\exp\left[i\left(k''_{S1}x_1 - \omega''_S t\right)\right] - k''^2_{S3}B''\exp\left[i\left(k''_{S1}x_1 - \omega''_S t\right)\right]\}. \end{aligned} \tag{3.3.46d}$$

In order for Eqs. (3.3.46a, 3.3.46b, 3.3.46c, 3.3.46d) to hold for any x_1 and t, there must be

$$\omega_S = \omega'_S = \omega''_S = \omega'_P = \omega''_P, \tag{3.3.47}$$

$$k_{S1} = k'_{S1} = k''_{S1} = k'_{P1} = k''_{P1}. \tag{3.3.48}$$

Equation (3.3.48) can also be written as

$$\frac{\sin\beta}{c_S} = \frac{\sin\beta'}{c_S} = \frac{\sin\beta''}{c'_S} = \frac{\sin\alpha'}{c_P} = \frac{\sin\alpha''}{c'_P}. \tag{3.3.49}$$

It can be seen from Eqs. (3.3.47) to (3.3.49) that the frequency of incident wave, reflected wave and transmitted wave are the same, and the reflection angles of each reflected waves and transmission angles of each transmitted waves are related with the incident angle by Snell's law. Since the longitudinal wave velocity is always greater than the transverse wave velocity, the reflection angle of P wave is greater than that of SV wave. The transmission angle of the transmitted P wave is greater than that of the transmitted SV wave.

The simplification Eqs. (3.3.46a, 3.3.46b, 3.3.46c, 3.3.46d) leads to:

$$A' + p_2(B - B') - A'' - p_4 B'' = 0, \tag{3.3.50a}$$

$$p_1 A' + B + B' + p_3 A'' - B'' = 0, \tag{3.3.50b}$$

$$\begin{aligned}&[\lambda(1 + p_1^2) + 2\mu p_1^2]A' - 2\mu p_2(B - B')\\&\quad - [\lambda'(1 + p_3^2) + 2\mu' p_3^2]A'' + 2\mu' p_4 B'' = 0,\end{aligned} \tag{3.3.50c}$$

$$2\mu p_1 A' + \mu(1 - p_2^2)(B + B') + 2\mu' p_3 A'' - \mu'(1 - p_4^2)B'' = 0. \tag{3.3.50d}$$

where

$$p_1 = \frac{k'_{P3}}{k'_{P1}} = \cot\alpha', \quad p_2 = \frac{k_{S3}}{k_{S1}} = \cot\beta,$$

$$p_3 = \frac{k''_{P3}}{k''_{P1}} = \cot\alpha'', \quad p_4 = \frac{k''_{S3}}{k''_{S1}} = \cot\beta''.$$

Let

$$P = B - B', \quad Q = B + B'. \tag{3.3.51}$$

From Eqs. (3.3.50a) and (3.3.50c), we obtain

$$P = -l_2 A'' - h_2 B'', \quad A' = l_1 A'' + h_1 B''. \tag{3.3.52}$$

From Eqs. (3.3.50b) and (3.3.50d), we obtain

$$Q = -l_4 A'' - h_4 B'', \quad A' = -l_3 A'' - h_3 B''. \tag{3.3.53}$$

where l_i and h_i have the same meaning as l_i and h_i in Sect. 3.3.1.

Inserting Eq. (3.3.51) into Eqs. (3.3.52) and (3.3.53) leads to

$$-\frac{A'}{B} + l_1\frac{A''}{B} + h_1\frac{B''}{B} = 0, \tag{3.3.54a}$$

$$\frac{B'}{B} - l_2\frac{A''}{B} - h_2\frac{B''}{B} = 1, \tag{3.3.54b}$$

$$\frac{A'}{B} + l_3 \frac{A''}{B} + h_3 \frac{B''}{B} = 0, \tag{3.3.54c}$$

$$-\frac{B'}{B} - l_4 \frac{A''}{B} - h_4 \frac{B''}{B} = 1. \tag{3.3.54d}$$

The reflection coefficients of the potential function of reflected SV wave and P wave are defined as the ratio of the amplitudes of the potential function of reflected SV wave and P wave to the amplitude of the potential function of incident SV wave, i.e.

$$F_{\mathrm{VV}} = \frac{B'}{B} = \frac{(l_2 - l_4)(h_1 + h_3) - (l_1 + l_3)(h_2 - h_4)}{(l_1 + l_3)(h_2 + h_4) - (l_2 + l_4)(h_1 + h_3)}, \tag{3.3.55}$$

$$F_{\mathrm{VP}} = \frac{A'}{B} = \frac{2(l_1 h_3 - l_3 h_1)}{(l_1 + l_3)(h_2 + h_4) - (l_2 + l_4)(h_1 + h_3)}. \tag{3.3.56}$$

The transmission coefficients of the potential functions of transmitted SV wave and P wave are defined as the ratio of the amplitudes of the potential function of transmitted SV wave and P wave to the amplitude of the potential function of incident P wave, i.e.

$$E_{\mathrm{VV'}} = \frac{B''}{B} = \frac{-2(l_1 + l_3)}{(l_1 + l_3)(h_2 + h_4) - (l_2 + l_4)(h_1 + h_3)}, \tag{3.3.57a}$$

$$E_{\mathrm{VP'}} = \frac{A''}{B} = \frac{2(h_1 + h_3)}{(l_1 + l_3)(h_2 + h_4) - (l_2 + l_4)(h_1 + h_3)}. \tag{3.3.57b}$$

According to Eqs. (3.3.55) and (3.3.57b), the following conclusions can be drawn.

(1) Vertical incidence

When SV wave is incident vertically, according to Snell's law, there is

$$\beta = \beta' = \beta'' = \alpha' = \alpha'' = 0. \tag{3.3.58}$$

Then, p_1, p_2, p_3 and p_4 tend to infinity. In this case, we obtain from Eqs. (3.3.56) and (3.3.57b)

$$F_{\mathrm{VP}} = 0, \quad E_{\mathrm{VP'}} = 0. \tag{3.3.59}$$

It indicates that there is no mode conversion when SV wave is incident vertically; namely, there is no longitudinal wave component.

(2) Total reflection

(a) When the wave velocities in the two media satisfy the relationship as follows

$$c'_S < c'_P < c_S < c_P. \tag{3.3.60}$$

According to Snell's law, we know that $\beta'' < \alpha'' < \beta < \alpha'$. When the incident wave is SV wave, there is a critical angle

$$\beta_{\text{cr1}} = \arcsin\left(\frac{c_S}{c_P}\right). \tag{3.3.61}$$

When the incident angle $\beta > \beta_{\text{cr}}$, the reflected P wave does not exist and it becomes an inhomogeneous wave propagating along the interface. At this time, the transmitted P wave and SV wave still exist.

(b) When the wave velocities in the two media satisfy

$$c_S < c_P < c'_S < c'_P. \tag{3.3.62}$$

According to Snell's law, we know that $\beta < \alpha' < \beta'' < \alpha''$; for this situation, there are three critical angles

$$\beta_{\text{cr1}} = \arcsin\left(\frac{c_S}{c'_P}\right), \tag{3.3.63a}$$

$$\beta_{\text{cr2}} = \arcsin\left(\frac{c_S}{c'_S}\right), \tag{3.3.63b}$$

$$\beta_{\text{cr3}} = \arcsin\left(\frac{c_S}{c_P}\right). \tag{3.3.63c}$$

When $\beta = \beta_{\text{cr1}}$, we get $\alpha'' = \frac{\pi}{2}$; when $\beta = \beta_{\text{cr2}}$, we get $\beta'' = \frac{\pi}{2}$; when $\beta = \beta_{\text{cr3}}$, we get $\alpha' = \frac{\pi}{2}$. If the incident angle $\beta > \beta_{\text{cr3}}$, then α'', β'' and α' must be complex. In this case, the transmitted P wave and SV wave as well as the reflected P wave do not exist, they all become inhomogeneous waves propagating along the interface, and only the reflected SV wave exists, which is called the **total reflection** phenomenon.

The displacement reflection coefficients of reflected SV wave and P wave are defined as the ratio of the displacement amplitudes of reflected SV wave and P wave to the displacement amplitude of incident SV wave, i.e.

$$R_{\text{VV}} = \frac{B'}{B}\frac{c_S}{c_S} = \frac{(l_2 - l_4)(h_1 + h_3) - (l_1 + l_3)(h_2 - h_4)}{(l_1 + l_3)(h_2 + h_4) - (l_2 + l_4)(h_1 + h_3)}, \tag{3.3.64a}$$

$$R_{\text{VP}} = \frac{A'}{B}\frac{c_S}{c_P} = \frac{2c_S(l_1 h_3 - l_3 h_1)}{c_P[(l_1 + l_3)(h_2 + h_4) - (l_2 + l_4)(h_1 + h_3)]}. \tag{3.3.64b}$$

The displacement transmission coefficients of transmitted SV and P wave are defined as the ratio of the displacement amplitudes of transmitted SV wave and P wave to the displacement amplitude of incident SV wave, i.e.

$$T_{VV'} = \frac{B''}{B}\frac{c_S}{c'_S} = \frac{-2c_S(l_1 + l_3)}{c'_S[(l_1 + l_3)(h_2 + h_4) - (l_2 + l_4)(h_1 + h_3)]}, \tag{3.3.65a}$$

$$T_{VP'} = \frac{A''}{B}\frac{c_S}{c'_P} = \frac{2c_S(h_1 + h_3)}{c'_P[(l_1 + l_3)(h_2 + h_4) - (l_2 + l_4)(h_1 + h_3)]}. \tag{3.3.65b}$$

3.3.4 P Wave and SV Wave Incidence Simultaneously

In this section, we will discuss the reflection and transmission of plane waves at the interface when P wave and SV wave incidence occur at the same time.

1. P wave and SV wave incidence simultaneously on one side

As shown in Fig. 3.21, the density and Lame coefficients of medium 1 and medium 2 are (ρ, λ, μ) and (ρ', λ', μ'), respectively. In the medium 1, P wave and SV wave incidence occur simultaneously along the plane x_1x_3. The plane x_1x_2 is the boundary plane. The reflected and transmitted P wave and SV wave are formed in medium 2. Since P wave and SV wave are coupled in the same medium, the angle between them is determined by the parameters of the medium and cannot be arbitrary. Therefore, the reflected wave is a set of P and SV waves, not two sets; similarly, the transmitted waves are also a set of P and SV waves.

The wave functions of incident P wave and SV wave, reflected P wave and SV wave, transmitted P wave and SV wave are defined as, respectively,

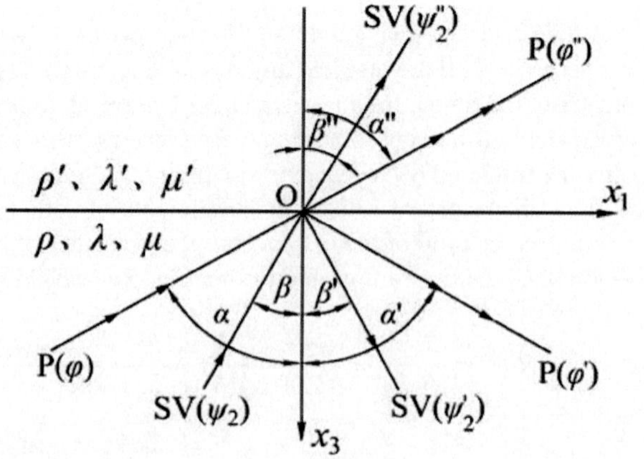

Fig. 3.21 P wave and SV wave incidence simultaneously on one side

$$\varphi = A\exp[i(k_{P1}x_1 - k_{P3}x_3 - \omega_P t)], \tag{3.3.66a}$$

$$\psi_2 = B\exp[i(k_{S1}x_1 - k_{S3}x_3 - \omega_S t)], \tag{3.3.66b}$$

$$\varphi' = A'\exp\left[i\left(k'_{P1}x_1 + k'_{P3}x_3 - \omega'_P t\right)\right], \tag{3.3.66c}$$

$$\psi'_2 = B'\exp\left[i\left(k'_{S1}x_1 + k'_{S3}x_3 - \omega'_S t\right)\right], \tag{3.3.66d}$$

$$\varphi'' = A''\exp\left[i\left(k''_{P1}x_1 - k''_{P3}x_3 - \omega''_P t\right)\right], \tag{3.3.66e}$$

$$\psi''_2 = B''\exp\left[i\left(k''_{S1}x_1 - k''_{S3}x_3 - \omega''_S t\right)\right]. \tag{3.3.66f}$$

where A, A', A'', B, B', B'' and ω_P, ω'_P, ω''_P, ω_S, ω'_S, ω''_S are the amplitudes and angular frequencies of the corresponding waves, and

$$k_{P1} = \frac{\omega_P}{c_P}\sin\alpha,\quad k_{P3} = \frac{\omega_P}{c_P}\cos\alpha,\quad k'_{P1} = \frac{\omega'_P}{c_P}\sin\alpha',\quad k'_{P3} = \frac{\omega'_P}{c_P}\cos\alpha',$$

$$k''_{P1} = \frac{\omega''_P}{c'_P}\sin\alpha'',\quad k''_{P3} = \frac{\omega''_P}{c'_P}\cos\alpha'',\quad k_{S1} = \frac{\omega_S}{c_S}\sin\beta,\quad k_{S3} = \frac{\omega_S}{c_S}\cos\beta,$$

$$k'_{S1} = \frac{\omega'_S}{c_S}\sin\beta',\quad k'_{S3} = \frac{\omega'_S}{c_S}\cos\beta',\quad k''_{S1} = \frac{\omega''_S}{c'_S}\sin\beta'',\quad k'_{S3} = \frac{\omega''_S}{c'_S}\cos\beta''.$$

The displacements and stresses in both medium can be calculated as follows

$$\begin{aligned} u_1 &= \frac{\partial(\varphi + \varphi')}{\partial x_1} - \frac{\partial(\psi_2 + \psi'_2)}{\partial x_3} \\ &= ik_{P1}A\exp[i(k_{P1}x_1 - k_{P3}x_3 - \omega_P t)] \\ &\quad + ik'_{P1}A'\exp\left[i\left(k'_{P1}x_1 + k'_{P3}x_3 - \omega'_P t\right)\right] \\ &\quad + ik_{S3}B\exp[i(k_{S1}x_1 - k_{S3}x_3 - \omega_S t)] \\ &\quad - ik'_{S3}B'\exp\left[i\left(k'_{S1}x_1 + k'_{S3}x_3 - \omega'_S t\right)\right], \end{aligned} \tag{3.3.67a}$$

$$\begin{aligned} u_3 &= \frac{\partial(\varphi + \varphi')}{\partial x_3} + \frac{\partial(\psi_2 + \psi'_2)}{\partial x_1} \\ &= -ik_{P3}A\exp[i(k_{P1}x_1 - k_{P3}x_3 - \omega_P t)] \\ &\quad + ik'_{P3}A'\exp\left[i\left(k'_{P1}x_1 + k'_{P3}x_3 - \omega'_P t\right)\right] \\ &\quad + ik_{S1}B\exp[i(k_{S1}x_1 - k_{S3}x_3 - \omega_S t)] \\ &\quad + ik'_{S1}B'\exp\left[i\left(k'_{S1}x_1 + k'_{S3}x_3 - \omega'_S t\right)\right], \end{aligned} \tag{3.3.67b}$$

$$
\begin{aligned}
\sigma_{33} &= \lambda \frac{\partial^2(\varphi+\varphi')}{\partial x_1^2} + (\lambda+2\mu)\frac{\partial^2(\varphi+\varphi')}{\partial x_3^2} + 2\mu \frac{\partial^2(\psi_2+\psi_2')}{\partial x_1 \partial x_3} \\
&= \lambda\{-k_{P1}^2 A \exp[i(k_{P1}x_1 - k_{P3}x_3 - \omega_P t)] \\
&\quad -k_{P1}'^2 A' \exp[i(k_{P1}'x_1 + k_{P3}'x_3 - \omega_P' t)]\} \\
&\quad + (\lambda+2\mu)\{-k_{P3}^2 A \exp[i(k_{P1}x_1 - k_{P3}x_3 - \omega_P t)] \\
&\quad - k_{P3}'^2 A' \exp[i(k_{P1}'x_1 + k_{P3}'x_3 - \omega_P' t)] \\
&\quad + 2\mu\{k_{S1}k_{S3} B \exp[i(k_{S1}x_1 - k_{S3}x_3 - \omega_S t)] \\
&\quad -k_{S1}'k_{S3}' B' \exp[i(k_{S1}'x_1 + k_{S3}'x_3 - \omega_S' t)]\},
\end{aligned} \tag{3.3.67c}
$$

$$
\begin{aligned}
\sigma_{31} &= \mu\left[2\frac{\partial^2(\varphi+\varphi')}{\partial x_1 \partial x_3} + \frac{\partial^2(\psi_2+\psi_2')}{\partial x_1^2} - \frac{\partial^2(\psi_2+\psi_2')}{\partial x_3^2}\right] \\
&= \mu\{2k_{P1}k_{P3} A \exp[i(k_{P1}x_1 - k_{P3}x_3 - \omega_P t)] \\
&\quad - 2k_{P1}'k_{P3}' A' \exp[i(k_{P1}'x_1 + k_{P3}'x_3 - \omega_P' t)] \\
&\quad - k_{S1}^2 B \exp[i(k_{S1}x_1 - k_{S3}x_3 - \omega_S t)] - k_{S1}'^2 B' \exp[i(k_{S1}'x_1 + k_{S3}'x_3 - \omega_S' t)] \\
&\quad + k_{S3}^2 B \exp[i(k_{S1}x_1 - k_{S3}x_3 - \omega_S t)] + k_{S3}'^2 B' \exp[i(k_{S1}'x_1 + k_{S3}'x_3 - \omega_S' t)]\}.
\end{aligned} \tag{3.3.67d}
$$

Using the relationship between displacement, stress and potential function, we can get

$$
\begin{aligned}
u_1' &= \frac{\partial \varphi''}{\partial x_1} - \frac{\partial \psi_2''}{\partial x_3} \\
&= ik_{P1}'' A'' \exp[i(k_{P1}''x_1 - k_{P3}''x_3 - \omega_P'' t)] \\
&\quad + ik_{S3}'' B'' \exp[i(k_{S1}''x_1 - k_{S3}''x_3 - \omega_S'' t)],
\end{aligned} \tag{3.3.68a}
$$

$$
\begin{aligned}
u_3' &= \frac{\partial \varphi''}{\partial x_3} + \frac{\partial \psi_2''}{\partial x_1} \\
&= -ik_{P3}'' A'' \exp[i(k_{P1}''x_1 - k_{P3}''x_3 - \omega_P'' t)] \\
&\quad + ik_{S1}'' B'' \exp[i(k_{S1}''x_1 - k_{S3}''x_3 - \omega_S'' t)],
\end{aligned} \tag{3.3.68b}
$$

$$
\begin{aligned}
\sigma_{33}' &= \lambda' \frac{\partial^2 \varphi''}{\partial x_1^2} + (\lambda'+2\mu')\frac{\partial^2 \varphi''}{\partial x_3^2} + 2\mu' \frac{\partial^2 \psi_2''}{\partial x_1 \partial x_3} \\
&= -\lambda' k_{P1}''^2 A'' \exp[i(k_{P1}''x_1 - k_{P3}''x_3 - \omega_P'' t)] \\
&\quad - (\lambda'+2\mu')k_{P3}''^2 A'' \exp[i(k_{P1}''x_1 - k_{P3}''x_3 - \omega_P'' t)] \\
&\quad + 2\mu' k_{S1}''k_{S3}'' B'' \exp[i(k_{S1}''x_1 - k_{S3}''x_3 - \omega_S'' t)],
\end{aligned} \tag{3.3.68c}
$$

$$\sigma'_{31} = \mu'\left(\frac{\partial^2\varphi''}{\partial x_1 \partial x_3} + \frac{\partial^2\psi''_2}{\partial x_1^2} - \frac{\partial^2\psi''_2}{\partial x_3^2}\right)$$
$$= \mu'\{2k''_{P1}k''_{P3}A''\exp\left[i\left(k''_{P1}x_1 - k''_{P3}x_3 - \omega''_P t\right)\right]$$
$$- k''^2_{S1}B''\exp\left[i\left(k''_{S1}x_1 - k''_{S3}x_3 - \omega''_S t\right)\right]$$
$$+ k''^2_{S3}B''\exp\left[i\left(k''_{S1}x_1 - k''_{S3}x_3 - \omega''_S t\right)\right]. \quad (3.3.68d)$$

The displacement and stress components at the interface satisfy the continuity conditions, namely

$$u_1|_{x_3=0} = u'_1|_{x_3=0}, \quad (3.3.69a)$$

$$u_3|_{x_3=0} = u'_3|_{x_3=0}, \quad (3.3.69b)$$

$$\sigma_{33}|_{x_3=0} = \sigma'_{33}|_{x_3=0}, \quad (3.3.69c)$$

$$\sigma_{31}|_{x_3=0} = \sigma'_{31}|_{x_3=0}. \quad (3.3.69d)$$

Equations (3.3.69a, 3.3.69b, 3.3.69c, 3.3.69d) is required to hold for any x_1 and t, there must be

$$\omega_P = \omega'_P = \omega''_P = \omega_S = \omega'_S = \omega''_S, \quad (3.3.70)$$

$$k_{P1} = k'_{P1} = k''_{P1} = k_{S1} = k'_{S1} = k''_{S1}. \quad (3.3.71)$$

Equation (3.3.71) can also be written as

$$\frac{\sin\alpha}{c_P} = \frac{\sin\alpha'}{c_P} = \frac{\sin\alpha''}{c'_P} = \frac{\sin\beta}{c_S} = \frac{\sin\beta'}{c_S} = \frac{\sin\beta''}{c'_S}. \quad (3.3.72)$$

Therefore,

$$\alpha = \alpha', \quad \beta = \beta'. \quad (3.3.73)$$

According to Eqs. (3.3.70)–(3.3.72), we know

(1) The frequency of incident wave, reflected wave and transmitted wave are the same;
(2) The reflection angle of the reflected P wave is equal to the incident angle of incident P wave, and the reflection angle of the reflected SV wave is equal to the incident angle of incident SV wave. Moreover, the reflection angle of the

reflected SV wave is smaller than that of the reflected P wave, and the transmission angle of the transmitted SV wave is smaller than that of the transmitted P wave.

(3) The relationships between the reflection angle of each reflected wave as well as the transmission angle of the transmitted wave and the incident angle are determined by the parameters of two kinds of medium, which satisfies Snell's law.

Based on the above relations, the continuous equations of the interface reduce to

$$A + A' + p_2(B - B') - A'' - p_4 B'' = 0, \tag{3.3.74a}$$

$$p_1(A - A') - (B + B') - p_3 A'' + B'' = 0, \tag{3.3.74b}$$

$$\begin{aligned} &[\lambda(1 + p_1^2) + 2\mu p_1^2](A + A') - 2\mu p_2(B - B') \\ &- [\lambda'(1 + p_3^2) + 2\mu' p_3^2]A'' + 2\mu' p_4 B'' = 0, \end{aligned} \tag{3.3.74c}$$

$$\mu[2p_1(A' - A) + (1 - p_2^2)(B + B')] + \mu'[2p_3 A'' - (1 - p_4^2)B''] = 0. \tag{3.3.74d}$$

where

$$p_1 = \frac{k_{P3}}{k_{P1}} = \cot\alpha, \quad p_2 = \frac{k_{S3}}{k_{S1}} = \cot\beta,$$

$$p_3 = \frac{k''_{P3}}{k''_{P1}} = \cot\alpha'', \quad p_4 = \frac{k''_{S3}}{k''_{S1}} = \cot\beta''.$$

Let

$$P = A + A', \quad Q = A - A'. \tag{3.3.75}$$

Substituting P into Eqs. (3.3.74a) and (3.3.74c), P and $B - B'$ can be solved as

$$P = l_1 A'' + h_1 B'', \quad B - B' = -l_2 A'' - h_2 B''. \tag{3.3.76}$$

Substituting Q into Eqs. (3.3.74b) and (3.3.74d), Q and $B' + B$ can be solved as

$$Q = l_3 A'' + h_3 B'', \quad B' + B = -l_4 A'' - h_4 B''. \tag{3.3.77}$$

in which l_i and h_i have the same meaning as l_i and h_i in Sect. 3.3.1.

Inserting Eq. (3.3.75) into Eqs. (3.3.76) and (3.3.77), there are

$$A + A' = l_1 A'' + h_1 B'', \tag{3.3.78a}$$

$$B' - B = l_2 A'' + h_2 B'', \tag{3.3.78b}$$

$$A - A' = l_3 A'' + h_3 B'', \tag{3.3.78c}$$

$$B' + B = -l_4 A'' - h_4 B''. \tag{3.3.78d}$$

Solving Eqs. (3.3.78a, 3.3.78b, 3.3.78c, 3.3.78d) simultaneously, we get

$$A' = \frac{[(l_1 - l_3)(h_2 + h_4) - (l_2 + l_4)(h_1 - h_3)]A + 2(l_1 h_3 - h_1 l_3)B}{(l_1 + l_3)(h_2 + h_4) - (l_2 + l_4)(h_1 + h_3)}, \tag{3.3.79a}$$

$$B' = \frac{2(l_2 h_4 - l_4 h_2)A + [(l_2 - l_4)(h_1 + h_3) - (l_1 + l_3)(h_2 - h_4)]B}{(l_1 + l_3)(h_2 + h_4) - (l_2 + l_4)(h_1 + h_3)}, \tag{3.3.79b}$$

$$A'' = \frac{2(h_2 + h_4)A + 2(h_1 + h_3)B}{(l_1 + l_3)(h_2 + h_4) - (l_2 + l_4)(h_1 + h_3)}, \tag{3.3.79c}$$

$$B'' = \frac{-2(l_2 + l_4)A - 2(l_1 + l_3)B}{(l_1 + l_3)(h_2 + h_4) - (l_2 + l_4)(h_1 + h_3)}. \tag{3.3.79d}$$

Equations (3.3.79a, 3.3.79b, 3.3.79c, 3.3.79d) can be rewritten in matrix form

$$\begin{pmatrix} A' \\ B' \end{pmatrix} = \boldsymbol{F}\begin{pmatrix} A \\ B \end{pmatrix} = \begin{pmatrix} F_{11} & F_{12} \\ F_{21} & F_{22} \end{pmatrix}\begin{pmatrix} A \\ B \end{pmatrix}, \tag{3.3.80a}$$

$$\begin{pmatrix} A'' \\ B'' \end{pmatrix} = \boldsymbol{E}\begin{pmatrix} A \\ B \end{pmatrix} = \begin{pmatrix} E_{11} & E_{12} \\ E_{21} & E_{22} \end{pmatrix}\begin{pmatrix} A \\ B \end{pmatrix}. \tag{3.3.80b}$$

where

$$F_{11} = \frac{(l_1 - l_3)(h_2 + h_4) - (l_2 + l_4)(h_1 - h_3)}{(l_1 + l_3)(h_2 + h_4) - (l_2 + l_4)(h_1 + h_3)},$$

$$F_{12} = \frac{2(l_1 h_3 - l_3 h_1)}{(l_1 + l_3)(h_2 + h_4) - (l_2 + l_4)(h_1 + h_3)},$$

$$F_{21} = \frac{2(l_2 h_4 - l_4 h_2)}{(l_1 + l_3)(h_2 + h_4) - (l_2 + l_4)(h_1 + h_3)},$$

$$F_{22} = \frac{(l_2 - l_4)(h_1 + h_3) - (l_1 + l_3)(h_2 - h_4)}{(l_1 + l_3)(h_2 + h_4) - (l_2 + l_4)(h_1 + h_3)},$$

$$E_{11} = \frac{2(h_2 + h_4)}{(l_1 + l_3)(h_2 + h_4) - (l_2 + l_4)(h_1 + h_3)},$$

$$E_{12} = \frac{2(h_1 + h_3)}{(l_1 + l_3)(h_2 + h_4) - (l_2 + l_4)(h_1 + h_3)},$$

$$E_{21} = \frac{-2(l_2 + l_4)}{(l_1 + l_3)(h_2 + h_4) - (l_2 + l_4)(h_1 + h_3)},$$

$$E_{22} = \frac{-2(l_1 + l_3)}{(l_1 + l_3)(h_2 + h_4) - (l_2 + l_4)(h_1 + h_3)}.$$

The matrices $\boldsymbol{F}$ and $\boldsymbol{E}$ are called reflection matrix and transmission matrix of the interface. They relate the potential function amplitudes of reflected and transmitted waves with the potential function amplitudes of incident waves. The values of each element of the matrix depend on the material parameters of the media on both sides of the interface.

When P wave and SV wave incidence occur simultaneously, we can draw the following conclusions by analyzing the expression of reflection and transmission coefficients.

(1) According to Eqs. (3.3.80a, 3.3.80b), when $B = 0$, it degenerates into the case of P wave incidence alone, and F_{11} and F_{21} are, respectively, the reflection coefficients of the potential function of reflected P wave and reflected SV wave when P wave incidence occurs alone. E_{11} and E_{21} are the transmission coefficients of the potential function of the transmitted P wave and transmitted SV wave when P wave incidence occurs separately.
(2) When $A = 0$, it degenerates into the case of SV wave incidence alone, and F_{12} and F_{22} are, respectively, the reflection coefficients of the potential function of reflected P wave and reflected SV wave. E_{12} and E_{22} are, respectively, the transmission coefficients of the potential function of the transmitted P wave and transmitted SV wave.

Let the displacement amplitudes of the incident P wave and SV wave be C_1 and C_2. The displacement amplitudes of reflected P wave and SV wave are C_1' and C_2'. The displacement amplitudes of transmitted P wave and SV wave are C_1'' and C_2'', respectively. If P wave incidence occurs alone, the displacement reflection coefficients of P wave and SV wave are R_{11} and R_{21}, respectively, and the displacement transmission coefficients are T_{11} and T_{21}, respectively, then

$$R_{11} = F_{11} \cdot \frac{c_P}{c_P} = F_{11}, \quad R_{21} = F_{21} \cdot \frac{c_P}{c_S},$$

$$T_{11} = E_{11} \cdot \frac{c_P}{c_P'}, \quad T_{21} = E_{21} \cdot \frac{c_P}{c_S'}.$$

If SV wave incidence occurs alone, the displacement reflection coefficients of P wave and SV wave are denoted by R_{12} and R_{22}, respectively, and the displacement transmission coefficients are denoted by T_{12} and T_{22}, respectively, then

$$R_{12} = F_{12} \cdot \frac{c_S}{c_P}, \quad R_{22} = F_{22} \cdot \frac{c_S}{c_S} = F_{22},$$

$$T_{12} = E_{12} \cdot \frac{c_S}{c'_P}, \quad T_{22} = E_{22} \cdot \frac{c_S}{c'_S}.$$

thus

$$\begin{pmatrix} C'_1 \\ C'_2 \end{pmatrix} = \begin{pmatrix} R_{11} & R_{12} \\ R_{21} & R_{22} \end{pmatrix} \begin{pmatrix} C_1 \\ C_2 \end{pmatrix}, \tag{3.3.81a}$$

$$\begin{pmatrix} C''_1 \\ C''_2 \end{pmatrix} = \begin{pmatrix} T_{11} & T_{12} \\ T_{21} & T_{22} \end{pmatrix} \begin{pmatrix} C_1 \\ C_2 \end{pmatrix}. \tag{3.3.81b}$$

The matrix $\boldsymbol{R} = \begin{pmatrix} R_{11} & R_{12} \\ R_{21} & R_{22} \end{pmatrix}$ is defined as the displacement reflection matrix of reflected P wave and reflected SV wave when P wave and SV wave are simultaneously incident. And the matrix $\boldsymbol{T} = \begin{pmatrix} T_{11} & T_{12} \\ T_{21} & T_{22} \end{pmatrix}$ is defined as the displacement transmission matrix of transmitted P wave and transmitted SV wave when P wave and SV wave are incident simultaneously.

2. P wave and SV wave simultaneously incident on both sides

As shown in Fig. 3.22, the density and Lame coefficients of medium 1 and medium 2 are (ρ, λ, μ) and (ρ', λ', μ'), respectively. P wave and SV wave incidences occur at the same time along the plane x_1x_3 in the two mediums. The plane x_1x_2 is the interface, and the reflection and transmission are formed on both sides of the boundary.

P wave and SV wave are coupled to each other. Therefore, under the action of bilateral incident waves, a set of reflected P waves and reflected SV waves are excited, respectively, in the two mediums, and their propagation directions are away from the

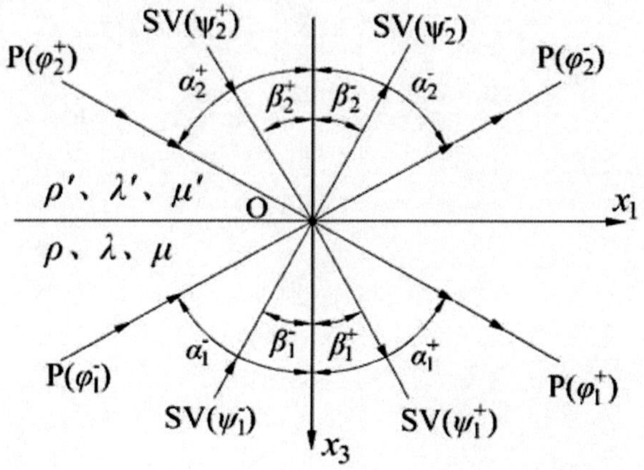

Fig. 3.22 P wave and SV wave are simultaneously incident on both sides

interface. Suppose the wave functions of incident P wave and SV wave as well as reflected P wave and SV wave of medium 1 and medium 2 be, respectively,

$$\varphi_1^- = A_1^- \exp\left[i\left(k_{P1}^{1-}x_1 - k_{P3}^{1-}x_3 - \omega_P^{1-}t\right)\right], \tag{3.3.82a}$$

$$\psi_1^- = B_1^- \exp\left[i\left(k_{S1}^{1-}x_1 - k_{S3}^{1-}x_3 - \omega_S^{1-}t\right)\right], \tag{3.3.82b}$$

$$\varphi_1^+ = A_1^+ \exp\left[i\left(k_{P1}^{1+}x_1 + k_{P3}^{1+}x_3 - \omega_P^{1+}t\right)\right], \tag{3.3.82c}$$

$$\psi_1^+ = B_1^+ \exp\left[i\left(k_{S1}^{1+}x_1 + k_{S3}^{1+}x_3 - \omega_S^{1+}t\right)\right], \tag{3.3.82d}$$

$$\varphi_2^+ = A_2^+ \exp\left[i\left(k_{P1}^{2+}x_1 + k_{P3}^{2+}x_3 - \omega_P^{2+}t\right)\right], \tag{3.3.82e}$$

$$\psi_2^+ = B_2^+ \exp\left[i\left(k_{S1}^{2+}x_1 + k_{S3}^{2+}x_3 - \omega_S^{2+}t\right)\right], \tag{3.3.82f}$$

$$\varphi_2^- = A_2^- \exp\left[i\left(k_{P1}^{2-}x_1 - k_{P3}^{2-}x_3 - \omega_P^{2-}t\right)\right], \tag{3.3.82g}$$

$$\psi_2^- = B_2^- \exp\left[i\left(k_{S1}^{2-}x_1 - k_{S3}^{2-}x_3 - \omega_S^{2-}t\right)\right], \tag{3.3.82h}$$

where A_1^-, B_1^-, A_1^+, B_1^+, A_2^+, B_2^+, A_2^-, B_2^- and ω_P^{1-}, ω_S^{1-}, ω_P^{1+}, ω_S^{1+}, ω_P^{2+}, ω_S^{2+}, ω_P^{2-}, ω_S^{2-} are the amplitudes and angular frequencies of the corresponding waves, and

$$k_{P1}^{1-} = \frac{\omega_P^{1-}}{c_P}\sin\alpha_1^-,\quad k_{P3}^{1-} = \frac{\omega_P^{1-}}{c_P}\cos\alpha_1^-,$$

$$k_{P1}^{1+} = \frac{\omega_P^{1+}}{c_P}\sin\alpha_1^+,\quad k_{P3}^{1+} = \frac{\omega_P^{1+}}{c_P}\cos\alpha_1^+,$$

$$k_{P1}^{2+} = \frac{\omega_P^{2+}}{c'_P}\sin\alpha_2^+,\quad k_{P3}^{2+} = \frac{\omega_P^{2+}}{c'_P}\cos\alpha_2^+,$$

$$k_{P1}^{2-} = \frac{\omega_P^{2-}}{c'_P}\sin\alpha_2^-,\quad k_{P3}^{2-} = \frac{\omega_P^{2-}}{c'_P}\cos\alpha_2^-,$$

$$k_{S1}^{1-} = \frac{\omega_S^{1-}}{c_S}\sin\beta_1^-,\quad k_{S3}^{1-} = \frac{\omega_S^{1-}}{c_S}\cos\beta_1^-,$$

$$k_{S1}^{1+} = \frac{\omega_S^{1+}}{c_S}\sin\beta_1^+,\quad k_{S3}^{1+} = \frac{\omega_S^{1+}}{c_S}\cos\beta_1^+,$$

$$k_{S1}^{2+} = \frac{\omega_S^{2+}}{c'_S}\sin\beta_2^+,\quad k_{S3}^{2+} = \frac{\omega_S^{2+}}{c'_S}\cos\beta_2^+,$$

$$k_{S1}^{2-} = \frac{\omega_S^{2-}}{c_S'} \sin\beta_2^-, \quad k_{S3}^{2-} = \frac{\omega_S^{2-}}{c_S'} \cos\beta_2^-.$$

The displacement and stress components in both mediums can be calculated as follows

$$\begin{aligned} u_1 &= \frac{\partial(\varphi_1^- + \varphi_1^+)}{\partial x_1} - \frac{\partial(\psi_1^- + \psi_1^+)}{\partial x_3} \\ &= ik_{P1}^{1-} A_1^- \exp\left[i\left(k_{P1}^{1-} x_1 - k_{P3}^{1-} x_3 - \omega_P^{1-} t\right)\right] \\ &\quad + ik_{P1}^{1+} A_1^+ \exp\left[i\left(k_{P1}^{1+} x_1 + k_{P3}^{1+} x_3 - \omega_P^{1+} t\right)\right] \\ &\quad + ik_{S3}^{1-} B_1^- \exp\left[i\left(k_{S1}^{1-} x_1 - k_{S3}^{1-} x_3 - \omega_S^{1-} t\right)\right] \\ &\quad - ik_{S3}^{1+} B_1^+ \exp\left[i\left(k_{S1}^{1+} x_1 + k_{S3}^{1+} x_3 - \omega_S^{1+} t\right)\right], \end{aligned} \tag{3.3.83a}$$

$$\begin{aligned} u_3 &= \frac{\partial(\varphi_1^- + \varphi_1^+)}{\partial x_3} + \frac{\partial(\psi_1^- + \psi_1^+)}{\partial x_1} \\ &= -ik_{P3}^{1-} A_1^- \exp\left[i\left(k_{P1}^{1-} x_1 - k_{P3}^{1-} x_3 - \omega_P^{1-} t\right)\right] \\ &\quad + ik_{P3}^{1+} A_1^+ \exp\left[i\left(k_{P1}^{1+} x_1 + k_{P3}^{1+} x_3 - \omega_P^{1+} t\right)\right] \\ &\quad + ik_{S1}^{1-} B_1^- \exp\left[i\left(k_{S1}^{1-} x_1 - k_{S3}^{1-} x_3 - \omega_S^{1-} t\right)\right] \\ &\quad + ik_{S1}^{1+} B_1^+ \exp\left[i\left(k_{S1}^{1+} x_1 + k_{S3}^{1+} x_3 - \omega_S^{1+} t\right)\right], \end{aligned} \tag{3.3.83b}$$

$$\begin{aligned} \sigma_{33} &= \lambda \frac{\partial^2(\varphi_1^- + \varphi_1^+)}{\partial x_1^2} + (\lambda + 2\mu) \frac{\partial^2(\varphi_1^- + \varphi_1^+)}{\partial x_3^2} + 2\mu \frac{\partial^2(\psi_1^- + \psi_1^+)}{\partial x_1 \partial x_3} \\ &= \lambda\Big\{-\left(k_{P1}^{1-}\right)^2 A_1^- \exp\left[i\left(k_{P1}^{1-} x_1 - k_{P3}^{1-} x_3 - \omega_P^{1-} t\right)\right] \\ &\quad -\left(k_{P1}^{1+}\right)^2 A_1^+ \exp\left[i\left(k_{P1}^{1+} x_1 + k_{P3}^{1+} x_3 - \omega_P^{1+} t\right)\right]\Big\} \\ &\quad + (\lambda + 2\mu)\Big\{-\left(k_{P3}^{1-}\right)^2 A_1^- \exp\left[i\left(k_{P1}^{1-} x_1 - k_{P3}^{1-} x_3 - \omega_P^{1-} t\right)\right] \\ &\quad -\left(k_{P3}^{1+}\right)^2 A_1^+ \exp\left[i\left(k_{P1}^{1+} x_1 + k_{P3}^{1+} x_3 - \omega_P^{1+} t\right)\right]\Big\} \\ &\quad + 2\mu\Big\{k_{S1}^{1-} k_{S3}^{1-} B_1^- \exp\left[i\left(k_{S1}^{1-} x_1 - k_{S3}^{1-} x_3 - \omega_S^{1-} t\right)\right] \\ &\quad - k_{S1}^{1+} k_{S3}^{1+} B_1^+ \exp\left[i\left(k_{S1}^{1+} x_1 + k_{S3}^{1+} x_3 - \omega_S^{1+} t\right)\right]\Big\}, \end{aligned} \tag{3.3.83c}$$

$$\begin{aligned} \sigma_{31} &= \mu\Big[2\frac{\partial^2(\varphi_1^- + \varphi_1^+)}{\partial x_1 \partial x_3} + \frac{\partial^2(\psi_1^- + \psi_1^+)}{\partial x_1^2} - \frac{\partial^2(\psi_1^- + \psi_1^+)}{\partial x_3^2} \\ &= \mu\Big\{2k_{P1}^{1-} k_{P3}^{1-} A_1^- \exp\left[i\left(k_{P1}^{1-} x_1 - k_{P3}^{1-} x_3 - \omega_P^{1-} t\right)\right] \\ &\quad - 2k_{P1}^{1+} k_{P3}^{1+} A_1^+ \exp\left[i\left(k_{P1}^{1+} x_1 + k_{P3}^{1+} x_3 - \omega_P^{1+} t\right)\right] \\ &\quad - \left(k_{S1}^{1-}\right)^2 B_1^- \exp\left[i\left(k_{S1}^{1-} x_1 - k_{S3}^{1-} x_3 - \omega_S^{1-} t\right)\right] \end{aligned}$$

$$
\begin{aligned}
&- \left(k_{S1}^{1+}\right)^2 B_1^+ \exp\left[i\left(k_{S1}^{1+} x_1 + k_{S3}^{1+} x_3 - \omega_S^{1+} t\right)\right] \\
&+ \left(k_{S3}^{1-}\right)^2 B_1^- \exp\left[i\left(k_{S1}^{1-} x_1 - k_{S3}^{1-} x_3 - \omega_S^{1-} t\right)\right] \\
&+ \left(k_{S3}^{1+}\right)^2 B_1^+ \exp\left[i\left(k_{S1}^{1+} x_1 + k_{S3}^{1+} x_3 - \omega_S^{1+} t\right)\right]\Big\},
\end{aligned} \tag{3.3.83d}
$$

$$
\begin{aligned}
u_1' &= \frac{\partial\left(\varphi_2^+ + \varphi_2^-\right)}{\partial x_1} - \frac{\partial\left(\psi_2^+ + \psi_2^-\right)}{\partial x_3} \\
&= ik_{P1}^{2+} A_2^+ \exp\left[i\left(k_{P1}^{2+} x_1 + k_{P3}^{2+} x_3 - \omega_P^{2+} t\right)\right] \\
&\quad + ik_{P1}^{2-} A_2^- \exp\left[i\left(k_{P1}^{2-} x_1 - k_{P3}^{2-} x_3 - \omega_P^{2-} t\right)\right] \\
&\quad - ik_{S3}^{2+} B_2^+ \exp\left[i\left(k_{S1}^{2+} x_1 + k_{S3}^{2+} x_3 - \omega_S^{2+} t\right)\right] \\
&\quad + ik_{S3}^{2-} B_2^- \exp\left[i\left(k_{S1}^{2-} x_1 - k_{S3}^{2-} x_3 - \omega_S^{2-} t\right)\right],
\end{aligned} \tag{3.3.83e}
$$

$$
\begin{aligned}
u_3' &= \frac{\partial\left(\varphi_2^+ + \varphi_2^-\right)}{\partial x_3} + \frac{\partial\left(\psi_2^+ + \psi_2^-\right)}{\partial x_1} \\
&= ik_{P3}^{2+} A_2^+ \exp\left[i\left(k_{P1}^{2+} x_1 + k_{P3}^{2+} x_3 - \omega_P^{2+} t\right)\right] \\
&\quad - ik_{P3}^{2-} A_2^- \exp\left[i\left(k_{P1}^{2-} x_1 - k_{P3}^{2-} x_3 - \omega_P^{2-} t\right)\right] \\
&\quad + ik_{S1}^{2+} B_2^+ \exp\left[i\left(k_{S1}^{2+} x_1 + k_{S3}^{2+} x_3 - \omega_S^{2+} t\right)\right] \\
&\quad + ik_{S1}^{2-} B_2^- \exp\left[i\left(k_{S1}^{2-} x_1 - k_{S3}^{2-} x_3 - \omega_S^{2-} t\right)\right],
\end{aligned} \tag{3.3.83f}
$$

$$
\begin{aligned}
\sigma_{33}' &= \lambda' \frac{\partial^2\left(\varphi_2^+ + \varphi_2^-\right)}{\partial x_1^2} + (\lambda' + 2\mu') \frac{\partial^2\left(\varphi_2^+ + \varphi_2^-\right)}{\partial x_3^2} + 2\mu' \frac{\partial^2\left(\psi_2^+ + \psi_2^-\right)}{\partial x_1 \partial x_3} \\
&= \lambda'\Big\{-\left(k_{P1}^{2+}\right)^2 A_2^+ \exp\left[i\left(k_{P1}^{2+} x_1 + k_{P3}^{2+} x_3 - \omega_P^{2+} t\right)\right] \\
&\quad -\left(k_{P1}^{2-}\right)^2 A_2^- \exp\left[i\left(k_{P1}^{2-} x_1 - k_{P3}^{2-} x_3 - \omega_P^{2-} t\right)\right]\Big\} \\
&\quad + (\lambda' + 2\mu')\Big\{-\left(k_{P3}^{2+}\right)^2 A_2^+ \exp\left[i\left(k_{P1}^{2+} x_1 + k_{P3}^{2+} x_3 - \omega_P^{2+} t\right)\right] \\
&\quad -\left(k_{P3}^{2-}\right)^2 A_2^- \exp\left[i\left(k_{P1}^{2-} x_1 - k_{P3}^{2-} x_3 - \omega_P^{2-} t\right)\right]\Big\} \\
&\quad + 2\mu'\{-k_{S1}^{2+} k_{S3}^{2+} B_2^+ \exp\left[i\left(k_{S1}^{2+} x_1 + k_{S3}^{2+} x_3 - \omega_S^{2+} t\right)\right] \\
&\quad + k_{S1}^{2-} k_{S3}^{2-} B_2^- \exp\left[i\left(k_{S1}^{2-} x_1 - k_{S3}^{2-} x_3 - \omega_S^{2-} t\right)\right]\},
\end{aligned} \tag{3.3.83g}
$$

$$
\begin{aligned}
\sigma_{31}' &= \mu'\left[2\frac{\partial^2\left(\varphi_2^+ + \varphi_2^-\right)}{\partial x_1 \partial x_3} + \frac{\partial^2\left(\psi_2^+ + \psi_2^-\right)}{\partial x_1^2} - \frac{\partial^2\left(\psi_2^+ + \psi_2^-\right)}{\partial x_3^2}\right] \\
&= \mu'\{-2k_{P1}^{2+} k_{P3}^{2+} A_2^+ \exp\left[i\left(k_{P1}^{2+} x_1 + k_{P3}^{2+} x_3 - \omega_P^{2+} t\right)\right] \\
&\quad + 2k_{P1}^{2-} k_{P3}^{2-} A_2^- \exp\left[i\left(k_{P1}^{2-} x_1 - k_{P3}^{2-} x_3 - \omega_P^{2-} t\right)\right] \\
&\quad - \left(k_{S1}^{2+}\right)^2 B_2^+ \exp\left[i\left(k_{S1}^{2+} x_1 + k_{S3}^{2+} x_3 - \omega_S^{2+} t\right)\right]
\end{aligned}
$$

$$
\begin{aligned}
& - \left(k_{S1}^{2-}\right)^2 B_2^- \exp\left[i\left(k_{S1}^{2-} x_1 - k_{S3}^{2-} x_3 - \omega_S^{2-} t\right)\right] \\
& + \left(k_{S3}^{2+}\right)^2 B_2^+ \exp\left[i\left(k_{S1}^{2+} x_1 + k_{S3}^{2+} x_3 - \omega_S^{2+} t\right)\right] \\
& + \left(k_{S3}^{2-}\right)^2 B_2^- \exp\left[i\left(k_{S1}^{2-} x_1 - k_{S3}^{2-} x_3 - \omega_S^{2-} t\right)\right]\Big\}.
\end{aligned} \tag{3.3.83h}
$$

The displacement and stress components at the interface satisfy the continuity condition, namely

$$
u_1|_{x_3=0} = u_1'|_{x_3=0}, \quad u_3|_{x_3=0} = u_3'|_{x_3=0}, \tag{3.3.84ab}
$$

$$
\sigma_{33}|_{x_3=0} = \sigma_{33}'|_{x_3=0}, \quad \sigma_{31}|_{x_3=0} = \sigma_{31}'|_{x_3=0} \tag{3.3.84cd}
$$

Eqs. (3.3.84ab, 3.3.84cd) is required to hold for any x_1 and t, there must be

$$
\omega_P^{1-} = \omega_P^{1+} = \omega_P^{2+} = \omega_P^{2-} = \omega_S^{1-} = \omega_S^{1+} = \omega_S^{2+} = \omega_S^{2-}, \tag{3.3.85}
$$

$$
k_{P1}^{1-} = k_{P1}^{1+} = k_{P1}^{2+} = k_{P1}^{2-} = k_{S1}^{1-} = k_{S1}^{1+} = k_{S1}^{2+} = k_{S1}^{2-}. \tag{3.3.86}
$$

Equation (3.3.86) can also be written as

$$
\frac{\sin\alpha_1^-}{c_P} = \frac{\sin\alpha_1^+}{c_P} = \frac{\sin\alpha_2^+}{c_P'} = \frac{\sin\alpha_2^-}{c_P'} = \frac{\sin\beta_1^-}{c_S} = \frac{\sin\beta_1^+}{c_S} = \frac{\sin\beta_2^+}{c_S'} = \frac{\sin\beta_2^-}{c_S'}. \tag{3.3.87}
$$

Obviously,

$$
\alpha_1^- = \alpha_1^+, \ \alpha_2^- = \alpha_2^+, \ \beta_1^- = \beta_1^+, \ \beta_2^- = \beta_2^+. \tag{3.3.88}
$$

It shows that the frequency of each incident wave is the same as that of each reflected wave from Eq. (3.3.85). In the same medium, the incident angle of P wave is equal to the reflected angle of P wave, and the incident angle of SV wave is equal to the reflected angle of SV wave. Moreover, the incident angle of the P wave is always greater than that of the SV wave.

Using the above relation, the continuous equation of the interface reduce to

$$
A_1^- + A_1^+ + p_2\left(B_1^- - B_1^+\right) - \left(A_2^- + A_2^+\right) - \left(B_2^- - B_2^+\right)p_4 = 0, \tag{3.3.89a}
$$

$$
p_1\left(A_1^- - A_1^+\right) - \left(B_1^- + B_1^+\right) - p_3\left(A_2^- - A_2^+\right) + \left(B_2^- + B_2^+\right) = 0, \tag{3.3.89b}
$$

$$
\begin{aligned}
& \left[\lambda\left(1 + p_1^2\right) + 2\mu p_1^2\right]\left(A_1^- + A_1^+\right) - 2\mu p_2\left(B_1^- B_1^+\right) \\
& \quad - \left[\lambda'\left(1 + p_3^2\right) + 2\mu' p_3^2\right]\left(A_2^- + A_2^+\right) + 2\mu' p_4\left(B_2^- - B_2^+\right) = 0,
\end{aligned} \tag{3.3.89c}
$$

$$2\mu p_1(A_1^- - A_1^+) - \mu(1 - p_2^2)(B_1^- + B_1^+)$$
$$- 2\mu' p_3(A_2^- - A_2^+) + \mu'(1 - p_4^2)(B_2^- + B_2^+) = 0, \tag{3.3.89d}$$

where

$$P_1 = \frac{k_{P3}^{1-}}{k_{p1}^{1-}} \cot\alpha_1^-, \quad P_2 = \frac{k_{S3}^{1-}}{k_{S1}^{1-}} \cot\beta_1^-,$$

$$P_3 = \frac{k_{P3}^{2-}}{k_{P1}^{2-}} \cot\alpha_2^-, \quad P_4 = \frac{k_{S3}^{2-}}{k_{S1}^{2-}} \cot\beta_2^-.$$

Let

$$P = A_1^- + A_1^+, \ Q = A_1^- - A_1^+. \tag{3.3.90}$$

From Eqs. (3.3.89a) and (3.3.89c), we get

$$P = l_1(A_2^- + A_2^+) + h_1(B_2^- - B_2^+), \quad B_1^+ - B_1^- = l_2(A_2^- + A_2^+) + h_2(B_2^- - B_2^+). \tag{3.3.91}$$

From Eqs. (3.3.89b) and (3.3.89d), we get

$$Q = l_3(A_2^- - A_2^+) + h_3(B_2^- + B_2^+),$$
$$B_1^+ + B_1^- = -l_4(A_2^- - A_2^+) - h_4(B_2^- + B_2^+). \tag{3.3.92}$$

In the formula, l_i and h_i have the same meaning as l_i and h_i in Sect. 3.3.1 The interface conditions (3.3.89a, 3.3.89b, 3.3.89c, 3.3.89d) is equivalent to

$$A_1^- + A_1^+ = l_1(A_2^- + A_2^+) + h_1(B_2^- - B_2^+), \tag{3.3.93a}$$

$$B_1^+ - B_1^- = l_2(A_2^- + A_2^+) + h_2(B_2^- - B_2^+), \tag{3.3.93b}$$

$$A_1^- - A_1^+ = l_3(A_2^- - A_2^+) + h_3(B_2^- + B_2^+), \tag{3.3.93c}$$

$$B_1^+ + B_1^- = -l_4(A_2^- - A_2^+) - h_4(B_2^- + B_2^+). \tag{3.3.93d}$$

Solving Eqs. (3.3.93a, 3.3.93b, 3.3.93c, 3.3.93d) simultaneously, we obtain

$$A_1^+ = \frac{[(l_1-l_3)(h_2+h_4)-(l_2+l_4)(h_1-h_3)]A_1^- + 2(l_1h_3-l_3h_1)B_1^-}{(l_1+l_3)(h_2+h_4)-(l_2+l_4)(h_1+h_3)}$$
$$+ \frac{2[l_1(l_3h_4-l_4h_3)-l_3(l_1h_2-l_2h_1)]A_2^+ + 2[h_3(l_2h_1-l_1h_2)+h_1(l_4h_3-l_3h_4)]B_2^+}{(l_1+l_3)(h_2+h_4)-(l_2+l_4)(h_1+h_3)},$$

$$B_1^+ = \frac{2(l_2h_4-l_4h_2)A_1^- + [(l_2-l_4)(h_1+h_3)-(l_1+l_3)(h_2-h_4)]B_1^-}{(l_1+l_3)(h_2+h_4)-(l_2+l_4)(h_1+h_3)} + \frac{2[l_2(l_3h_4-l_4h_3)+l_4(l_1h_2-l_2h_1)]A_2^+ + 2[h_4(l_2h_1-l_1h_2)+h_2(l_4h_3-l_3h_4)]B_2^+}{(l_1+l_3)(h_2+h_4)-(l_2+l_4)(h_1+h_3)},$$

$$A_2^- = \frac{2(h_2+h_4)A_1^- + 2(h_1+h_3)B_1^-}{(l_1+l_3)(h_2+h_4)-(l_2+l_4)(h_1+h_3)} + \frac{[(l_3-l_1)(h_2+h_4)+(l_2-l_4)(h_1+h_3)]A_2^+ + 2(h_1h_4-h_2h_3)B_2^+}{(l_1+l_3)(h_2+h_4)-(l_2+l_4)(h_1+h_3)},$$

$$B_2^- = \frac{-2(l_2+l_4)A_1^- - 2(l_1+l_3)B_1^-}{(l_1+l_3)(h_2+h_4)-(l_2+l_4)(h_1+h_3)} + \frac{2(l_1l_4-l_2l_3)A_2^+ + [(l_2+l_4)(h_3-h_1)+(l_1+l_3)(h_2-h_4)]B_2^+}{(l_1+l_3)(h_2+h_4)-(l_2+l_4)(h_1+h_3)}.$$

Thus, the four unknown amplitudes A_1^+, B_1^+, A_2^- and B_2^- can be represented by the known incident wave amplitudes A_1^-, B_1^-, A_2^+ and B_2^+ in matrix form

$$\begin{pmatrix} A_1^+ \\ B_1^+ \end{pmatrix} = (\boldsymbol{F}^{+-})\begin{pmatrix} A_1^- \\ B_1^- \end{pmatrix} + (\boldsymbol{E}^{++})\begin{pmatrix} A_2^+ \\ B_2^+ \end{pmatrix}, \tag{3.3.94a}$$

$$\begin{pmatrix} A_2^- \\ B_2^- \end{pmatrix} = (\boldsymbol{E}^{--})\begin{pmatrix} A_1^- \\ B_1^- \end{pmatrix} + (\boldsymbol{F}^{-+})\begin{pmatrix} A_2^+ \\ B_2^+ \end{pmatrix}, \tag{3.3.94b}$$

in which

$$F_{11}^{+-} = \frac{(l_1-l_2)(h_2+h_4)-(l_2+l_4)(h_1-h_3)}{(l_1+l_3)(h_2+h_4)-(l_2+l_4)(h_1+h_3)},$$

$$F_{12}^{+-} = \frac{2(l_1h_3-l_3h_1)}{(l_1+l_3)(h_2+h_4)-(l_2+l_4)(h_1+h_3)},$$

$$F_{21}^{+-} = \frac{2(l_2h_4-l_4h_2)}{(l_1+l_3)(h_2+h_4)-(l_2+l_4)(h_1+h_3)},$$

$$F_{22}^{+-} = \frac{(l_2-l_4)(h_1+h_3)-(l_1+l_3)(h_2-h_4)}{(l_1+l_3)(h_2+h_4)-(l_2+l_4)(h_1+h_3)},$$

$$F_{11}^{-+} = \frac{(l_3-l_1)(h_2+h_4)+(l_2-l_4)(h_1+h_3)}{(l_1+l_3)(h_2+h_4)-(l_2+l_4)(h_1+h_3)},$$

$$F_{12}^{-+} = \frac{2(h_1h_4-h_2h_3)}{(l_1+l_3)(h_2+h_4)-(l_2+l_4)(h_1+h_3)},$$

$$F_{21}^{-+} = \frac{2(l_1l_4-l_2l_3)}{(l_1+l_3)(h_2+h_4)-(l_2+l_4)(h_1+h_3)},$$

$$F_{22}^{-+} = \frac{(l_2 + l_4)(h_3 - h_1) + (l_1 + l_3)(h_2 - h_4)}{(l_1 + l_3)(h_2 + h_4) - (l_2 + l_4)(h_1 + h_3)},$$

$$E_{11}^{++} = \frac{2[l_1(l_3h_4 - l_4h_3) - l_3(l_1h_2 - l_2h_1)]}{(l_1 + l_3)(h_2 + h_4) - (l_2 + l_4)(h_1 + h_3)},$$

$$E_{12}^{++} = \frac{2[h_3(l_2h_1 - l_1h_2) + h_1(l_4h_3 - l_3h_4)]}{(l_1 + l_3)(h_2 + h_4) - (l_2 + l_4)(h_1 + h_3)},$$

$$E_{21}^{++} = \frac{2[l_2(l_3h_4 - l_4h_3) + l_4(l_1h_2 - l_2h_1)]}{(l_1 + l_3)(h_2 + h_4) - (l_2 + l_4)(h_1 + h_3)},$$

$$E_{22}^{++} = \frac{2[h_4(l_2h_1 - l_1h_2) + h_2(l_4h_3 - l_3h_4)]}{(l_1 + l_3)(h_2 + h_4) - (l_2 + l_4)(h_1 + h_3)},$$

$$E_{11}^{--} = \frac{2(h_2 + h_4)}{(l_1 + l_3)(h_2 + h_4) - (l_2 + l_4)(h_1 + h_3)},$$

$$E_{12}^{--} = \frac{2(h_1 + h_3)}{(l_1 + l_3)(h_2 + h_4) - (l_2 + l_4)(h_1 + h_3)},$$

$$E_{21}^{--} = \frac{-2(l_2 + l_4)}{(l_1 + l_3)(h_2 + h_4) - (l_2 + l_4)(h_1 + h_3)},$$

$$E_{22}^{--} = \frac{-2(l_1 + l_3)}{(l_1 + l_3)(h_2 + h_4) - (l_2 + l_4)(h_1 + h_3)}.$$

When P wave and SV wave are incident on both sides at the same time, we can get the conclusions as follows

(1) Taking $A_2^+ = B_2^+ = 0$, it degenerates into simultaneous incident P wave and SV wave from one side of interface, i.e. medium 1, F^{+-} represents the reflection matrix of the potential function, and E^{--} represents the transmission matrix of the potential function. If B_1^- is zero further, it means that the P wave is incident alone, and if A_1^- is zero, it means that the SV wave is incident alone.
(2) Taking $A_1^- = B_1^- = 0$, it degenerates into simultaneous incident P wave and SV wave from one side of interface, i.e. medium 2, E^{++} represents the potential function transmission matrix, and F^{-+} represents the potential function reflection matrix. Further, If $B_2^+ = 0$, it means P wave is incident alone, if $A_2^+ = 0$, it means SV wave is incident alone.
(3) F_{ij}^{+-} and F_{ij}^{-+} represent the reflection coefficient of the potential function when a single P wave or SV wave is incident, and E_{ij}^{++} and E_{ij}^{--} represent the transmission coefficient of the potential function as a single P wave or SV wave incident.

Now, let us study the relationship of displacement amplitudes when P wave and SV wave are incident from both sides at the same time. Assuming that the displacement amplitudes of incident P wave and SV wave as well as reflected P wave and SV wave in medium 1 and medium 2 be C_1^-, D_1^-, C_1^+, D_1^+ and C_2^+, D_2^+, C_2^-, D_2^-, respectively. And suppose

$$\begin{pmatrix} C_1^+ \\ D_1^+ \end{pmatrix} = (\boldsymbol{R}^{+-})\begin{pmatrix} C_1^- \\ D_1^- \end{pmatrix} + (\boldsymbol{T}^{++})\begin{pmatrix} C_2^+ \\ D_2^+ \end{pmatrix},$$

$$\begin{pmatrix} C_2^- \\ D_2^- \end{pmatrix} = (\boldsymbol{T}^{--})\begin{pmatrix} C_1^- \\ D_1^- \end{pmatrix} + (\boldsymbol{R}^{-+})\begin{pmatrix} C_2^+ \\ D_2^+ \end{pmatrix}.$$

Then, $\boldsymbol{R}^{+-}$ is the displacement reflection matrix when P wave and SV wave are simultaneously incident on the side of medium 1. Thus,

$$R_{11}^{+-} = F_{11}^{+-} \cdot \frac{c_P}{c_P} = F_{11}^{+-}, \quad R_{21}^{+-} = F_{21}^{+-} \cdot \frac{c_P}{c_S},$$

$$R_{12}^{+-} = F_{12}^{+-} \cdot \frac{c_S}{c_P}, \quad R_{22}^{+-} = F_{22}^{+-} \cdot \frac{c_S}{c_S} = F_{22}^{+-}.$$

$\boldsymbol{R}^{-+}$ is the displacement reflection matrix when P wave and SV wave are simultaneously incident on the side of medium 2. Thus,

$$R_{11}^{-+} = F_{11}^{-+} \cdot \frac{c'_P}{c'_P} = F_{11}^{-+}, \quad R_{21}^{-+} = F_{21}^{-+} \cdot \frac{c'_P}{c'_S},$$

$$R_{12}^{-+} = F_{12}^{-+} \cdot \frac{c'_S}{c'_P}, \quad R_{22}^{-+} = F_{22}^{-+} \cdot \frac{c'_S}{c'_S} = F_{22}^{-+}.$$

$\boldsymbol{T}^{++}$ is the displacement transmission matrix when P wave and SV wave are simultaneously incident on the side of medium 2. Thus,

$$T_{11}^{++} = E_{11}^{++} \cdot \frac{c'_P}{c_P}, \quad T_{21}^{++} = E_{21}^{++} \cdot \frac{c'_P}{c_S},$$

$$T_{12}^{++} = E_{12}^{++} \cdot \frac{c'_S}{c_P}, \quad T_{22}^{++} = E_{22}^{++} \cdot \frac{c'_S}{c_S}.$$

$\boldsymbol{T}^{--}$ is the displacement transmission matrix when P wave and SV wave are simultaneously incident on the side of medium 1. Thus

$$T_{11}^{--} = E_{11}^{--} \cdot \frac{c_P}{c'_P}, \quad T_{21}^{--} = E_{21}^{--} \cdot \frac{c_P}{c'_S},$$

$$T_{12}^{--} = E_{12}^{--} \cdot \frac{c_S}{c'_P}, \quad T_{22}^{--} = E_{22}^{--} \cdot \frac{c_S}{c'_S}.$$

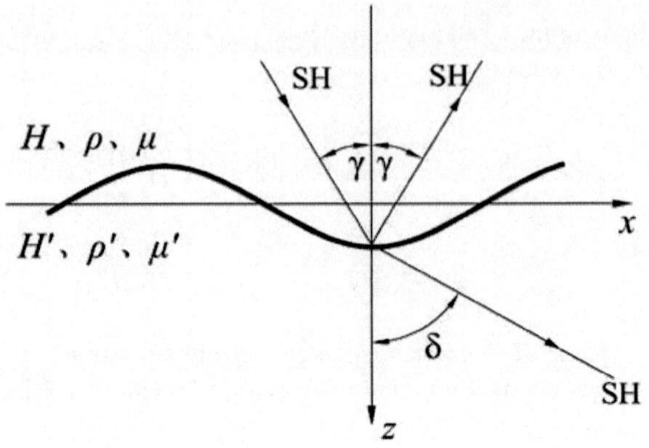

Fig. 3.23 Reflection and transmission of elastic waves at periodic corrugated interfaces

3.4 Reflection and Transmission of Waves at the Periodic Corrugated Interface

In the previous chapter, when discussing the reflection and transmission of elastic waves at the interface, we all assumed that the interface is a plane. In this section, we will discuss the reflection and transmission of elastic waves at periodic corrugated interfaces. The structure of the problem and the Cartesian coordinate axes are shown in Fig. 3.23. The upper and lower half-spaces are separated by periodic corrugated interfaces. H, H' represents two isotropic half-spaces. The equation of the corrugated interface is given by $z = \zeta(x)$. Two half-spaces are perfectly connected at the periodic corrugated interface [11,12].

The mean value of ζ is assumed to be zero. Fourier series representation of ζ is given by

$$\zeta(x) = \sum_{n=1}^{\infty} [c_n \cos(npx) + s_n \sin(npx)], \tag{3.4.1}$$

where c_n and s_n are the Fourier coefficient; $p = \frac{2\pi}{\Lambda}$, Λ is the structure period. Consider Euler formula $e^{i\theta} = \cos\theta + i\sin\theta$, Eq. (3.4.1) can be written as

$$\zeta(x) = \sum_{n\to-\infty(n\neq 0)}^{+\infty} [\zeta_n e^{inpx}], \tag{3.4.2}$$

where

$$\zeta_n = \begin{cases} \frac{c_n - is_n}{2}, & n > 0 \\ \frac{c_{-n} + is_{-n}}{2}, & n < 0 \end{cases} \tag{3.4.3}$$

For an SH wave propagating in the xz plane, all field variables are independent of y, and the governing equation of an isotropic material is expressed by:

$$\mu \frac{\partial^2 v}{\partial x^2} + \mu \frac{\partial^2 v}{\partial z^2} = \rho \frac{\partial^2 v}{\partial t^2}, \tag{3.4.4}$$

where μ is the shear modulus; v denotes the y-component of the displacement; ρ is the density; t is the time. The constitutive equation of an isotropic elastic material is:

$$\sigma_{xy} = \mu \frac{\partial v}{\partial x}, \quad \sigma_{yz} = \mu \frac{\partial v}{\partial z}. \tag{3.4.5}$$

Consider the time harmonic waves and let

$$v = X(x)Z(z)\mathrm{e}^{-i\omega t}, \tag{3.4.6}$$

where ω is the angular frequency. Insertion of Eq. (3.4.6) into Eq. (3.4.4) and separating the variables,

$$\begin{aligned} \frac{\mathrm{d}^2 X}{\mathrm{d}x^2} + k_x^2 X = 0, \\ \frac{\mathrm{d}^2 Z}{\mathrm{d}z^2} + k_z^2 Z = 0, \end{aligned} \tag{3.4.7}$$

where

$$k_x^2 + k_z^2 = k^2, \quad k^2 = \frac{\rho \omega^2}{\mu}. \tag{3.4.8}$$

where k_x is the wavenumber component in the x-direction; k_z is the wavenumber component in the z-direction. Using φ denotes the angle between the wave normal and the z-axis, namely

$$k_x = k \sin \varphi, \quad k_z = k \cos \varphi, \quad \tan \varphi = \frac{k_x}{k_z}. \tag{3.4.9}$$

The solution of Eq. (3.4.7) can be expressed as

$$X(x) = A_1 \mathrm{e}^{ik_x x} + A_2 \mathrm{e}^{-ik_x x}, \tag{3.4.10a}$$

$$Z(z) = B_1 \mathrm{e}^{ik_z z} + B_2 \mathrm{e}^{-ik_z z}. \tag{3.4.10b}$$

Considering the existence of periodic corrugated interface along the x-direction and using Bloch–Floquet principle, the $X(x)$ in Eq. (3.4.10a) will be modulated by periodic corrugated interface and is given by

$$X(x) = A_1(x)\mathrm{e}^{ik_x x} + A_2(x)\mathrm{e}^{-ik_x x}, \tag{3.4.11}$$

where $A_1(x)$ and $A_2(x)$ are functions with period Λ, namely

$$A_1(x+\Lambda) = A_1(x), \quad A_2(x+\Lambda) = A_2(x). \tag{3.4.12}$$

These two periodic functions can be expressed by Fourier series as

$$A_1(x) = \sum_{n\to-\infty}^{+\infty} A_{1n}\mathrm{e}^{inpx}, \quad A_2(x) = \sum_{n\to-\infty}^{+\infty} A_{2n}\mathrm{e}^{-inpx}. \tag{3.4.13}$$

Inserting Eq. (3.4.13) into Eq. (3.4.11) leads to

$$X(x) = \sum_{n\to-\infty}^{+\infty} A_{1n}\mathrm{e}^{i(k_x+np)x} + \sum_{n\to-\infty}^{+\infty} A_{2n}\mathrm{e}^{-i(k_x+np)x}. \tag{3.4.14}$$

Let $k_{nx} = k_x + np$, then

$$X(x) = \sum_{n\to-\infty}^{+\infty} A_{1n}\mathrm{e}^{ik_{nx}x} + \sum_{n\to-\infty}^{+\infty} A_{2n}\mathrm{e}^{-ik_{nx}x}. \tag{3.4.15}$$

Combining Eqs. (3.4.6) and (3.4.15), the general form of displacement can be expressed as

$$\begin{aligned} v(x,z,t) = \sum_{n\to-\infty}^{+\infty} &\{[A_{1n}^{+}\mathrm{e}^{ik_{nz}z} + A_{1n}^{-}\mathrm{e}^{-ik_{nz}z}]\mathrm{e}^{ik_{nx}x} \\ &+[A_{2n}^{+}\mathrm{e}^{ik_{nz}z} + A_{2n}^{-}\mathrm{e}^{-ik_{nz}z}]\mathrm{e}^{-ik_{nx}x}\}\mathrm{e}^{-i\omega t}, \end{aligned} \tag{3.4.16}$$

where $A_{1n}^{\pm}$ and $A_{2n}^{\pm}$ are the displacement amplitudes of the nth harmonic propagating in different directions (namely the different combinations of k_{nz} and k_{nx}). Determined by

$$k_{nz}^2 + k_{nx}^2 = \frac{\rho\omega^2}{\mu} = k^2, \tag{3.4.17}$$

namely $k_{nz} = \pm\sqrt{k^2 - k_{nx}^2}$. Considering $\alpha_n = \sqrt{k^2 - k_{nx}^2}$, the set of possible displacement modes in the upper half-space is

$$\begin{aligned} v(x,z,t) = \sum_{n\to-\infty}^{+\infty} &\{[A_{1n}^{+}\mathrm{e}^{i\alpha_n z} + A_{1n}^{-}\mathrm{e}^{-i\alpha_n z}]\mathrm{e}^{ik_{nx}x} \\ &+ [A_{2n}^{+}\mathrm{e}^{i\alpha_n z} + A_{2n}^{-}\mathrm{e}^{-i\alpha_n z}]\mathrm{e}^{-ik_{nx}x}\}\mathrm{e}^{-i\omega t}. \end{aligned} \tag{3.4.18}$$

Similarly, the set of possible displacement modes in the lower half of the space is

$$v'(x,z,t) = \sum_{n\to-\infty}^{+\infty} \Big\{\Big[A'^{+}_{1n}\mathrm{e}^{i\alpha'_n z} + A'^{-}_{1n}\mathrm{e}^{-i\alpha'_n z}\Big]\mathrm{e}^{ik'_{nx}x} + \Big[A'^{+}_{2n}\mathrm{e}^{i\alpha'_n z} + A'^{-}_{2n}\mathrm{e}^{-i\alpha'_n z}\Big]\mathrm{e}^{-ik'_{nx}x}\Big\}\mathrm{e}^{-i\omega t}, \quad (3.4.19)$$

where $k'_{nx} = k'_x + np$, $\alpha'_n = \sqrt{k'^2 - k'^2_{nx}}$, $k'^2 = \rho'\omega^2/\mu'$. According to Snell's theorem, it can be seen that $k_x = k'_x$; namely, $k_{0x} = k'_{0x}$ and $k_{nx} = k'_{nx}$ can be further obtained. Introducing $k_{nx} = k'_{nx} = \beta_n$, Eqs. (3.4.18) and (3.4.19) can be simplified as

$$v(x,z,t) = \sum_{n=-\infty}^{+\infty} \{[A^{+}_{1n}\mathrm{e}^{i\alpha_n z} + A^{-}_{1n}\mathrm{e}^{-i\alpha_n z}]\mathrm{e}^{i\beta_n x} + [A^{+}_{2n}\mathrm{e}^{i\alpha_n z} + A^{-}_{2n}\mathrm{e}^{-i\alpha_n z}]\mathrm{e}^{-i\beta_n x}\}\mathrm{e}^{-i\omega t}, \quad (3.4.20)$$

$$v'(x,z,t) = \sum_{n=-\infty}^{+\infty} \Big\{\Big[A'^{+}_{1n}\mathrm{e}^{i\alpha'_n z} + A'^{-}_{1n}\mathrm{e}^{-i\alpha'_n z}\Big]\mathrm{e}^{i\beta_n x} + \Big[A'^{+}_{2n}\mathrm{e}^{i\alpha'_n z} + A'^{-}_{2n}\mathrm{e}^{-i\alpha'_n z}\Big]\mathrm{e}^{-i\beta_n x}\Big\}\mathrm{e}^{-i\omega t}. \quad (3.4.21)$$

Consider an SH wave with a unit amplitude frequency of ω and an angle of γ with the z-axis from the upper half of the space, as shown in Fig. 3.23. In the upper half-space, only incident waves propagate in the positive direction of the z-axis, which means

$$A^{+}_{0} = 1, \quad A^{+}_{n} = 0\,(n \neq 0). \quad (3.4.22)$$

For the reflected wave, which propagates along the negative direction of the z-axis, all harmonics propagating in the negative direction of the z-axis should be retained. When the projection of the incident wave vector on the x-axis is positive, all the harmonic terms containing $\mathrm{e}^{i\beta_n x}$ should be retained, and the harmonic terms containing $\mathrm{e}^{-i\beta_n x}$ should be discarded. Therefore, the total displacement in the upper half-space (incident wave + reflected wave) can be expressed as

$$v = \mathrm{e}^{i\alpha_0 z}\mathrm{e}^{i\beta_0 x} + \sum_{n\to-\infty}^{+\infty} A^{-}_{n}\mathrm{e}^{-i\alpha_n z}\mathrm{e}^{i\beta_n x}. \quad (3.4.23)$$

The transmitted wave propagates in the positive direction of the x-axis, so for the lower half of the space, all waves propagating in the positive direction of the z-axis should be retained. Therefore, the displacement in the lower half-space is expressed as

$$v' = \sum_{n=-\infty}^{+\infty} A_n'^{+} e^{i\alpha_n' z} e^{i\beta_n x}, \tag{3.4.24}$$

where $\beta_0 = k_{0x} = k \sin\gamma$, $\beta_n = k \sin\gamma + np$.

The interface condition of a perfect interface requires continuous displacement and stress at the interface $z = \zeta(x)$. The surface force component P_i at any point on the interface $z = \zeta(x)$ can be expressed as

$$P_i = \sigma_{ij} n_j, \tag{3.4.25}$$

where n_j is the normal vector on the interface. Then, the corrugated interface $z = \zeta(x)$ can be expressed as

$$n_x = -\frac{\zeta'}{\sqrt{1+\zeta'^2}}, \quad n_y = 0, \quad n_z = \frac{1}{\sqrt{1+\zeta'^2}}, \tag{3.4.26}$$

where ζ' denotes $\frac{\partial\zeta}{\partial x}$. Substituting Eqs. (3.4.5) and (3.4.26) into Eq. (3.4.25), we can obtain

$$P_x = 0, \quad P_y = \frac{\mu_1\left(\frac{\partial v}{\partial z} - \frac{\partial v}{\partial x}\zeta'\right)}{\sqrt{1+\zeta'^2}}, \quad P_z = 0. \tag{3.4.27}$$

Therefore, the continuity condition of the interface can be expressed as:

$$v = v', \tag{3.4.28a}$$

$$\mu_1\left(\frac{\partial v}{\partial z} - \frac{\partial v}{\partial x}\zeta'\right) = \mu_2\left(\frac{\partial v'}{\partial z} - \frac{\partial v'}{\partial x}\zeta'\right). \tag{3.4.28b}$$

Inserting Eqs. (3.4.23) and (3.4.24) into Eqs. (3.4.28a, 3.4.28b) and omitting the common factor $e^{i\beta_0 x}$, we obtain the following equation:

$$e^{i\alpha_0\zeta} + \sum_{n=-\infty}^{+\infty}\left[A_n^{-} e^{-i\alpha_n\zeta} e^{inpx}\right] = \sum_{n=-\infty}^{+\infty}\left[A_n'^{+} e^{i\alpha_n'\zeta} e^{inpx}\right], \tag{3.4.29a}$$

$$\begin{aligned} &\mu\left[(-\beta_0\zeta' + \alpha_0)e^{i\alpha_0\zeta} + \sum_{n=-\infty}^{+\infty}\left[(-\beta_n\zeta' - \alpha_n)A_n^{-} e^{-i\alpha_n\zeta} e^{inpx}\right]\right] \\ &= \mu' \sum_{n=-\infty}^{+\infty}\left[(-\beta_n\zeta' + \alpha_n')A_n'^{+} e^{i\alpha_n'\zeta} e^{inpx}\right]. \end{aligned} \tag{3.4.29b}$$

Assuming the corrugation of the interface $z = \zeta(x)$ is small, the Taylor series expansion is performed on $\mathrm{e}^{-i\alpha_n\zeta}$ at $z = 0$, and only the first term of ζ is retained; therefore, we have

$$\mathrm{e}^{-i\alpha_n\zeta} = 1 - i\alpha_n\zeta. \tag{3.4.30}$$

Inserting Eq. (3.4.30) into Eqs. (3.4.29a, 3.4.29b), we have

$$1 + i\alpha_0\zeta + \sum_{n=-\infty}^{+\infty}\left[A_n^-(1 - i\alpha_n\zeta)\mathrm{e}^{inpx}\right] = \sum_{n=-\infty}^{+\infty}\left[A_n'^+(1 + i\alpha_n'\zeta)\mathrm{e}^{inpx}\right], \tag{3.4.31a}$$

$$\mu\left[(-\beta_0\zeta' + \alpha_0)(1 + i\alpha_0\zeta) + \sum_{n=-\infty}^{+\infty}\left[(-\beta_n\zeta' - \alpha_n)(1 - i\alpha_n\zeta)A_n^-\mathrm{e}^{inpx}\right]\right] = \mu'\sum_{n=-\infty}^{+\infty}\left[(-\beta_n\zeta' + \alpha_n')(1 + i\alpha_n'\zeta)A_n'^+\mathrm{e}^{inpx}\right]. \tag{3.4.31b}$$

Further, Eqs. (3.4.31a, 3.4.31b) can be written as

$$\begin{aligned} &1 + A_0^- + i\alpha_0\zeta - i\alpha_0\zeta A_0^- + \sum_{n=-\infty(n\neq 0)}^{+\infty}\left[A_n^-\mathrm{e}^{inpx}\right] \\ &+ \sum_{n=-\infty(n\neq 0)}^{+\infty}\left[-i\alpha_n\zeta A_n^-\mathrm{e}^{inpx}\right] \\ &= A_0'^+ + \sum_{n=-\infty(n\neq 0)}^{+\infty}\left[A_n'^+\mathrm{e}^{inpx}\right] + i\alpha_0'\zeta A_0'^+ \\ &+ \sum_{n=-\infty(n\neq 0)}^{+\infty}\left[i\alpha_n'\zeta A_n'^+\mathrm{e}^{inpx}\right], \end{aligned} \tag{3.4.32a}$$

$$\begin{aligned} &\mu\Big\{\alpha_0 - \beta_0\zeta' + i\alpha_0^2\zeta - i\alpha_0\beta_0\zeta\zeta' + (-\alpha_0 - \beta_0\zeta' + i\alpha_0^2\zeta + \alpha_0\beta_0\zeta\zeta')A_0^- \\ &+ \sum_{n=-\infty(n\neq 0)}^{+\infty}\left[(-\alpha_n - \beta_n\zeta' + i\alpha_n^2\zeta + \alpha_n\beta_n\zeta\zeta')A_n^-\mathrm{e}^{inpx}\right]\Big\} \\ &= \mu'\Big\{(\alpha_0' - \beta_0\zeta' + i\alpha_0'^2\zeta - \alpha_0'\beta_0\zeta\zeta')A_0'^+ \\ &+ \sum_{n=-\infty(n\neq 0)}^{+\infty}\left[(\alpha_0' - \beta_0\zeta' + i\alpha_0'^2\zeta - \alpha_0'\beta_0\zeta\zeta')A_n'^+\mathrm{e}^{inpx}\right]\Big\}. \end{aligned} \tag{3.4.32b}$$

It should be noted that A_n^-, $A_n'^+ (n \neq 0)$ are caused by periodic corrugation and are considered to be small amounts of the first order of ζ. In Eqs. (3.4.32a, 3.4.32b), only the first term of ζ is retained, thus obtaining

$$1 + A_0^- - i\alpha_0\zeta(A_0^- - 1) + \sum_{\substack{n=-\infty \\ (n \neq 0)}}^{+\infty} \left[A_n^- e^{inpx}\right]$$

$$= A_0'^+ + i\alpha_0'\zeta A_0'^+ + \sum_{\substack{n=-\infty \\ (n \neq 0)}}^{+\infty} \left[A_n'^+ e^{inpx}\right], \tag{3.4.33a}$$

$$\mu\left\{\alpha_0 - \beta_0\zeta' + i\alpha_0^2\zeta + (-\alpha_0 - \beta_0\zeta' + i\alpha_0^2\zeta)A_0^- + \sum_{\substack{n=-\infty \\ (n \neq 0)}}^{+\infty} \left[-\alpha_n A_n^- e^{inpx}\right]\right\}$$

$$= \mu'\left\{(\alpha_0' - \beta_0\zeta' + i\alpha_0'^2\zeta)A_0'^+ + \sum_{\substack{n=-\infty \\ (n \neq 0)}}^{+\infty} \left[\alpha_n' A_n'^+ e^{inpx}\right]\right\}. \tag{3.4.33b}$$

By extracting the terms that do not contain x and ζ from Eqs. (3.4.33a, 3.4.33b), the relational expression about A_0^-, $A_0'^+$ can be written as

$$1 + A_0^- = A_0'^+, \tag{3.4.34a}$$

$$\mu\left[\alpha_0 - \alpha_0 A_0^-\right] = \mu'\alpha_0' A_0'^+. \tag{3.4.34b}$$

In order to find the first-order approximation of A_n^-, $A_n'^+ (n \neq 0)$, substituting Eq. (3.4.2) into Eqs. (3.4.33a, 3.4.33b), we have

$$1 + A_0^- - i\alpha_0(A_0^- - 1) \sum_{n \to -\infty (n \neq 0)}^{+\infty} \left[\zeta_n e^{inpx}\right]$$

$$+ \sum_{n \to -\infty (n \neq 0)}^{+\infty} \left[A_n^- e^{inpx}\right]$$

$$= A_0'^{+} + i\alpha_0' A_0'^{+} \sum_{n \to -\infty (n \neq 0)}^{+\infty} \left[\zeta_n e^{inpx}\right] + \sum_{n \to -\infty (n \neq 0)}^{+\infty} \left[A_n'^{+} e^{inpx}\right], \tag{3.4.35a}$$

$$\mu \Bigg\{ \alpha_0 - \beta_0 \sum_{n \to -\infty (n \neq 0)}^{+\infty} \left[inp\zeta_n e^{inpx}\right] + i\alpha_0^2 \sum_{n \to -\infty (n \neq 0)}^{+\infty} \left[\zeta_n e^{inpx}\right] + \left(-\alpha_0 - \beta_0 \sum_{n \to -\infty (n \neq 0)}^{+\infty} \left[inp\zeta_n e^{inpx}\right] + i\alpha_0^2 \sum_{\substack{n \to -\infty \\ (n \neq 0)}}^{+\infty} \left[\zeta_n e^{inpx}\right] \right) A_0^- + \sum_{\substack{n = -\infty \\ (n \neq 0)}}^{+\infty} \left[-\alpha_n A_n^- e^{inpx}\right] \Bigg\}$$

$$= \mu' \Bigg\{ \left(\alpha_0' - \beta_0 \sum_{\substack{n \to -\infty \\ (n \neq 0)}}^{+\infty} \left[inp\zeta_n e^{inpx}\right] + i\alpha_0'^2 \sum_{\substack{n \to -\infty \\ (n \neq 0)}}^{+\infty} \left[\zeta_n e^{inpx}\right] \right) A_0'^{+} + \sum_{\substack{n = -\infty \\ (n \neq 0)}}^{+\infty} \left[\alpha_n' A_n'^{+} e^{inpx}\right] \Bigg\}. \tag{3.4.35b}$$

Extracting the first-order small term containing e^{inpx} from Eqs. (3.4.35a, 3.4.35b) and eliminating the common term e^{inpx}, we can get

$$i\alpha_0\zeta_n - i\alpha_0\zeta_n A_0^- + A_n^- = i\alpha_0'\zeta_n A_0'^{+} + A_n'^{+}, \tag{3.4.36a}$$

$$\mu\left[(-inp\beta_0 + i\alpha_0^2)\zeta_n + (-inp\beta_0 + i\alpha_0^2)\zeta_n A_0^- - \alpha_n A_n^-\right] = \mu'\left[(-inp\beta_0 + i\alpha_0'^2)\zeta_n A_0'^{+} + \alpha_0' A_n'^{+}\right]. \tag{3.4.36b}$$

Arranging A_n^-, $A_n'^+ (n \neq 0)$ in Eqs. (3.4.36a, 3.4.36b) to the left side of the equation, we can get

$$A_n^- - A_n'^+ = -i\alpha_0\zeta_n(1 - A_0^-) + i\alpha'\zeta_n A_0'^+, \tag{3.4.37a}$$

$$\begin{aligned}\mu\alpha_n A_n^- + \mu'\alpha_0' A_n'^+ &= i\mu\zeta_n[-(np\beta_0 - \alpha_0^2)(1 + A_0^-)] \\ &\quad + i\mu'\zeta_n(np\beta_0 - \alpha_0'^2)A_0'^+.\end{aligned} \tag{3.4.37b}$$

By solving Eqs. (3.4.34a, 3.4.34b), the first-order approximate solution of A_0^-, $A_0'^+$ can be written as

$$A_0^- = \frac{\mu\alpha_0 - \mu'\alpha_0'}{\mu\alpha_0 + \mu'\alpha_0'}, \tag{3.4.38a}$$

$$A_0'^+ = \frac{2\mu\alpha_0}{\mu\alpha_0 + \mu'\alpha_0'}. \tag{3.4.38b}$$

Similarly, by solving Eqs. (3.4.37a, 3.4.37b), the first-order approximate solution of A_n^-, $A_n'^+ (n \neq 0)$ can be written as

$$A_n^- = \frac{b_n}{a_n},\ A_n'^+ = \frac{b_n'}{a_n}, \tag{3.4.39}$$

where

$$\begin{aligned}b_n &= -i\zeta_n\{[\mu(np\beta_0 - \alpha_0^2)(1 + A_0^-) - \mu'(np\beta_0 - \alpha_0'^2)A_0'^+] \\ &\quad + \mu'\alpha_n'[\alpha_0(1 - A_0^-) - \alpha_0' A_0'^+]\}, \\ b_n' &= -i\zeta_n\{[\mu(np\beta_0 - \alpha_0^2)(1 + A_0^-) - \mu'(np\beta_0 - \alpha_0'^2)A_0'^+] \\ &\quad - \mu\alpha_n[\alpha_0(1 - A_0^-) - \alpha_0' A_0'^+]\}, \\ a_n &= \mu\alpha_n + \mu'\alpha_n'.\end{aligned}$$

As a special case, consider the periodic fluctuation interface as a cosine interface, expressed as

$$z = c\cos px. \tag{3.4.40}$$

In this case, the value corresponding to ζ_n in formula (3.4.2) can be written as

$$\zeta_1 = \zeta_{-1} = \frac{c}{2}. \tag{3.4.41}$$

Inserting Eq. (3.4.41) into Eq. (3.4.39), the expressions of A_1^-, $A_1'^+$, A_{-1}^-, $A_{-1}'^+$ can be obtained

$$A_1^- = \frac{b_1}{a_1},\quad A_1'^+ = \frac{b_1'}{a_1},\quad A_{-1}^- = \frac{b_{-1}}{a_{-1}},\quad A_{-1}'^+ = \frac{b_{-1}'}{a_{-1}},\tag{3.4.42}$$

where

$$b_1 = -\frac{ic}{2}\{[\mu(p\beta_0 - \alpha_0^2)(1 + A_0^-) - \mu'(p\beta_0 - \alpha_0'^2)A_0'^+] \\ +\mu'\alpha_1'[\alpha_0(1 - A_0^-) - \alpha_0'A_0'^+]\},$$

$$b_1' = -\frac{ic}{2}\{[\mu(p\beta_0 - \alpha_0^2)(1 + A_0^-) - \mu'(p\beta_0 - \alpha_0'^2)A_0'^+] \\ -\mu\alpha_1[\alpha_0(1 - A_0^-) - \alpha_0'A_0'^+]\},$$

$$b_{-1} = \frac{ic}{2}\{[\mu(p\beta_0 + \alpha_0^2)(1 + A_0^-) - \mu'(p\beta_0 + \alpha_0'^2)A_0'^+] \\ -\mu'\alpha_{-1}'[\alpha_0(1 - A_0^-) - \alpha_0'A_0'^+]\},$$

$$b_{-1}' = \frac{ic}{2}\{[\mu(p\beta_0 + \alpha_0^2)(1 + A_0^-) - \mu'(p\beta_0 + \alpha_0'^2)A_0'^+] \\ +\mu\alpha_{-1}[\alpha_0(1 - A_0^-) - \alpha_0'A_0'^+]\},$$

$$a_1 = \mu\alpha_1 + \mu'\alpha_1',\ a_{-1} = \mu\alpha_{-1} + \mu'\alpha_{-1}'.$$

When the incident wave is incident perpendicularly ($\gamma = 0$), we have $\beta_0 = 0$, $\alpha_0 = k$, $\alpha_0' = k'$, $\alpha_1 = \alpha_{-1} = \sqrt{k^2 - p^2}$, $\alpha_1' = \alpha_{-1}' = \sqrt{k'^2 - p^2}$. Then, the expression of $b_{\pm 1}, b_{\pm 1}', a_{\pm 1}$ can be expressed as

$$b_1 = b_{-1} = -\frac{ic}{2}\{[-\mu k^2(1 + A_0^-) + \mu' k'^2] \\ +\mu'\sqrt{k'^2 - p^2}[k(1 - A_0^-) - k'A_0'^+]\},\tag{3.4.43a}$$

$$b_1' = b_{-1}' = -\frac{ic}{2}\{[-\mu k^2(1 + A_0^-) + \mu' k'^2] \\ -\mu\sqrt{k^2 - p^2}[k(1 - A_0^-) - k'A_0'^+]\},\tag{3.4.43b}$$

$$a_1 = a_{-1} = \mu\sqrt{k^2 - p^2} + \mu'\sqrt{k'^2 - p^2},\tag{3.4.43c}$$

therefore

$$A_1^- = A_{-1}^-,\ A_1'^+ = A_{-1}'^+.\tag{3.4.44}$$

So far, the expressions of the reflection and transmission coefficients for the oblique incidence and normal incidence of SH waves on the periodic corrugated interface $z = \zeta(x)$ have been obtained. Here, the numerical calculation is performed only for the case of normal incidence and the cosine interface ($z = c\cos px$). The shear bulk wave velocity ratio of the upper and lower half-spaces is taken as $V_s/V_s' = 3/4$ and the shear modulus ratio $\mu'/\mu = 2.1$. At normal incidence, $A_1^- = A_{-1}^-$, $A_1'^+ = A_{-1}'^+$. Therefore, only the law that the values of $|A_1^-|$ and $|A_1'^+|$ vary with the period Λ of the periodic corrugated interface and the amplitude c are given.

In Fig. 3.24, λ and λ' are the wavelengths of the shear bulk wave in the upper and lower half-space, respectively. From Fig. 3.24, it is seen that when keeping c unchanged, with the increase of $\frac{\Lambda}{\lambda}$, the value of $|A_1^-|$ first increases and reaches the

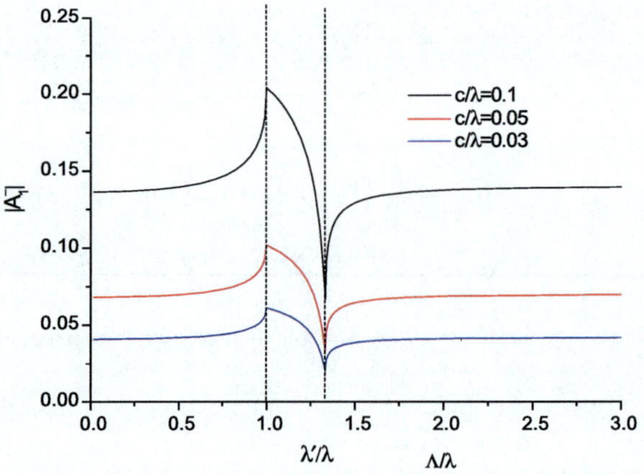

Fig. 3.24 Reflection coefficient $|A_1^-|$ versus the period Λ and amplitude $\boldsymbol{c}$

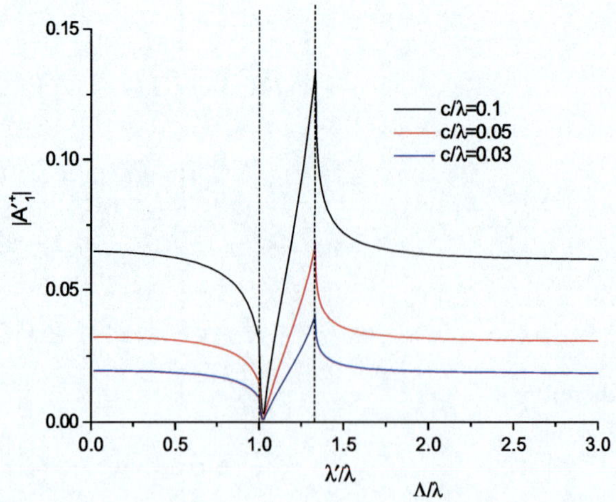

Fig. 3.25 Reflection coefficient $|A_1'^+|$ versus the period Λ and amplitude $\boldsymbol{c}$

Fig. 3.26 Reflection and transmission of P and SV wave at periodic corrugated interfaces

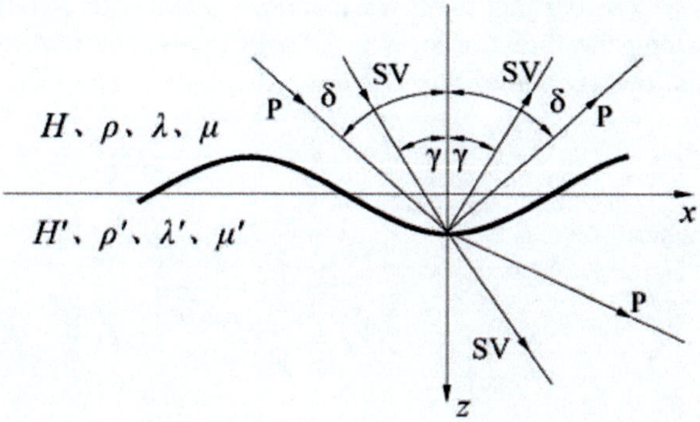

maximum value when $\frac{\Lambda}{\lambda} = 1$, and then begins to decrease and when $\frac{\Lambda}{\lambda}$ is equal to $\frac{\Lambda}{\lambda}$ takes the minimum value, and finally starts to increase and stabilize.

Figure 3.25 plots the transmission coefficient $\left|A_1'^{+}\right|$ against the corrugated period Λ and the amplitude c of the corrugated interface. It can be seen that when keeping c unchanged, with the increase of $\frac{\Lambda}{\lambda}$, the value of $\left|A_1'^{+}\right|$ decreases first and obtains the minimum value when $\frac{\Lambda}{\lambda} = 1$ and then starts to increase. The maximum value is reached when $\frac{\Lambda}{\lambda}$ is equal to $\frac{\lambda'}{\lambda}$, and finally begins to decrease and stabilize. Comparing Figs. 3.24 and 3.25, it can be found that when the reflection coefficient reaches the maximum value, the transmission coefficient takes the minimum value; when the transmission coefficient reaches the maximum value, the reflection coefficient takes the minimum value, which reflects the law of conservation of energy. In addition, it can be seen that when the value of $\frac{\Lambda}{\lambda}$ remains unchanged, the values of the reflection coefficient $\left|A_1^{-}\right|$ and the transmission coefficient $\left|A_1'^{+}\right|$ are all proportional to the amplitude c.

As shown in Fig. 3.26, now discuss the reflection and transmission of P wave and SV wave at the periodic corrugated interface. For isotropic elastic materials, the governing differential equations of plane P wave and SV wave can be written as

$$(\lambda + 2\mu)\nabla^2\phi = \rho\frac{\partial^2\phi}{\partial t^2}, \tag{3.4.45a}$$

$$\mu\nabla^2\psi = \rho\frac{\partial^2\psi}{\partial t^2}, \tag{3.4.45b}$$

where $\nabla^2 = \frac{\partial^2}{\partial x^2} + \frac{\partial^2}{\partial z^2}$, λ and μ are the Lame constant, ρ is the density. The relationship between the potential functions ϕ and ψ and the displacement u in the x-direction and the displacement w in the z-direction is

$$u = \frac{\partial\phi}{\partial x} - \frac{\partial\psi}{\partial z}, \tag{3.4.46a}$$

$$w = \frac{\partial\phi}{\partial z} + \frac{\partial\psi}{\partial x}. \tag{3.4.46b}$$

Consider that the P wave and SV wave with frequency ω are incident obliquely along the direction of δ or γ with the z-axis, and the angles δ and γ satisfy the following relationship

$$k_P \sin\delta = k_S \sin\gamma \equiv \beta_0, \tag{3.4.47}$$

where

$$k_P = \sqrt{\frac{\rho\omega^2}{\lambda + 2\mu}}, \quad k_S = \sqrt{\frac{\rho\omega^2}{\mu}}.$$

Considering the existence of the periodic corrugated interface along the x-direction, applying the Bloch–Floquet principle, the solution of the potential functions ϕ and ψ can be written as

$$\phi = \sum_{n\to-\infty}^{+\infty} A_n e^{ik_{Pnz}z} e^{i\beta_n x} e^{-i\omega t}, \tag{3.4.48a}$$

$$\psi = \sum_{n\to-\infty}^{+\infty} B_n e^{ik_{Snz}z} e^{i\beta_n x} e^{-i\omega t}, \tag{3.4.48b}$$

where A_n and B_n are the displacement amplitudes of the P wave and SV wave of the nth mode in the upper half-space, respectively. $\beta_n = \beta_0 + np$ denotes the wavenumber component of the nth modal in the x-direction; k_{Pnz} and k_{Snz} are the P wave and SV wave wavenumber component of the nth mode in the z-direction, respectively. Substituting Eqs. (3.4.48a, 3.4.48b) into Eqs. (3.4.46a, 3.4.46b), we can get

$$k_{Pnz}^2 + \beta_n^2 = k_P^2, \quad k_{Snz}^2 + \beta_n^2 = k_S^2. \tag{3.4.49}$$

$k_{Pnz} = \pm\sqrt{k_P^2 - \beta_n^2}$ and $k_{Snz} = \pm\sqrt{k_S^2 - \beta_n^2}$ denote a pair of P waves and SV waves propagating along the positive and negative directions of the z-axis, respectively. Let $\alpha_n = \sqrt{k_P^2 - \beta_n^2}$ and $\chi_n = \sqrt{k_S^2 - \beta_n^2}$, the set of potential functions that may exist in the upper half-space are

$$\phi = \sum_{n\to-\infty}^{+\infty} \left[A_n^+ e^{i\alpha_n z} e^{i\beta_n x} + A_n^- e^{-i\alpha_n z} e^{i\beta_n x}\right] e^{-i\omega t}, \tag{3.4.50a}$$

$$\psi = \sum_{n\to-\infty}^{+\infty} \left[B_n^+ e^{i\chi_n z} e^{i\beta_n x} + B_n^- e^{-i\chi_n z} e^{i\beta_n x}\right] e^{-i\omega t}. \tag{3.4.50b}$$

Similar to Eqs. (3.4.50a, 3.4.50b), the set of potential functions that may exist in the lower half-space can be obtained as

$$\phi' = \sum_{n\to-\infty}^{+\infty} \left[A_n'^{+} e^{i\alpha_n' z} e^{i\beta_n' x} + A_n'^{-} e^{-i\alpha_n' z} e^{i\beta_n' x} \right] e^{-i\omega t}, \tag{3.4.51a}$$

$$\psi' = \sum_{n\to-\infty}^{+\infty} \left[B_n'^{+} e^{i\chi_n' z} e^{i\beta_n' x} + B_n'^{-} e^{-i\chi_n' z} e^{i\beta_n' x} \right] e^{-i\omega t}, \tag{3.4.51b}$$

where $\beta_n' = \beta_0' + np; \alpha_n' = \sqrt{k_P'^2 - \beta_n'^2}, \chi_n' = \sqrt{k_S'^2 - \beta_n'^2}, k_P'^2 = \frac{\rho'\omega^2}{(\lambda'+2\mu')}, k_S'^2 = \frac{\rho'\omega^2}{\mu'}$. According to Snell's law, $\beta_0 = \beta_0'$ can be obtained, and further obtain $\beta_n = \beta_n'$, $\alpha_n' = \sqrt{k_P'^2 - \beta_n^2}$, $\chi_n' = \sqrt{k_S'^2 - \beta_n^2}$.

Considering that the P wave with amplitude a and the SV wave with amplitude b are incident obliquely along the direction of the z-axis with an angle of δ or γ. In the upper half-space, only incident waves propagate in the positive direction of the z-axis, namely

$$\begin{aligned} A_n^+ &= 0(n \neq 0), \quad A_0^+ = a, \\ B_n^+ &= 0(n \neq 0), \quad B_0^+ = b. \end{aligned} \tag{3.4.52}$$

All harmonics propagating in the negative direction of the z-axis should be retained. Therefore, the general solution of the ϕ and ψ in the upper half-space is expressed as

$$\phi = a e^{i\alpha_0 z} e^{i\beta_0 x} + \sum_{n\to-\infty}^{+\infty} A_n^- e^{-i\alpha_n z} e^{i\beta_n x}, \tag{3.4.53a}$$

$$\psi = b e^{i\chi_0 z} e^{i\beta_0 x} + \sum_{n\to-\infty}^{+\infty} B_n^- e^{-i\chi_n z} e^{i\beta_n x}. \tag{3.4.53b}$$

Here, we omit the time factor $e^{-i\omega t}$. For the lower half-space, there should be no harmonics propagating in the negative direction of the z-axis, and all harmonics propagating in the positive direction of the z-axis should be retained. The general solution of the displacement in the lower half-space is expressed as (omit the time factor $e^{-i\omega t}$)

$$\phi' = \sum_{n\to-\infty}^{+\infty} A_n'^{+} e^{i\alpha_n' z} e^{i\beta_n x}, \tag{3.4.54a}$$

$$\psi' = \sum_{n\to-\infty}^{+\infty} B_n'^{+} e^{i\chi_n' z} e^{i\beta_n x}. \tag{3.4.54b}$$

It is assumed that the two half-spaces are perfectly bonded at the periodic corrugated interface. Therefore, the displacement and stress remain continuous at the interface $z = \zeta(x)$. The normal vector at any point on the interface $z = \zeta(x)$ can be expressed as

$$n_x = -\frac{\zeta'}{\sqrt{1+\zeta'^2}}, \quad n_y = 0, \quad n_z = \frac{1}{\sqrt{1+\zeta'^2}}. \tag{3.4.55}$$

Take Eq. (3.4.55) into consideration; the expression of the face force component $P_i = \sigma_{ij} n_j$ can be written as

$$P_x = \frac{(-\sigma_{xx}\zeta' + \sigma_{xz})}{\sqrt{1+\zeta'^2}}, \quad P_y = 0, \quad P_z = \frac{(-\sigma_{zx}\zeta' + \sigma_{zz})}{\sqrt{1+\zeta'^2}}. \tag{3.4.56}$$

Thus, the continuity condition of the interface can be expressed as:

$$u = u', \tag{3.4.57a}$$

$$w = w', \tag{3.4.57b}$$

$$-\sigma_{xx}\zeta' + \sigma_{xz} = -\sigma'_{xx}\zeta' + \sigma'_{xz}, \tag{3.4.57c}$$

$$-\sigma_{zx}\zeta' + \sigma_{zz} = -\sigma'_{zx}\zeta' + \sigma'_{zz}. \tag{3.4.57d}$$

The constitutive equations of an isotropic elastic material are

$$\sigma_{xx} = (\lambda + 2\mu)\frac{\partial u}{\partial x} + \lambda\frac{\partial w}{\partial z}, \tag{3.4.58a}$$

$$\sigma_{zz} = \lambda\frac{\partial u}{\partial x} + (\lambda + 2\mu)\frac{\partial w}{\partial z}, \tag{3.4.58b}$$

$$\sigma_{xz} = \mu\left(\frac{\partial u}{\partial z} + \frac{\partial w}{\partial x}\right). \tag{3.4.58c}$$

Combing Eqs. (3.4.46a, 3.4.46b) with Eqs. (3.4.58a, 3.4.58b, 3.4.58c) and using the boundary conditions, i.e., Eqs. (3.4.57a, 3.4.57b, 3.4.57c, 3.4.57d), the interface conditions expressed by the ϕ and ψ can be obtained, such as

$$\frac{\partial \phi}{\partial x} - \frac{\partial \psi}{\partial z} = \frac{\partial \phi'}{\partial x} - \frac{\partial \psi'}{\partial z}, \tag{3.4.59a}$$

$$\frac{\partial \phi}{\partial z} + \frac{\partial \psi}{\partial x} = \frac{\partial \phi'}{\partial z} + \frac{\partial \psi'}{\partial x}, \tag{3.4.59b}$$

$$-\zeta'\left[(\lambda+2\mu)\left(\frac{\partial^2\phi}{\partial x^2}-\frac{\partial^2\psi}{\partial x\partial z}\right)+\lambda\left(\frac{\partial^2\phi}{\partial z^2}+\frac{\partial^2\psi}{\partial x\partial z}\right)\right]$$
$$+\mu\left(\frac{\partial^2\psi}{\partial x^2}-\frac{\partial^2\psi}{\partial z^2}+2\frac{\partial^2\phi}{\partial x\partial z}\right)$$
$$=-\zeta'\left[(\lambda'+2\mu')\left(\frac{\partial^2\phi'}{\partial x^2}-\frac{\partial^2\psi'}{\partial x\partial z}\right)+\lambda'\left(\frac{\partial^2\phi'}{\partial z^2}+\frac{\partial^2\psi'}{\partial x\partial z}\right)\right]$$
$$+\mu'\left(\frac{\partial^2\psi'}{\partial x^2}-\frac{\partial^2\psi'}{\partial z^2}+2\frac{\partial^2\phi'}{\partial x\partial z}\right), \tag{3.4.59c}$$

$$-\zeta'\mu\left(\frac{\partial^2\psi}{\partial x^2}-\frac{\partial^2\psi}{\partial z^2}+2\frac{\partial^2\phi}{\partial x\partial z}\right)+(\lambda+2\mu)\left(\frac{\partial^2\phi}{\partial z^2}+\frac{\partial^2\psi}{\partial x\partial z}\right)+\lambda\left(\frac{\partial^2\phi}{\partial x^2}-\frac{\partial^2\psi}{\partial x\partial z}\right)$$
$$=-\zeta'\mu'\left(\frac{\partial^2\psi'}{\partial x^2}-\frac{\partial^2\psi'}{\partial z^2}+2\frac{\partial^2\phi'}{\partial x\partial z}\right)+(\lambda'+2\mu')\left(\frac{\partial^2\phi'}{\partial z^2}+\frac{\partial^2\psi'}{\partial x\partial z}\right)$$
$$+\lambda'\left(\frac{\partial^2\phi'}{\partial x^2}-\frac{\partial^2\psi'}{\partial x\partial z}\right). \tag{3.4.59d}$$

Substituting Eqs. (3.4.53a, 3.4.53b) and (3.4.54a, 3.4.54b) into the above equation and eliminating the common term $e^{i\beta_0 x}$, we can get

$$i\left(a\beta_0 e^{i\alpha_0\zeta}+\sum_{n\to-\infty}^{+\infty}\left(\beta_n A_n^- e^{-i\alpha_n\zeta}e^{inpx}\right)-b\chi_0 e^{i\chi_0\zeta}+\sum_{n\to-\infty}^{+\infty}\left(\chi_n B_n^- e^{-i\chi_n\zeta}e^{inpx}\right)\right)$$
$$=i\left(\sum_{n\to-\infty}^{+\infty}\left(\beta_n A_n'^+ e^{i\alpha_n'\zeta}e^{inpx}\right)-\sum_{n\to-\infty}^{+\infty}\left(\chi_n' B_n'^+ e^{i\chi_n'\zeta}e^{inpx}\right)\right), \tag{3.4.60a}$$

$$i\left(a\alpha_0 e^{i\alpha_0\zeta}-\sum_{n\to-\infty}^{+\infty}\alpha_n A_n^- e^{-i\alpha_n\zeta}e^{inpx}+b\beta_0 e^{i\chi_0\zeta}+\sum_{n\to-\infty}^{+\infty}\beta_n B_n^- e^{-i\chi_n\zeta}e^{inpx}\right)$$
$$=i\left(\sum_{n\to-\infty}^{+\infty}\alpha_n' A_n'^+ e^{i\alpha_n'\zeta}e^{inpx}+\sum_{n\to-\infty}^{+\infty}\beta_n B_n'^+ e^{i\chi_n'\zeta}e^{inpx}\right), \tag{3.4.60b}$$

$$-\zeta'\left\{(\lambda+2\mu)\left[-\left(a\beta_0^2 e^{i\alpha_0\zeta}+\sum_{n\to-\infty}^{+\infty}\beta_n^2 A_n^- e^{-i\alpha_n\zeta}e^{inpx}\right)\right.\right.$$
$$\left.+\left(b\chi_0\beta_0 e^{i\chi_0\zeta}+\sum_{n\to-\infty}^{+\infty}-\chi_n\beta_n B_n^- e^{-i\chi_n\zeta}e^{inpx}\right)\right]$$
$$+\lambda\left[-\left(a\alpha_0^2 e^{i\alpha_0\zeta}+\sum_{n\to-\infty}^{+\infty}\alpha_n^2 A_n^- e^{-i\alpha_n\zeta}e^{inpx}\right)\right.$$
$$\left.\left.-\left(b\chi_0\beta_0 e^{i\chi_0\zeta}+\sum_{n\to-\infty}^{+\infty}-\chi_n\beta_n B_n^- e^{-i\chi_n\zeta}e^{inpx}\right)\right]\right\}$$

$$
\begin{aligned}
&+\mu\Bigg[-\Bigg(b\beta_0^2 e^{i\chi_0\zeta} + \sum_{n\to-\infty}^{+\infty} \beta_n^2 B_n^- e^{-i\chi_n\zeta} e^{inpx}\Bigg) \\
&+\Bigg(b\chi_0^2 e^{i\chi_0\zeta} + \sum_{n\to-\infty}^{+\infty} \chi_n^2 B_n^- e^{-i\chi_n\zeta} e^{inpx}\Bigg) \\
&-2\Bigg(a\alpha_0\beta_0 e^{i\alpha_0\zeta} - \sum_{n\to-\infty}^{+\infty} \alpha_n\beta_n A_n^- e^{-i\alpha_n\zeta} e^{inpx}\Bigg)\Bigg] \\
&= -\zeta'\Bigg\{(\lambda'+2\mu')\Bigg[\sum_{n\to-\infty}^{+\infty}\Big(-\beta_n^2 A_n'^+ e^{i\alpha_n'\zeta} e^{inpx}\Big) \\
&-\sum_{n\to-\infty}^{+\infty}\Big(-\beta_n\chi_n' B_n'^+ e^{i\chi_n'\zeta} e^{inpx}\Big)\Bigg] \\
&+\lambda'\Bigg(\sum_{n\to-\infty}^{+\infty}\Big(-\alpha_n'^2 A_n'^+ e^{i\alpha_n'\zeta} e^{inpx}\Big) + \sum_{n\to-\infty}^{+\infty}\Big(-\beta_n\chi_n' B_n'^+ e^{i\chi_n'\zeta} e^{inpx}\Big)\Bigg)\Bigg\} \\
&+\mu'\Bigg[\sum_{n\to-\infty}^{+\infty}\Big(-\beta_n^2 B_n'^+ e^{i\chi_n'\zeta} e^{inpx}\Big) - \sum_{n\to-\infty}^{+\infty}\Big(-\chi_n'^2 B_n'^+ e^{i\chi_n'\zeta} e^{inpx}\Big) \\
&+2\sum_{n\to-\infty}^{+\infty}\Big(-\alpha_n'\beta_n A_n'^+ e^{i\alpha_n'\zeta} e^{inpx}\Big)\Bigg],
\end{aligned}
\tag{3.4.60c}
$$

$$
\begin{aligned}
&-\zeta'\mu\Bigg[-\Bigg(b\beta_0^2 e^{i\chi_0\zeta} + \sum_{n\to-\infty}^{+\infty} \beta_n^2 B_n^- e^{-i\chi_n\zeta} e^{inpx}\Bigg) \\
&+\Bigg(b\chi_0^2 e^{i\chi_0\zeta} + \sum_{n\to-\infty}^{+\infty} \chi_n^2 B_n^- e^{-i\chi_n\zeta} e^{inpx}\Bigg) \\
&-2\Bigg(a\alpha_0\beta_0 e^{i\alpha_0\zeta} - \sum_{n\to-\infty}^{+\infty} \alpha_n\beta_n A_n^- e^{-i\alpha_n\zeta} e^{inpx}\Bigg)\Bigg] \\
&+(\lambda+2\mu)\Bigg[-\Bigg(a\alpha_0^2 e^{i\alpha_0\zeta} + \sum_{n\to-\infty}^{+\infty} \alpha_n^2 A_n^- e^{-i\alpha_n\zeta} e^{inpx}\Bigg) \\
&-\Bigg(b\chi_0\beta_0 e^{i\chi_0\zeta} + \sum_{n\to-\infty}^{+\infty} -\chi_n\beta_n B_n^- e^{-i\chi_n\zeta} e^{inpx}\Bigg)\Bigg] \\
&+\lambda\Bigg[-\Bigg(a\beta_0^2 e^{i\alpha_0\zeta} + \sum_{n\to-\infty}^{+\infty} \beta_n^2 A_n^- e^{-i\alpha_n\zeta} e^{inpx}\Bigg) \\
&+\Bigg(b\chi_0\beta_0 e^{i\chi_0\zeta} + \sum_{n\to-\infty}^{+\infty} -\chi_n\beta_n B_n^- e^{-i\chi_n\zeta} e^{inpx}\Bigg)\Bigg]
\end{aligned}
$$

$$= -\zeta'\mu'\left[\sum_{n\to-\infty}^{+\infty}\left(-\beta_n^2 B_n'^{+}\mathrm{e}^{i\chi_n'\zeta}\mathrm{e}^{inpx}\right) - \sum_{n\to-\infty}^{+\infty}\left(-\chi_n'^2 B_n'^{+}\mathrm{e}^{i\chi_n'\zeta}\mathrm{e}^{inpx}\right)\right.$$
$$\left.+2\sum_{n\to-\infty}^{+\infty}\left(-\alpha_n'\beta_n A_n'^{+}\mathrm{e}^{i\alpha_n'\zeta}\mathrm{e}^{inpx}\right)\right]$$
$$+(\lambda'+2\mu')\left[\sum_{n\to-\infty}^{+\infty}\left(-\alpha_n'^2 A_n'^{+}\mathrm{e}^{i\alpha_n'\zeta}\mathrm{e}^{inpx}\right) + \sum_{n\to-\infty}^{+\infty}\left(-\beta_n\chi_n' B_n'^{+}\mathrm{e}^{i\chi_n'\zeta}\mathrm{e}^{inpx}\right)\right]$$
$$+\lambda'\left[\sum_{n\to-\infty}^{+\infty}\left(-\beta_n^2 A_n'^{+}\mathrm{e}^{i\alpha_n'\zeta}\mathrm{e}^{inpx}\right) - \sum_{n\to-\infty}^{+\infty}\left(-\beta_n\chi_n' B_n'^{+}\mathrm{e}^{i\chi_n'\zeta}\mathrm{e}^{inpx}\right)\right]. \quad (3.4.60\text{d})$$

Assuming that the periodic corrugation $z = \zeta(x)$ of the interface is very small, perform Taylor series expansion on $\mathrm{e}^{-i\alpha_n\zeta}$ at $z = 0$, and only retain the first-order term of ζ, namely

$$\mathrm{e}^{-i\alpha_n\zeta} \approx 1 - \mathrm{i}\alpha_n\zeta. \quad (3.4.61)$$

Substituting Eq. (3.4.61) into Eqs. (3.4.60a, 3.4.60b, 3.4.60c, 3.4.60d) and retaining only the first-order term of ζ, we get

$$a\beta_0(1+i\alpha_0\zeta) + \beta_0 A_0^-(1-i\alpha_0\zeta) + \sum_{n\to-\infty(n\neq0)}^{+\infty}\left(\beta_n A_n^-\mathrm{e}^{inpx}\right) - b\chi_0(1+i\chi_0\zeta)$$
$$+\chi_0 B_0^-(1-i\chi_0\zeta) + \sum_{n\to-\infty(n\neq0)}^{+\infty}\left(\chi_n B_n^-\mathrm{e}^{inpx}\right)$$
$$= \beta_0 A_0'^{+}(1+i\alpha_0'\zeta) + \sum_{n\to-\infty(n\neq0)}^{+\infty}\left(\beta_n A_n'^{+}\mathrm{e}^{inpx}\right)$$
$$-\left[\chi_0' B_0'^{+}(1+i\chi_0'\zeta) + \sum_{n\to-\infty(n\neq0)}^{+\infty}\left(\chi_n' B_n'^{+}\mathrm{e}^{inpx}\right)\right], \quad (3.4.62\text{a})$$

$$a\alpha_0(1+i\alpha_0\zeta) - \alpha_0 A_0^-(1-i\alpha_0\zeta)$$
$$-\sum_{n\to-\infty(n\neq0)}^{+\infty}\left(\alpha_n A_n^-\mathrm{e}^{inpx}\right) + b\beta_0(1+i\chi_0\zeta)$$
$$+\beta_0 B_0^-(1-i\chi_0\zeta) + \sum_{n\to-\infty(n\neq0)}^{+\infty}\left(\beta_n B_n^-\mathrm{e}^{inpx}\right)$$

$$= \alpha_0' A_0'^{+}(1 + i\alpha_0'\zeta) + \sum_{\substack{n \to -\infty \\ (n \neq 0)}}^{+\infty} (\alpha_n' A_n'^{+} \mathrm{e}^{inpx}) + \beta_0 B_0'^{+}(1 + i\chi_0'\zeta)$$

$$+ \sum_{\substack{n \to -\infty \\ (n \neq 0)}}^{+\infty} (\beta_n B_n'^{+} \mathrm{e}^{inpx}), \tag{3.4.62b}$$

$$- \zeta' \begin{bmatrix} (\lambda + 2\mu)(-(a\beta_0^2 + \beta_0^2 A_0^-) + (b\chi_0\beta_0 - \chi_0\beta_0 B_0^-)) \\ +\lambda(-(a\alpha_0^2 + \alpha_0^2 A_0^-) - (b\chi_0\beta_0 - \chi_0\beta_0 B_0^-)) \end{bmatrix}$$

$$- \mu \left(b\beta_0^2(1 + i\chi_0\zeta) + \beta_0^2 B_0^-(1 - i\chi_0\zeta) + \sum_{n \to -\infty (n \neq 0)}^{+\infty} \beta_n^2 B_n^- \mathrm{e}^{inpx} \right)$$

$$+ \mu \left(b\chi_0^2(1 + i\chi_0\zeta) + \chi_0^2 B_0^-(1 - i\chi_0\zeta) + \sum_{n \to -\infty (n \neq 0)}^{+\infty} \chi_n^2 B_n^- \mathrm{e}^{inpx} \right)$$

$$- 2\mu \left(a\alpha_0\beta_0(1 + i\alpha_0\zeta) - \alpha_0\beta_0 A_0^-(1 - i\alpha_0\zeta) - \sum_{n \to -\infty (n \neq 0)}^{+\infty} \alpha_n\beta_n A_n^- \mathrm{e}^{inpx} \right)$$

$$= -\zeta'[(\lambda' + 2\mu')(-\beta_0^2 A_0'^{+} + \beta_0\chi_0' B_0'^{+}) + \lambda'(-\alpha_0'^2 A_0'^{+} - \beta_0\chi_0' B_0'^{+})]$$

$$- \mu'\beta_0^2 B_0'^{+}(1 + i\chi_0'\zeta) + \mu' \sum_{\substack{n \to -\infty \\ (n \neq 0)}}^{+\infty} (-\beta_n^2 B_n'^{+} \mathrm{e}^{inpx}) + \mu'\chi_0'^2 B_0'^{+}(1 + i\chi_0'\zeta)$$

$$+ \mu' \sum_{\substack{n \to -\infty \\ (n \neq 0)}}^{+\infty} (\chi_n'^2 B_n'^{+} \mathrm{e}^{inpx})$$

$$- 2\mu' \left(\alpha_0'\beta_0 A_0'^{+}(1 + i\alpha_0'\zeta) + \sum_{\substack{n \to -\infty \\ (n \neq 0)}}^{+\infty} (\alpha_n'\beta_n A_n'^{+} \mathrm{e}^{inpx}) \right), \tag{3.4.62c}$$

$$- \zeta'\mu\left(-\left(b\beta_0^2 + \beta_0^2 B_0^-\right) + \left(b\chi_0^2 + \chi_0^2 B_0^-\right) - 2\left(a\alpha_0\beta_0 - \alpha_0\beta_0 A_0^-\right)\right)$$

$$+ (\lambda + 2\mu) \begin{pmatrix} -\left(a\alpha_0^2(1 + i\alpha_0\zeta) + \alpha_0^2 A_0^-(1 - i\alpha_0\zeta) + \sum_{n \to -\infty (n \neq 0)}^{+\infty} \alpha_n^2 A_n^- \mathrm{e}^{inpx}\right) \\ -\left(b\chi_0\beta_0(1 + i\chi_0\zeta) - \chi_0\beta_0 B_0^-(1 - i\chi_0\zeta) - \sum_{n \to -\infty (n \neq 0)}^{+\infty} \chi_n\beta_n B_n^- \mathrm{e}^{inpx}\right) \end{pmatrix}$$

$$
\begin{aligned}
&+\lambda\left(\begin{array}{l}
-\left(a\beta_0^2(1+i\alpha_0\zeta)+\beta_0^2A_0^-(1-i\alpha_0\zeta)+\sum\limits_{n\to-\infty(n\neq0)}^{+\infty}\beta_n^2A_n^-\mathrm{e}^{inpx}\right)\\
+\left(b\chi_0\beta_0(1+i\chi_0\zeta)-\chi_0\beta_0B_0^-(1-i\chi_0\zeta)-\sum\limits_{n\to-\infty(n\neq0)}^{+\infty}\chi_n\beta_nB_n^-\mathrm{e}^{inpx}\right)
\end{array}\right)\\
&=-\zeta'\mu'\left(-\beta_0^2B_0'^{+}+\chi_0'^2B_0'^{+}-2\alpha_0'\beta_0A_0'^{+}\right)\\
&+(\lambda'+2\mu')\left(\begin{array}{l}
-\alpha_0'^2A_0'^{+}(1+i\alpha_0'\zeta)+\sum\limits_{n\to-\infty(n\neq0)}^{+\infty}\left(-\alpha_n'^2A_n'^{+}\mathrm{e}^{inpx}\right)\\
-\beta_0\chi_0'B_0'^{+}(1+i\chi_0'\zeta)+\sum\limits_{n\to-\infty(n\neq0)}^{+\infty}\left(-\beta_n\chi_n'B_n'^{+}\mathrm{e}^{inpx}\right)
\end{array}\right)\\
&+\lambda'\left(\begin{array}{l}
-\beta_0^2A_0'^{+}(1+i\alpha_0'\zeta)+\sum\limits_{n\to-\infty(n\neq0)}^{+\infty}\left(-\beta_n^2A_n'^{+}\mathrm{e}^{inpx}\right)\\
+\beta_0\chi_0'B_0'^{+}(1+i\chi_0'\zeta)+\sum\limits_{n\to-\infty(n\neq0)}^{+\infty}\left(\beta_n\chi_n'B_n'^{+}\mathrm{e}^{inpx}\right)
\end{array}\right).
\end{aligned}
\tag{3.4.62d}
$$

Extracting terms that do not contain x and ζ from Eqs. (3.4.62a, 3.4.62b, 3.4.62c, 3.4.62d), the relational expressions about A_0^-, $A_0'^{+}$, B_0^-, $B_0'^{+}$ can be obtained as

$$
a\beta_0+\beta_0A_0^- - b\chi_0+\chi_0B_0^- = \beta_0A_0'^{+}-\chi_0'B_0'^{+}, \tag{3.4.63a}
$$

$$
a\alpha_0-\alpha_0A_0^- + b\beta_0+\beta_0B_0^- = \alpha_0'A_0'^{+}+\beta_0B_0'^{+}, \tag{3.4.63b}
$$

$$
\begin{aligned}
&\mu\left(-\left(b\beta_0^2+\beta_0^2B_0^-\right)+\left(b\chi_0^2+\chi_0^2B_0^-\right)-2\left(a\alpha_0\beta_0-\alpha_0\beta_0A_0^-\right)\right)\\
&\quad=\mu'\left(-\beta_0^2B_0'^{+}+\chi_0'^2B_0'^{+}-2\left(\alpha_0'\beta_0A_0'^{+}\right)\right),
\end{aligned}
\tag{3.4.63c}
$$

$$
\begin{aligned}
&(\lambda+2\mu)\left(-a\alpha_0^2-\alpha_0^2A_0^- - b\chi_0\beta_0+\chi_0\beta_0B_0^-\right)\\
&\quad+\lambda\left(-a\beta_0^2-\beta_0^2A_0^- + b\chi_0\beta_0-\chi_0\beta_0B_0^-\right)\\
&\quad=(\lambda'+2\mu')\left(-\alpha_0'^2A_0'^{+}-\beta_0\chi_0'B_0'^{+}\right)+\lambda'\left(-\beta_0^2A_0'^{+}+\beta_0\chi_0'B_0'^{+}\right).
\end{aligned}
\tag{3.4.63d}
$$

In order to find the first-order approximation of A_n^-, $A_n'^{+}$, B_n^-, $B_n'^{+}(n\neq0)$, substitute Eq. (3.4.2) into Eqs. (3.4.62a, 3.4.62b, 3.4.62c, 3.4.62d) and extract the first-order small items containing e^{inpx} and eliminate the common item e^{inpx}, we obtain

$$
\begin{aligned}
&i\beta_0\alpha_0\zeta_n\left(a-A_0^-\right)+\beta_nA_n^- - ib\chi_0^2\zeta_n-i\chi_0^2B_0^-\zeta_n+\chi_nB_n^-\\
&\quad=i\alpha'\beta_0A_0'^{+}\zeta_n+\beta_nA_n'^{+}-\left(i\chi_0'^2B_0'^{+}\zeta_n+\chi_n'B_n'^{+}\right),
\end{aligned}
\tag{3.4.64a}
$$

$$
\begin{aligned}
&-i\alpha_0^2\zeta_n\left(a+A_0^-\right)+\alpha_nA_n^- + i\chi_0\beta_0\zeta_n\left(B_0^- - b\right)-\beta_nB_n^-\\
&\quad=-i\alpha_0'^2A_0'^{+}-\alpha_n'A_n'^{+}-i\chi_0'\beta_0B_0'^{+}\zeta_n-\beta_nB_n'^{+},
\end{aligned}
\tag{3.4.64b}
$$

$$
-inp\zeta_n\left[\begin{array}{c}
(\lambda+2\mu)\left(-\left(a\beta_0^2+\beta_0^2A_0^-\right)+\left(b\chi_0\beta_0-\chi_0\beta_0B_0^-\right)\right)\\
+\lambda\left(-\left(a\alpha_0^2+\alpha_0^2A_0^-\right)-\left(b\chi_0\beta_0-\chi_0\beta_0B_0^-\right)\right)
\end{array}\right]
$$

$$
\begin{aligned}
&+\mu\begin{pmatrix} -(-i\chi_0\beta_0^2\zeta_n(B_0^- - b) + \beta_n^2 B_n^-) + (-i\chi_0^3\zeta_n(B_0^- - b) + \chi_n^2 B_n^-) \\ -2(i\alpha_0^2\beta_0\zeta_n(A_0^- + a) - \alpha_n\beta_n A_n^-) \end{pmatrix} \\
&= -inp\zeta_n\left[(\lambda' + 2\mu')(-\beta_0^2 A_0'^+ + \beta_0\chi_0' B_0'^+) + \lambda'(-\alpha_0'^2 A_0'^+ - \beta_0\chi_0' B_0'^+)\right] \\
&+ \mu'(-i\beta_0^2 B_0'^+\chi_0'\zeta_n - \beta_n^2 B_n'^+ + i\chi_0'^3 B_0'^+\zeta_n + \chi_n'^2 B_n'^+ \\
&-2(i\alpha_0'^2\beta_0\zeta_n A_0'^+ + \alpha_n'\beta_n A_n'^+)),
\end{aligned} \tag{3.4.64c}
$$

$$
\begin{aligned}
&- inp\zeta_n\mu(-(b\beta_0^2 + \beta_0^2 B_0^-) + (b\chi_0^2 + \chi_0^2 B_0^-) - 2(a\alpha_0\beta_0 - \alpha_0\beta_0 A_0^-)) \\
&+ (\lambda + 2\mu)(-(-i\alpha_0^3\zeta_n(A_0^- - a) + \alpha_n^2 A_n^-) + (-i\chi_0^2\beta_0\zeta_n(\mathrm{b} + B_0^-) + \chi_n\beta_n B_n^-)) \\
&+ \lambda(-(-i\alpha_0\beta_0^2\zeta_n(A_0^- - a) + \beta_n^2 A_n^-) - (-i\chi_0^2\beta_0\zeta_n(b + B_0^-) + \chi_n\beta_n B_n^-)) \\
&= -inp\zeta_n\mu'(-\beta_0^2 B_0'^+ + \chi_0'^2 B_0'^+ - 2\alpha_0'\beta_0 A_0'^+) \\
&+ (\lambda' + 2\mu')(-i\alpha_0'^3\zeta_n A_0'^+ - \alpha_n'^2 A_n'^+ - i\beta_0\chi_0'^2\zeta_n B_0'^+ - \beta_n\chi_n' B_n'^+) \\
&+ \lambda'(-i\alpha_0'\beta_0^2\zeta_n A_0'^+ - \beta_n^2 A_n'^+ + i\beta_0\chi_0'^2\zeta_n B_0'^+ + \beta_n\chi_n' B_n'^+).
\end{aligned} \tag{3.4.64d}
$$

Arranging the terms containing A_0^-, $A_0'^+$, B_0^-, $B_0'^+$ in Eqs. (3.4.63a, 3.4.63b, 3.4.63c, 3.4.63d) to the left side of the equation, we have

$$
\beta_0 A_0^- - \beta_0 A_0'^+ + \chi_0 B_0^- + \chi_0' B_0'^+ = -a\beta_0 + b\chi_0, \tag{3.4.65a}
$$

$$
\alpha_0 A_0^- + \alpha_0' A_0'^+ - \beta_0 B_0^- + \beta_0 B_0'^+ = a\alpha_0 + b\beta_0, \tag{3.4.65b}
$$

$$
\begin{aligned}
&2\mu\alpha_0\beta_0 A_0^- + 2\mu'\alpha_0'\beta_0 A_0'^+ + \mu(\chi_0^2 - \beta_0^2)B_0^- - \mu'(\chi_0'^2 - \beta_0^2)B_0'^+ \\
&= 2\mu\alpha_0\beta_0 a - \mu(\chi_0^2 - \beta_0^2)b,
\end{aligned} \tag{3.4.65c}
$$

$$
\begin{aligned}
&-\left[(\lambda + 2\mu)\alpha_0^2 + \lambda\beta_0^2\right]A_0^- + \left[(\lambda' + 2\mu')\alpha_0'^2 + \lambda'\beta_0^2\right]A_0'^+ \\
&+ [(\lambda + 2\mu) - \lambda]\chi_0\beta_0 B_0^- + \left[(\lambda' + 2\mu') - \lambda\right]\beta_0\chi_0' B_0'^+ \\
&= \left[(\lambda + 2\mu)\alpha_0^2 + \lambda\beta_0^2\right]a + [(\lambda + 2\mu) - \lambda]\chi_0\beta_0 b.
\end{aligned} \tag{3.4.65d}
$$

Similarly, arranging the terms containing $A_n^-, A_n'^+, B_n^-, B_n'^+ (n \neq 0)$ in Eqs. (3.4.64a, 3.4.64b, 3.4.64c, 3.4.64d) to the left side of the equation, we have

$$
\begin{aligned}
&\beta_n A_n^- - \beta_n A_n'^+ + \chi_n B_n^- + \chi_n' B_n'^+ = -i\beta_0\alpha_0\zeta_n(a - A_0^-) \\
&+ i\alpha'\beta_0 A_0'^+\zeta_n + i\chi_0^2\zeta_n(B_0^- + b) - i\chi_0'^2 B_0'^+\zeta_n,
\end{aligned} \tag{3.4.66a}
$$

$$
\begin{aligned}
&\alpha_n A_n^- + \alpha_n' A_n'^+ - \beta_n B_n^- + \beta_n B_n'^+ = i\alpha_0^2\zeta_n(a + A_0^-) \\
&- i\alpha_0'^2 A_0'^+ - i\chi_0\beta_0\zeta_n(B_0^- - b) - i\chi_0'\beta_0 B_0'^+\zeta_n,
\end{aligned} \tag{3.4.66b}
$$

$$
2\mu\alpha_n\beta_n A_n^- + 2\mu'\alpha_n'\beta_n A_n'^+ + (\chi_n^2 - \beta_n^2)\mu B_n^- + (\beta_n^2 - \chi_n'^2)\mu' B_n'^+
$$

$$
= -inp\zeta_n \begin{bmatrix} (\lambda+2\mu)\left(a\beta_0^2+\beta_0^2 A_0^- - b\chi_0\beta_0 + \chi_0\beta_0 B_0^-\right) \\ +\lambda\left(a\alpha_0^2+\alpha_0^2 A_0^- + b\chi_0\beta_0 - \chi_0\beta_0 B_0^-\right) \\ +(\lambda'+2\mu')\left(-\beta_0^2 A_0'^+ + \beta_0\chi_0' B_0'^+\right) + \lambda'\left(-\alpha_0'^2 A_0'^+ - \beta_0\chi_0' B_0'^+\right) \end{bmatrix}
$$

$$
- i\zeta_n \begin{bmatrix} \mu\left(\chi_0\beta_0^2(B_0^- - b) - \chi_0^3(B_0^- - b) - 2\alpha_0^2\beta_0(A_0^- + a)\right) \\ +\mu'\left(\beta_0^2 B_0'^+\chi_0' - \chi_0'^3 B_0'^+ - 2(-\alpha_0'^2\beta_0 A_0'^+)\right) \end{bmatrix}, \tag{3.4.66c}
$$

$$
\begin{aligned}
& -\left[(\lambda+2\mu)\alpha_n^2+\lambda\beta_n^2\right]A_n^- + \left[(\lambda'+2\mu')\alpha_n'^2+\lambda'\beta_n^2\right]A_n'^+ \\
& + 2\mu\chi_n\beta_n B_n^- + 2\mu'\beta_n\chi_n' B_n'^+
\end{aligned}
$$

$$
= -inp\zeta_n \begin{bmatrix} \mu\left((b\beta_0^2+\beta_0^2 B_0^-) - (b\chi_0^2+\chi_0^2 B_0^-) + 2(a\alpha_0\beta_0 - \alpha_0\beta_0 A_0^-)\right) \\ +\mu'\left(-\beta_0^2 B_0'^+ + \chi_0'^2 B_0'^+ - 2\alpha_0'\beta_0 A_0'^+\right) \end{bmatrix}
$$

$$
- i\zeta_n \begin{bmatrix} (\lambda+2\mu)\left(\alpha_0^3(A_0^- - a) - \chi_0^2\beta_0(b+B_0^-)\right) \\ +\lambda\left(\alpha_0\beta_0^2(A_0^- - a) + \chi_0^2\beta_0(b+B_0^-)\right) \\ +(\lambda'+2\mu')\left(\alpha_0'^3 A_0'^+ + \beta_0\chi_0'^2 B_0'^+\right) \\ +\lambda'\left(\alpha_0'\beta_0^2 A_0'^+ - \beta_0\chi_0'^2 B_0'^+\right) \end{bmatrix}. \tag{3.4.66d}
$$

In matrix form, Eqs. (3.4.65a, 3.4.65b, 3.4.65c, 3.4.65d) can be written as

$$
\begin{bmatrix} \beta_0 & -\beta_0 & \chi_0 & \chi_0' \\ \alpha_0 & \alpha_0' & -\beta_0 & \beta_0 \\ G_1 & G_1' & G_2 & -G_2' \\ -L_1 & L_1' & L_2 & L_2' \end{bmatrix} \begin{pmatrix} A_0^- \\ A_0'^+ \\ B_0^- \\ B_0'^+ \end{pmatrix} = \begin{pmatrix} -\beta_0 \\ \alpha_0 \\ G_1 \\ L_1 \end{pmatrix} a + \begin{pmatrix} \chi_0 \\ \beta_0 \\ -G_2 \\ L_2 \end{pmatrix} b \tag{3.4.67}
$$

where

$G_1 = 2\mu\alpha_0\beta_0$, $G_1' = 2\mu'\alpha_0'\beta_0$, $G_2 = \mu(\chi_0^2 - \beta_0^2)$, $G_2' = \mu'(\chi_0'^2 - \beta_0^2)$.

$L_1 = (\lambda+2\mu)\alpha_0^2 + \lambda\beta_0^2$, $L_1' = (\lambda'+2\mu')\alpha_0'^2 + \lambda'\beta_0^2$.

$L_2 = 2\mu\chi_0\beta_0$, $L_2' = 2\mu'\beta_0\chi_0'$.

The first-order approximate solution of A_0^-, $A_0'^+$, B_0^-, $B_0'^+$ can be obtained by Eq. (3.4.67).

Similarly, Eqs. (3.4.66a, 3.4.66b, 3.4.66c, 3.4.66d) can be written in matrix form as

$$
\begin{bmatrix} \beta_n & -\beta_n & \chi_n & \chi_n' \\ \alpha_n & \alpha_n' & -\beta_n & \beta_n \\ G_{1n} & G_{1n}' & G_{2n} & -G_{2n}' \\ -L_{1n} & L_{1n}' & L_{2n} & L_{2n}' \end{bmatrix} \begin{pmatrix} A_n^- \\ A_n'^+ \\ B_n^- \\ B_n'^+ \end{pmatrix} = \begin{pmatrix} d_1 \\ d_2 \\ d_3 \\ d_4 \end{pmatrix}. \tag{3.4.68}
$$

where

$G_{1n} = 2\mu\alpha_n\beta_n$, $G_{1n}' = 2\mu'\alpha_n'\beta_n$, $G_{2n} = \mu(\chi_n^2 - \beta_n^2)$, $G_{2n}' = \mu'(\chi_n'^2 - \beta_n^2)$,

$L_{1n} = (\lambda+2\mu)\alpha_n^2 + \lambda\beta_n^2$, $L_{1n}' = (\lambda'+2\mu')\alpha_n'^2 + \lambda'\beta_n^2$,

$L_{2n} = 2\mu\chi_n\beta_n$, $L_2' = 2\mu'\beta_n\chi_n'$,

$$d_1 = -i\beta_0\alpha_0\zeta_n(a - A_0^-) + i\alpha'\beta_0 A_0'^+\zeta_n + i\chi_0^2\zeta_n(B_0^- + b) - i\chi_0'^2 B_0'^+\zeta_n,$$

$$d_2 = i\alpha_0^2\zeta_n(\mathrm{a} + A_0^-) - i\alpha_0'^2 A_0'^+ - i\chi_0\beta_0\zeta_n(B_0^- - b) - i\chi_0'\beta_0 B_0'^+\zeta_n,$$

$$d_3 = -inp\zeta_n\begin{bmatrix}(\lambda + 2\mu)(a\beta_0^2 + \beta_0^2 A_0^- - b\chi_0\beta_0 + \chi_0\beta_0 B_0^-)\\ +\lambda(a\alpha_0^2 + \alpha_0^2 A_0^- + b\chi_0\beta_0 - \chi_0\beta_0 B_0^-)\\ +(\lambda' + 2\mu')(-\beta_0^2 A_0'^+ + \beta_0\chi_0' B_0'^+) + \lambda'(-\alpha_0'^2 A_0'^+ - \beta_0\chi_0' B_0'^+)\end{bmatrix}$$
$$- i\zeta_n\begin{bmatrix}\mu(\chi_0\beta_0^2(B_0^- - b) - \chi_0^3(B_0^- - b) - 2\alpha_0^2\beta_0(A_0^- + a))\\ +\mu'(\beta_0^2 B_0'^+\chi_0' - \chi_0'^3 B_0'^+ - 2(-\alpha_0'^2\beta_0 A_0'^+))\end{bmatrix},$$

$$d_4 = -inp\zeta_n\begin{bmatrix}\mu((b\beta_0^2 + \beta_0^2 B_0^-) - (b\chi_0^2 + \chi_0^2 B_0^-) + 2(a\alpha_0\beta_0 - \alpha_0\beta_0 A_0^-))\\ +\mu'(-\beta_0^2 B_0'^+ + \chi_0'^2 B_0'^+ - 2\alpha_0'\beta_0 A_0'^+)\end{bmatrix}$$
$$- i\zeta_n\begin{bmatrix}(\lambda + 2\mu)(\alpha_0^3(A_0^- - a) - \chi_0^2\beta_0(b + B_0^-))\\ +\lambda(\alpha_0\beta_0^2(A_0^- - a) + \chi_0^2\beta_0(b + B_0^-))\\ +(\lambda' + 2\mu')(\alpha_0'^3 A_0'^+ + \beta_0\chi_0'^2 B_0'^+)\\ +\lambda'(\alpha_0'\beta_0^2 A_0'^+ - \beta_0\chi_0'^2 B_0'^+)\end{bmatrix}.$$

The first-order approximate solution of A_n^-, $A_n'^+$, B_n^-, $B_n'^+$ can be obtained by Eq. (3.4.68).

Chapter 4
Reflection and Transmission of Elastic Waves in Multilayer Media

We have studied the reflection and transmission of elastic waves through the interface of two media in Chap. 3. In this chapter, we will further study the reflection and transmission of elastic waves through single-layer medium and multilayer medium. One of the actual backgrounds of this problem is the study of seismic wave intensity when the seismic wave propagates from the source to the ground through different types of strata [7]. We also need to study the reflection and transmission coefficients of elastic waves through multilayered media to solve the problem of the vibration reduction and noise reduction of construction machinery and structures, and the acoustic structure design of sound-absorbing materials. Since the reflection and transmission of elastic waves through a single-layer medium are the basis for further research on the reflection and transmission of elastic waves through multilayer media, this chapter first introduces the five methods to treat with the reflection and transmission of elastic waves through a single-layer medium and further discusses how to use these methods to study the reflection and transmission of elastic waves through multilayer media.

4.1 Simultaneous Interface Conditions Method

In this section, we will study the reflection and transmission of elastic waves through sandwiched slab. The slab with finite thickness h is in the middle of two semi-infinite media. λ_1, μ_1, ρ_1 and λ_3, μ_3, ρ_3 are the material parameters of two semi-infinite media. The incident wave from the medium 1 cause the reflected waves in the medium 1 and the transmitted waves in the medium 3, as shown in Fig. 4.1.

Firstly, consider the reflection and transmission of the incident SH wave through the interlayer of finite thickness. Suppose the displacement fields of incident SH wave, reflected SH wave, and transmitted SH wave (downward wave in the interlayer) at interface 2 ($x_3 = 0$) are:

© Science Press 2022

P. Wei, *Theory of Elastic Waves*,
https://doi.org/10.1007/978-981-19-5662-1_4

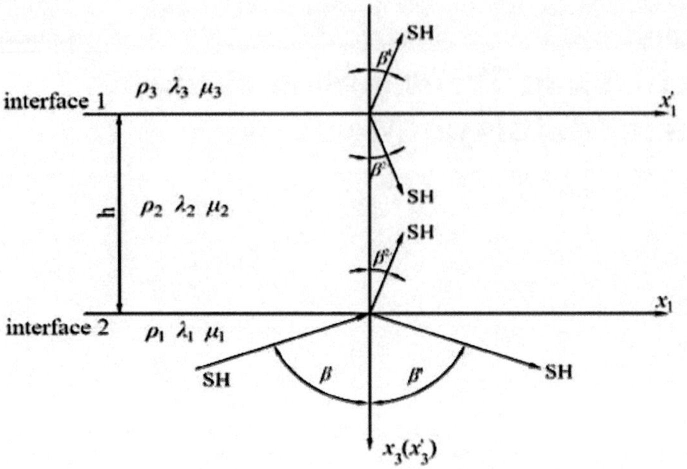

Fig. 4.1 Reflection and transmission of SH wave through the sandwiched slab

$$u_2^i = C^i \exp\left[i\left(\xi x_1 - k_{s3}^{(1)} x_3 - \omega_s t\right)\right], \tag{4.1.1a}$$

$$u_2^r = C^r \exp\left[i\left(\xi x_1 + k_{s3}^{(1)} x_3 - \omega_s t\right)\right], \tag{4.1.1b}$$

$$u_2^{2-} = C^{2-} \exp\left[i\left(\xi x_1 - k_{s3}^{(2)} x_3 - \omega_s t\right)\right]. \tag{4.1.1c}$$

We also suppose the displacement fields of the reflected SH wave which is the downward wave in the interlayer and the transmitted SH wave at the interface 1 ($x_3 = -h$) are

$$u_2^{2+} = C^{2+} \exp\left\{i\left[\xi x_1 + k_{s3}^{(2)} (x_3 + h) - \omega_s t\right]\right\}, \tag{4.1.1d}$$

$$u_2^t = C^t \exp\left\{i\left[\xi x_1 - k_{s3}^{(3)} (x_3 + h) - \omega_s t\right]\right\}. \tag{4.1.1e}$$

Equation (4.1.1) already implies that the reflected wave, the transmitted wave and the incident wave have the same frequency and apparent wave number (Snell's law), where

$$\begin{aligned}
&\xi = \frac{\omega_s}{c_s} \sin\beta, k_{s3}^{(1)} = \frac{\omega_s}{c_s} \cos\beta, k_{s3}^{(1)} = \frac{\omega_s}{c_s} \cos\beta^r, \\
&k_{s3}^{(2)} = \frac{\omega_s}{c_s^{(2)}} \cos\beta^{2-}, k_{s3}^{(2)} = \frac{\omega_s}{c_s^{(2)}} \cos\beta^{2+}, k_{s3}^{(3)} = \frac{\omega_s}{c_s^{(3)}} \cos\beta^t, \\
&\beta^r = \beta, \beta^{2-} = \beta^{2+}.
\end{aligned}$$

The continuity conditions on both sides of the interface 2 ($x_3 = 0$) and interface 1 ($x_3 = -h$) are

$$u_2^+\big|_{x_3=0} = u_2^-\big|_{x_3=0},\ \sigma_{32}^+\big|_{x_3=0} = \sigma_{32}^-\big|_{x_3=0}. \tag{4.1.2a,b}$$

$$u_2^+\big|_{x_3=-h} = u_2^-\big|_{x_3=-h},\ \sigma_{32}^+\big|_{x_3=-h} = \sigma_{32}^-\big|_{x_3=-h}. \tag{4.1.2c,d}$$

They also can be written as

$$\begin{aligned} &C^i \exp[i(\xi x_1 - \omega_s t)] + C^r \exp[i(\xi x_1 - \omega_s t)] \\ &\quad = C^{2-} \exp[i(\xi x_1 - \omega_s t)] + C^{2+} \exp\left[i\left(\xi x_1 + k_{s3}^{(2)} h - \omega_s t\right)\right], \end{aligned} \tag{4.1.3a}$$

$$\begin{aligned} &\mu_1\left\{-ik_{s3}^{(1)} C^i \exp[i(\xi x_1 - \omega_s t)] + ik_{s3}^{(1)} C^r \exp[i(\xi x_1 - \omega_s t)]\right\} \\ &\quad = \mu_2\left\{-ik_{s3}^{(2)} C^{2-} \exp[i(\xi x_1 - \omega_s t)] + ik_{s3}^{(2)} C^{2+} \exp\left[i\left(\xi x_1 + k_{s3}^{(2)} h - \omega_s t\right)\right]\right\}, \end{aligned} \tag{4.1.3b}$$

$$C^{2-} \exp\left[i\left(\xi x_1 + k_{s3}^{(2)} h - \omega_s t\right)\right] + C^{2+} \exp[i(\xi x_1 - \omega_s t)] = C^t \exp[i(\xi x_1 - \omega_s t)], \tag{4.1.3c}$$

$$\begin{aligned} &\mu_2\left\{-ik_{s3}^{(2)} C^{2-} \exp\left[i\left(\xi x_1 + k_{s3}^{(2)} h - \omega_s t\right)\right] + ik_{s3}^{(2)} C^{2+} \exp[i(\xi x_1 - \omega_s t)]\right\} \\ &\quad = \mu_3\left\{-ik_{s3}^{(3)} C^t \exp[i(\xi x_1 - \omega_s t)]\right\}. \end{aligned} \tag{4.1.3d}$$

After the simplification of Eq. (4.1.3), we get

$$C^i + C^r = C^{2-} + C^{2+} e^{ik_{s3}^{(2)} h}, \tag{4.1.4a}$$

$$\mu_1\left(k_{s3}^{(1)} C^i - k_{s3}^{(1)} C^r\right) = \mu_2\left(k_{s3}^{(2)} C^{2-} - k_{s3}^{(2)} C^{2+} e^{ik_{s3}^{(2)} h}\right), \tag{4.1.4b}$$

$$C^{2-} e^{ik_{s3}^{(2)} h} + C^{2+} = C^t, \tag{4.1.4c}$$

$$\mu_2\left(k_{s3}^{(2)} C^{2-} e^{ik_{s3}^{(2)} h} - k_{s3}^{(2)} C^{2+}\right) = \mu_3 k_{s3}^{(3)} C^t. \tag{4.1.4d}$$

Equation (4.1.4) can be written in matrix form as

$$\begin{pmatrix} -1 & 1 & e^{ik_{s3}^{(2)}h} & 0 \\ 1 & \frac{\mu_2 k_{s3}^{(2)}}{\mu_1 k_{s3}^{(1)}} & -\frac{\mu_2 k_{s3}^{(2)}}{\mu_1 k_{s3}^{(1)}} e^{ik_{s3}^{(2)}h} & 0 \\ 0 & e^{ik_{s3}^{(2)}h} & 1 & -1 \\ 0 & \frac{\mu_2 k_{s3}^{(2)}}{\mu_3 k_{s3}^{(3)}} e^{ik_{s3}^{(2)}h} & -\frac{\mu_2 k_{s3}^{(2)}}{\mu_3 k_{s3}^{(3)}} & -1 \end{pmatrix} \begin{Bmatrix} C^r \\ C^{2-} \\ C^{2+} \\ C^t \end{Bmatrix} = C^i \begin{Bmatrix} 1 \\ 1 \\ 0 \\ 0 \end{Bmatrix}. \tag{4.1.5}$$

When the incident wave is P wave or SV wave, the reflected P wave and reflected SV wave will be generated in medium 1, transmitted P wave and transmitted SV wave will be generated in medium 3 and P waves and SV waves along the positive and negative directions of the interface normal will be generated in medium 2, respectively, as shown in Fig. 4.2 [13].

We suppose the wave functions of P waves and SV waves in medium 1, medium 2 and medium 3 are, respectively,

$$\varphi^i = A^i \exp\Big[i\Big(\xi x_1 - k_{p3}^{(1)} x_3 - \omega t\Big)\Big], \tag{4.1.6a}$$

$$\psi^i = B^i \exp\Big[i\Big(\xi x_1 - k_{s3}^{(1)} x_3 - \omega t\Big)\Big], \tag{4.1.6b}$$

$$\varphi^r = A^r \exp\Big[i\Big(\xi x_1 + k_{p3}^{(1)} x_3 - \omega t\Big)\Big], \tag{4.1.6c}$$

$$\psi^r = B^r \exp\Big[i\Big(\xi x_1 + k_{s3}^{(1)} x_3 - \omega t\Big)\Big], \tag{4.1.6d}$$

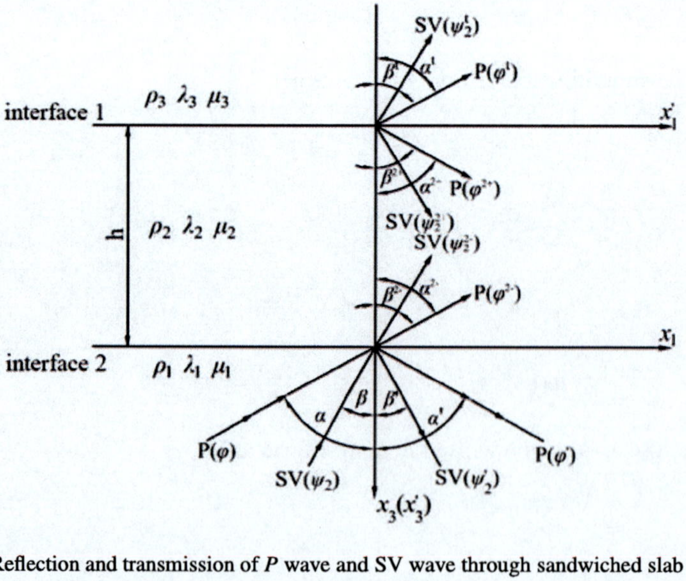

Fig. 4.2 Reflection and transmission of P wave and SV wave through sandwiched slab

$$\varphi^{2-} = A^{2-} \exp\left[i\left(\xi x_1 - k_{p3}^{(2)} x_3 - \omega t\right)\right], \tag{4.1.6e}$$

$$\psi^{2-} = B^{2-} \exp\left[i\left(\xi x_1 - k_{s3}^{(2)} x_3 - \omega t\right)\right], \tag{4.1.6f}$$

$$\varphi^{2+} = A^{2+} \exp\left[i\left(\xi x_1 + k_{p3}^{(2)} (x_3 + h) - \omega t\right)\right], \tag{4.1.6g}$$

$$\psi^{2+} = B^{2+} \exp\left[i\left(\xi x_1 + k_{s3}^{(2)} (x_3 + h) - \omega t\right)\right], \tag{4.1.6h}$$

$$\varphi^{t} = A^{t} \exp\left[i\left(\xi x_1 - k_{p3}^{(3)} (x_3 + h) - \omega t\right)\right], \tag{4.1.6i}$$

$$\psi^{t} = B^{t} \exp\left[i\left(\xi x_1 - k_{s3}^{(3)} (x_3 + h) - \omega t\right)\right]. \tag{4.1.6j}$$

The displacements and stresses in medium 1 can be written as

$$\begin{aligned} u_1 &= \frac{\partial(\varphi^i + \varphi^r)}{\partial x_1} - \frac{\partial(\psi^i + \psi^r)}{\partial x_3} \\ &= i\xi A^i \exp\left[i(\xi x_1 - k_{p3}^{(1)} x_3 - \omega t)\right] \\ &+ i\xi A^r \exp\left[i(\xi x_1 + k_{p3}^{(1)} x_3 - \omega t)\right] \\ &+ ik_{s3}^{(1)} B^i \exp\left[i(\xi x_1 - k_{s3}^{(1)} x_3 - \omega t)\right] \\ &- ik_{s3}^{(1)} B^r \exp\left[i(\xi x_1 + k_{s3}^{(1)} x_3 - \omega t)\right], \end{aligned} \tag{4.1.7a}$$

$$\begin{aligned} u_3 &= \frac{\partial(\varphi^i + \varphi^r)}{\partial x_3} + \frac{\partial(\psi^i + \psi^r)}{\partial x_1} \\ &= -ik_{p3}^{(1)} A^i \exp\left[i(\xi x_1 - k_{p3}^{(1)} x_3 - \omega t)\right] + ik_{p3}^{(1)} A^r \exp\left[i(\xi x_1 + k_{p3}^{(1)} x_3 - \omega t)\right] \\ &+ i\xi B^i \exp\left[i(\xi x_1 - k_{s3}^{(1)} x_3 - \omega t)\right] + i\xi B^r \exp\left[i(\xi x_1 + k_{s3}^{(1)} x_3 - \omega t)\right], \end{aligned} \tag{4.1.7b}$$

$$\begin{aligned} \sigma_{33} &= \lambda_1 \frac{\partial^2(\varphi^i + \varphi^r)}{\partial x_1^2} + (\lambda_1 + 2\mu_1) \frac{\partial^2(\varphi^i + \varphi^r)}{\partial x_3^2} + 2\mu_1 \frac{\partial^2(\psi^i + \psi^r)}{\partial x_1 \partial x_3} \\ &= \lambda_1 \left\{-\xi^2 A^i \exp\left[i(\xi x_1 - k_{p3}^{(1)} x_3 - \omega t)\right] - \xi^2 A^r \exp\left[i(\xi x_1 + k_{p3}^{(1)} x_3 - \omega t)\right]\right\} \\ &+ (\lambda_1 + 2\mu_1)\left\{-(k_{p3}^{(1)})^2 A^i \exp\left[i(\xi x_1 - k_{p3}^{(1)} x_3 - \omega t)\right]\right. \\ &\left.-(k_{p3}^{(1)})^2 A^r \exp\left[i(\xi x_1 + k_{p3}^{(1)} x_3 - \omega t)\right]\right\} \\ &+ 2\mu_1 \left\{\xi k_{s3}^{(1)} B^i \exp\left[i(\xi x_1 - k_{s3}^{(1)} x_3 - \omega t)\right]\right. \end{aligned}$$

$$-\xi k_{s3}^{(1)} B^r \exp\Big[i(\xi x_1 + k_{s3}^{(1)} x_3 - \omega t)\Big]\Big\}, \tag{4.1.7c}$$

$$\begin{aligned}
\sigma_{31} &= \mu_1 \left[2\frac{\partial^2(\varphi^i + \varphi^r)}{\partial x_1 \partial x_3} + \frac{\partial^2(\psi^i + \psi^r)}{\partial x_1^2} - \frac{\partial^2(\psi^i + \psi^r)}{\partial x_3^2} \right] \\
&= \mu_1 \Big\{ 2\xi k_{p3}^{(1)} A^i \exp\Big[i(\xi x_1 - k_{p3}^{(1)} x_3 - \omega t)\Big] - 2\xi k_{p3}^{(1)} A^r \exp\Big[i(\xi x_1 + k_{p3}^{(1)} x_3 - \omega t)\Big] \\
&\quad - \xi^2 B^i \exp\Big[i(\xi x_1 - k_{s3}^{(1)} x_3 - \omega t)\Big] - \xi^2 B^r \exp\Big[i(\xi x_1 + k_{s3}^{(1)} x_3 - \omega t)\Big] \\
&\quad + (k_{s3}^{(1)})^2 B^i \exp\Big[i(\xi x_1 - k_{s3}^{(1)} x_3 - \omega t)\Big] + (k_{s3}^{(1)})^2 B^r \exp\Big[i(\xi x_1 + k_{s3}^{(1)} x_3 - \omega t)\Big]\Big\}.
\end{aligned} \tag{4.1.7d}$$

The displacements and stresses in medium 2 can be written as

$$\begin{aligned}
u_1' &= \frac{\partial(\varphi^{2-} + \varphi^{2+})}{\partial x_1} - \frac{\partial(\psi^{2-} + \psi^{2+})}{\partial x_3} \\
&= i\xi A^{2-} \exp\Big[i(\xi x_1 - k_{p3}^{(2)} x_3 - \omega t)\Big] + i\xi A^{2+} \exp\Big[i(\xi x_1 + k_{p3}^{(2)} (x_3 + h) - \omega t)\Big] \\
&\quad + i k_{s3}^{(2)} B^{2-} \exp\Big[i(\xi x_1 - k_{s3}^{(2)} x_3 - \omega t)\Big] - i k_{s3}^{(2)} B^{2+} \exp\Big[i(\xi x_1 + k_{s3}^{(2)} (x_3 + h) - \omega t)\Big],
\end{aligned} \tag{4.1.8a}$$

$$\begin{aligned}
u_3' &= \frac{\partial(\varphi^{2-} + \varphi^{2+})}{\partial x_3} + \frac{\partial(\psi^{2-} + \psi^{2+})}{\partial x_1} \\
&= -i k_{p3}^{(2)} A^{2-} \exp\Big[i(\xi x_1 - k_{p3}^{(2)} x_3 - \omega t)\Big] + i k_{p3}^{(2)} A^{2+} \exp\Big[i(\xi x_1 + k_{p3}^{(2)} (x_3 + h) - \omega t)\Big] \\
&\quad + i\xi B^{2-} \exp\Big[i(\xi x_1 - k_{s3}^{(2)} x_3 - \omega t)\Big] + i\xi B^{2+} \exp\Big[i(\xi x_1 + k_{s3}^{(2)} (x_3 + h) - \omega t)\Big],
\end{aligned} \tag{4.1.8b}$$

$$\begin{aligned}
\sigma_{33}' &= \lambda_2 \frac{\partial^2(\varphi^{2-} + \varphi^{2+})}{\partial x_1^2} + (\lambda_2 + 2\mu_2) \frac{\partial^2(\varphi^{2-} + \varphi^{2+})}{\partial x_3^2} + 2\mu_2 \frac{\partial^2(\psi^{2-} + \psi^{2+})}{\partial x_1 \partial x_3} \\
&= \lambda_2 \Big\{ -\xi^2 A^{2-} \exp\Big[i(\xi x_1 - k_{p3}^{(2)} x_3 - \omega t)\Big] - \xi^2 A^{2+} \\
&\quad \exp\Big[i(\xi x_1 + k_{p3}^{(2)} (x_3 + h) - \omega t)\Big]\Big\} \\
&\quad + (\lambda_2 + 2\mu_2) \Big\{ -(k_{p3}^{(2)})^2 A^{2-} \exp\Big[i(\xi x_1 - k_{p3}^{(2)} x_3 - \omega t)\Big] - (k_{p3}^{(2)})^2 A^{2+} \\
&\quad \exp\Big[i(\xi x_1 + k_{p3}^{(2)} (x_3 + h) - \omega t)\Big]\Big\} \\
&\quad + 2\mu_2 \Big\{ \xi k_{s3}^{(2)} B^{2-} \exp\Big[i(\xi x_1 - k_{s3}^{(2)} x_3 - \omega t)\Big] \\
&\quad - \xi k_{s3}^{(2)} B^{2+} \exp\Big[i(\xi x_1 + k_{s3}^{(2)} (x_3 + h) - \omega t)\Big]\Big\},
\end{aligned} \tag{4.1.8c}$$

$$\begin{aligned}
\sigma_{31}' &= \mu_2 \left[2\frac{\partial^2(\varphi^{2-} + \varphi^{2+})}{\partial x_1 \partial x_3} + \frac{\partial^2(\psi^{2-} + \psi^{2+})}{\partial x_1^2} - \frac{\partial^2(\psi^{2-} + \psi^{2+})}{\partial x_3^2} \right] \\
&= \mu_2 \Big\{ 2\xi k_{p3}^{(2)} A^{2-} \exp\Big[i(\xi x_1 - k_{p3}^{(2)} x_3 - \omega t)\Big]
\end{aligned}$$

$$- 2\xi k_{p3}^{(2)} A^{2+} \exp\Big[i(\xi x_1 + k_{p3}^{(2)}(x_3 + h) - \omega t)\Big]$$
$$- \xi^2 B^{2-} \exp\Big[i(\xi x_1 - k_{s3}^{(2)} x_3 - \omega t)\Big]$$
$$- \xi^2 B^{2+} \exp\Big[i(\xi x_1 + k_{s3}^{(2)}(x_3 + h) - \omega t)\Big]$$
$$+ (k_{s3}^{(2)})^2 B^{2-} \exp\Big[i(\xi x_1 - k_{s3}^{(2)} x_3 - \omega t)\Big]$$
$$+ (k_{s3}^{(2)})^2 B^{2+} \exp\Big[i(\xi x_1 + k_{s3}^{(2)}(x_3 + h) - \omega t)\Big]\Big\}. \tag{4.1.8d}$$

The displacements and stresses in medium 3 can be written as

$$\begin{aligned} u_1'' &= \frac{\partial \varphi^t}{\partial x_1} - \frac{\partial \psi_2^t}{\partial x_3} \\ &= i\xi A^t \exp\Big[i(\xi x_1 - k_{p3}^{(3)}(x_3 + h) - \omega t)\Big] \\ &\quad + ik_{s3}^{(3)} B^t \exp\Big[i(\xi x_1 - k_{s3}^{(3)}(x_3 + h) - \omega t)\Big], \end{aligned} \tag{4.1.9a}$$

$$\begin{aligned} u_3'' &= \frac{\partial \varphi^t}{\partial x_3} + \frac{\partial \psi_2^t}{\partial x_1} \\ &= -ik_{p3}^{(3)} A^t \exp\Big[i(\xi x_1 - k_{p3}^{(3)}(x_3 + h) - \omega t)\Big] \\ &\quad + i\xi B^t \exp\Big[i(\xi x_1 - k_{s3}^{(3)}(x_3 + h) - \omega t)\Big], \end{aligned} \tag{4.1.9b}$$

$$\begin{aligned} \sigma_{33}'' &= \lambda_3 \frac{\partial^2 \varphi^t}{\partial x_1^2} + (\lambda_3 + 2\mu_3)\frac{\partial^2 \varphi^t}{\partial x_3^2} + 2\mu_3 \frac{\partial^2 \psi^t}{\partial x_1 \partial x_3} \\ &= -\lambda_3 \xi^2 A^t \exp\Big[i(\xi x_1 - k_{p3}^{(3)}(x_3 + h) - \omega t)\Big] \\ &\quad - (\lambda_3 + 2\mu_3)(k_{p3}^{(3)})^2 A^t \exp\Big[i(\xi x_1 - k_{p3}^{(3)}(x_3 + h) - \omega t)\Big] \\ &\quad + 2\mu_3 \xi k_{s3}^{(3)} B^t \exp\Big[i(\xi x_1 - k_{s3}^{(3)}(x_3 + h) - \omega t)\Big], \end{aligned} \tag{4.1.9c}$$

$$\begin{aligned} \sigma_{31}'' &= \mu_3 \left(2\frac{\partial^2 \varphi^t}{\partial x_1 \partial x_3} + \frac{\partial^2 \psi^t}{\partial x_1^2} - \frac{\partial^2 \psi^t}{\partial x_3^2}\right) \\ &= \mu_3 \Big\{2\xi k_{p3}^{(3)} A^t \exp\Big[i(\xi x_1 - k_{p3}^{(3)}(x_3 + h) - \omega t)\Big] \\ &\quad - \xi^2 B^t \exp\Big[i(\xi x_1 - k_{s3}^{(3)}(x_3 + h) - \omega t)\Big] \\ &\quad + (k_{s3}^{(3)})^2 B^t \exp\Big[i(\xi x_1 - k_{s3}^{(3)}(x_3 + h) - \omega t)\Big]\Big\}. \end{aligned} \tag{4.1.9d}$$

The continuity conditions of the displacement and stress at interface 1 and interface 2 can be written as

$$u_1|_{x_3=0} = u_1'\big|_{x_3=0}, \quad u_3|_{x_3=0} = u_3'\big|_{x_3=0}, \tag{4.1.10a}$$

$$\sigma_{33}|_{x_3=0} = \sigma_{33}'\big|_{x_3=0}, \quad \sigma_{31}|_{x_3=0} = \sigma_{31}'\big|_{x_3=0}, \tag{4.1.10b}$$

$$u_1'\big|_{x_3=-h} = u_1''\big|_{x_3=-h}, \quad u_3'\big|_{x_3=-h} = u_3''\big|_{x_3=-h}, \tag{4.1.11a}$$

$$\sigma_{33}'\big|_{x_3=-h} = \sigma_{33}''\big|_{x_3=-h}, \quad \sigma_{31}'\big|_{x_3=-h} = \sigma_{31}''\big|_{x_3=-h}. \tag{4.1.11b}$$

At interface 2, it can be seen from $u_1|_{x_3=0} = u_1'\big|_{x_3=0}$ that

$$i\xi A^i + i\xi A^r + ik_{s3}^{(1)} B^i - ik_{s3}^{(1)} B^r = i\xi A^{2-} + i\xi A^{2+} e^{ik_{p3}^{(2)} h} + ik_{s3}^{(2)} B^{2-} \\ -ik_{s3}^{(2)} B^{2+} e^{ik_{s3}^{(2)} h}.$$

After simplification, we get

$$-A^r + P_2 B^r + A^{2-} + P_4 B^{2-} + e^{ik_{p3}^{(2)} h} A^{2+} - P_4 e^{ik_{s3}^{(2)} h} B^{2+} = A^i + P_2 B^i. \tag{4.1.12a}$$

It can be seen from $u_3|_{x_3=0} = u_3'\big|_{x_3=0}$ that

$$-ik_{p3}^{(1)} A^i + ik_{p3}^{(1)} A^r + i\xi B^i + i\xi B^r = -ik_{p3}^{(2)} A^{2-} \\ + ik_{p3}^{(2)} A^{2+} e^{ik_{p3}^{(2)} h} + i\xi B^{2-} + i\xi B^{2+} e^{ik_{s3}^{(2)} h}.$$

After simplification, we get

$$-P_1 A^r - B^r - P_3 A^{2-} + B^{2-} + P_3 e^{ik_{p3}^{(2)} h} A^{2+} + e^{ik_{s3}^{(2)} h} B^{2+} = -P_1 A^i + B^i. \tag{4.1.12b}$$

It can be seen from $\sigma_{33}|_{x_3=0} = \sigma_{33}'\big|_{x_3=0}$ that

$$\lambda_1\left(-\xi^2 A^i - \xi^2 A^r\right) + (\lambda_1 + 2\mu_1)\left[-(k_{p3}^{(1)})^2 A^i - (k_{p3}^{(1)})^2 A^r\right] + 2\mu_1\left(\xi k_{s3}^{(1)} B^i - \xi k_{s3}^{(1)} B^r\right) \\ = \lambda_2\left(-\xi^2 A^{2-} - \xi^2 A^{2+} e^{ik_{p3}^{(2)} h}\right) + (\lambda_2 + 2\mu_2)\left[-(k_{p3}^{(2)})^2 A^{2-} - (k_{p3}^{(2)})^2 A^{2+} e^{ik_{p3}^{(2)} h}\right] \\ + 2\mu_2\left(\xi k_{s3}^{(2)} B^{2-} - \xi k_{s3}^{(2)} B^{2+} e^{ik_{s3}^{(2)} h}\right).$$

After simplification, we get

$$\left[\lambda_1 + (\lambda_1 + 2\mu_1) P_1^2\right] A^r + 2\mu_1 P_2 B^r \\ - \left[\lambda_2 + (\lambda_2 + 2\mu_2) P_3^2\right] A^{2-} + 2\mu_2 P_4 B^{2-}$$

$$- \left[\lambda_2 + (\lambda_2 + 2\mu_2) P_3^2\right] e^{ik_{p3}^{(2)}h} A^{2+} - 2\mu_2 P_4 e^{ik_{s3}^{(2)}h} B^{2+}$$
$$= -\left[\lambda_1 + (\lambda_1 + 2\mu_1) P_1^2\right] A^i + 2\mu_1 P_2 B^i. \tag{4.1.12c}$$

It can be seen from $\sigma_{31}|_{x_3=0} = \sigma'_{31}\big|_{x_3=0}$ that

$$\mu_1\left[2\xi k_{p3}^{(1)} A^i - 2\xi k_{p3}^{(1)} A^r - \xi^2 B^i - \xi^2 B^r + (k_{s3}^{(1)})^2 B^i + (k_{s3}^{(1)})^2 B^r\right]$$
$$= \mu_2 (2\xi k_{p3}^{(2)} A^{2-} - 2\xi k_{p3}^{(2)} A^{2+} e^{ik_{p3}^{(2)}h} - \xi^2 B^{2-} - \xi^2 B^{2+} e^{ik_{s3}^{(2)}h}$$
$$+ (k_{s3}^{(2)})^2 B^{2-} + (k_{s3}^{(2)})^2 B^{2+} e^{ik_{s3}^{(2)}h}).$$

After simplification, we get

$$2\mu_1 P_1 A^r - \mu_1 (P_2^2 - 1) B^r + 2\mu_2 P_3 A^{2-} + \mu_2 (P_4^2 - 1) B^{2-} - 2\mu_2 P_3 e^{ik_{p3}^{(2)}h} A^{2+}$$
$$+ \mu_2 (P_4^2 - 1) e^{ik_{s3}^{(2)}h} B^{2+} = 2\mu_1 P_1 A^i + \mu_1 (P_2^2 - 1) B^i. \tag{4.1.12d}$$

At interface 1, it can be seen from $u'_1\big|_{x_3=-h} = u''_1\big|_{x_3=-h}$ that

$$i\xi e^{ik_{p3}^{(2)}h} A^{2-} + i\xi A^{2+} + ik_{s3}^{(2)} B^{2-} e^{ik_{s3}^{(2)}h} - ik_{s3}^{(2)} B^{2+} = i\xi A^t + ik_{s3}^{(3)} B^t.$$

After simplification, we get

$$e^{ik_{p3}^{(2)}h} A^{2-} + P_4 e^{ik_{s3}^{(2)}h} B^{2-} + A^{2+} - P_4 B^{2+} - A^t - P_6 B^t = 0. \tag{4.1.13a}$$

It can be seen from $u'_3\big|_{x_3=-h} = u''_3\big|_{x_3=-h}$ that

$$-ik_{p3}^{(2)} A^{2-} e^{ik_{p3}^{(2)}h} + ik_{p3}^{(2)} A^{2+} + i\xi B^{2-} e^{ik_{s3}^{(2)}h} + i\xi B^{2+} = -ik_{p3}^{(3)} A^t + i\xi B^t.$$

After simplification, we get

$$-P_3 e^{ik_{p3}^{(2)}h} A^{2-} + e^{ik_{s3}^{(2)}h} B^{2-} + P_3 A^{2+} + B^{2+} + P_5 A^t - B^t = 0. \tag{4.1.13b}$$

It can be seen from $\sigma'_{33}\big|_{x_3=-h} = \sigma''_{33}\big|_{x_3=-h}$ that

$$\lambda_2[-\xi^2 A^{2-} e^{ik_{p3}^{(2)}h} - \xi^2 A^{2+}] + (\lambda_2 + 2\mu_2)[-(k_{p3}^{(2)})^2 A^{2-} e^{ik_{p3}^{(2)}h} - (k_{p3}^{(2)})^2 A^{2+}]$$
$$+2\mu_2(\xi k_{s3}^{(2)} B^{2-} e^{ik_{s3}^{(2)}h} - \xi k_{s3}^{(2)} B^{2+}) = -\lambda_3 \xi^2 A^t - (\lambda_3 + 2\mu_3)(k_{p3}^{(3)})^2 A^t + 2\mu_3 \xi k_{s3}^{(3)} B^t.$$

After simplification, we get

$$- \left[\lambda_2 + (\lambda_2 + 2\mu_2) P_3^2\right] e^{ik_{p3}^{(2)}h} A^{2-} + 2\mu_2 P_4 e^{ik_{s3}^{(2)}h} B^{2-}$$

$$
\begin{aligned}
&-\left[\lambda_2+(\lambda_2+2\mu_2)P_3^2\right]A^{2+}-2\mu_2P_4B^{2+}\\
&+\left[\lambda_3+(\lambda_3+2\mu_3)P_5^2\right]A^t-2\mu_3P_6B^t=0.
\end{aligned}
\tag{4.1.13c}
$$

It can be seen from $\sigma'_{31}\big|_{x_3=-h}=\sigma''_{31}\big|_{x_3=-h}$ that

$$
\begin{aligned}
&\mu_2\Big[2\xi k_{p3}^{(2)}A^{2-}\mathrm{e}^{ik_{p3}^{(2)}h}-2\xi k_{p3}^{(2)}A^{2+}-\xi^2B^{2-}\mathrm{e}^{ik_{s3}^{(2)}h}\\
&-\xi^2B^{2+}+(k_{s3}^{(2)})^2B^{2-}\mathrm{e}^{ik_{s3}^{(2)}h}+(k_{s3}^{(2)})^2B^{2+}\Big]\\
&=\mu_3\Big[2\xi k_{p3}^{(3)}A^t-\xi^2B^t+(k_{s3}^{(3)})^2B^t\Big].
\end{aligned}
$$

After simplification, we get

$$
\begin{aligned}
&2\mu_2P_3A^{2-}\mathrm{e}^{ik_{p3}^{(2)}h}+\mu_2(P_4^2-1)B^{2-}\mathrm{e}^{ik_{s3}^{(2)}h}-2\mu_2P_3A^{2+}\\
&+\mu_2(P_4^2-1)B^{2+}-2\mu_3P_5A^t-\mu_3(P_6^2-1)B^t=0.
\end{aligned}
\tag{4.1.13d}
$$

Equations (4.1.12) and (4.1.13) can be written in matrix form as

$$
\begin{bmatrix}
-1 & P_2 & 1 & P_4 & \mathrm{e}^{ik_{p3}^{(2)}h} & -P_4\mathrm{e}^{ik_{s3}^{(2)}h} & 0 & 0\\
-P_1 & -1 & -P_3 & 1 & P_3\mathrm{e}^{ik_{p3}^{(2)}h} & \mathrm{e}^{ik_{s3}^{(2)}h} & 0 & 0\\
\lambda_1+(\lambda_1+2\mu_1)P_1^2 & 2\mu_1P_2 & -\lambda_2-(\lambda_2+2\mu_2)P_3^2 & 2\mu_2P_4 & \left[-\lambda_2-(\lambda_2+2\mu_2)P_3^2\right]\mathrm{e}^{ik_{p3}^{(2)}h} & -2\mu_2P_4\mathrm{e}^{ik_{s3}^{(2)}h} & 0 & 0\\
2\mu_1P_1 & -\mu_1\left(P_2^2-1\right) & 2\mu_2P_3 & \mu_2\left(P_4^2-1\right) & -2\mu_2P_3\mathrm{e}^{ik_{p3}^{(2)}h} & \mu_2\left(P_4^2-1\right)\mathrm{e}^{ik_{s3}^{(2)}h} & 0 & 0\\
0 & 0 & \mathrm{e}^{ik_{p3}^{(2)}h} & P_4\mathrm{e}^{ik_{s3}^{(2)}h} & 1 & -P_4 & -1 & -P_6\\
0 & 0 & -P_3\mathrm{e}^{ik_{p3}^{(2)}h} & \mathrm{e}^{ik_{s3}^{(2)}h} & P_3 & 1 & P_5 & -1\\
0 & 0 & -\left[\lambda_2+(\lambda_2+2\mu_2)P_3^2\right]\mathrm{e}^{ik_{p3}^{(2)}h} & 2\mu_2P_4\mathrm{e}^{ik_{s3}^{(2)}h} & -\left[\lambda_2+(\lambda_2+2\mu_2)P_3^2\right] & -2\mu_2P_4 & \lambda_3+(\lambda_3+2\mu_3)P_5^2 & -2\mu_3P_6\\
0 & 0 & 2\mu_2P_3\mathrm{e}^{ik_{p3}^{(2)}h} & \mu_2\left(P_4^2-1\right)\mathrm{e}^{ik_{s3}^{(2)}h} & -2\mu_2P_3 & \mu_2\left(P_4^2-1\right) & -2\mu_3P_5 & -\mu_3\left(P_6^2-1\right)
\end{bmatrix}
\begin{bmatrix} A^r\\ B^r\\ A^{2-}\\ B^{2-}\\ A^{2+}\\ B^{2+}\\ A^t\\ B^t \end{bmatrix}
$$

$$
=\begin{bmatrix} 1\\ -P_1\\ -\left[\lambda_1+(\lambda_1+2\mu_1)P_1^2\right]\\ 2\mu_1P_1\\ 0\\ 0\\ 0\\ 0 \end{bmatrix}A^i+\begin{bmatrix} P_2\\ 1\\ 2\mu_1P_2\\ \mu_1\left(P_2^2-1\right)\\ 0\\ 0\\ 0\\ 0 \end{bmatrix}B^i,
\tag{4.1.14}
$$

where

$$P_1 = \frac{k_{p3}^{(1)}}{\xi} = \cot\alpha,\ P_2 = \frac{k_{s3}^{(1)}}{\xi} = \cot\beta,$$
$$P_3 = \frac{k_{p3}^{(2)}}{\xi} = \cot\alpha^{2-},\ P_4 = \frac{k_{s3}^{(2)}}{\xi} = \cot\beta^{2-},$$
$$P_5 = \frac{k_{p3}^{(3)}}{\xi} = \cot\alpha^{t},\ P_6 = \frac{k_{s3}^{(3)}}{\xi} = \cot\beta^{t}.$$

When $A^i = 0$ and $B^i \neq 0$, it means SV wave is incident; when $A^i \neq 0$ and $B^i = 0$, it means P wave is incident; when $A^i \neq 0$ and $B^i \neq 0$, it means P wave and SV wave are incident at the same time.

The method of solving interface conditions simultaneously can also be extended to the case of multilayer media. Generally, the interface conditions can be written as a system of linear algebraic equation

$$\boldsymbol{D} \cdot \boldsymbol{X} = \boldsymbol{b}_1 A^i + \boldsymbol{b}_2 B^i, \tag{4.1.15}$$

where $\boldsymbol{X}$ is the amplitude of the reflected wave, the transmitted wave, the upward and downward waves in the interlayer. A^i and B^i are the amplitudes of the incident P wave and SV wave, respectively. When the number of layers N increases, the number of interfaces, $N+1$, increases accordingly. The number of interfacial conditions, $(N+1)\times 4$ (corresponding to the incidence of P wave and SV wave), also increases, which will make the order of the coefficient matrix $\mathbf{D}$ of linear algebraic equation set to become very large. Not only will the amount of calculation be greatly increased, the calculation error and the algorithm stability are also not guaranteed.

4.2 Transfer Matrix Method

Firstly, let us consider that the incident wave is SH wave. In medium 2,

$$\begin{aligned} u_2 &= C^{2+}\exp\left[i\left(\xi x_1 + k_{s3}^{(2)}x_3 - \omega_s t\right)\right] + C^{2-}\exp\left[i\left(\xi x_1 - k_{s3}^{(2)}x_3 - \omega_s t\right)\right] \\ &= \left(C^{2+}\mathrm{e}^{ik_{s3}^{(2)}x_3} + C^{2-}\mathrm{e}^{-ik_{s3}^{(2)}x_3}\right)\exp[i(\xi x_1 - \omega_s t)], \end{aligned} \tag{4.2.1}$$

$$\sigma_{32} = \mu_2\left(ik_{s3}^{(2)}C^{2+}\mathrm{e}^{ik_{s3}^{(2)}x_3} - ik_{s3}^{(2)}C^{2-}\mathrm{e}^{-ik_{s3}^{(2)}x_3}\right)\exp[i(\xi x_1 - \omega_s t)],$$
$$\begin{Bmatrix} u_2 \\ \sigma_{32} \end{Bmatrix} = \begin{pmatrix} 1 & 1 \\ i\mu_2 k_{s3}^{(2)} & -i\mu_2 k_{s3}^{(2)} \end{pmatrix}\begin{pmatrix} \mathrm{e}^{ik_{s3}^{(2)}x_3} & 0 \\ 0 & \mathrm{e}^{-ik_{s3}^{(2)}x_3} \end{pmatrix}\begin{Bmatrix} C^{2+} \\ C^{2-} \end{Bmatrix}\exp[i(\xi x_1 - \omega_s t)]. \tag{4.2.2}$$

Let

$$\boldsymbol{P}=\begin{Bmatrix} u_2 \\ \sigma_{32} \end{Bmatrix},\ \boldsymbol{X}=\begin{pmatrix} 1 & 1 \\ i\mu_2 k_{s3}^{(2)} & -i\mu_2 k_{s3}^{(2)} \end{pmatrix},\ \boldsymbol{D}=\begin{pmatrix} \mathrm{e}^{ik_{s3}^{(2)}x_3} & 0 \\ 0 & \mathrm{e}^{-ik_{s3}^{(2)}x_3} \end{pmatrix},\ \boldsymbol{U}=\begin{Bmatrix} C^{2+} \\ C^{2-} \end{Bmatrix}. \tag{4.2.3}$$

Equation (4.2.2) can be written as

$$\boldsymbol{P}=\boldsymbol{X}\cdot\boldsymbol{D}\cdot\boldsymbol{U}\exp[i(\xi x_1-\omega_s t)]. \tag{4.2.4}$$

The column vector $\boldsymbol{P}$ is usually called the state vector. Equation (4.2.4) gives the calculation formulas of state vectors in different positions in a certain medium. Generally, the state vector is a position-dependent function [14,15,17].

At the upper and lower boundaries of medium 2,

$$\begin{Bmatrix} u_2 \\ \sigma_{32} \end{Bmatrix}_{x_3=0^-}=\begin{pmatrix} 1 & 1 \\ i\mu_2 k_{s3}^{(2)} & -i\mu_2 k_{s3}^{(2)} \end{pmatrix}\begin{Bmatrix} C^{2+} \\ C^{2-} \end{Bmatrix}\exp[i(\xi x_1-\omega_s t)], \tag{4.2.5a}$$

$$\begin{Bmatrix} u_2 \\ \sigma_{32} \end{Bmatrix}_{x_3=-h^+}=\begin{pmatrix} 1 & 1 \\ i\mu_2 k_{s3}^{(2)} & -i\mu_2 k_{s3}^{(2)} \end{pmatrix}\begin{pmatrix} \mathrm{e}^{-ik_{s3}^{(2)}h} & 0 \\ 0 & \mathrm{e}^{ik_{s3}^{(2)}h} \end{pmatrix}\begin{Bmatrix} C^{2+} \\ C^{2-} \end{Bmatrix}\exp[i(\xi x_1-\omega_s t)]. \tag{4.2.5b}$$

According to

$$\boldsymbol{P}(x_3=0^-)=\boldsymbol{X}\cdot\boldsymbol{U}\ \exp[i(\xi x_1-\omega_s t)], \tag{4.2.6}$$

we can obtain

$$\boldsymbol{X}^{-1}\cdot\boldsymbol{P}(x_3=0^-)=\boldsymbol{U}\ \exp[i(\xi x_1-\omega_s t)]. \tag{4.2.7}$$

Thereby

$$\begin{aligned}\boldsymbol{P}(x_3=-h^+)&=\boldsymbol{X}\cdot\boldsymbol{D}(x_3=-h^+)\cdot\boldsymbol{U}\exp[i(\xi x_1-\omega_s t)]\\&=\boldsymbol{X}\cdot\boldsymbol{D}(x_3=-h^+)\cdot\boldsymbol{X}^{-1}\cdot\boldsymbol{P}(x_3=0^-).\end{aligned} \tag{4.2.8}$$

Let

$$\begin{aligned}\boldsymbol{T}(h)&=\boldsymbol{X}\cdot\boldsymbol{D}(x_3=-h^+)\cdot\boldsymbol{X}^{-1}\\&=\begin{pmatrix} 1 & 1 \\ i\mu_2 k_{s3}^{(2)} & -i\mu_2 k_{s3}^{(2)} \end{pmatrix}\begin{pmatrix} \mathrm{e}^{-ik_{s3}^{(2)}h} & 0 \\ 0 & \mathrm{e}^{ik_{s3}^{(2)}h} \end{pmatrix}\begin{pmatrix} 1 & 1 \\ i\mu_2 k_{s3}^{(2)} & -i\mu_2 k_{s3}^{(2)} \end{pmatrix}^{-1},\end{aligned}$$

then, the state vectors at the upper and lower boundaries of medium 2 have the following relationship

$$\boldsymbol{P}(x_3=-h^+)=\boldsymbol{T}(h)\cdot\boldsymbol{P}(x_3=0^-), \tag{4.2.9}$$

where $\boldsymbol{T}(h)$ is usually called the transfer matrix of the state vector in the medium.

Let us calculate the state vectors in medium 1 and medium 3. The displacement field at interface 2 $\left(x_3 = 0^+\right)$ is

$$u_2|_{x_3=0^+} = C^i \mathrm{e}^{i(\xi x_1 - \omega_s t)} + C^r \mathrm{e}^{i(\xi x_1 - \omega_s t)} = \left(C^i + C^r\right)\mathrm{e}^{i(\xi x_1 - \omega_s t)}, \tag{4.2.10}$$

and the stress field at interface 2 $\left(x_3 = 0^+\right)$ is

$$\begin{aligned}\sigma_{32}|_{x_3=0^+} &= \mu_1\left(-ik_{s3}^{(1)}C^i \mathrm{e}^{i(\xi x_1 - \omega_s t)} + ik_{s3}^r C^r \mathrm{e}^{i(\xi x_1 - \omega_s t)}\right) \\ &= i\mu_1 k_{s3}^{(1)}\left(C^r - C^i\right)\mathrm{e}^{i(\xi x_1 - \omega_s t)},\end{aligned} \tag{4.2.11}$$

thereby

$$\begin{aligned}\begin{Bmatrix} u_2 \\ \sigma_{32} \end{Bmatrix}_{x_3=0^+} &= \begin{pmatrix} C^i + C^r \\ i\mu_1 k_{s3}^{(1)}\left(C^r - C^i\right) \end{pmatrix} \cdot \mathrm{e}^{i(\xi x_1 - \omega_s t)} \\ &= \begin{pmatrix} 1 & 1 \\ i\mu_1 k_{s3}^{(1)} & -i\mu_1 k_{s3}^{(1)} \end{pmatrix}\begin{Bmatrix} C^r \\ C^i \end{Bmatrix} \cdot \mathrm{e}^{i(\xi x_1 - \omega_s t)} \\ &= \boldsymbol{X}_1\begin{Bmatrix} C^r \\ C^i \end{Bmatrix} \cdot \mathrm{e}^{i(\xi x_1 - \omega_s t)}.\end{aligned} \tag{4.2.12}$$

The displacement field at interface 1 $\left(x_3 = -h^-\right)$ is

$$u_2|_{x_3=-h^-} = C^t \mathrm{e}^{i(\xi x_1 - \omega_s t)}, \tag{4.2.13}$$

and the stress field at interface 1 $\left(x_3 = -h^-\right)$ is

$$\sigma_{32}|_{x_3=-h^-} = -i\mu_3 k_{s3}^{(3)} C^t \mathrm{e}^{i(\xi x_1 - \omega_s t)}, \tag{4.2.14}$$

thereby

$$\begin{Bmatrix} u_2 \\ \sigma_{32} \end{Bmatrix}_{x_3=-h^-} = \begin{Bmatrix} 1 \\ -i\mu_3 k_{s3}^{(3)} \end{Bmatrix} C^t \mathrm{e}^{i(\xi x_1 - \omega_s t)} = \boldsymbol{X}_3 C^t \mathrm{e}^{i(\xi x_1 - \omega_s t)}. \tag{4.2.15}$$

For the perfect interface, the state vectors on both sides of the interface are continuous, i.e.

$$\begin{Bmatrix} u_2 \\ \sigma_{32} \end{Bmatrix}_{x_3=-h^+} = \begin{Bmatrix} u_2 \\ \sigma_{32} \end{Bmatrix}_{x_3=-h^-}, \tag{4.2.16a}$$

$$\begin{Bmatrix} u_2 \\ \sigma_{32} \end{Bmatrix}_{x_3=0^-} = \begin{Bmatrix} u_2 \\ \sigma_{32} \end{Bmatrix}_{x_3=0^+}. \tag{4.2.16b}$$

Therefore,

$$\begin{Bmatrix} u_2 \\ \sigma_{32} \end{Bmatrix}_{x_3=-h^-} = \boldsymbol{T}(h)\begin{Bmatrix} u_2 \\ \sigma_{32} \end{Bmatrix}_{x_3=0^+}, \tag{4.2.17a}$$

that is

$$\boldsymbol{X}_3 C^t = \boldsymbol{T}(h)\cdot \boldsymbol{X}_1 \begin{Bmatrix} C^r \\ C^i \end{Bmatrix}. \tag{4.2.17b}$$

Through matrix partitioning, the dependence of reflection and transmission amplitude on incident amplitude can be obtained from the above formula

$$\begin{Bmatrix} C^r \\ C^t \end{Bmatrix} = \begin{Bmatrix} R \\ T \end{Bmatrix} C^i,$$

where R and T are the reflection and transmission coefficients of SH wave, respectively.

Now, let us study the reflection and transmission of P wave and SV wave through the sandwiched slab by using the transfer matrix method. According to Eq. (4.1.8), the state vector composed of displacement and stress in medium 2 can be written in matrix form as

$$\boldsymbol{P} = \boldsymbol{X}\cdot\boldsymbol{D}\cdot\boldsymbol{U}e^{i(\xi x_1-\omega t)}, \tag{4.2.18}$$

where

$$\boldsymbol{X} = \begin{bmatrix} i\xi & i\xi & ik_{s3}^{(2)} & -ik_{s3}^{(2)} \\ -ik_{p3}^{(2)} & ik_{p3}^{(2)} & i\xi & i\xi \\ -\lambda_2\xi^2-(\lambda_2+2\mu_2)(k_{p3}^{(2)})^2 & -\lambda_2\xi^2-(\lambda_2+2\mu_2)(k_{p3}^{(2)})^2 & 2\mu_2\xi k_{s3}^{(2)} & -2\mu_2\xi k_{s3}^{(2)} \\ 2\mu_2\xi k_{p3}^{(2)} & -2\mu_2\xi k_{p3}^{(2)} & \mu_2\left[\left(k_{s3}^{(2)}\right)^2-\xi^2\right] & \mu_2\left[\left(k_{s3}^{(2)}\right)^2-\xi^2\right] \end{bmatrix}$$

$$\boldsymbol{D} = \begin{bmatrix} e^{-ik_{p3}^{(2)}x_3} & 0 & 0 & 0 \\ 0 & e^{ik_{p3}^{(2)}x_3} & 0 & 0 \\ 0 & 0 & e^{-ik_{s3}^{(2)}x_3} & 0 \\ 0 & 0 & 0 & e^{ik_{s3}^{(2)}x_3} \end{bmatrix},\ \boldsymbol{P} = \begin{bmatrix} u_1 \\ u_3 \\ \sigma_{33} \\ \sigma_{31} \end{bmatrix},\ \boldsymbol{U} = \begin{bmatrix} A^{2-} \\ A^{2+} \\ B^{2-} \\ B^{2+} \end{bmatrix}. \tag{4.2.19}$$

Similarly, the state vectors in medium 1 and medium 3 can be expressed as

$$\boldsymbol{P}_1 = \boldsymbol{X}_1\cdot\boldsymbol{D}_1\cdot\boldsymbol{U}_1 e^{i(\xi x_1-\omega t)}, \tag{4.2.20a}$$

$$\boldsymbol{P}_3 = \boldsymbol{X}_3\cdot\boldsymbol{D}_3\cdot\boldsymbol{U}_3 e^{i(\xi x_1-\omega t)}. \tag{4.2.20b}$$

For medium 2, at the upper and lower boundaries of the medium, we can obtain

$$P(x_3 = 0^-) = X \cdot U e^{i(\xi x_1 - \omega t)}, \tag{4.2.21}$$

$$X^{-1} \cdot P(x_3 = 0^-) = U e^{i(\xi x_1 - \omega t)}, \tag{4.2.22}$$

$$P(x_3 = -h^+) = X \cdot D(x_3 = -h^+) \cdot U e^{i(\xi x_1 - \omega t)}, \tag{4.2.23}$$

thereby

$$P(x_3 = -h^+) = X \cdot D(x_3 = -h^+) \cdot X^{-1} P(x_3 = 0^-) = T(h) P(x_3 = 0^-), \tag{4.2.24}$$

where,

$$T(h) = X \cdot D(x_3 = -h^+) \cdot X^{-1}. \tag{4.2.25}$$

At interface 2, the displacement field and stress field at $x_3 = 0^+$ are

$$P(x_3 = 0^+) = X_1 \cdot \begin{bmatrix} A^i \\ A^r \\ B^i \\ B^r \end{bmatrix} e^{i(\xi x_1 - \omega t)}. \tag{4.2.26}$$

At interface 1, the displacement field and stress field at $x_3 = -h^-$ are

$$P(x_3 = -h^-) = X_3 \cdot \begin{bmatrix} A^t \\ B^t \end{bmatrix} e^{i(\xi x_1 - \omega t)}. \tag{4.2.27}$$

When the interface is perfect, we can obtain the following relation

$$P(x_3 = -h^+) = P(x_3 = -h^-), \; P(x_3 = 0^+) = P(x_3 = 0^-), \tag{4.2.28}$$

thereby

$$P(x_3 = -h^-) = T(h) \cdot P(x_3 = 0^+). \tag{4.2.29}$$

Further,

$$X_3 \begin{bmatrix} A^t \\ B^t \end{bmatrix} = X \cdot D_{x_3 = -h^+} \cdot X^{-1} \cdot X_1 \cdot \begin{bmatrix} A^i \\ A^r \\ B^i \\ B^r \end{bmatrix}. \tag{4.2.30}$$

Through the matrix partitioning processing, the above formula can be written as

$$\begin{Bmatrix} A^r \\ B^r \end{Bmatrix} = \begin{bmatrix} R_{AA} & R_{AB} \\ R_{BA} & R_{BB} \end{bmatrix} \begin{Bmatrix} A^i \\ B^i \end{Bmatrix}, \quad (4.2.31a)$$

$$\begin{Bmatrix} A^t \\ B^t \end{Bmatrix} = \begin{bmatrix} T_{AA} & T_{AB} \\ T_{BA} & T_{BB} \end{bmatrix} \begin{Bmatrix} A^i \\ B^i \end{Bmatrix}, \quad (4.2.31b)$$

where $\boldsymbol{R}$ and $\boldsymbol{T}$ are called the reflection matrix and the transmission matrix of the sandwiched slab. When $A^i = 0$ and $B^i \neq 0$, it means SV wave is incident; when $A^i \neq 0$ and $B^i = 0$, it means P wave is incident; when $A^i \neq 0$ and $B^i \neq 0$, it means P wave and SV wave are incident at the same time. Thus, R_{AA} represents the reflection coefficient of the reflected P wave when the P wave is incident alone; R_{AB} represents the reflection coefficient of the reflected P wave when the SV wave is incident alone; T_{AA} represents the transmission coefficient of the transmitted P wave when the P wave is incident alone; T_{AB} represents the transmission coefficient of the transmitted P wave when the SV wave is incident alone. The physical meaning of other coefficients can be understood similarly.

Now, let us consider the case that there are multiple layers of medium existing between two half-spaces. Assuming that the transfer matrix of the first layer of media is $\boldsymbol{T}_1(h)$, the transfer matrix of the second layer of media is $\boldsymbol{T}_2(h)$, and so on, the transfer matrix of the n-th layer of media is $\boldsymbol{T}_n(h)$. When the interface is perfect, there are

$$\boldsymbol{P}_{x_3=-nh^+} = \boldsymbol{P}_{x_3=-nh^-}, \ldots, \boldsymbol{P}_{x_3=0^+} = \boldsymbol{P}_{x_3=0^-}. \quad (4.2.32)$$

Therefore,

$$\boldsymbol{P}_{x_3=-nh^-} = \boldsymbol{T}(h)\boldsymbol{P}_{x_3=0^+}, \quad (4.2.33)$$

where,

$$\boldsymbol{T}(h) = \boldsymbol{T}_1(h) \cdot \boldsymbol{T}_2(h) \cdot \boldsymbol{T}_3(h) \cdots\cdots \boldsymbol{T}_n(h) \quad (4.2.34)$$

is the total transfer matrix of N-layer media. Once the total transfer matrix is obtained, the reflection and transmission of the elastic wave through the N-layer medium are the same as those through one layer of the medium. When SH wave is incident, the total transfer matrix is 2×2 orders, and when the incident wave is P wave or SV wave, the total transfer matrix is 4×4 orders. Compared with the simultaneous interfacial condition method, the order of the final linear algebraic equations caused by the transfer matrix method is greatly reduced, which greatly reduce the storage space of the computer and improve the calculation efficiency. This is the main reason why the transfer matrix method is widely used [14–18]. However, the transfer matrix method itself has some shortcomings. As the number of layers of the layered medium increases, the cumulative error of the transfer matrix increases rapidly, which makes the numerical algorithm based on the transfer matrix method unstable [14].

4.3 Stiffness Matrix Method

Let us firstly consider the case that the SH wave is incident. In medium 2,

$$\begin{bmatrix} u_2|_{x_3=0^-} \\ u_2|_{x_3=-h^+} \end{bmatrix} = \begin{bmatrix} 1 & 1 \\ \mathrm{e}^{-ik_{s3}^{(2)}h} & \mathrm{e}^{ik_{s3}^{(2)}h} \end{bmatrix} \begin{bmatrix} C^{2+} \\ C^{2-} \end{bmatrix} \exp[i(\xi x_1 - \omega_s t)], \tag{4.3.1a}$$

$$\begin{bmatrix} \sigma_{32}|_{x_3=0^-} \\ \sigma_{32}|_{x_3=-h^+} \end{bmatrix} = \begin{bmatrix} i\mu_2 k_{s3}^{(2)} & -i\mu_2 k_{s3}^{(2)} \\ i\mu_2 k_{s3}^{(2)} \mathrm{e}^{-ik_{s3}^{(2)}h} & -i\mu_2 k_{s3}^{(2)} \mathrm{e}^{ik_{s3}^{(2)}h} \end{bmatrix} \begin{bmatrix} C^{2+} \\ C^{2-} \end{bmatrix} \exp[i(\xi x_1 - \omega_s t)]. \tag{4.3.1b}$$

Therefore

$$\begin{aligned} \begin{bmatrix} \sigma_{32}|_{x_3=0^-} \\ \sigma_{32}|_{x_3=-h^+} \end{bmatrix} &= \begin{bmatrix} i\mu_2 k_{s3}^{(2)} & -i\mu_2 k_{s3}^{(2)} \\ i\mu_2 k_{s3}^{(2)} \mathrm{e}^{-ik_{s3}^{(2)}h} & -i\mu_2 k_{s3}^{(2)} \mathrm{e}^{ik_{s3}^{(2)}h} \end{bmatrix} \begin{bmatrix} 1 & 1 \\ \mathrm{e}^{-ik_{s3}^{(2)}h} & \mathrm{e}^{ik_{s3}^{(2)}h} \end{bmatrix}^{-1} \begin{bmatrix} u_2|_{x_3=0^-} \\ u_2|_{x_3=-h^+} \end{bmatrix} \\ &= \mathbf{K} \cdot \begin{bmatrix} u_2|_{x_3=0^-} \\ u_2|_{x_3=-h^+} \end{bmatrix}, \end{aligned} \tag{4.3.2}$$

where

$$\mathbf{K} = \begin{bmatrix} i\mu_2 k_{s3}^{(2)} & -i\mu_2 k_{s3}^{(2)} \\ i\mu_2 k_{s3}^{(2)} \mathrm{e}^{-ik_{s3}^{(2)}h} & -i\mu_2 k_{s3}^{(2)} \mathrm{e}^{ik_{s3}^{(2)}h} \end{bmatrix} \begin{bmatrix} 1 & 1 \\ \mathrm{e}^{-ik_{s3}^{(2)}h} & \mathrm{e}^{ik_{s3}^{(2)}h} \end{bmatrix}^{-1} \tag{4.3.3}$$

is the stiffness matrix of medium 2, which establishes the relationship between the stress and displacement at the upper and lower boundaries of medium 2.

In medium 1 and medium 3, the displacement fields at interface 1 ($x_3 = -h^-$) and interface 2 ($x_3 = 0^+$) are

$$\begin{bmatrix} u_2|_{x_3=0^+} \\ u_2|_{x_3=-h^-} \end{bmatrix} = \begin{bmatrix} C^i + C^r \\ C^t \end{bmatrix} \mathrm{e}^{i(\xi x_1 - \omega_s t)}. \tag{4.3.4}$$

The stress fields at interface 1 ($x_3 = -h^-$) and interface 2 ($x_3 = 0^+$) are

$$\begin{aligned} \begin{bmatrix} \sigma_{32}|_{x_3=0^+} \\ \sigma_{32}|_{x_3=-h^-} \end{bmatrix} &= \begin{bmatrix} i\mu_1 k_{s3}^{(1)} & 0 \\ 0 & -i\mu_3 k_{s3}^{(3)} \end{bmatrix} \begin{bmatrix} C^r - C^i \\ C^t \end{bmatrix} \mathrm{e}^{i(\xi x_1 - \omega_s t)} \\ &= \boldsymbol{X} \begin{bmatrix} C^r - C^i \\ C^t \end{bmatrix} \mathrm{e}^{i(\xi x_1 - \omega_s t)}. \end{aligned} \tag{4.3.5}$$

When the interface is perfect, the displacement and stress satisfy the continuous condition, namely

$$\begin{bmatrix} u_2|_{x_3=0^+} \\ u_2|_{x_3=-h^-} \end{bmatrix} = \begin{bmatrix} u_2|_{x_3=0^-} \\ u_2|_{x_3=-h^+} \end{bmatrix}, \tag{4.3.6a}$$

$$\begin{bmatrix} \sigma_{32}|_{x_3=0^+} \\ \sigma_{32}|_{x_3=-h^-} \end{bmatrix} = \begin{bmatrix} \sigma_{32}|_{x_3=0^-} \\ \sigma_{32}|_{x_3=-h^+} \end{bmatrix}. \tag{4.3.6b}$$

Therefore

$$\begin{bmatrix} \sigma_{32}|_{x_3=0^+} \\ \sigma_{32}|_{x_3=-h^-} \end{bmatrix} = \begin{bmatrix} i\mu_2 k_{s3}^{(2)} & -i\mu_2 k_{s3}^{(2)} \\ i\mu_2 k_{s3}^{(2)} \mathrm{e}^{-ik_{s3}^{(2)}h} & -i\mu_2 k_{s3}^{(2)}{}_{s3}\mathrm{e}^{ik_{s3}^{(2)}h} \end{bmatrix} \begin{bmatrix} 1 & 1 \\ \mathrm{e}^{-ik_{s3}^{(2)}h} & \mathrm{e}^{ik_{s3}^{(2)}h} \end{bmatrix}^{-1} \begin{bmatrix} u_2|_{x_3=0^+} \\ u_2|_{x_3=-h^-} \end{bmatrix}. \tag{4.3.7}$$

Further, inserting Eqs. (4.3.4) and (4.3.5) into Eq. (4.3.7) leads to

$$\begin{bmatrix} i\mu_1 k_{s3}^{(1)} & \\ & -i\mu_3 k_{s3}^{(3)} \end{bmatrix} \begin{bmatrix} C^r - C^i \\ C^t \end{bmatrix} =$$
$$\begin{bmatrix} i\mu_2 k_{s3}^{(2)} & -i\mu_2 k_{s3}^{(2)} \\ i\mu_2 k_{s3}^{(2)} \mathrm{e}^{-ik_{s3}^{(2)}h} & -i\mu_2 k_{s3}^{(2)}{}_{s3}\mathrm{e}^{ik_{s3}^{(2)}h} \end{bmatrix} \begin{bmatrix} 1 & 1 \\ \mathrm{e}^{-ik_{s3}^{(2)}h} & \mathrm{e}^{ik_{s3}^{(2)}h} \end{bmatrix}^{-1} \begin{bmatrix} C^i + C^r \\ C^t \end{bmatrix}. \tag{4.3.8}$$

Equation (4.3.8) can be abbreviated as

$$\mathbf{X} \cdot \begin{bmatrix} C^r - C^i \\ C^t \end{bmatrix} = \mathbf{K} \cdot \begin{bmatrix} C^i + C^r \\ C^t \end{bmatrix}, \tag{4.3.9}$$

or

$$(\boldsymbol{X} - \boldsymbol{K}) \cdot \begin{bmatrix} C^r \\ C^t \end{bmatrix} = (\boldsymbol{X} + \boldsymbol{K}) \cdot \begin{bmatrix} C^i \\ 0 \end{bmatrix}. \tag{4.3.10}$$

Thus, the reflection coefficient and transmission coefficient can be obtained. When there are two interlayers between two half-spaces, we can suppose that

$$\begin{bmatrix} \sigma_{32}|_{x_3=0^-} \\ \sigma_{32}|_{x_3=-h^+} \end{bmatrix} = \begin{bmatrix} k_{11}^{(1)} & k_{12}^{(1)} \\ k_{21}^{(1)} & k_{22}^{(1)} \end{bmatrix} \begin{bmatrix} u_2|_{x_3=0^-} \\ u_2|_{x_3=-h^+} \end{bmatrix}, \tag{4.3.11a}$$

$$\begin{bmatrix} \sigma_{32}|_{x_3=-h^-} \\ \sigma_{32}|_{x_3=-2h^+} \end{bmatrix} = \begin{bmatrix} k_{11}^{(2)} & k_{12}^{(2)} \\ k_{21}^{(2)} & k_{22}^{(2)} \end{bmatrix} \begin{bmatrix} u_2|_{x_3=h^-} \\ u_2|_{x_3=-2h^+} \end{bmatrix}. \tag{4.3.11b}$$

When the interface is perfect, by using the continuous conditions

$$\begin{bmatrix} \sigma_{32}|_{x_3=-h^+} \\ u_2|_{x_3=-h^+} \end{bmatrix} = \begin{bmatrix} \sigma_{32}|_{x_3=-h^-} \\ u_2|_{x_3=-h^-} \end{bmatrix}, \tag{4.3.12}$$

we get

$$\begin{bmatrix} \sigma_{32}|_{x_3=0^-} \\ \sigma_{32}|_{x_3=-2h^+} \end{bmatrix} = \begin{bmatrix} m_{11}^{(2)} & m_{12}^{(2)} \\ m_{21}^{(2)} & m_{22}^{(2)} \end{bmatrix} \begin{bmatrix} u_2|_{x_3=0^-} \\ u_2|_{x_3=-2h^+} \end{bmatrix}, \tag{4.3.13}$$

where

$$m_{11}^{(2)} = k_{11}^{(1)} + k_{12}^{(1)}(k_{11}^{(2)} - k_{22}^{(1)})^{-1}k_{21}^{(1)}, m_{12}^{(2)} = -k_{12}^{(1)}(k_{11}^{(2)} - k_{22}^{(1)})^{-1}k_{12}^{(2)},$$
$$m_{21}^{(2)} = k_{21}^{(2)}(k_{11}^{(2)} - k_{22}^{(1)})^{-1}k_{21}^{(1)}, m_{22}^{(2)} = k_{22}^{(2)} - k_{21}^{(2)}(k_{11}^{(2)} - k_{22}^{(1)})^{-1}k_{12}^{(2)}.$$

The matrix **M** in Eq. (4.3.13) can be considered as the equivalent stiffness matrix of bilayer medium, which is determined completely by the stiffness matrix of each layer.

When there are n interlayers between two half-spaces, we can suppose that

$$\begin{bmatrix} \sigma_{32}|_{x_3=-(n-1)h^-} \\ \sigma_{32}|_{x_3=-nh^+} \end{bmatrix} = \begin{bmatrix} k_{11}^{(n)} & k_{12}^{(n)} \\ k_{21}^{(n)} & k_{22}^{(n)} \end{bmatrix} \begin{bmatrix} u_2|_{x_3=-(n-1)h^-} \\ u_2|_{x_3=-nh^+} \end{bmatrix}. \tag{4.3.14}$$

For the perfect interface, the continuous conditions hold, i.e.

$$\begin{bmatrix} \sigma_{32}|_{x_3=-(n-1)h^+} \\ u_2|_{x_3=-(n-1)h^+} \end{bmatrix} = \begin{bmatrix} \sigma_{32}|_{x_3=-(n-1)h^-} \\ u_2|_{x_3=-(n-1)h^-} \end{bmatrix}. \tag{4.3.15}$$

The application of the continuous conditions onto Eq. (4.3.14) leads to

$$\begin{bmatrix} \sigma_{32}|_{x_3=0^-} \\ \sigma_{32}|_{x_3=-nh^+} \end{bmatrix} = \begin{bmatrix} m_{11}^{(n)} & m_{12}^{(n)} \\ m_{21}^{(n)} & m_{22}^{(n)} \end{bmatrix} \begin{bmatrix} u_2|_{x_3=0^-} \\ u_2|_{x_3=-nh^+} \end{bmatrix}, \tag{4.3.16}$$

where

$$\begin{aligned} m_{11}^{(n)} &= m_{11}^{(n-1)} + m_{12}^{(n-1)}(k_{11}^{(n)} - m_{22}^{(n-1)})^{-1}m_{21}^{(n-1)}, \\ m_{12}^{(n)} &= -m_{12}^{(n-1)}(k_{11}^{(n)} - m_{22}^{(n-1)})^{-1}k_{12}^{(n)}, \\ m_{21}^{(n)} &= k_{21}^{(n)}(k_{11}^{(n)} - m_{22}^{(n-1)})^{-1}m_{21}^{(n-1)}, \\ m_{22}^{(n)} &= k_{22}^{(n)} - k_{21}^{(n)}(k_{11}^{(n)} - m_{22}^{(n-1)})^{-1}k_{12}^{(n)}. \end{aligned}$$

In the above equations, $k_{ij}^{(n)}(i = 1, 2;\ j = 1, 2)$ is the coefficient of stiffness matrix of n-th layer medium. $m_{ij}^{(n)}(i = 1, 2;\ j = 1, 2)$ is the coefficient of overall stiffness matrix of total n layer. The recursion equation can be used to obtain the equivalent stiffness matrix of multiple layers media. Once the stiffness matrix of the N-layer medium is obtained, the reflection and transmission of the elastic wave through the N-layer medium are the same as those through a layer of medium.

When the incident wave is P wave or SV wave, the displacement fields at interface 1 ($x_3 = -h^+$) and interface 2 ($x_3 = 0^-$) are

$$u_1|_{x_3=0^-} = (i\xi A^{2-} + i\xi A^{2+} + ik_{s3}^{(2)} B^{2-} - ik_{s3}^{(2)} B^{2+})\mathrm{e}^{i(\xi x_1-\omega t)}, \tag{4.3.17a}$$

$$u_3|_{x_3=0^-} = (-ik_{p3}^{(2)} A^{2-} + ik_{p3}^{(2)} A^{2+} + i\xi B^{2-} + i\xi B^{2+})\mathrm{e}^{i(\xi x_1-\omega t)}, \tag{4.3.17b}$$

$$\begin{aligned} u_1|_{x_3=-h^+} = {} & (i\xi \mathrm{e}^{ik_{p3}^{(2)}h} A^{2-} + i\xi \mathrm{e}^{-ik_{p3}^{(2)}h} A^{2+} + ik_{s3}^{(2)} B^{2-} \mathrm{e}^{ik_{s3}^{(2)}h} \\ & - ik_{s3}^{(2)} \mathrm{e}^{-ik_{s3}^{(2)}h} B^{2+})\mathrm{e}^{i(\xi x_1-\omega t)}, \end{aligned} \tag{4.3.17c}$$

$$\begin{aligned} u_3|_{x_3=-h^+} = {} & (-ik_{p3}^{(2)} A^{2-} \mathrm{e}^{ik_{p3}^{(2)}h} + ik_{p3}^{(2)} \mathrm{e}^{-ik_{p3}^{(2)}h} A^{2+} \\ & + i\xi B^{2-} \mathrm{e}^{ik_{s3}^{(2)}h} + i\xi \mathrm{e}^{-ik_{s3}^{(2)}h} B^{2+})\mathrm{e}^{i(\xi x_1-\omega t)}. \end{aligned} \tag{4.3.17d}$$

Equation (4.3.17) can be written in the form of a matrix as

$$\begin{bmatrix} u_1|_{x_3=0^-} \\ u_3|_{x_3=0^-} \\ u_1|_{x_3=-h^+} \\ u_3|_{x_3=-h^+} \end{bmatrix} = \mathbf{M} \begin{bmatrix} A^{2-} \\ A^{2+} \\ B^{2-} \\ B^{2+} \end{bmatrix} \mathrm{e}^{i(\xi x_1-\omega t)}, \tag{4.3.18}$$

where

$$\mathbf{M} = \begin{bmatrix} i\xi & i\xi & ik_{s3}^{(2)} & -ik_{s3}^{(2)} \\ -ik_{p3}^{(2)} & ik_{p3}^{(2)} & i\xi & i\xi \\ i\xi \mathrm{e}^{ik_{p3}^{(2)}h} & i\xi \mathrm{e}^{-ik_{p3}^{(2)}h} & ik_{s3}^{(2)} \mathrm{e}^{ik_{s3}^{(2)}h} & -ik_{s3}^{(2)} \mathrm{e}^{-ik_{s3}^{(2)}h} \\ -ik_{p3}^{(2)} \mathrm{e}^{ik_{p3}^{(2)}h} & ik_{p3}^{(2)} \mathrm{e}^{-ik_{p3}^{(2)}h} & i\xi \mathrm{e}^{ik_{s3}^{(2)}h} & i\xi \mathrm{e}^{-ik_{s3}^{(2)}h} \end{bmatrix}. \tag{4.3.19}$$

The stress fields at interface 1 ($x_3 = -h^+$) and interface 2 ($x_3 = 0^-$) are

$$\begin{aligned} \sigma_{33}|_{x_3=0^-} = {} & [(-\lambda_2\xi^2 - (\lambda_2 + 2\mu_2)(k_{p3}^{(2)})^2)A^{2-} + (-\lambda_2\xi^2 - (\lambda_2 + 2\mu_2)(k_{p3}^{(2)})^2)A^{2+} \\ & + 2\mu_2\xi k_{s3}^{(2)} B^{2-} - 2\mu_2\xi k_{s3}^{(2)} B^{2+}]\mathrm{e}^{i(\xi x_1-\omega t)}, \end{aligned}$$

$$\begin{aligned} \sigma_{31}|_{x_3=0^-} = {} & [2\mu_2\xi k_{p3}^{(2)} A^{2-} - 2\mu_2\xi k_{p3}^{(2)} A^{2+} + \mu_2((k_{s3}^{(2)})^2 - \xi^2)B^{2-} \\ & + \mu_2((k_{s3}^{(2)})^2 - \xi^2)B^{2+}]\mathrm{e}^{i(\xi x_1-\omega t)}, \end{aligned}$$

$$\begin{aligned} \sigma_{33}|_{x_3=-h^+} = {} & [(-\lambda_2\xi^2 - (\lambda_2 + 2\mu_2)(k_{p3}^{(2)})^2)\mathrm{e}^{ik_{p3}^{(2)}h} A^{2-} + (-\lambda_2\xi^2 \\ & - (\lambda_2 + 2\mu_2)(k_{p3}^{(2)})^2)\mathrm{e}^{-ik_{p3}^{(2)}h} A^{2+} \\ & + 2\mu_2\xi k_{s3}^{(2)} \mathrm{e}^{ik_{s3}^{(2)}h} B^{2-} - 2\mu_2\xi k_{s3}^{(2)} \mathrm{e}^{-ik_{s3}^{(2)}h} B^{2+}]\mathrm{e}^{i(\xi x_1-\omega t)}, \end{aligned}$$

$$\begin{aligned} \sigma_{31}|_{x_3=-h^+} = {} & [2\mu_2\xi k_{p3}^{(2)} \mathrm{e}^{ik_{p3}^{(2)}h} A^{2-} - 2\mu_2\xi k_{p3}^{(2)} \mathrm{e}^{-ik_{p3}^{(2)}h} A^{2+} \\ & + \mu_2((k_{s3}^{(2)})^2 - \xi^2)\mathrm{e}^{ik_{s3}^{(2)}h} B^{2-} \end{aligned}$$

$$+\mu_2((k_{s3}^{(2)})^2-\xi^2)\mathrm{e}^{-ik_{s3}^{(2)}h}B^{2+}]\mathrm{e}^{i(\xi x_1-\omega t)}. \tag{4.3.20a,b,c,d}$$

Equation (4.3.20) can be written in the form of a matrix as

$$\begin{bmatrix}\sigma_{33}|_{x_3=0^-}\\ \sigma_{31}|_{x_3=0^-}\\ \sigma_{33}|_{x_3=-h^+}\\ \sigma_{31}|_{x_3=-h^+}\end{bmatrix}=\mathbf{N}\begin{bmatrix}A^{2-}\\ A^{2+}\\ B^{2-}\\ B^{2+}\end{bmatrix}\mathrm{e}^{i(\xi x_1-\omega t)}, \tag{4.3.21}$$

where

$$\mathbf{N}=\begin{bmatrix}-\lambda_2\xi^2-(\lambda_2+2\mu_2)(k_{p3}^{(2)})^2 & -\lambda_2\xi^2-(\lambda_2+2\mu_2)(k_{p3}^{(2)})^2 & 2\mu_2\xi k_{s3}^{(2)} & -2\mu_2\xi k_{s3}^{(2)}\\ 2\mu_2\xi k_{p3}^{(2)} & -2\mu_2\xi k_{p3}^{(2)} & \mu_2((k_{s3}^{(2)})^2-\xi^2) & \mu_2((k_{s3}^{(2)})^2-\xi^2)\\ (-\lambda_2\xi^2-(\lambda_2+2\mu_2)(k_{p3}^{(2)})^2)\mathrm{e}^{ik_{p3}^{(2)}h} & (-\lambda_2\xi^2-(\lambda_2+2\mu_2)(k_{p3}^{(2)})^2)\mathrm{e}^{-ik_{p3}^{(2)}h} & 2\mu_2\xi k_{s3}^{(2)}\mathrm{e}^{ik_{s3}^{(2)}h} & -2\mu_2\xi k_{s3}^{(2)}\mathrm{e}^{-ik_{s3}^{(2)}h}\\ 2\mu_2\xi k_{p3}^{(2)}\mathrm{e}^{ik_{p3}^{(2)}h} & -2\mu_2\xi k_{p3}^{(2)}\mathrm{e}^{-ik_{p3}^{(2)}h} & \mu_2((k_{s3}^{(2)})^2-\xi^2)\mathrm{e}^{ik_{s3}^{(2)}h} & \mu_2((k_{s3}^{(2)})^2-\xi^2)\mathrm{e}^{-ik_{s3}^{(2)}h}\end{bmatrix}$$

Therefore

$$\begin{bmatrix}\sigma_{33}|_{x_3=0^-}\\ \sigma_{31}|_{x_3=0^-}\\ \sigma_{33}|_{x_3=-h^+}\\ \sigma_{31}|_{x_3=-h^+}\end{bmatrix}=\mathbf{N}\cdot\mathbf{M}^{-1}\begin{bmatrix}u_1|_{x_3=0^-}\\ u_3|_{x_3=0^-}\\ u_1|_{x_3=-h^+}\\ u_3|_{x_3=-h^+}\end{bmatrix}=\mathbf{K}\cdot\begin{bmatrix}u_1|_{x_3=0^-}\\ u_3|_{x_3=0^-}\\ u_1|_{x_3=-h^+}\\ u_3|_{x_3=-h^+}\end{bmatrix}. \tag{4.3.22}$$

The displacement fields at interface 1 ($x_3=-h^-$) and interface 2 ($x_3=0^+$) are

$$u_1|_{x_3=0^+}=(i\xi A^i+i\xi A^r+ik_{s3}^{(1)}B^i-ik_{s3}^{(1)}B^r)\mathrm{e}^{i(\xi x_1-\omega t)}, \tag{4.3.23a}$$

$$u_3|_{x_3=0^+}=(-ik_{p3}^{(1)}A^i+ik_{p3}^{(1)}A^r+i\xi B^i+i\xi B^r)\mathrm{e}^{i(\xi x_1-\omega t)}, \tag{4.3.23b}$$

$$u_1|_{x_3=-h^-}=(i\xi A^t+ik_{s3}^{(3)}B^t)\mathrm{e}^{i(\xi x_1-\omega t)}, \tag{4.3.23c}$$

$$u_3|_{x_3=-h^-}=(-ik_{p3}^{(3)}A^t+i\xi B^t)\mathrm{e}^{i(\xi x_1-\omega t)}. \tag{4.3.23d}$$

Equation (4.3.23) can be written in the form of a matrix as

$$\begin{bmatrix}u_1|_{x_3=0^+}\\ u_3|_{x_3=0^+}\\ u_1|_{x_3=-h^-}\\ u_3|_{x_3=-h^-}\end{bmatrix}=\mathbf{P}\begin{bmatrix}A^i\\ A^r\\ B^i\\ B^r\\ A^t\\ B^t\end{bmatrix}\mathrm{e}^{i(\xi x_1-\omega t)}, \tag{4.3.24}$$

where

$$\mathbf{P} = \begin{bmatrix} i\xi & i\xi & ik_{s3}^{(1)} & -ik_{s3}^{(1)} & 0 & 0 \\ -ik_{p3}^{(1)} & ik_{p3}^{(1)} & i\xi & i\xi & 0 & 0 \\ 0 & 0 & 0 & 0 & i\xi & ik_{s3}^{(3)} \\ 0 & 0 & 0 & 0 & -ik_{p3}^{(3)} & i\xi \end{bmatrix}. \tag{4.3.25}$$

The stress fields at interface 1 ($x_3 = -h^-$) and interface 2 ($x_3 = 0^+$) are

$$\sigma_{33}|_{x_3=0^+} = [(-\lambda_1\xi^2 - (\lambda_1 + 2\mu_1)(k_{p3}^{(1)})^2)A^i + (-\lambda_1\xi^2 - (\lambda_1 + 2\mu_1)(k_{p3}^{(1)})^2)A^r + 2\mu_1\xi k_{s3}^{(1)} B^i - 2\mu_1\xi k_{s3}^{(1)} B^r]\mathrm{e}^{i(\xi x_1 - \omega t)}, \tag{4.3.26a}$$

$$\sigma_{31}|_{x_3=0^+} = [2\mu_1\xi k_{p3}^{(1)} A^i - 2\mu_1\xi k_{p3}^{(1)} A^r + \mu_1((k_{s3}^{(1)})^2 - \xi^2)B^i + \mu_1((k_{s3}^{(1)})^2 - \xi^2)B^r]\mathrm{e}^{i(\xi x_1 - \omega t)}, \tag{4.3.26b}$$

$$\sigma_{33}|_{x_3=-h^-} = [(-\lambda_3\xi^2 - (\lambda_3 + 2\mu_3)(k_{p3}^{(3)})^2)A^t + 2\mu_3\xi k_{s3}^{(3)} B^t]\mathrm{e}^{i(\xi x_1 - \omega t)}, \tag{4.3.26c}$$

$$\sigma_{31}|_{x_3=-h^-} = [2\mu_3\xi k_{p3}^{(3)} A^t + \mu_3((k_{s3}^{(3)})^2 - \xi^2)B^t]\mathrm{e}^{i(\xi x_1 - \omega t)}. \tag{4.3.26d}$$

Equation (4.3.26) can be written in the form of a matrix as

$$\begin{bmatrix} \sigma_{33}|_{x_3=0^+} \\ \sigma_{31}|_{x_3=0^+} \\ \sigma_{33}|_{x_3=-h^-} \\ \sigma_{31}|_{x_3=-h^-} \end{bmatrix} = \mathbf{Q} \begin{bmatrix} A^i \\ A^r \\ B^i \\ B^r \\ A^t \\ B^t \end{bmatrix} \mathrm{e}^{i(\xi x_1 - \omega t)} \tag{4.3.27}$$

where

$Q_{11} = Q_{12} = -\lambda_1\xi^2 - (\lambda_1 + 2\mu_1)(k_{p3}^{(1)})^2$, $Q_{13} = -Q_{14} = 2\mu_1\xi k_{s3}^{(1)}$, $Q_{21} = -Q_{22} = 2\mu_1\xi k_{p3}^{(1)}$, $Q_{23}=-Q_{24}=\mu_1((k_{s3}^{(1)})^2-\xi^2)$, $Q_{35}=-\lambda_3\xi^2-(\lambda_3+2\mu_3)(k_{p3}^{(3)})^2$, $Q_{36}=2\mu_3\xi k_{s3}^{(3)}$, $Q_{45} = 2\mu_3\xi k_{p3}^{(3)}$, $Q_{46} = \mu_3((k_{s3}^{(3)})^2 - \xi^2)$. Other elements are zero.

For the perfect interface, the displacement and stress satisfy the continuous conditions, i.e.

$$\begin{bmatrix} u_1|_{x_3=0^+} \\ u_3|_{x_3=0^+} \\ u_1|_{x_3=-h^-} \\ u_3|_{x_3=-h^-} \end{bmatrix} = \begin{bmatrix} u_1|_{x_3=0^-} \\ u_3|_{x_3=0^-} \\ u_1|_{x_3=-h^+} \\ u_3|_{x_3=-h^+} \end{bmatrix}, \begin{bmatrix} \sigma_{33}|_{x_3=0^+} \\ \sigma_{31}|_{x_3=0^+} \\ \sigma_{33}|_{x_3=-h^-} \\ \sigma_{31}|_{x_3=-h^-} \end{bmatrix} = \begin{bmatrix} \sigma_{33}|_{x_3=0^-} \\ \sigma_{31}|_{x_3=0^-} \\ \sigma_{33}|_{x_3=-h^+} \\ \sigma_{31}|_{x_3=-h^+} \end{bmatrix}. \tag{4.3.28}$$

The application of the continuous conditions leads to

$$\begin{bmatrix} \sigma_{33}|_{x_3=0^+} \\ \sigma_{31}|_{x_3=0^+} \\ \sigma_{33}|_{x_3=-h^-} \\ \sigma_{31}|_{x_3=-h^-} \end{bmatrix} = \mathbf{N} \cdot \mathbf{M}^{-1} \begin{bmatrix} u_1|_{x_3=0^+} \\ u_3|_{x_3=0^+} \\ u_1|_{x_3=-h^-} \\ u_3|_{x_3=-h^-} \end{bmatrix}. \tag{4.3.29}$$

Further, inserting Eqs. (4.3.24) and (4.3.27) into Eq. (4.3.29) leads to

$$\mathbf{Q} \begin{bmatrix} A^i \\ A^r \\ B^i \\ B^r \\ A^t \\ B^t \end{bmatrix} = \mathbf{N} \cdot \mathbf{M}^{-1} \cdot \mathbf{P} \begin{bmatrix} A^i \\ A^r \\ B^i \\ B^r \\ A^t \\ B^t \end{bmatrix}. \tag{4.3.30}$$

Therefore

$$(\mathbf{Q} - \mathbf{N} \cdot \mathbf{M}^{-1} \cdot \mathbf{P}) \begin{bmatrix} A^i \\ A^r \\ B^i \\ B^r \\ A^t \\ B^t \end{bmatrix} = 0. \tag{4.3.31}$$

Through the matrix partitioning processing, Eq. (4.3.31) can be rewritten as

$$\mathbf{D} \cdot \begin{Bmatrix} A^r \\ B^r \\ A^t \\ B^t \end{Bmatrix} = \mathbf{b}_1 A^i + \mathbf{b}_2 B^i. \tag{4.3.32}$$

When $A^i = 0$ and $B^i \neq 0$, it means SV wave is incident alone; when $A^i \neq 0$ and $B^i = 0$, it means P wave is incident alone; when $A^i \neq 0$ and $B^i \neq 0$, it means the P wave and SV wave are incident at the same time.

Now, let us consider the situation when there are two interlayers sandwiched between half-spaces. We can suppose that

$$\begin{bmatrix} \sigma_{33}|_{x_3=0^-} \\ \sigma_{31}|_{x_3=0^-} \\ \sigma_{33}|_{x_3=-h^+} \\ \sigma_{31}|_{x_3=-h^+} \end{bmatrix} = \begin{bmatrix} \mathbf{K}_{11}^{(1)} & \mathbf{K}_{12}^{(1)} \\ \mathbf{K}_{21}^{(1)} & \mathbf{K}_{22}^{(1)} \end{bmatrix} \begin{bmatrix} u_1|_{x_3=0^-} \\ u_3|_{x_3=0^-} \\ u_1|_{x_3=-h^+} \\ u_3|_{x_3=-h^+} \end{bmatrix}, \tag{4.3.33a}$$

$$\begin{bmatrix} \sigma_{33}|_{x_3=-h^-} \\ \sigma_{31}|_{x_3=-h^-} \\ \sigma_{33}|_{x_3=-2h^+} \\ \sigma_{31}|_{x_3=-2h^+} \end{bmatrix} = \begin{bmatrix} \mathbf{K}_{11}^{(2)} & \mathbf{K}_{12}^{(2)} \\ \mathbf{K}_{21}^{(2)} & \mathbf{K}_{22}^{(2)} \end{bmatrix} \begin{bmatrix} u_1|_{x_3=-h^-} \\ u_3|_{x_3=-h^-} \\ u_1|_{x_3=-2h^+} \\ u_3|_{x_3=-2h^+} \end{bmatrix}. \tag{4.3.33b}$$

When the interface is perfect, the continuous conditions hold, i.e.

$$\begin{bmatrix} \sigma_{33}|_{x_3=-h^+} \\ \sigma_{31}|_{x_3=-h^+} \\ u_1|_{x_3=-h^+} \\ u_3|_{x_3=-h^+} \end{bmatrix} = \begin{bmatrix} \sigma_{33}|_{x_3=-h^-} \\ \sigma_{31}|_{x_3=-h^-} \\ u_1|_{x_3=-h^-} \\ u_3|_{x_3=-h^-} \end{bmatrix}. \tag{4.3.34}$$

The application of the continuous conditions leads to

$$\begin{bmatrix} \sigma_{33}|_{x_3=0^-} \\ \sigma_{31}|_{x_3=0^-} \\ \sigma_{33}|_{x_3=-2h^+} \\ \sigma_{31}|_{x_3=-2h^+} \end{bmatrix} = \begin{bmatrix} \mathbf{M}_{11}^{(2)} & \mathbf{M}_{12}^{(2)} \\ \mathbf{M}_{21}^{(2)} & \mathbf{M}_{22}^{(2)} \end{bmatrix} \begin{bmatrix} u_1|_{x_3=0^-} \\ u_3|_{x_3=0^-} \\ u_1|_{x_3=-2h^+} \\ u_3|_{x_3=-2h^+} \end{bmatrix}, \tag{4.3.35}$$

where

$$\mathbf{M}_{11}^{(2)} = \mathbf{K}_{11}^{(1)} + \mathbf{K}_{12}^{(1)}(\mathbf{K}_{11}^{(2)} - \mathbf{K}_{22}^{(1)})^{-1}\mathbf{K}_{21}^{(1)}, \mathbf{M}_{12}^{(2)} = -\mathbf{K}_{12}^{(1)}(\mathbf{K}_{11}^{(2)} - \mathbf{K}_{22}^{(1)})^{-1}\mathbf{K}_{12}^{(2)},$$
$$\mathbf{M}_{21}^{(2)} = \mathbf{K}_{21}^{(2)}(\mathbf{K}_{11}^{(2)} - \mathbf{K}_{22}^{(1)})^{-1}\mathbf{K}_{21}^{(1)}, \mathbf{M}_{22}^{(2)} = \mathbf{K}_{22}^{(2)} - \mathbf{K}_{21}^{(2)}(\mathbf{K}_{11}^{(2)} - \mathbf{K}_{22}^{(1)})^{-1}\mathbf{K}_{12}^{(2)}.$$

The matrix **M** in Eq. (4.3.35) is the equivalent stiffness matrix which is determined completely by the stiffness matrix of each layer.

When there are n interlayers sandwiched between two half-spaces, we can suppose that the relation in the n-th layer is

$$\begin{bmatrix} \sigma_{33}|_{x_3=-(n-1)h^-} \\ \sigma_{31}|_{x_3=-(n-1)h^-} \\ \sigma_{33}|_{x_3=-nh^+} \\ \sigma_{31}|_{x_3=-nh^+} \end{bmatrix} = \begin{bmatrix} \mathbf{K}_{11}^{(n)} & \mathbf{K}_{12}^{(n)} \\ \mathbf{K}_{21}^{(n)} & \mathbf{K}_{22}^{(n)} \end{bmatrix} \begin{bmatrix} u_1|_{x_3=-(n-1)h^-} \\ u_3|_{x_3=-(n-1)h^-} \\ u_1|_{x_3=-nh^+} \\ u_3|_{x_3=-nh^+} \end{bmatrix}. \tag{4.3.36}$$

For the perfect interface,

$$\begin{bmatrix} \sigma_{33}|_{x_3=-(n-1)h^+} \\ \sigma_{31}|_{x_3=-(n-1)h^+} \\ u_1|_{x_3=-(n-1)h^+} \\ u_3|_{x_3=-(n-1)h^+} \end{bmatrix} = \begin{bmatrix} \sigma_{33}|_{x_3=-(n-1)h^-} \\ \sigma_{31}|_{x_3=-(n-1)h^-} \\ u_1|_{x_3=-(n-1)h^-} \\ u_3|_{x_3=-(n-1)h^-} \end{bmatrix}. \tag{4.3.37}$$

By using the continuous conditions, i.e. Eq. (4.3.37), we get

$$\begin{bmatrix} \sigma_{33}|_{x_3=0^-} \\ \sigma_{31}|_{x_3=0^-} \\ \sigma_{33}|_{x_3=-nh^+} \\ \sigma_{31}|_{x_3=-nh^+} \end{bmatrix} = \begin{bmatrix} \mathbf{M}_{11}^{(n)} & \mathbf{M}_{12}^{(n)} \\ \mathbf{M}_{21}^{(n)} & \mathbf{M}_{22}^{(n)} \end{bmatrix} \begin{bmatrix} u_1|_{x_3=0^-} \\ u_3|_{x_3=0^-} \\ u_1|_{x_3=-nh^+} \\ u_3|_{x_3=-nh^+} \end{bmatrix}, \tag{4.3.38}$$

where

$$\begin{aligned}
\mathbf{M}_{11}^{(n)} &= \mathbf{M}_{11}^{(n-1)} + \mathbf{M}_{12}^{(n-1)}(\mathbf{K}_{11}^{(n)} - \mathbf{M}_{22}^{(n-1)})^{-1}\mathbf{M}_{21}^{(n-1)}, \\
\mathbf{M}_{12}^{(n)} &= -\mathbf{M}_{12}^{(n-1)}(\mathbf{K}_{11}^{(n)} - \mathbf{M}_{22}^{(n-1)})^{-1}\mathbf{K}_{12}^{(n)}, \\
\mathbf{M}_{21}^{(n)} &= \mathbf{K}_{21}^{(n)}(\mathbf{K}_{11}^{(n)} - \mathbf{M}_{22}^{(n-1)})^{-1}\mathbf{M}_{21}^{(n-1)}, \\
\mathbf{M}_{22}^{(n)} &= \mathbf{K}_{22}^{(n)} - \mathbf{K}_{21}^{(n)}(\mathbf{K}_{11}^{(n)} - \mathbf{M}_{22}^{(n-1)})^{-1}\mathbf{K}_{12}^{(n)}.
\end{aligned}$$

In the above formula, $\boldsymbol{K}_{ij}^{(n)}(i = 1, 2;\ j = 1, 2)$ is the coefficient of stiffness matrix of the n-th layer. $\boldsymbol{M}_{ij}^{(n)}(i = 1, 2;\ j = 1, 2)$ is the coefficients of overall stiffness matrix of the first n layers. Once the stiffness matrix of the total n-layer medium is obtained, the reflection and transmission of the elastic wave through the n-layer medium are the same as those through a layer of medium. Compared with the transfer matrix method, obtaining the overall or global stiffness matrix is more complicated than obtaining the overall or global transfer matrix, and the calculation workload is also greatly increased. When the number of layers is very large, the transfer matrix method may become unstable. However, the stiffness matrix can suppress the propagation of numerical errors and the stiffness matrix method still keep good stability [14, 16]. In other words, the stiffness matrix method is more robust than the transfer matrix method.

4.4 Multiple Reflection/Transmission Method

Firstly, let us consider the situation of incident SH wave. According to the process of wave propagation, when SH wave is incident on the interface between the incident medium and the interlayer, reflected SH wave and transmitted SH wave will be generated. When the transmitted SH wave travels to the interface between the interlayer and the transmission medium, the reflection and the transmission will occur again. The waves reflected back into the interlayer will be repeatedly reflected at the upper and lower interfaces of the interlayer which form SH^+ waves (upward waves) and SH^- waves (downward waves). These waves are superimposed on each other and finally form a guided wave that propagates along the sandwiched slab. In the incident medium (denoted by medium 1), the waves that have been reflected and transmitted multiple times into the incident medium (the half-space that the incident wave exists) are superimposed on each other and finally form the reflected wave of the interlayer. In the transmission medium (the half-space on the other side of the interlayer, and denoted by medium 3 in this section), the waves that have been transmitted multiple times into the transmission medium are superimposed on each other and finally form a transmitted wave through the interlayer. When the SH wave is incident, the multiple reflection/transmission process of the wave at the upper and lower interfaces of the interlayer is shown in Fig. 4.3. The essence of the multiple reflection and transmission method is to consider the phase difference between the multiple reflection/transmission waves according to the wave propagation process and superimpose the multiple reflection waves to obtain the reflected wave of the sandwiched slab while superimpose the multiple transmission waves on the other side of interlayer to obtain the transmission wave of sandwiched slab [20].

When the SH wave is incident, there is no mode conversion of wave, and the propagation process is relatively simple. Therefore, the result of multiple reflection and transmission of SH wave can be written as

$$H^R = R_{12}H^I + T_{21}R_{23}T_{12}H^I\exp(i2k_y^{sh}h) + T_{21}R_{23}R_{21}R_{23}T_{12}H^I\exp(i4k_y^{sh}h) + \cdots, \tag{4.4.1}$$

$$\begin{aligned} H^T &= T_{23}T_{12}H^I\exp(ik_y^{sh}h) + T_{23}R_{21}R_{23}T_{12}H^I\exp(i3k_y^{sh}h) \\ &\quad + T_{23}R_{21}R_{23}R_{21}R_{23}T_{12}H^I\exp(i5k_y^{sh}h) + \cdots, \end{aligned} \tag{4.4.2}$$

where H^R, H^T and H^I are the amplitude of the reflected wave, transmitted wave and incident wave, respectively. When the incident wave is incident from the medium i, R_{ij} and T_{ij}, respectively, represent the reflection and transmission coefficients at the interface between the medium i and the medium j. The first term on the right-hand side of Eq. (4.4.1) represents the reflection wave at the interface 1 (between medium 1 and medium 2), namely the first reflection wave. The second term represents the second reflected wave which is first transmitted from interface 1 to medium 2, and then reflected on interface 2 (the interface between medium 2 and medium 3) to

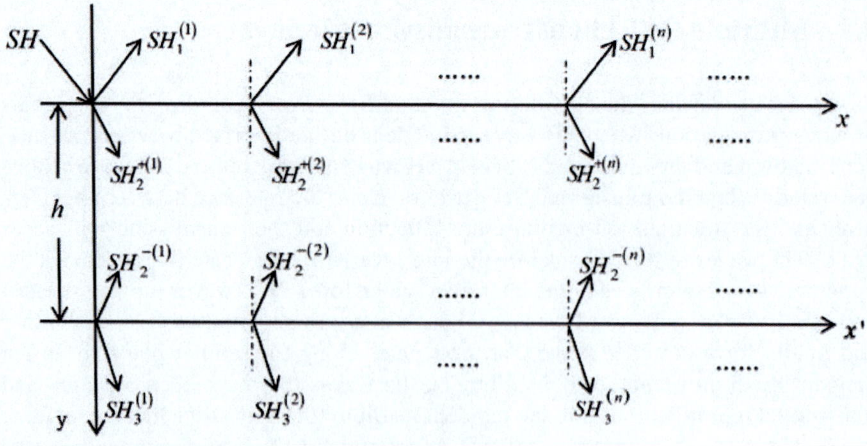

Fig. 4.3 Multiple reflection and transmission of SH wave

interface 1, then transmitted from interface 1. The phase shift of the second reflection wave during the entire propagation process is $\exp(i2k_y^{sh}h)$. The third term represents the third reflection wave which is first transmission from interface 1 to medium 2, next, reflection on interface 2, then, reflection on interface 1 again, then, reflection from interface 2 again, and finally is transmitted into medium 1 at the interface 1. It is reflected in the interlayer for a total of 3 times and experience round-trip 2 times. Each item on the right-hand side of Eq. (4.4.2) is similar to the understanding of Eq. (4.4.1). The first term represents the wave transmitted from interface 1 to medium 2, and then transmitted from interface 2 again; the second term represents the wave transmitted from interface 1 to medium 2, and experience one round-trip and one one-way travel, and finally transmitted from interface 2 into medium 3. Equations (4.4.1) and (4.4.2) are geometric series, which can be written as

$$\begin{aligned} H^R &= R_{12}H^I + T_{21}R_{23}T_{12}H^I \exp(i2k_y^{sh}h) \\ &\quad + T_{21}(R_{23}R_{21})R_{23}T_{12}H^I \exp(i4k_y^{sh}h) \\ &\quad + T_{21}(R_{23}R_{21}R_{23}R_{21})R_{23}T_{12}H^I \exp(i6k_y^{sh}h) + \cdots, \end{aligned} \tag{4.4.3}$$

$$\begin{aligned} H^T &= T_{23}T_{12}H^I \exp(ik_y^{sh}h) + T_{23}(R_{21}R_{23})T_{12}H^I \exp(i3k_y^{sh}h) \\ &\quad + T_{23}(R_{21}R_{23}R_{21}R_{23})T_{12}H^I \exp(i5k_y^{sh}h) + \cdots. \end{aligned} \tag{4.4.4}$$

According to the summation formula of the geometric series, we get

$$H^R = \left(R_{12} + \frac{T_{21}R_{23}T_{12}\exp(i2k_y^{sh}h)}{I - R_{23}R_{21}\exp(i2k_y^{sh}h)}\right)H^I, \tag{4.4.5}$$

$$H^T = \frac{T_{23}T_{12}\exp(ik_y^{sh}h)}{I - R_{21}R_{23}\exp(i2k_y^{sh}h)}H^I. \tag{4.4.6}$$

As long as the medium parameters are given, the reflection coefficient and the transmission coefficient in the above formula are known. Thus, the reflection coefficient and transmission coefficient of the sandwiched slab can be obtained from the above formula. Note that formulas Eqs. (4.4.5) and (4.4.6) contain the factor $\exp(i2k_y^{sh}h)$. It is noted that the reflection and transmission coefficients of the sandwiched slab do not change, as long as $2k_y^{sh}h = 2\pi n$, or, $h = \frac{\lambda}{2}n$, namely the thickness of the interlayer is the integer multiples of half-wavelength. It can also be said that the reflection and transmission coefficients of the sandwiched slab are periodic functions of the thickness and the period is the half-wavelength.

When P wave or SV wave is incident, the multiple reflection and transmission process is more complicated due to the mode conversion, as shown in Fig. 4.4. The reflection and transmission processes of various waves at the upper and lower interfaces of the interlayer can be represented by the reflection matrix and transmission matrix, and the multiple reflection and transmission processes can be expressed more clearly and concisely.

Let $\boldsymbol{A}^I = (A_0, B_0)^T, \boldsymbol{A}^R = (A_1, B_1)^T$ and $\boldsymbol{A}^T = (A_3, B_3)^T$ denote the amplitude matrix of incident waves, reflected waves and transmitted waves, respectively. When the incident wave is incident from the medium i, $\boldsymbol{R}_{ij}$ and $\boldsymbol{T}_{ij}$ represent the reflection and transmission matrix between the medium i and the medium j, respectively, and the multiple reflection and transmission processes of the P wave and SV wave can be expressed as [20]

$$\begin{aligned}\boldsymbol{A}^R &= \boldsymbol{R}_{12}\boldsymbol{A}^I + \boldsymbol{T}_{21}\boldsymbol{\Lambda}\boldsymbol{R}_{23}\boldsymbol{\Lambda}\boldsymbol{T}_{12}\boldsymbol{A}^I + \boldsymbol{T}_{21}(\boldsymbol{\Lambda}\boldsymbol{R}_{23}\boldsymbol{\Lambda}\boldsymbol{R}_{21})\boldsymbol{\Lambda}\boldsymbol{R}_{23}\boldsymbol{\Lambda}\boldsymbol{T}_{12}\boldsymbol{A}^I \\ &\quad + \boldsymbol{T}_{21}(\boldsymbol{\Lambda}\boldsymbol{R}_{23}\boldsymbol{\Lambda}\boldsymbol{R}_{21}\boldsymbol{\Lambda}\boldsymbol{R}_{23}\boldsymbol{\Lambda}\boldsymbol{R}_{21})\boldsymbol{\Lambda}\boldsymbol{R}_{23}\boldsymbol{\Lambda}\boldsymbol{T}_{12}\boldsymbol{A}^I + \cdots,\end{aligned} \tag{4.4.7}$$

$$\begin{aligned}\boldsymbol{A}^T &= \boldsymbol{T}_{23}\boldsymbol{\Lambda}\boldsymbol{T}_{12}\boldsymbol{A}^I + \boldsymbol{T}_{23}(\boldsymbol{\Lambda}\boldsymbol{R}_{21}\boldsymbol{\Lambda}\boldsymbol{R}_{23})\boldsymbol{\Lambda}\boldsymbol{T}_{12}\boldsymbol{A}^I \\ &\quad + \boldsymbol{T}_{23}(\boldsymbol{\Lambda}\boldsymbol{R}_{21}\boldsymbol{\Lambda}\boldsymbol{R}_{23}\boldsymbol{\Lambda}\boldsymbol{R}_{21}\boldsymbol{\Lambda}\boldsymbol{R}_{23})\boldsymbol{\Lambda}\boldsymbol{T}_{12}\boldsymbol{A}^I + \cdots,\end{aligned} \tag{4.4.8}$$

where

$$\boldsymbol{\Lambda} = \begin{pmatrix} \exp(ik_y^p h) & 0 \\ 0 & \exp(ik_y^s h) \end{pmatrix}$$

is the phase shift matrix generated when P wave and SV wave propagate between interface 1 and interface 2. According to the summation formula of geometric series, we get

$$\boldsymbol{A}^R = \left[\boldsymbol{R}_{12} + \boldsymbol{T}_{21}(\boldsymbol{I} - \boldsymbol{\Lambda}\boldsymbol{R}_{23}\boldsymbol{\Lambda}\boldsymbol{R}_{21})^{-1}\boldsymbol{\Lambda}\boldsymbol{R}_{23}\boldsymbol{\Lambda}\boldsymbol{T}_{12}\right]\boldsymbol{A}^I, \tag{4.4.9}$$

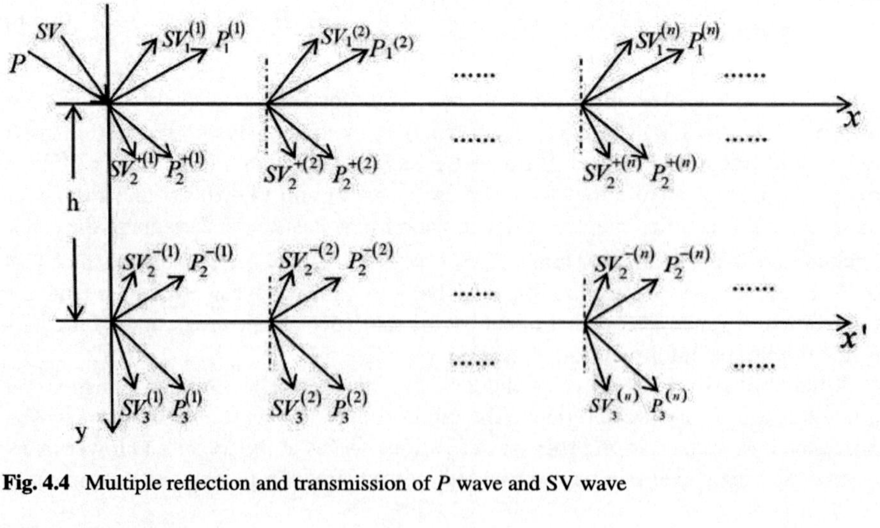

Fig. 4.4 Multiple reflection and transmission of P wave and SV wave

$$A^T = T_{23}(I - \Lambda R_{21}\Lambda R_{23})^{-1}\Lambda T_{12}A^I. \tag{4.4.10}$$

From $\exp(ik_y^p h)$ and $\exp(ik_y^s h)$, it can be seen that the reflected P wave and the transmitted P wave of the interlayer are periodic functions of the thickness and the period is the half-wavelength of the P wave. The reflected SV wave and the transmitted SV wave of the interlayer are also periodic functions of the thickness and the period is the half-wavelength of the SV wave. Because the reflection (transmission) P waves and the reflection (transmission) SV waves have different period, the reflection and transmission coefficients are actually the quasi-periodic function of thickness.

Compared with the methods in the previous sections, the method of multiple reflection and transmission can reflect the wave propagation and the formation process of reflection and transmission in the interlayer. The physical image of the wave propagation through the sandwiched slab (interlayer) is clearly displayed, which makes people have a deeper understanding of the physical process of waves propagating through sandwiched slab. However, this method is more cumbersome and not suitable for generalization to multilayer media.

4.5 Super-Interface Method

Firstly, let us discuss the reflection and transmission matrices when the SH wave is incident simultaneously on both sides of the interface, as shown in Fig. 4.5.

Suppose that the displacement of the SH waves on both sides of the interface are

$$u^{1+} = C^{1+}(0,0)\exp\left[i\left(-\xi x_1 + k_{S3}^{(1)}x_3 - \omega t\right)\right], \tag{4.5.1a}$$

$$u^{1-} = C^{1-}(0,0)\exp\left[i\left(-\xi x_1 - k_{S3}^{(1)}x_3 - \omega t\right)\right], \tag{4.5.1b}$$

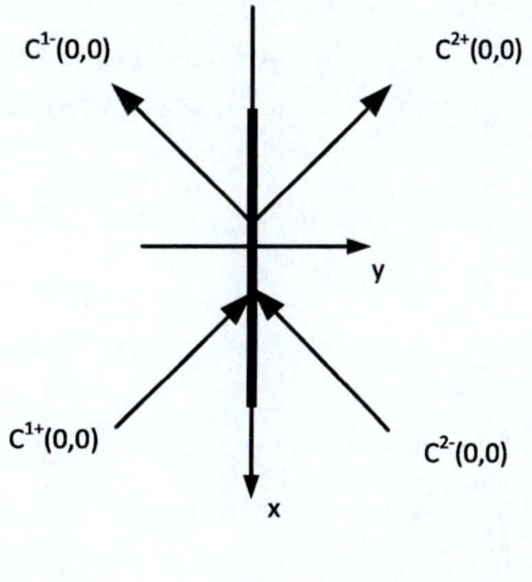

Fig. 4.5 Reflection and transmission problems in a single interface when the SH waves are incident simultaneously on both sides of the interface

$$u^{2-} = C^{2-}(0,0)\exp\left[i\left(-\xi x_1 - k_{S3}^{(2)}x_3 - \omega t\right)\right], \tag{4.5.1c}$$

$$u^{2+} = C^{2+}(0,0)\exp\left[i\left(-\xi x_1 + k_{S3}^{(2)}x_3 - \omega t\right)\right]. \tag{4.5.1d}$$

R_{ij} is defined as the reflection coefficient of the elastic wave incident from the medium i into the medium j. T_{ij} is defined as the transmission coefficient of the elastic wave incident from the medium i into the medium j. At $y = 0$, the displacement satisfies

$$\begin{cases} u^{1-} = R_{12}u^{1+} + T_{21}u^{2-} \\ u^{2+} = T_{12}u^{1+} + R_{21}u^{2-} \end{cases}, \tag{4.5.2}$$

or equivalently

$$\begin{cases} C^{1-}(0,0) = R_{12}(0,0)C^{1+}(0,0) + T_{21}(0,0)C^{2-}(0,0) \\ C^{2+}(0,0) = T_{12}(0,0)C^{1+}(0,0) + R_{21}(0,0)C^{2-}(0,0) \end{cases}. \tag{4.5.3}$$

Write Eq. (4.5.3) in matrix form, namely

$$\begin{bmatrix} C^{1-}(0,0) \\ C^{2+}(0,0) \end{bmatrix} = \begin{bmatrix} R_{12}(0,0) & T_{21}(0,0) \\ T_{12}(0,0) & R_{21}(0,0) \end{bmatrix} \cdot \begin{bmatrix} C^{1+}(0,0) \\ C^{2-}(0,0) \end{bmatrix}. \tag{4.5.4}$$

Next, let us discuss the reflection and transmission of elastic waves at the interface with two wings, as shown in Fig. 4.6. The thicknesses of the wings on both sides of the interface are a and b, respectively. The amplitudes of incident SH waves at left and right flanks are written as $C^{1+}(a,b)$ and $C^{2-}(a,b)$. The amplitudes of the reflection and transmission waves are written as $C^{1-}(a,b)$ and $C^{2+}(a,b)$. . The amplitudes at the interface ($x_3 = 0$) are written as $C^{1+}(0,0)$, $C^{2-}(0,0)$, $C^{1-}(0,0)$, $C^{2+}(0,0)$. According to Eq. (4.5.1), we can obtain the displacement expression of incident, reflection and transmission waves

$$\begin{aligned} u^{1+}|_{x_3=-a} &= C^{1+}(0,0)\exp\left[i\left(-\xi x_1 - k_{S3}^{(1)}a - \omega t\right)\right] \\ &= \exp(-ik_{S3}^{(1)}a)C^{1+}(0,0)\exp[i(-\xi x_1 - \omega t)] \\ &= C^{1+}(a,b)\exp[i(-\xi x_1 - \omega t)], \end{aligned} \tag{4.5.5a}$$

$$\begin{aligned} u^{1-}|_{x_3=-a} &= C^{1-}(0,0)\exp\left[i\left(-\xi x_1 + k_{S3}^{(1)}a - \omega t\right)\right] \\ &= \exp(ik_{S3}^{(1)}a)C^{1-}(0,0)\exp[i(-\xi x_1 - \omega t)] \\ &= C^{1-}(a,b)\exp[i(-\xi x_1 - \omega t)], \end{aligned} \tag{4.5.5b}$$

$$u^{2+}|_{x_3=b} = C^{2+}(0,0)\exp\left[i\left(-\xi x_1 + k_{S3}^{(2)}b - \omega t\right)\right]$$

$$= \exp(ik_{S3}^{(2)}b)C^{2+}(0,0)\exp[i(-\xi x_1 - \omega t)]$$
$$= C^{2+}(a,b)\exp[i(-\xi x_1 - \omega t)], \quad (4.5.5c)$$

$$u^{2-}|_{x_3=b} = C^{2-}(0,0)\exp\left[i\left(-\xi x_1 - k_{S3}^{(2)}b - \omega t\right)\right]$$
$$= \exp(-ik_{S3}^{(2)}b)C^{2-}(0,0)\exp[i(-\xi x_1 - \omega t)]$$
$$= C^{2-}(a,b)\exp[i(-\xi x_1 - \omega t)]. \quad (4.5.5d)$$

From Eq. (4.5.5), we know

$$\begin{aligned} C^{1+}(a,b) &= \mathrm{e}^{-ik_{S3}^{(1)}a}C^{1+}(0,0), C^{1-}(a,b) = \mathrm{e}^{ik_{S3}^{(1)}a}C^{1-}(0,0), \\ C^{2+}(a,b) &= \mathrm{e}^{ik_{S3}^{(2)}b}C^{2+}(0,0), C^{2-}(a,b) = \mathrm{e}^{-ik_{S3}^{(2)}b}C^{2-}(0,0). \end{aligned} \quad (4.5.6)$$

Write Eq. (4.5.6) in matrix form, namely

$$\begin{pmatrix} C^{1-}(a,b) \\ C^{2+}(a,b) \end{pmatrix} = \begin{pmatrix} \mathrm{e}^{ik_{S3}^{(1)}a} & 0 \\ 0 & \mathrm{e}^{ik_{S3}^{(2)}b} \end{pmatrix} \begin{pmatrix} C^{1-}(0,0) \\ C^{2+}(0,0) \end{pmatrix}, \quad (4.5.7a)$$

$$\begin{pmatrix} C^{1+}(a,b) \\ C^{2-}(a,b) \end{pmatrix} = \begin{pmatrix} \mathrm{e}^{-ik_{S3}^{(1)}a} & 0 \\ 0 & \mathrm{e}^{-ik_{S3}^{(2)}b} \end{pmatrix} \begin{pmatrix} C^{1+}(0,0) \\ C^{2-}(0,0) \end{pmatrix}. \quad (4.5.7b)$$

We can assume the amplitudes at the interface satisfy

$$\begin{aligned} C^{1-}(a,b) &= R_{12}(a,b)C^{1+}(a,b) + T_{21}(a,b)C^{2-}(a,b), \\ C^{2+}(a,b) &= T_{12}(a,b)C^{1+}(a,b) + R_{21}(a,b)C^{2-}(a,b). \end{aligned} \quad (4.5.8)$$

Write Eq. (4.5.8) in matrix form, namely

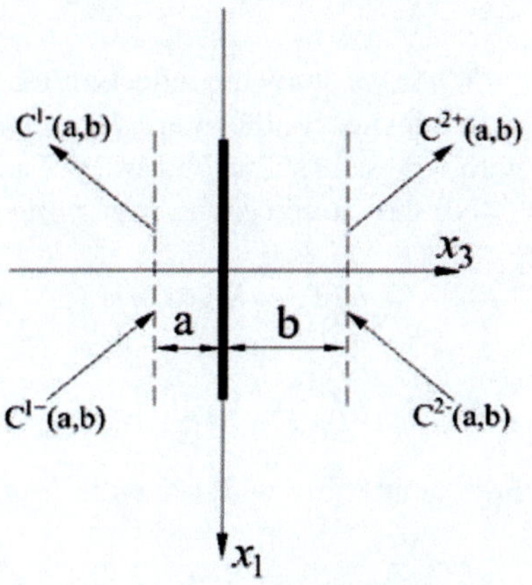

Fig. 4.6 Reflection and transmission of interface with two wings

$$\begin{pmatrix} C^{1-}(a,b) \\ C^{2+}(a,b) \end{pmatrix} = \begin{pmatrix} R_{12}(a,b) & T_{21}(a,b) \\ T_{12}(a,b) & R_{21}(a,b) \end{pmatrix} \begin{pmatrix} C^{1+}(a,b) \\ C^{2-}(a,b) \end{pmatrix}. \tag{4.5.9}$$

Inserting Eq. (4.5.7) into Eq. (4.5.9) leads to

$$\begin{aligned} &\begin{pmatrix} e^{ik_{S3}^{(1)}a} & 0 \\ 0 & e^{ik_{S3}^{(2)}b} \end{pmatrix} \begin{pmatrix} C^{1-}(0,0) \\ C^{2+}(0,0) \end{pmatrix} \\ &= \begin{pmatrix} R_{12}(a,b) & T_{21}(a,b) \\ T_{12}(a,b) & R_{21}(a,b) \end{pmatrix} \begin{pmatrix} e^{-ik_{S3}^{(1)}a} & 0 \\ 0 & e^{-ik_{S3}^{(2)}b} \end{pmatrix} \begin{pmatrix} C^{1+}(0,0) \\ C^{2-}(0,0) \end{pmatrix}. \end{aligned} \tag{4.5.10}$$

The amplitudes of elastic waves at the interface without wings satisfy

$$\begin{pmatrix} C^{1-}(0,0) \\ C^{2+}(0,0) \end{pmatrix} = \begin{pmatrix} R_{12}(0,0) & T_{21}(0,0) \\ T_{12}(0,0) & R_{21}(0,0) \end{pmatrix} \begin{pmatrix} C^{1+}(0,0) \\ C^{2-}(0,0) \end{pmatrix}. \tag{4.5.11}$$

Therefore, the relationship of the reflection/transmission coefficient between the interface with two wings and the interface without wings can be obtained as

$$\begin{pmatrix} R_{12}(0,0) & T_{21}(0,0) \\ T_{12}(0,0) & R_{21}(0,0) \end{pmatrix} = \begin{pmatrix} e^{ik_{S3}^{(1)}a} & 0 \\ 0 & e^{ik_{S3}^{(2)}b} \end{pmatrix}^{-1} \begin{pmatrix} R_{12}(a,b) & T_{21}(a,b) \\ T_{12}(a,b) & R_{21}(a,b) \end{pmatrix} \begin{pmatrix} e^{-ik_{S3}^{(1)}a} & 0 \\ 0 & e^{-ik_{S3}^{(2)}b} \end{pmatrix}, \tag{4.5.12}$$

$$\begin{pmatrix} R_{12}(a,b) & T_{21}(a,b) \\ T_{12}(a,b) & R_{21}(a,b) \end{pmatrix} = \begin{pmatrix} e^{ik_{S3}^{(1)}a} & 0 \\ 0 & e^{ik_{S3}^{(2)}b} \end{pmatrix} \begin{pmatrix} R_{12}(0,0) & T_{21}(0,0) \\ T_{12}(0,0) & R_{21}(0,0) \end{pmatrix} \begin{pmatrix} e^{-ik_{S3}^{(1)}a} & 0 \\ 0 & e^{-ik_{S3}^{(2)}b} \end{pmatrix}^{-1}, \tag{4.5.13}$$

or equivalently

$$\begin{pmatrix} R_{12}(a,b) & T_{21}(a,b) \\ T_{12}(a,b) & R_{21}(a,b) \end{pmatrix} = \begin{pmatrix} e^{ik_{S3}^{(1)}a} R_{12}(0,0) e^{ik_{S3}^{(1)}a} & e^{ik_{S3}^{(1)}a} T_{21}(0,0) e^{ik_{S3}^{(2)}b} \\ e^{ik_{S3}^{(2)}b} T_{12}(0,0) e^{ik_{S3}^{(1)}a} & e^{ik_{S3}^{(2)}b} R_{21}(0,0) e^{ik_{S3}^{(2)}b} \end{pmatrix}. \tag{4.5.14}$$

When we study the reflection/transmission of elastic waves through the interlayer with thickness h, the interlayer can be regarded as the combination of two interfaces with wings, as shown in Figs. 4.7 and 4.8.

For the interface with right wing (Fig. 4.8a), the amplitudes satisfy

$$C^{1-}(0,h/2) = R_{12}(0,h/2)C^{1+}(0,h/2) + T_{21}(0,h/2)C^{2-}(0,h/2), \tag{4.5.15a}$$

$$C^{2+}(0,h/2) = T_{12}(0,h/2)C^{1+}(0,h/2) + R_{21}(0,h/2)C^{2-}(0,h/2). \tag{4.5.15b}$$

For the interface with left wing (Fig. 4.8b), the amplitudes satisfy

$$C^{2-}(h/2,0) = R_{23}(h/2,0)C^{2+}(h/2,0) + T_{32}(h/2,0)C^{3-}(h/2,0), \tag{4.5.16a}$$

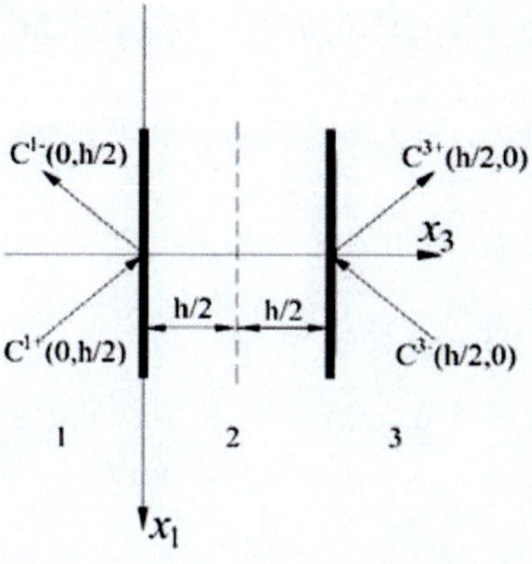

Fig. 4.7 Reflection and transmission of the elastic wave through an interlayer sandwiched between two half-spaces

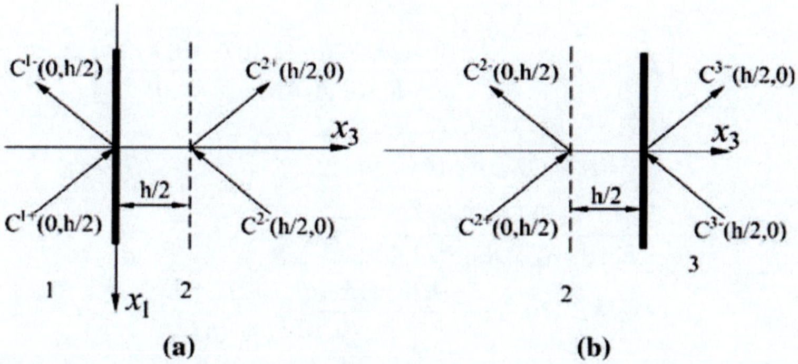

Fig. 4.8 Interlayer is divided into two interfaces with wings. In other words, the interlayer is a combination of two interfaces with wings. **a** two wings are $a = 0, b = h/2$; **b** two wings are $a = h/2, b = 0$

$$C^{3+}(h/2, 0) = T_{23}(h/2, 0)C^{2+}(h/2, 0) + R_{32}(h/2, 0)C^{3-}(h/2, 0). \quad (4.5.16b)$$

Inserting

$$C^{2-}(h/2, 0) = C^{2-}(0, h/2) = C^{2-}, C^{2+}(h/2, 0) = C^{2+}(0, h/2) = C^{2+}, \quad (4.5.17)$$

into Eqs. (4.5.15) and (4.5.16) leads to

$$C^{1-}(0, h/2) = R_{12}(0, h/2)C^{1+}(0, h/2) + T_{21}(0, h/2)C^{2-}, \quad (4.5.18a)$$

$$C^{2+} = T_{12}(0, h/2)C^{1+}(0, h/2) + R_{21}(0, h/2)C^{2-}, \tag{4.5.18b}$$

$$C^{2-} = R_{23}(h/2, 0)C^{2+} + T_{32}(h/2, 0)C^{3-}(h/2, 0), \tag{4.5.18c}$$

$$C^{3+}(h/2, 0) = T_{23}(h/2, 0)C^{2+} + R_{32}(h/2, 0)C^{3-}(h/2, 0). \tag{4.5.18d}$$

From Eqs. (4.5.18b) and (4.5.18c), we get

$$C^{2+} = \frac{T_{12}(0, h/2)C^{1+}(0, h/2) + R_{21}(0, h/2)T_{32}(h/2, 0)C^{3-}(h/2, 0)}{I - R_{21}(0, h/2)R_{23}(h/2, 0)}, \tag{4.5.19a}$$

$$C^{2-} = \frac{R_{23}(h/2, 0)T_{12}(0, h/2)C^{1+}(0, h/2) + T_{32}(h/2, 0)C^{3-}(h/2, 0)}{I - R_{21}(0, h/2)R_{23}(h/2, 0)}. \tag{4.5.19b}$$

Inserting Eq. (4.5.19) into Eqs. (4.5.18a) and (4.5.18d) leads to

$$\begin{aligned} C^{1-}(0, h/2) = {} & \left[R_{12}(0, h/2) + \frac{T_{21}(0, h/2)R_{23}(h/2, 0)T_{12}(0, h/2)}{I - R_{21}(0, h/2)R_{23}(h/2, 0)}\right]C^{1+}(0, h/2) \\ & + \frac{T_{21}(0, h/2)T_{32}(h/2, 0)}{I - R_{21}(0, h/2)R_{23}(h/2, 0)}C^{3-}(h/2, 0), \end{aligned} \tag{4.5.20a}$$

$$\begin{aligned} C^{3+}(h/2, 0) = {} & \frac{T_{23}(h/2, 0)T_{12}(0, h/2)}{I - R_{21}(0, h/2)R_{23}(h/2, 0)}C^{1+}(0, h/2) \\ & + \left[R_{32}(h/2, 0) + \frac{T_{23}(h/2, 0)R_{21}(0, h/2)T_{32}(h/2, 0)}{I - R_{21}(0, h/2)R_{23}(h/2, 0)}\right]C^{3-}(h/2, 0). \end{aligned} \tag{4.5.20b}$$

Considering Eq. (4.5.13), the relationships between the amplitudes of the reflection/transmission waves and the incident wave through an interlayer of thickness h are

$$\begin{aligned} C^{1-} = {} & \left[R_{12} + \frac{\left(T_{21}e^{ik_{s3}^{(2)}\frac{h}{2}}\right)\cdot\left(e^{ik_{s3}^{(2)}\frac{h}{2}}R_{23}e^{ik_{s3}^{(2)}\frac{h}{2}}\right)\cdot\left(e^{ik_{s3}^{(2)}\frac{h}{2}}T_{12}\right)}{I - \left(e^{ik_{s3}^{(2)}\frac{h}{2}}R_{21}e^{ik_{s3}^{(2)}\frac{h}{2}}\right)\cdot\left(e^{ik_{s3}^{(2)}\frac{h}{2}}R_{23}e^{ik_{s3}^{(2)}\frac{h}{2}}\right)}\right]C^{1+} \\ & + \frac{\left(T_{21}e^{ik_{s3}^{(2)}\frac{h}{2}}\right)\cdot\left(e^{ik_{s3}^{(2)}\frac{h}{2}}T_{32}\right)}{I - \left(e^{ik_{s3}^{(2)}\frac{h}{2}}R_{21}e^{ik_{s3}^{(2)}\frac{h}{2}}\right)\cdot\left(e^{ik_{s3}^{(2)}\frac{h}{2}}R_{23}e^{ik_{s3}^{(2)}\frac{h}{2}}\right)}C^{3-}, \end{aligned} \tag{4.5.21a}$$

$$C^{3+} = \frac{\left(T_{23}e^{ik_{s3}^{(2)}\frac{h}{2}}\right)\cdot\left(e^{ik_{s3}^{(2)}\frac{h}{2}}T_{12}\right)}{I - \left(e^{ik_{s3}^{(2)}\frac{h}{2}}R_{21}e^{ik_{s3}^{(2)}\frac{h}{2}}\right)\cdot\left(e^{ik_{s3}^{(2)}\frac{h}{2}}R_{23}e^{ik_{s3}^{(2)}\frac{h}{2}}\right)}C^{1+}$$

$$+\left[R_{32}+\frac{\left(T_{23}\mathrm{e}^{ik_{S3}^{(2)}\frac{h}{2}}\right)\cdot\left(\mathrm{e}^{ik_{S3}^{(2)}\frac{h}{2}}R_{21}\mathrm{e}^{ik_{S3}^{(2)}\frac{h}{2}}\right)\cdot\left(\mathrm{e}^{ik_{S3}^{(2)}\frac{h}{2}}T_{32}\right)}{I-\left(\mathrm{e}^{ik_{S3}^{(2)}\frac{h}{2}}R_{21}\mathrm{e}^{ik_{S3}^{(2)}\frac{h}{2}}\right)\cdot\left(\mathrm{e}^{ik_{S3}^{(2)}\frac{h}{2}}R_{23}\mathrm{e}^{ik_{S3}^{(2)}\frac{h}{2}}\right)}\right]C^{3-}. \quad (4.5.21b)$$

From Eq. (4.5.21), we can obtain

$$C^{1-}=\left[R_{12}+\frac{T_{21}R_{23}T_{12}\mathrm{e}^{ik_{S3}^{(2)}\cdot 2h}}{I-R_{21}R_{23}\mathrm{e}^{ik_{S3}^{(2)}\cdot 2h}}\right]C^{1+}+\frac{T_{21}T_{32}\mathrm{e}^{ik_{S3}^{(2)}\cdot h}}{I-R_{21}R_{23}\mathrm{e}^{ik_{S3}^{(2)}\cdot 2h}}C^{3-}, \quad (4.5.22a)$$

$$C^{3+}=\frac{T_{23}T_{12}\mathrm{e}^{ik_{S3}^{(2)}\cdot h}}{I-R_{21}R_{23}\mathrm{e}^{ik_{S3}^{(2)}\cdot 2h}}C^{1+}+\left[R_{32}+\frac{T_{23}R_{21}T_{32}\mathrm{e}^{ik_{S3}^{(2)}\cdot 2h}}{I-R_{21}R_{23}\mathrm{e}^{ik_{S3}^{(2)}\cdot 2h}}\right]C^{3-}. \quad (4.5.22b)$$

Write Eq. (4.5.22) in matrix form

$$\begin{pmatrix} C^{1-} \\ C^{3+} \end{pmatrix}=\begin{pmatrix} R_{13} & T_{31} \\ T_{13} & R_{31} \end{pmatrix}\begin{pmatrix} C^{1+} \\ C^{3-} \end{pmatrix}, \quad (4.5.23)$$

where

$$R_{13}=R_{12}+\frac{T_{21}R_{23}T_{12}\mathrm{e}^{ik_{S3}^{(2)}\cdot 2h}}{I-R_{21}R_{23}\mathrm{e}^{ik_{S3}^{(2)}\cdot 2h}}, \quad (4.5.24a)$$

$$T_{31}=\frac{T_{21}T_{32}\mathrm{e}^{ik_{S3}^{(2)}h}}{I-R_{21}R_{23}\mathrm{e}^{ik_{S3}^{(2)}\cdot 2h}}, \quad (4.5.24b)$$

$$T_{13}=\frac{T_{23}T_{12}\mathrm{e}^{ik_{S3}^{(2)}\cdot h}}{I-R_{21}R_{23}\mathrm{e}^{ik_{S3}^{(2)}\cdot 2h}}, \quad (4.5.24c)$$

$$R_{31}=R_{32}+\frac{T_{23}R_{21}T_{32}\mathrm{e}^{ik_{S3}^{(2)}\cdot 2h}}{I-R_{21}R_{23}\mathrm{e}^{ik_{S3}^{(2)}\cdot 2h}}. \quad (4.5.24d)$$

In this way, the reflection and transmission of elastic waves through the bi-interfaces (both sides of the interlayer) are regarded as the reflection and transmission through an equivalent single interface. This equivalent interface is called a **super-interface**. The reflection and transmission of elastic waves through n layers ($n+1$ interfaces) can also be regarded as the reflection and transmission through a super-interface. Now we consider the problem of reflection and transmission of SH waves through a double-layer medium (3 interfaces). The first and the second interfaces are equivalent to a super-interface, and the reflection/transmission matrixes (i.e. R_{13}, R_{31}, T_{13}, T_{31}) of the super-interface are obtained. Then, the problem of the tri-interfaces is regarded as the combination of a super-interface and an interface with left wing, which can be equivalent to a new super-interface. It is assumed that the final super-interface has the following reflection and transmission relationship

$$\begin{pmatrix} C^{1-} \\ C^{4+} \end{pmatrix} = \begin{pmatrix} R_{14} & T_{41} \\ T_{14} & R_{41} \end{pmatrix} \begin{pmatrix} C^{1+} \\ C^{4-} \end{pmatrix}. \tag{4.5.25}$$

From Eq. (4.5.25), we can obtain

$$\begin{pmatrix} R_{14} & T_{41} \\ T_{14} & R_{41} \end{pmatrix} = \begin{pmatrix} R_{13} + \frac{T_{31} R_{34} T_{13} e^{ik_{S3}^{(3)} \cdot 2h}}{I - R_{31} R_{34} e^{ik_{S3}^{(3)} \cdot 2h}} & \frac{T_{31} T_{43} e^{ik_{S3}^{(3)} \cdot h}}{I - R_{31} R_{34} e^{ik_{S3}^{(3)} \cdot 2h}} \\ \frac{T_{34} T_{13} e^{ik_{S3}^{(3)} \cdot h}}{I - R_{31} R_{34} e^{ik_{S3}^{(3)} \cdot 2h}} & R_{43} + \frac{T_{34} R_{31} T_{43} e^{ik_{S3}^{(3)} \cdot 2h}}{I - R_{31} R_{34} e^{ik_{S3}^{(3)} \cdot 2h}} \end{pmatrix}. \tag{4.5.26}$$

For the N-layer medium ($N = n - 2$), it is assumed that the amplitude of the reflection wave and incident wave can be related by the reflection/transmission matrix of the super-interface

$$\begin{pmatrix} C^{1-} \\ C^{n+} \end{pmatrix} = \begin{pmatrix} R_{1\mathrm{n}} & T_{\mathrm{n}1} \\ T_{1\mathrm{n}} & R_{\mathrm{n}1} \end{pmatrix} \begin{pmatrix} C^{1+} \\ C^{n-} \end{pmatrix}. \tag{4.5.27}$$

Then, we can obtain the recursion relation

$$\begin{pmatrix} R_{1\mathrm{n}} & T_{\mathrm{n}1} \\ T_{1\mathrm{n}} & R_{\mathrm{n}1} \end{pmatrix} = \begin{pmatrix} R_{1,n-1} + \frac{T_{n-1,1} R_{n-1,n} T_{1,n-1} e^{ik_{S3}(n-1) \cdot 2h}}{I - R_{n-1,1} R_{n-1,n} e^{ik_{S3}(n-1) \cdot 2h}} & \frac{T_{n-1,1} T_{n,n-1} e^{ik_{S3}(n-1) \cdot h}}{I - R_{n-1,1} R_{n-1,n} e^{ik_{S3}(n-1) \cdot 2h}} \\ \frac{T_{n-1,n} T_{1,n-1} e^{ik_{S3}(n-1) \cdot h}}{I - R_{n-1,1} R_{n-1,n} e^{ik_{S3}(n-1) \cdot 2h}} & R_{n,n-1} + \frac{T_{n-1,n} R_{n-1,1} T_{n,n-1} e^{ik_{S3}(n-1) \cdot 2h}}{I - R_{n-1,1} R_{n-1,n} e^{ik_{S3}(n-1) \cdot 2h}} \end{pmatrix}. \tag{4.5.28}$$

The reflection and transmission coefficients of SH wave through N-layer medium can be obtained by this recursion relation.

The reflection/transmission of P and SV waves through the N-layer medium is now considered. Firstly, we consider the reflection and transmission of the P and SV waves through a single interface. The potential function on both sides of the interface is assumed to be

$$\varphi^{1+} = A^{1+}(0,0) \exp\left[i\left(-\xi x_1 + k_{P3}^{(1)} x_3 - \omega t\right)\right], \tag{4.5.29a}$$

$$\psi^{1+} = B^{1+}(0,0) \exp\left[i\left(-\xi x_1 + k_{S3}^{(1)} x_3 - \omega t\right)\right], \tag{4.5.29b}$$

$$\varphi^{1-} = A^{1-}(0,0) \exp\left[i\left(-\xi x_1 - k_{P3}^{(1)} x_3 - \omega t\right)\right], \tag{4.5.29c}$$

$$\psi^{1-} = B^{1-}(0,0) \exp\left[i\left(-\xi x_1 - k_{S3}^{(1)} x_3 - \omega t\right)\right], \tag{4.5.29d}$$

$$\varphi^{2+} = A^{2+}(0,0) \exp\left[i\left(-\xi x_1 + k_{P3}^{(2)} x_3 - \omega t\right)\right], \tag{4.5.29e}$$

$$\psi^{2+} = B^{2+}(0,0) \exp\left[i\left(-\xi x_1 + k_{S3}^{(2)} x_3 - \omega t\right)\right], \tag{4.5.29f}$$

$$\varphi^{2-} = A^{2-}(0,0)\exp\left[i\left(-\xi x_1 - k_{P3}^{(2)}x_3 - \omega t\right)\right], \tag{4.5.29g}$$

$$\psi^{2-} = B^{2-}(0,0)\exp\left[i\left(-\xi x_1 - k_{S3}^{(2)}x_3 - \omega t\right)\right]. \tag{4.5.29h}$$

Let $\boldsymbol{R}_{ij}$ is the reflection coefficient of the elastic wave incident from the medium i into the medium j, and $\boldsymbol{T}_{ij}$ is the transmission coefficient of the elastic wave incident from the medium i into the medium j. The amplitudes of the incident, reflection, and transmission waves are assumed to be

$$\boldsymbol{C}^{1-} = \begin{pmatrix} A^{1-} \\ B^{1-} \end{pmatrix}, \boldsymbol{C}^{1+} = \begin{pmatrix} A^{1+} \\ B^{1+} \end{pmatrix}, \boldsymbol{C}^{2-} = \begin{pmatrix} A^{2-} \\ B^{2-} \end{pmatrix}, \boldsymbol{C}^{2+} = \begin{pmatrix} A^{2+} \\ B^{2+} \end{pmatrix}. \tag{4.5.30}$$

The amplitudes satisfy

$$\boldsymbol{C}^{1-}(0,0) = \boldsymbol{R}_{12}(0,0)\boldsymbol{C}^{1+}(0,0) + \boldsymbol{T}_{21}(0,0)\boldsymbol{C}^{2-}(0,0), \tag{4.5.31a}$$

$$\boldsymbol{C}^{2+}(0,0) = \boldsymbol{T}_{12}(0,0)\boldsymbol{C}^{1+}(0,0) + \boldsymbol{R}_{21}(0,0)\boldsymbol{C}^{2-}(0,0). \tag{4.5.31b}$$

For interfaces with the wings, similarly, the amplitudes satisfy

$$\boldsymbol{C}^{1-}(a,b) = \boldsymbol{R}_{12}(a,b)\boldsymbol{C}^{1+}(a,b) + \boldsymbol{T}_{21}(a,b)\boldsymbol{C}^{2-}(a,b), \tag{4.5.32a}$$

$$\boldsymbol{C}^{2+}(a,b) = \boldsymbol{T}_{12}(a,b)\boldsymbol{C}^{1+}(a,b) + \boldsymbol{R}_{21}(a,b)\boldsymbol{C}^{2-}(a,b). \tag{4.5.32b}$$

The potential functions at the interfaces can be expressed as

$$\begin{aligned} \varphi^{1+}|_{x_3=-a} &= A^{1+}(0,0)\exp\left[i\left(-\xi x_1 - k_{P3}^{(1)}a - \omega t\right)\right] \\ &= \exp(-ik_{P3}^{(1)}a)A^{1+}(0,0)\exp[i(-\xi x_1 - \omega t)] \\ &= A^{1+}(a,b)\exp[i(-\xi x_1 - \omega t)], \end{aligned} \tag{4.5.33a}$$

$$\begin{aligned} \psi^{1+}|_{x_3=-a} &= B^{1+}(0,0)\exp\left[i\left(-\xi x_1 - k_{S3}^{(1)}a - \omega t\right)\right] \\ &= \exp(-ik_{S3}^{(1)}a)B^{1+}(0,0)\exp[i(-\xi x_1 - \omega t)] \\ &= B^{1+}(a,b)\exp[i(-\xi x_1 - \omega t)], \end{aligned} \tag{4.5.33b}$$

$$\begin{aligned} \varphi^{1-}|_{x_3=-a} &= A^{1-}(0,0)\exp\left[i\left(-\xi x_1 + k_{P3}^{(1)}a - \omega t\right)\right] \\ &= \exp(ik_{P3}^{(1)}a)A^{1-}(0,0)\exp[i(-\xi x_1 - \omega t)] \\ &= A^{1-}(a,b)\exp[i(-\xi x_1 - \omega t)], \end{aligned} \tag{4.5.33c}$$

$$\begin{aligned}\psi^{1-}|_{x_3=-a} &= B^{1-}(0,0)\exp\left[i\left(-\xi x_1 + k_{S3}^{(1)}a - \omega t\right)\right] \\ &= \exp(ik_{S3}^{(1)}a)B^{1-}(0,0)\exp[i(-\xi x_1 - \omega t)] \\ &= B^{1-}(a,b)\exp[i(-\xi x_1 - \omega t)],\end{aligned} \tag{4.5.33d}$$

$$\begin{aligned}\varphi^{2+}|_{x_3=b} &= A^{1+}(0,0)\exp\left[i\left(-\xi x_1 + k_{P3}^{(1)}b - \omega t\right)\right] \\ &= \exp(ik_{P3}^{(1)}b)A^{2+}(0,0)\exp[i(-\xi x_1 - \omega t)] \\ &= A^{2+}(a,b)\exp[i(-\xi x_1 - \omega t)],\end{aligned} \tag{4.5.33e}$$

$$\begin{aligned}\psi^{2+}|_{x_3=b} &= B^{2+}(0,0)\exp\left[i\left(-\xi x_1 + k_{S3}^{(2)}b - \omega t\right)\right] \\ &= \exp(ik_{S3}^{(2)}b)B^{2+}(0,0)\exp[i(-\xi x_1 - \omega t)] \\ &= B^{2+}(a,b)\exp[i(-\xi x_1 - \omega t)],\end{aligned} \tag{4.5.33f}$$

$$\begin{aligned}\varphi^{2-}|_{x_3=b} &= A^{2-}(0,0)\exp\left[i\left(-\xi x_1 - k_{P3}^{(1)}b - \omega t\right)\right] \\ &= \exp(-ik_{P3}^{(2)}b)A^{2-}(0,0)\exp[i(-\xi x_1 - \omega t)] \\ &= A^{2-}(a,b)\exp[i(-\xi x_1 - \omega t)],\end{aligned} \tag{4.5.33g}$$

$$\begin{aligned}\psi^{2-}|_{x_3=b} &= B^{2-}(0,0)\exp\left[i\left(-\xi x_1 - k_{S3}^{(1)}b - \omega t\right)\right] \\ &= \exp(-ik_{S3}^{(2)}b)B^{2-}(0,0)\exp[i(-\xi x_1 - \omega t)] \\ &= B^{2-}(a,b)\exp[i(-\xi x_1 - \omega t)].\end{aligned} \tag{4.5.33h}$$

The relationship of amplitudes of elastic waves through the interface with wings and the interface without wings can be written as

$$\begin{aligned}\boldsymbol{C}^{1-}(a,b) &= \begin{pmatrix} A^{1-}(a,b) \\ B^{1-}(a,b) \end{pmatrix} = \begin{pmatrix} \exp(ik_{P3}^{(1)}a) & 0 \\ 0 & \exp(ik_{S3}^{(1)}a) \end{pmatrix}\begin{pmatrix} A^{1-}(0,0) \\ B^{1-}(0,0) \end{pmatrix} \\ &= \begin{pmatrix} \exp(ik_{P3}^{(1)}a) & 0 \\ 0 & \exp(ik_{S3}^{(1)}a) \end{pmatrix}\boldsymbol{C}^{1-}(0,0),\end{aligned} \tag{4.5.34a}$$

$$\begin{aligned}\boldsymbol{C}^{1+}(a,b) &= \begin{pmatrix} A^{1+}(a,b) \\ B^{1+}(a,b) \end{pmatrix} = \begin{pmatrix} \exp(-ik_{P3}^{(1)}a) & 0 \\ 0 & \exp(-ik_{S3}^{(1)}a) \end{pmatrix}\begin{pmatrix} A^{1+}(0,0) \\ B^{1+}(0,0) \end{pmatrix} \\ &= \begin{pmatrix} \exp(-ik_{P3}^{(1)}a) & 0 \\ 0 & \exp(-ik_{S3}^{(1)}a) \end{pmatrix}\boldsymbol{C}^{1+}(0,0),\end{aligned} \tag{4.5.34b}$$

$$
\begin{aligned}
\boldsymbol{C}^{2-}(a,b) = \begin{pmatrix} A^{2-}(a,b) \\ B^{2-}(a,b) \end{pmatrix} &= \begin{pmatrix} \exp(-ik_{P3}^{(2)}b) & 0 \\ 0 & \exp(-ik_{S3}^{(2)}b) \end{pmatrix} \begin{pmatrix} A^{2-}(0,0) \\ B^{2-}(0,0) \end{pmatrix} \\
&= \begin{pmatrix} \exp(-ik_{P3}^{(2)}b) & 0 \\ 0 & \exp(-ik_{S3}^{(2)}b) \end{pmatrix} \boldsymbol{C}^{2-}(0,0),
\end{aligned}
\tag{4.5.34c}
$$

$$
\begin{aligned}
\boldsymbol{C}^{2+}(a,b) = \begin{pmatrix} A^{2+}(a,b) \\ B^{2+}(a,b) \end{pmatrix} &= \begin{pmatrix} \exp(ik_{P3}^{(2)}b) & 0 \\ 0 & \exp(ik_{S3}^{(2)}b) \end{pmatrix} \begin{pmatrix} A^{2+}(0,0) \\ B^{2+}(0,0) \end{pmatrix} \\
&= \begin{pmatrix} \exp(ik_{P3}^{(2)}b) & 0 \\ 0 & \exp(ik_{S3}^{(2)}b) \end{pmatrix} \boldsymbol{C}^{2+}(0,0).
\end{aligned}
\tag{4.5.34d}
$$

Inserting Eq. (4.5.34) into Eq. (4.5.32) leads to

$$
\boldsymbol{R}_{12}(a,b) = \boldsymbol{A}^{(1)}(+a) \cdot \boldsymbol{R}_{12}(0,0) \cdot [\boldsymbol{A}^{(1)}]^{-1}(-a), \tag{4.5.35a}
$$

$$
\boldsymbol{T}_{21}(a,b) = \boldsymbol{A}^{(1)}(+a) \cdot \boldsymbol{T}_{21}(0,0) \cdot [\boldsymbol{A}^{(2)}]^{-1}(-b), \tag{4.5.35b}
$$

$$
\boldsymbol{R}_{21}(a,b) = \boldsymbol{A}^{(2)}(+b) \cdot \boldsymbol{R}_{12}(0,0) \cdot [\boldsymbol{A}^{(2)}]^{-1}(-b), \tag{4.5.35c}
$$

$$
\boldsymbol{T}_{12}(a,b) = \boldsymbol{A}^{(2)}(+b) \cdot \boldsymbol{T}_{21}(0,0) \cdot [\boldsymbol{A}^{(1)}]^{-1}(-a), \tag{4.5.35d}
$$

where

$$
\boldsymbol{A}^{(1)}(s) = \begin{pmatrix} e^{ik_{P3}^{(1)}s} & 0 \\ 0 & e^{ik_{S3}^{(1)}s} \end{pmatrix}, \tag{4.5.36a}
$$

$$
\boldsymbol{A}^{(2)}(s) = \begin{pmatrix} e^{ik_{P3}^{(2)}s} & 0 \\ 0 & e^{ik_{S3}^{(2)}s} \end{pmatrix}. \tag{4.5.36b}
$$

When studying the problem of reflection and transmission of elastic waves through an interlayer with a thickness of h, the interlayer can be regarded as a combination of two interfaces with wings. For the interface with right wing, the amplitudes satisfy

$$
\boldsymbol{C}^{1-}(0,h/2) = \boldsymbol{R}_{12}(0,h/2) \cdot \boldsymbol{C}^{1+}(0,h/2) + \boldsymbol{T}_{21}(0,h/2) \cdot \boldsymbol{C}^{2-}(0,h/2), \tag{4.5.37a}
$$

$$
\boldsymbol{C}^{1-}(0,h/2) = \boldsymbol{R}_{12}(0,h/2) \cdot \boldsymbol{C}^{1+}(0,h/2) + \boldsymbol{T}_{21}(0,h/2) \cdot \boldsymbol{C}^{2-}(0,h/2). \tag{4.5.37b}
$$

For the interface with left wing, the amplitudes satisfy

$$\boldsymbol{C}^{2-}(h/2,0)=\boldsymbol{R}_{23}(h/2,0)\cdot\boldsymbol{C}^{2+}(h/2,0)+\boldsymbol{T}_{32}(h/2,0)\cdot\boldsymbol{C}^{3-}(h/2,0), \tag{4.5.38a}$$

$$\boldsymbol{C}^{3+}(h/2,0)=\boldsymbol{T}_{23}(h/2,0)\cdot\boldsymbol{C}^{2+}(h/2,0)+\boldsymbol{R}_{32}(h/2,0)\cdot\boldsymbol{C}^{3-}(h/2,0). \tag{4.5.38b}$$

Considering

$$\boldsymbol{C}^{2-}(h/2,0)=\boldsymbol{C}^{2-}(0,h/2)=\boldsymbol{C}^{2-},\ \boldsymbol{C}^{2+}(h/2,0)=\boldsymbol{C}^{2+}(0,h/2)=\boldsymbol{C}^{2+}, \tag{4.5.39}$$

we can obtain

$$\boldsymbol{C}^{1-}(0,h/2)=\boldsymbol{R}_{12}(0,h/2)\cdot\boldsymbol{C}^{1+}(0,h/2)+\boldsymbol{T}_{21}(0,h/2)\cdot\boldsymbol{C}^{2-}, \tag{4.5.40a}$$

$$\boldsymbol{C}^{2+}=\boldsymbol{T}_{12}(0,h/2)\cdot\boldsymbol{C}^{1+}(0,h/2)+\boldsymbol{R}_{21}(0,h/2)\cdot\boldsymbol{C}^{2-}, \tag{4.5.40b}$$

$$\boldsymbol{C}^{2-}=\boldsymbol{R}_{23}(h/2,0)\cdot\boldsymbol{C}^{2+}+\boldsymbol{T}_{32}(h/2,0)\cdot\boldsymbol{C}^{3-}(h/2,0), \tag{4.5.40c}$$

$$\boldsymbol{C}^{3-}(h/2,0)=\boldsymbol{T}_{23}(h/2,0)\cdot\boldsymbol{C}^{2+}+\boldsymbol{R}_{32}(h/2,0)\cdot\boldsymbol{C}^{3-}(h/2,0). \tag{4.5.40d}$$

According to Eqs. (4.5.40b) and (4.5.40c), we can obtain

$$\begin{aligned}\boldsymbol{C}^{2+}=&[\boldsymbol{E}-\boldsymbol{R}_{21}(0,h/2)\cdot\boldsymbol{R}_{23}(h/2,0)]^{-1}\cdot[\mathbf{T}_{12}(0,h/2)\cdot\boldsymbol{C}^{1+}(0,h/2)\\&+\boldsymbol{R}_{21}(0,h/2)\cdot\boldsymbol{T}_{32}(h/2,0)\cdot\boldsymbol{C}^{3-}(h/2,0)],\end{aligned} \tag{4.5.41a}$$

$$\begin{aligned}\boldsymbol{C}^{2-}=&[\boldsymbol{E}-\boldsymbol{R}_{23}(h/2,0)\cdot\boldsymbol{R}_{21}(0,h/2)]^{-1}\cdot[\mathbf{R}_{23}(h/2,0)\cdot\boldsymbol{T}_{12}(0,h/2)\\&\cdot\boldsymbol{C}^{1+}(0,h/2)+\boldsymbol{T}_{32}(h/2,0)\cdot\boldsymbol{C}^{3-}(h/2,0)].\end{aligned} \tag{4.5.41b}$$

Inserting it into Eqs. (4.5.40a) and (4.5.40d) leads to

$$\begin{aligned}\boldsymbol{C}^{1-}(0,h/2)=&\{\boldsymbol{R}_{12}(0,h/2)+\boldsymbol{T}_{21}(0,h/2)\cdot\\&[\mathbf{E}-\mathbf{R}_{23}(h/2,0)\cdot\boldsymbol{R}_{21}(0,h/2)]^{-1}\cdot\boldsymbol{R}_{23}(h/2,0)\cdot\boldsymbol{T}_{12}(0,h/2)\}\cdot\boldsymbol{C}^{1+}(0,h/2)\\&+\mathbf{T}_{21}(0,h/2)\cdot\left[\boldsymbol{E}-\boldsymbol{R}_{23}(h/2,0)\cdot\boldsymbol{R}_{21}(0,h/2)\right]^{-1}\cdot\boldsymbol{T}_{32}(h/2,0)\cdot\boldsymbol{C}^{3-}(h/2,0)\right],\end{aligned} \tag{4.5.42a}$$

$$\begin{aligned}\boldsymbol{C}^{3+}(h/2,0)=&\boldsymbol{T}_{23}(h/2,0)\cdot[\boldsymbol{E}-\boldsymbol{R}_{21}(0,h/2)\cdot\boldsymbol{R}_{23}(h/2,0)]^{-1}\cdot\\&\mathbf{T}_{12}(0,h/2)\cdot\boldsymbol{C}^{1+}(0,h/2)+\{\boldsymbol{R}_{32}(h/2,0)+\boldsymbol{T}_{23}(h/2,0)\cdot\\&[\mathbf{E}-\mathbf{R}_{21}(0,h/2)\cdot\boldsymbol{R}_{23}(h/2,0)]^{-1}\cdot\boldsymbol{R}_{21}(0,h/2)\cdot\boldsymbol{T}_{32}(h/2,0)\}\cdot\boldsymbol{C}^{3-}(h/2,0).\end{aligned} \tag{4.5.42b}$$

Taking into account the Eq. (4.5.35), Eq. (4.5.42) can be rewritten as

$$\boldsymbol{C}^{1-}(0, h/2) = \{\mathbf{R}_{12} + \boldsymbol{T}_{21}\boldsymbol{\Lambda}'[\boldsymbol{E} - \boldsymbol{\Lambda}'\boldsymbol{R}_{23}\boldsymbol{\Lambda}'\boldsymbol{\Lambda}'\boldsymbol{R}_{21}\boldsymbol{\Lambda}']^{-1}\boldsymbol{\Lambda}'\boldsymbol{R}_{23}\boldsymbol{\Lambda}'\boldsymbol{\Lambda}'\boldsymbol{T}_{12}\}\boldsymbol{C}^{1+}(0, h/2) + \mathbf{T}_{21}\boldsymbol{\Lambda}'[\boldsymbol{E} - \boldsymbol{\Lambda}'\boldsymbol{R}_{23}\boldsymbol{\Lambda}'\boldsymbol{\Lambda}'\boldsymbol{R}_{21}\boldsymbol{\Lambda}']^{-1}\boldsymbol{\Lambda}'\boldsymbol{T}_{32}\boldsymbol{C}^{3-}(h/2, 0)], \tag{4.5.43a}$$

$$\boldsymbol{C}^{3+}(h/2, 0) = \mathbf{T}_{23}\boldsymbol{\Lambda}'[\boldsymbol{E} - \boldsymbol{\Lambda}'\boldsymbol{R}_{21}\boldsymbol{\Lambda}'\boldsymbol{\Lambda}'\boldsymbol{R}_{23}\boldsymbol{\Lambda}']^{-1}\boldsymbol{\Lambda}'\boldsymbol{T}_{12}\boldsymbol{C}^{1+}(0, h/2) + \{\mathbf{R}_{32} + \boldsymbol{T}_{23}\boldsymbol{\Lambda}'[\boldsymbol{E} - \boldsymbol{\Lambda}'\boldsymbol{R}_{21}\boldsymbol{\Lambda}'\boldsymbol{\Lambda}'\boldsymbol{R}_{23}\boldsymbol{\Lambda}']^{-1}\boldsymbol{\Lambda}'\boldsymbol{R}_{21}\boldsymbol{\Lambda}'\boldsymbol{\Lambda}'\boldsymbol{T}_{32}\}\boldsymbol{C}^{3-}(h/2, 0), \tag{4.5.43b}$$

where

$$\boldsymbol{\Lambda}' = \begin{pmatrix} \mathrm{e}^{ik_{P3}^{(2)}\cdot\frac{h}{2}} & 0 \\ 0 & \mathrm{e}^{ik_{S3}^{(2)}\cdot\frac{h}{2}} \end{pmatrix}.$$

After rearrangement, we can obtain

$$\boldsymbol{C}^{1-}(0, h/2) = \{\mathbf{R}_{12} + \boldsymbol{T}_{21} \cdot [[\boldsymbol{\Lambda}'']^{-1} - \boldsymbol{R}_{23} \cdot \boldsymbol{\Lambda}'' \cdot \boldsymbol{R}_{21}]^{-1} \cdot \boldsymbol{R}_{23} \cdot \boldsymbol{\Lambda}'' \cdot \boldsymbol{T}_{12}\} \cdot \boldsymbol{C}^{1+}(0, h/2) + \mathbf{T}_{21} \cdot [[\boldsymbol{\Lambda}'']^{-1} - \boldsymbol{R}_{23} \cdot \boldsymbol{\Lambda}'' \cdot \boldsymbol{R}_{21}]^{-1} \cdot \boldsymbol{T}_{32} \cdot \boldsymbol{C}^{3-}(h/2, 0)], \tag{4.5.44a}$$

$$\boldsymbol{C}^{3+}(h/2, 0) = \mathbf{T}_{23} \cdot [[\boldsymbol{\Lambda}'']^{-1} - \boldsymbol{R}_{21} \cdot \boldsymbol{\Lambda}'' \cdot \boldsymbol{R}_{23}]^{-1} \cdot \boldsymbol{T}_{12} \cdot \boldsymbol{C}^{1+}(0, h/2) + \{\mathbf{R}_{32} + \boldsymbol{T}_{23} \cdot [[\boldsymbol{\Lambda}'']^{-1} - \boldsymbol{R}_{21} \cdot \boldsymbol{\Lambda}'' \cdot \boldsymbol{R}_{23}]^{-1} \cdot \boldsymbol{R}_{21} \cdot \boldsymbol{\Lambda}'' \cdot \boldsymbol{T}_{32}\} \cdot \boldsymbol{C}^{3-}(h/2, 0), \tag{4.5.44b}$$

where

$$\boldsymbol{\Lambda}'' = \boldsymbol{\Lambda}' \cdot \boldsymbol{\Lambda}' = \begin{pmatrix} \mathrm{e}^{ik_{P3}^{(2)}\cdot h} & 0 \\ 0 & \mathrm{e}^{ik_{S3}^{(2)}\cdot h} \end{pmatrix}.$$

Write it in matrix form

$$\begin{pmatrix} \boldsymbol{C}^{1-} \\ \boldsymbol{C}^{3+} \end{pmatrix} = \begin{pmatrix} \boldsymbol{R}_{13} & \boldsymbol{T}_{31} \\ \boldsymbol{T}_{13} & \boldsymbol{R}_{31} \end{pmatrix} \begin{pmatrix} \boldsymbol{C}^{1+} \\ \boldsymbol{C}^{3-} \end{pmatrix},$$

where

$$\boldsymbol{R}_{13} = \boldsymbol{R}_{12} + \boldsymbol{T}_{21} \cdot [[\boldsymbol{\Lambda}'']^{-1} - \boldsymbol{R}_{23} \cdot \boldsymbol{\Lambda}'' \cdot \boldsymbol{R}_{21}]^{-1} \cdot \boldsymbol{R}_{23} \cdot \boldsymbol{\Lambda}'' \cdot \boldsymbol{T}_{12}, \tag{4.5.45a}$$

$$\boldsymbol{T}_{31} = \boldsymbol{T}_{21} \cdot [[\boldsymbol{\Lambda}'']^{-1} - \boldsymbol{R}_{23} \cdot \boldsymbol{\Lambda}'' \cdot \boldsymbol{R}_{21}]^{-1} \cdot \boldsymbol{T}_{32}, \tag{4.5.45b}$$

$$\boldsymbol{T}_{13} = \boldsymbol{T}_{23} \cdot [[\boldsymbol{\Lambda}'']^{-1} - \boldsymbol{R}_{21} \cdot \boldsymbol{\Lambda}'' \cdot \boldsymbol{R}_{23}]^{-1} \cdot \boldsymbol{T}_{12}, \tag{4.5.45c}$$

$$\boldsymbol{R}_{31} = \boldsymbol{R}_{32} + \boldsymbol{T}_{23} \cdot [[\boldsymbol{\Lambda}'']^{-1} - \boldsymbol{R}_{21} \cdot \boldsymbol{\Lambda}'' \cdot \boldsymbol{R}_{23}]^{-1} \cdot \boldsymbol{R}_{21} \cdot \boldsymbol{\Lambda}'' \cdot \boldsymbol{T}_{32}. \tag{4.5.45d}$$

It is noted that the Eqs. (4.5.45a) and (4.5.45c) are identical to the Eqs. (4.4.9) and (4.4.10), which indicates that the method of super-interface has the same results as the method of multiple reflection/transmission. But the ideas of two methods dealing with the reflection/transmission problems are completely different.

For the case of P and SV waves incident simultaneously to bi-interlayers, it is assumed that the reflection and transmission amplitudes satisfy

$$\begin{pmatrix} \boldsymbol{C}^{1-} \\ \boldsymbol{C}^{4+} \end{pmatrix} = \begin{pmatrix} \boldsymbol{R}_{14} & \boldsymbol{T}_{41} \\ \boldsymbol{T}_{14} & \boldsymbol{R}_{41} \end{pmatrix} \begin{pmatrix} \boldsymbol{C}^{1+} \\ \boldsymbol{C}^{4-} \end{pmatrix}, \tag{4.5.46}$$

where

$$\boldsymbol{R}_{14} = \boldsymbol{R}_{13} + \boldsymbol{T}_{31} \cdot [[\boldsymbol{\Lambda}'']^{-1} - \boldsymbol{R}_{34} \cdot \boldsymbol{\Lambda}'' \cdot \boldsymbol{R}_{31}]^{-1} \boldsymbol{R}_{34} \cdot \boldsymbol{\Lambda}'' \cdot \boldsymbol{T}_{13}, \tag{4.5.47a}$$

$$\boldsymbol{T}_{41} = \boldsymbol{T}_{31} \cdot [[\boldsymbol{\Lambda}'']^{-1} - \boldsymbol{R}_{34} \cdot \boldsymbol{\Lambda}'' \cdot \boldsymbol{R}_{31}]^{-1} \cdot \boldsymbol{T}_{43}, \tag{4.5.47b}$$

$$\boldsymbol{T}_{14} = \boldsymbol{T}_{34} \cdot [[\boldsymbol{\Lambda}'']^{-1} - \boldsymbol{R}_{31} \cdot \boldsymbol{\Lambda}'' \cdot \boldsymbol{R}_{34}]^{-1} \cdot \boldsymbol{T}_{13}, \tag{4.5.47c}$$

$$\boldsymbol{R}_{41} = \boldsymbol{R}_{43} + \boldsymbol{T}_{34} \cdot [[\boldsymbol{\Lambda}'']^{-1} - \boldsymbol{R}_{31} \cdot \boldsymbol{\Lambda}'' \cdot \boldsymbol{R}_{34}]^{-1} \cdot \boldsymbol{R}_{31} \cdot \boldsymbol{\Lambda}'' \cdot \boldsymbol{T}_{43}. \tag{4.5.47d}$$

$$\boldsymbol{\Lambda}'' = \begin{pmatrix} e^{ik_{P3}^{(3)} \cdot h} & 0 \\ 0 & e^{ik_{S3}^{(3)} \cdot h} \end{pmatrix}.$$

For the case of n-2 interlayers (corresponding to n-1 interfaces and n kinds of media). The amplitudes of reflection and transmission waves in n-2 layers are assumed as

$$\begin{pmatrix} \boldsymbol{C}^{1-} \\ \boldsymbol{C}^{n+} \end{pmatrix} = \begin{pmatrix} \boldsymbol{R}_{1,\mathrm{n}} & \boldsymbol{T}_{\mathrm{n},1} \\ \boldsymbol{T}_{1,\mathrm{n}} & \boldsymbol{R}_{\mathrm{n},1} \end{pmatrix} \begin{pmatrix} \boldsymbol{C}^{1+} \\ \boldsymbol{C}^{n-} \end{pmatrix}, \tag{4.5.48}$$

where

$$\boldsymbol{R}_{1,n} = \boldsymbol{R}_{1,n-1} + \boldsymbol{T}_{n-1,1} \cdot [[\boldsymbol{\Lambda}'']^{-1} - \boldsymbol{R}_{n-1,n} \cdot \boldsymbol{\Lambda}'' \cdot \boldsymbol{R}_{n-1,1}]^{-1} \cdot \boldsymbol{R}_{n-1,n} \cdot \boldsymbol{\Lambda}'' \cdot \boldsymbol{T}_{1,n-1}, \tag{4.5.49a}$$

$$\boldsymbol{T}_{n,1} = \boldsymbol{T}_{n-1,1} \cdot [[\boldsymbol{\Lambda}'']^{-1} - \boldsymbol{R}_{n-1,n} \cdot \boldsymbol{\Lambda}'' \cdot \boldsymbol{R}_{n-1,1}]^{-1} \cdot \boldsymbol{T}_{n,n-1}, \tag{4.5.49b}$$

$$\boldsymbol{T}_{1,n} = \boldsymbol{T}_{n-1,n} \cdot [[\boldsymbol{\Lambda}'']^{-1} - \boldsymbol{R}_{n-1,1} \cdot \boldsymbol{\Lambda}'' \cdot \boldsymbol{R}_{n-1,n}]^{-1} \cdot \boldsymbol{T}_{1,n-1}, \tag{4.5.49c}$$

$$\boldsymbol{R}_{n,1} = \boldsymbol{R}_{n,n-1} + \boldsymbol{T}_{n-1,n} \cdot [[\boldsymbol{\Lambda}'']^{-1} - \boldsymbol{R}_{n-1,1} \cdot \boldsymbol{\Lambda}'' \cdot \boldsymbol{R}_{n-1,n}]^{-1} \cdot \boldsymbol{R}_{n-1,1} \boldsymbol{\Lambda}'' \cdot \boldsymbol{T}_{n,n-1}, \tag{4.5.49d}$$

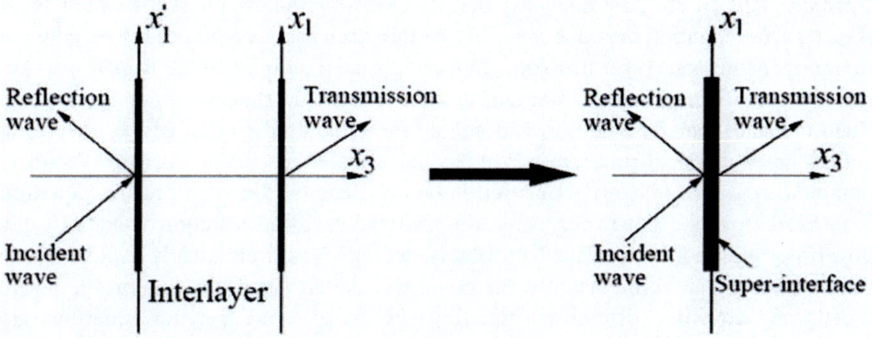

Fig. 4.9 Diagram of equating a single interlayer to a super-interface

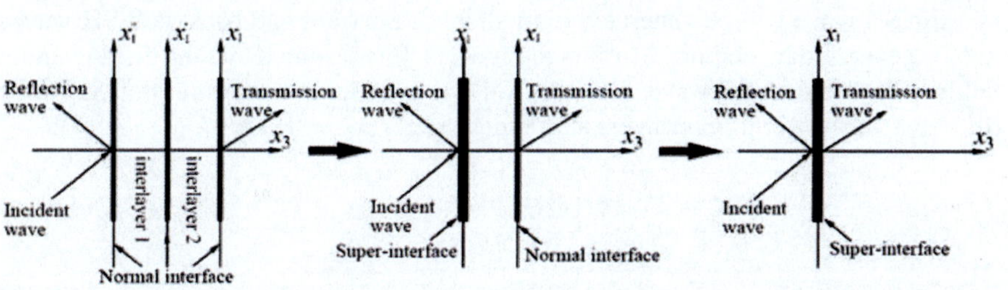

Fig. 4.10 Diagram of equivalent process making multiple interlayers equivalent to a super-interface

where

$$\boldsymbol{\Lambda}'' = \begin{pmatrix} e^{ik_{P3}(n-1)\cdot h} & 0 \\ 0 & e^{ik_{S3}(n-1)\cdot h} \end{pmatrix}.$$

The reflection and transmission coefficients of elastic waves through any number of layers can be obtained by Eq. (4.5.49).

The super-interface method equates the reflection/transmission of an interlayer with the reflection/transmission of a super-interface. The reflection/transmission of multilayer medium can also be equivalent to the reflection/transmission of a super-interface by repeatedly using the recursion relation, i.e. Eqs. (4.5.28) and (4.5.49). This process is shown in Figs. 4.9 and 4.10.

4.6 The State Transfer Equation Method

These methods discussed in previous sections have no difficulty in solving the reflection and transmission problem of homogeneous interlayer, but they are incapable to cope with inhomogeneous interlayer, of which mechanical parameters (elastic constant, density, etc.) are unevenly distributed along the thickness. The method of state transfer equation introduced in this section has great advantages for solving the inhomogeneous interlayer problem. In this approach, similar to the transfer matrix method and stiffness matrix method, the concept of the state vector is introduced. Then the state transfer equation is obtained by reducing the order of the governing equations of physical problems. Further, the transfer matrix of the state vector is obtained by solving the state transfer equation. Because the state transfer equation is derived from the governing equation, the state transfer equation is actually the governing equation of physical problems, which is mathematically equivalent to the governing equation. Formally, the governing equations of physical problems are mostly second-order differential equations, while the state transfer equations are first-order matrix equations [15–19].

Reflection and transmission of incident SH wave through a finite thickness interlayer are considered firstly. Reflection SH wave will be generated in medium 1. Transmission SH wave will be generated in medium 3. Forward and backward SH waves will be generated in medium 2 (refers to Fig. 4.1). It is assumed that the displacement fields of the incident SH wave, the reflection SH wave and the transmission SH wave (the backward wave in interlayer) at the interface 2 ($x_3 = 0$) are

$$u_2^i = C^i \exp\left[i\left(\xi x_1 - k_{S3}^{(1)} x_3 - \omega_S t\right)\right], \tag{4.6.1a}$$

$$u_2^r = C^r \exp\left[i\left(\xi x_1 + k_{S3}^{(1)} x_3 - \omega_S t\right)\right], \tag{4.6.1b}$$

$$u_2^{2-} = C^{2-} \exp\left[i\left(\xi x_1 - k_{S3}^{(2)} x_3 - \omega_S t\right)\right]. \tag{4.6.1c}$$

It is assumed that the displacement fields of the reflection SH wave (the forward wave in interlayer) and transmission SH wave at interface 1 ($x_3 = -h$) are

$$u_2^{2+} = C^{2+} \exp\left\{i\left[\xi x_1 + k_{S3}^{(2)}(x_3 + h) - \omega_S t\right]\right\}, \tag{4.6.1d}$$

$$u_2^t = C^t \exp\left\{i\left[\xi x_1 - k_{S3}^{(3)}(x_3 + h) - \omega_S t\right]\right\}. \tag{4.6.1e}$$

The displacement and stress in medium 1 can be written as

$$u_2 = u_2^i + u_2^r = C^i \exp\left[i\left(\xi x_1 - k_{S3}^{(1)} x_3 - \omega_S t\right)\right] + C^r \exp\left[i\left(\xi x_1 + k_{S3}^{(1)} x_3 - \omega_S t\right)\right], \tag{4.6.2a}$$

$$\begin{aligned}\sigma_{32} = \mu_1 u_{2,3} = \mu_1\Big\{&-ik_{S3}^{(1)} C^i \exp\left[i\left(\xi x_1 - k_{S3}^{(1)} x_3 - \omega_S t\right)\right]\\ &+ik_{S3}^{(1)} C^r \exp\left[i\left(\xi x_1 + k_{S3}^{(1)} x_3 - \omega_S t\right)\right]\Big\}.\end{aligned} \tag{4.6.2b}$$

The displacement and stress in medium 2 can be written as

$$u_2' = u_2^{2-} + u_2^{2+} = C^{2-}\exp\left[i\left(\xi x_1 - k_{S3}^{(2)} x_3 - \omega_S t\right)\right]$$
$$+ C^{2+}\exp\left\{i\left[\xi x_1 + k_{S3}^{(2)}(x_3 + h) - \omega_S t\right]\right\}, \quad (4.6.3a)$$

$$\sigma_{32}' = \mu_2 u_{2,3}' = \mu_2\left\{-ik_{S3}^{(2)} C^{2-}\exp\left[i\left(\xi x_1 - k_{S3}^{(2)} x_3 - \omega_S t\right)\right] + ik_{S3}^{(2)} C^{2+}\exp\right.$$
$$\left.\left\{i\left[\xi x_1 + k_{S3}^{(2)}(x_3 + h) - \omega_S t\right]\right\}\right\}. \quad (4.6.3b)$$

The displacement and stress in medium 3 can be written as

$$u_2'' = u_2^t = C^t exp\left\{i\left[\xi x_1 - k_{S3}^{(3)}(x_3 + h) - \omega_S t\right]\right\}, \quad (4.6.4a)$$

$$\sigma_{32}'' = \mu_3 u_{2,3}'' = \mu_3\left\{-ik_{S3}^{(3)} C^t exp\left\{i\left[\xi x_1 - k_{S3}^{(3)}(x_3 + h) - \omega_S t\right]\right\}\right\}. \quad (4.6.4b)$$

The governing equations can be written as

$$\rho \ddot{u}_i = \sigma_{ij,j}. \quad (4.6.5)$$

The first-order differential equations are obtained by reducing the order of the governing equations of the second-order differential equation, namely

$$\begin{cases} \mu u_{2,3} = \sigma_{32} \\ \sigma_{32,3} = -\rho\omega^2 u_2 \end{cases}. \quad (4.6.6)$$

Define the state vector

$$\xi = \begin{bmatrix} u_2(x_3) \\ \sigma_{32}(x_3) \end{bmatrix}. \quad (4.6.7)$$

According to Eq. (4.6.6), we can obtain

$$\begin{bmatrix} \mu & 0 \\ 0 & 1 \end{bmatrix}\begin{bmatrix} u_{2,3} \\ \sigma_{32,3} \end{bmatrix} = \begin{bmatrix} 0 & 1 \\ -\rho\omega^2 & 0 \end{bmatrix}\begin{bmatrix} u_2 \\ \sigma_{32} \end{bmatrix}, \quad (4.6.8a)$$

or equivalently

$$\begin{bmatrix} u_{2,3} \\ \sigma_{32,3} \end{bmatrix} = \begin{bmatrix} \mu & 0 \\ 0 & 1 \end{bmatrix}^{-1}\begin{bmatrix} 0 & 1 \\ -\rho\omega^2 & 0 \end{bmatrix}\begin{bmatrix} u_2 \\ \sigma_{32} \end{bmatrix}, \quad (4.6.8b)$$

$$\begin{bmatrix} u_{2,3} \\ \sigma_{32,3} \end{bmatrix} = \begin{bmatrix} 0 & \frac{1}{\mu} \\ -\rho\omega^2 & 0 \end{bmatrix}\begin{bmatrix} u_2 \\ \sigma_{32} \end{bmatrix}. \quad (4.6.8c)$$

For inhomogeneous medium, i.e. $\mu = \mu(x_3)$ and $\lambda = \lambda(x_3)$, the above derivation process is also valid. The first-order differential equation of the following form

$$\frac{\mathrm{d}\boldsymbol{\xi}}{\mathrm{d}x_3} = \boldsymbol{A}(x_3)\boldsymbol{\xi}, \tag{4.6.9}$$

is generally called the state transfer differential equation of isotropic elastic interlayer. Wherein the expression of the coefficient matrix is

$$\boldsymbol{A}(x_3) = \begin{bmatrix} 0 & \frac{1}{\mu(x_3)} \\ -\rho(x_3)\omega^2 & 0 \end{bmatrix}. \tag{4.6.10}$$

Assume that the state vectors at x_3^0 and x_3 satisfy

$$\boldsymbol{\xi}(x_3) = \boldsymbol{Q}(x_3, x_3^0) \cdot \boldsymbol{\xi}(x_3^0). \tag{4.6.11}$$

The matrix $\boldsymbol{Q}(x_3, x_3^0)$ is usually called the transfer matrix of state vector. Obviously, the transfer matrix satisfies $\boldsymbol{Q}(x_3^0, x_3^0) = \boldsymbol{I}$. Inserting Eq. (4.6.11) into the state transfer differential equation, i.e. Eq. (4.6.9) leads to

$$\frac{\mathrm{d}\boldsymbol{Q}(x_3, x_3^0)}{\mathrm{d}x_3} = \boldsymbol{A}(x_3) \cdot \boldsymbol{Q}(x_3, x_3^0),\ \boldsymbol{Q}(x_3^0, x_3^0) = \boldsymbol{I}. \tag{4.6.12}$$

The transfer matrix $\boldsymbol{Q}(x_3, x_3^0)$ can be generally solved from Eq. (4.6.12).

When the interlayer is homogeneous, $\boldsymbol{A}(x_3)$ is a constant matrix. The solution of Eq. (4.6.12) can be expressed as

$$\boldsymbol{Q}(x_3, x_3^0) = \mathrm{e}^{\boldsymbol{A}(x_3 - x_3^0)}. \tag{4.6.13}$$

Assuming that the thickness of the interlayer is *h*, the transfer matrix corresponding to the state vectors on the left and right sides of the interlayer is

$$\boldsymbol{Q}(h) = \mathrm{e}^{\boldsymbol{A}h}. \tag{4.6.14}$$

Assuming the eigenvalues of **A** are $\lambda_n (n = 1,\ 2)$, its corresponding eigenvectors constitute matrix **W** of 2 × 2 order. The two eigenvalues correspond to the waves propagating forward and backward in the interlayer. The eigenvector corresponding to the eigenvalue less than zero (denotes by λ^-) is $\boldsymbol{W}^- = \begin{bmatrix} P^- \\ D^- \end{bmatrix}$. The eigenvector corresponding to the eigenvalue greater than zero (denotes by λ^+) is $\boldsymbol{W}^+ = \begin{bmatrix} P^+ \\ D^+ \end{bmatrix}$. The coefficient matrix can be expressed as

$$A = \begin{bmatrix} P^- & P^+ \\ D^- & D^+ \end{bmatrix} \cdot \begin{bmatrix} \lambda^- & 0 \\ 0 & \lambda^+ \end{bmatrix} \cdot \begin{bmatrix} P^- & P^+ \\ D^- & D^+ \end{bmatrix}^{-1}. \tag{4.6.15}$$

Thus, the transfer matrix can be expressed as

$$\begin{aligned} Q(h) &= \mathrm{e}^{Ah} = I + Ah + \frac{1}{2!}(Ah)^2 + \frac{1}{3!}(Ah)^3 + \cdots + \frac{1}{n!}(Ah)^n + \cdots \\ &= I + hW \cdot \Lambda \cdot W^{-1} + \frac{1}{2!}h^2(W \cdot \Lambda \cdot W^{-1})(W \cdot \Lambda \cdot W^{-1}) \\ &\quad + \frac{1}{3!}h^3(W \cdot \Lambda \cdot W^{-1})(W \cdot \Lambda \cdot W^{-1})(W \cdot \Lambda \cdot W^{-1}) + \cdots \\ &\quad + \frac{1}{n!}h^n \underbrace{(W \cdot \Lambda \cdot W^{-1})(W \cdot \Lambda \cdot W^{-1})(W \cdot \Lambda \cdot W^{-1}) \cdots}_{n\uparrow} + \cdots \\ &= W\left(I + h\Lambda + \frac{1}{2!}h^2\Lambda^2 + \frac{1}{3!}h^3\Lambda^3 + \cdots + \frac{1}{n!}h^n\Lambda^n + \cdots\right)W^{-1} \\ &= W \cdot \mathrm{e}^{\Lambda h} \cdot W^{-1} = W \cdot \begin{bmatrix} \mathrm{e}^{\lambda^- h} & 0 \\ 0 & \mathrm{e}^{\lambda^+ h} \end{bmatrix} \cdot W^{-1}, \end{aligned} \tag{4.6.16}$$

where $\Lambda = \begin{bmatrix} \lambda^- & 0 \\ 0 & \lambda^+ \end{bmatrix}$.

The displacement and stress at interface 2 ($x_3 = 0^+$) can be written as

$$\begin{aligned} \xi_{x_3=0^+} &= \begin{bmatrix} u_2(x_3 = 0^+) \\ \sigma_{32}(x_3 = 0^+) \end{bmatrix} = \begin{bmatrix} 1 & 1 \\ -i\mu_1 k_{s3}^{(1)} & i\mu_1 k_{s3}^{(1)} \end{bmatrix} \cdot \begin{bmatrix} C^i \\ C^r \end{bmatrix} \mathrm{e}^{i(\xi x_1 - \omega t)} \\ &= X_1 \begin{bmatrix} C^i \\ C^r \end{bmatrix} \mathrm{e}^{i(\xi x_1 - \omega t)}. \end{aligned} \tag{4.6.17}$$

The displacement and stress at interface 1 ($x_3 = -h^-$) can be written as

$$\xi''_{x_3=-h^-} = \begin{bmatrix} u_2(x_3 = -h^-) \\ \sigma_{32}(x_3 = -h^-) \end{bmatrix} = \begin{bmatrix} 1 \\ -i\mu_3 k_{s3}^{(3)} \end{bmatrix} C^t \mathrm{e}^{i(\xi x_1 - \omega t)} = X_3 C^t \mathrm{e}^{i(\xi x_1 - \omega t)}. \tag{4.6.18}$$

At interface 1 ($x_3 = -h^-$) and interface 2 ($x_3 = 0^+$), the displacement and stress satisfy the continuous conditions on both sides of the interface, namely

$$u_2|_{x_3=0} = u'_2|_{x_3=0}, \quad \sigma_{32}|_{x_3=0} = \sigma'_{32}|_{x_3=0}, \tag{4.6.19a}$$

$$u'_2|_{x_3=-h} = u''_2|_{x_3=-h}, \quad \sigma'_{32}|_{x_3=-h} = \sigma''_{32}|_{x_3=-h}, \tag{4.6.19b}$$

or equivalently

$$\boldsymbol{\xi}'_{x_3=-h^+} = \boldsymbol{\xi}''_{x_3=-h^-}, \boldsymbol{\xi}_{x_3=0^+} = \boldsymbol{\xi}'_{x_3=0^-}, \tag{4.6.20}$$

where

$$\boldsymbol{\xi} = \begin{bmatrix} u_2(x_3) \\ \sigma_{32}(x_3) \end{bmatrix}, \boldsymbol{\xi}' = \begin{bmatrix} u'_2(x_3) \\ \sigma'_{32}(x_3) \end{bmatrix}, \boldsymbol{\xi}'' = \begin{bmatrix} u''_2(x_3) \\ \sigma''_{32}(x_3) \end{bmatrix}. \tag{4.6.21}$$

Thus, the state vectors in medium 1 and 3 satisfy

$$\boldsymbol{\xi}''(x_3 = -h^-) = \boldsymbol{Q}(h) \cdot \boldsymbol{\xi}(x_3 = 0^+). \tag{4.6.22}$$

According to Eqs. (4.6.17), (4.6.18), and (4.6.22), we can obtain

$$\boldsymbol{X}_3 C^t = \boldsymbol{Q}(h) \cdot \boldsymbol{X}_1 \cdot \begin{bmatrix} C^i \\ C^r \end{bmatrix}. \tag{4.6.23}$$

So the amplitudes of reflection and transmission waves are

$$C^r = RC^i, \tag{4.6.24a}$$

$$C^t = TC^i, \tag{4.6.24b}$$

where R and T are the reflection and transmission coefficients of SH wave, respectively.

In the case of incident P or SV wave, reflection P and SV waves will be generated in medium 1. Transmission P and SV waves will be generated in medium 3. Forward and backward P and SV waves will be generated in medium 2 (refer to Fig. 4.2). It is assumed that the wave functions of the P and SV waves in medium 1, medium 2 and medium 3 are, respectively

$$\varphi^i = A^i \exp\left[i\left(\xi x_1 - k_{P3}^{(1)} x_3 - \omega t\right)\right], \tag{4.6.25a}$$

$$\psi^i = B^i \exp\left[i\left(\xi x_1 - k_{S3}^{(1)} x_3 - \omega t\right)\right], \tag{4.6.25b}$$

$$\varphi^r = A^r \exp\left[i\left(\xi x_1 + k_{P3}^{(1)} x_3 - \omega t\right)\right], \tag{4.6.25c}$$

$$\psi^r = B^r \exp\left[i\left(\xi x_1 + k_{S3}^{(1)} x_3 - \omega t\right)\right], \tag{4.6.25d}$$

$$\varphi^{2-} = A^{2-} \exp\left[i\left(\xi x_1 - k_{P3}^{(2)} x_3 - \omega t\right)\right], \tag{4.6.25e}$$

$$\psi^{2-} = B^{2-} \exp\left[i\left(\xi x_1 - k_{S3}^{(2)} x_3 - \omega t\right)\right], \tag{4.6.25f}$$

$$\varphi^{2+} = A^{2+}\exp\left[i\left(\xi x_1 + k_{P3}^{(2)}(x_3+h) - \omega t\right)\right], \tag{4.6.25g}$$

$$\psi^{2+} = B^{2+}\exp\left[i\left(\xi x_1 + k_{S3}^{(2)}(x_3+h) - \omega t\right)\right], \tag{4.6.25h}$$

$$\varphi^{t} = A^{t}\exp\left[i\left(\xi x_1 - k_{P3}^{(3)}(x_3+h) - \omega t\right)\right], \tag{4.6.25i}$$

$$\psi^{t} = B^{t}\exp\left[i\left(\xi x_1 - k_{S3}^{(3)}(x_3+h) - \omega t\right)\right]. \tag{4.6.25j}$$

The displacement and stress in medium 1 can be written as

$$\begin{aligned} u_1 &= \frac{\partial(\varphi^i+\varphi^r)}{\partial x_1} - \frac{\partial(\psi^i+\psi^r)}{\partial x_3} = i\xi A^i\exp\left[i\left(\xi x_1 - k_{P3}^{(1)}x_3 - \omega t\right)\right] \\ &+ i\xi A^r\exp\left[i\left(\xi x_1 + k_{P3}^{(1)}x_3 - \omega t\right)\right] + ik_{S3}^{(1)}B^i\exp\left[i\left(\xi x_1 - k_{S3}^{(1)}x_3 - \omega t\right)\right] \\ &- ik_{S3}^{(1)}B^r\exp\left[i\left(\xi x_1 + k_{S3}^{(1)}x_3 - \omega t\right)\right], \end{aligned} \tag{4.6.26a}$$

$$\begin{aligned} u_3 &= \frac{\partial(\varphi^i+\varphi^r)}{\partial x_3} + \frac{\partial(\psi^i+\psi^r)}{\partial x_1} = -ik_{P3}^{(1)}A^i\exp\left[i\left(\xi x_1 - k_{P3}^{(1)}x_3 - \omega t\right)\right] \\ &+ ik_{P3}^{(1)}A^r\exp\left[i\left(\xi x_1 + k_{P3}^{(1)}x_3 - \omega t\right)\right] + i\xi B^i\exp\left[i\left(\xi x_1 - k_{S3}^{(1)}x_3 - \omega t\right)\right] \\ &+ i\xi B^r\exp\left[i\left(\xi x_1 + k_{S3}^{(1)}x_3 - \omega t\right)\right], \end{aligned} \tag{4.6.26b}$$

$$\begin{aligned} \sigma_{33} &= \lambda_1\frac{\partial^2(\varphi^i+\varphi^r)}{\partial x_1^2} + (\lambda_1+2\mu_1)\frac{\partial^2(\varphi^i+\varphi^r)}{\partial x_3^2} + 2\mu_1\frac{\partial^2(\psi^i+\psi^r)}{\partial x_1 x_3} \\ &= -\lambda_1\left\{\xi^2A^i\exp\left[i\left(\xi x_1 - k_{P3}^{(1)}x_3 - \omega t\right)\right] + \xi^2A^r\exp\left[i\left(\xi x_1 + k_{P3}^{(1)}x_3 - \omega t\right)\right]\right\} \\ &- (\lambda_1+2\mu_1)\left\{\left(k_{P3}^{(1)}\right)^2A^i\exp\left[i\left(\xi x_1 - k_{P3}^{(1)}x_3 - \omega t\right)\right]\right. \\ &\left.+\left(k_{P3}^{(1)}\right)^2A^r\exp\left[i\left(\xi x_1 + k_{P3}^{(1)}x_3 - \omega t\right)\right]\right\} \\ &+ 2\mu_1\left\{\xi k_{S3}^{(1)}B^i\exp\left[i\left(\xi x_1 - k_{S3}^{(1)}x_3 - \omega t\right)\right]\right. \\ &\left.-\xi k_{S3}^{(1)}B^r\exp\left[i\left(\xi x_1 + k_{S3}^{(1)}x_3 - \omega t\right)\right]\right\}, \end{aligned} \tag{4.6.26c}$$

$$\begin{aligned} \sigma_{31} &= \mu_1\left[2\frac{\partial^2(\varphi^i+\varphi^r)}{\partial x_1 x_3} + \frac{\partial^2(\psi^i+\psi^r)}{\partial x_1^2} - \frac{\partial^2(\psi^i+\psi^r)}{\partial x_3^2}\right] \\ &= \mu_1\left\{2\xi k_{P3}^{(1)}A^i\exp\left[i\left(\xi x_1 - k_{P3}^{(1)}x_3 - \omega t\right)\right]\right. \end{aligned}$$

$$
\begin{aligned}
&- 2\xi k_{P3}^{(1)} A^r \exp\left[i\left(\xi x_1 + k_{P3}^{(1)} x_3 - \omega t\right)\right] - \xi^2 B^i \exp\left[i\left(\xi x_1 - k_{S3}^{(1)} x_3 - \omega t\right)\right] \\
&- \xi^2 B^r \exp\left[i\left(\xi x_1 + k_{S3}^{(1)} x_3 - \omega t\right)\right] + \left(k_{S3}^{(1)}\right)^2 B^i \exp\left[i\left(\xi x_1 - k_{S3}^{(1)} x_3 - \omega t\right)\right] \\
&+ \left(k_{S3}^{(1)}\right)^2 B^r \exp\left[i\left(\xi x_1 + k_{S3}^{(1)} x_3 - \omega t\right)\right]\Big\}. \qquad (4.6.26d)
\end{aligned}
$$

The displacement and stress in medium 2 can be written as

$$
\begin{aligned}
u_1' = {} & \frac{\partial\left(\varphi^{2-} + \varphi^{2+}\right)}{\partial x_1} - \frac{\partial\left(\psi^{2-} + \psi^{2+}\right)}{\partial x_3} = i\xi A^{2-} \exp\left[i\left(\xi x_1 - k_{P3}^{(2)} x_3 - \omega t\right)\right] \\
&+ i\xi A^{2+} \exp\left[i\left(\xi x_1 + k_{P3}^{(2)} (x_3 + h) - \omega t\right)\right] + i k_{S3}^{(2)} B^{2-} \exp\left[i\left(\xi x_1 - k_{S3}^{(2)} x_3 - \omega t\right)\right] \\
&- i k_{S3}^{(2)} B^{2+} \exp\left[\left(\xi x_1 + k_{S3}^{(2)} (x_3 + h) - \omega t\right)\right], \qquad (4.6.27a)
\end{aligned}
$$

$$
\begin{aligned}
u_3' = {} & \frac{\partial\left(\varphi^{2-} + \varphi^{2+}\right)}{\partial x_3} + \frac{\partial\left(\psi^{2-} + \psi^{2+}\right)}{\partial x_1} = -i k_{P3}^{(2)} A^{2-} \exp\left[i\left(\xi x_1 - k_{P3}^{(2)} x_3 - \omega t\right)\right] \\
&+ i k_{P3}^{(2)} A^{2+} \exp\left[i\left(\xi x_1 + k_{P3}^{(2)} (x_3 + h) - \omega t\right)\right] + i\xi B^{2-} \exp\left[i\left(\xi x_1 - k_{S3}^{(2)} x_3 - \omega t\right)\right] \\
&+ i\xi B^{2+} \exp\left[i\left(\xi x_1 + k_{S3}^{(2)} (x_3 + h) - \omega t\right)\right], \qquad (4.6.27b)
\end{aligned}
$$

$$
\begin{aligned}
\sigma_{33}' = {} & \lambda_2 \frac{\partial^2\left(\varphi^{2-} + \varphi^{2+}\right)}{\partial x_1^2} + (\lambda_2 + 2\mu_2) \frac{\partial^2\left(\varphi^{2-} + \varphi^{2+}\right)}{\partial x_3^2} + 2\mu_2 \frac{\partial^2\left(\psi^{2-} + \psi^{2+}\right)}{\partial x_1 x_3} = \\
&- \lambda_2 \left\{\xi^2 A^{2-} \exp\left[i\left(\xi x_1 - k_{P3}^{(2)} x_3 - \omega t\right)\right] + \xi^2 A^{2+} \exp\left[i\left(\xi x_1 + k_{P3}^{(2)} (x_3 + h) - \omega t\right)\right]\right\} \\
&- (\lambda_2 + 2\mu_2) \left\{\left(k_{P3}^{(2)}\right)^2 A^{2-} \exp\left[i\left(\xi x_1 - k_{P3}^{(2)} x_3 - \omega t\right)\right] \right. \\
&\left. + \left(k_{P3}^{(2)}\right)^2 A^{2+} \exp\left[i\left(\xi x_1 + k_{P3}^{(2)} (x_3 + h) - \omega t\right)\right]\right\} \\
&+ 2\mu_2 \left\{\xi k_{S3}^{(2)} B^{2-} \exp\left[i\left(\xi x_1 - k_{S3}^{(2)} x_3 - \omega t\right)\right] \right. \\
&\left. - \xi k_{S3}^{(2)} B^{2+} \exp\left[i\left(\xi x_1 + k_{S3}^{(2)} (x_3 + h) - \omega t\right)\right]\right\}, \qquad (4.6.27c)
\end{aligned}
$$

$$
\begin{aligned}
\sigma_{31}' = {} & \mu_2 \left[2 \frac{\partial^2\left(\varphi^{2-} + \varphi^{2+}\right)}{\partial x_1 x_3} + \frac{\partial^2\left(\psi^{2-} + \psi^{2+}\right)}{\partial x_1^2} - \frac{\partial^2\left(\psi^{2-} + \psi^{2+}\right)}{\partial x_3^2}\right] \\
= {} & \mu_2 \left\{2\xi k_{P3}^{(2)} A^{2-} \exp\left[i\left(\xi x_1 - k_{P3}^{(2)} x_3 - \omega t\right)\right] \right. \\
&- 2\xi k_{P3}^{(2)} A^{2+} \exp\left[i\left(\xi x_1 + k_{P3}^{(2)} (x_3 + h) - \omega t\right)\right] \\
&- \xi^2 B^{2-} \exp\left[i\left(\xi x_1 - k_{S3}^{(2)} x_3 - \omega t\right)\right] \\
&- \xi^2 B^{2+} \exp\left[i\left(\xi x_1 + k_{S3}^{(2)} (x_3 + h) - \omega t\right)\right] \\
&+ \left(k_{S3}^{(2)}\right)^2 B^{2-} \exp\left[i\left(\xi x_1 - k_{S3}^{(2)} x_3 - \omega t\right)\right]
\end{aligned}
$$

$$+\left(k_{S3}^{(2)}\right)^2 B^{2+}\exp\left[i\left(\xi x_1 + k_{S3}^{(2)}(x_3+h)-\omega t\right)\right]\Big\}. \tag{4.6.27d}$$

The displacement and stress in medium 3 can be written as

$$\begin{aligned} u_1'' = \frac{\partial \varphi^t}{\partial x_1} - \frac{\partial \psi^t}{\partial x_3} &= i\xi A^t \exp\left[i\left(\xi x_1 - k_{P3}^{(3)}(x_3+h)-\omega t\right)\right] \\ &+ ik_{S3}^{(3)} B^t \exp\left[i\left(\xi x_1 - k_{S3}^{(3)}(x_3+h)-\omega t\right)\right], \end{aligned} \tag{4.6.28a}$$

$$\begin{aligned} u_3'' = \frac{\partial \varphi^t}{\partial x_3} + \frac{\partial \psi^t}{\partial x_1} &= -ik_{P3}^{(3)} A^t \exp\left[i\left(\xi x_1 - k_{P3}^{(3)}(x_3+h)-\omega t\right)\right] \\ &+ i\xi B^t \exp\left[i\left(\xi x_1 - k_{S3}^{(3)}(x_3+h)-\omega t\right)\right], \end{aligned} \tag{4.6.28b}$$

$$\begin{aligned} \sigma_{33}'' = \lambda_3\frac{\partial^2 \varphi^t}{\partial x_1^2} + (\lambda_3+2\mu_3)\frac{\partial^2 \varphi^t}{\partial x_3^2} + 2\mu_3\frac{\partial^2 \psi^t}{\partial x_1 x_3} &= -\lambda_3\xi^2 A^t \exp\left[i\left(\xi x_1 - k_{P3}^{(3)}(x_3+h)-\omega t\right)\right] \\ -(\lambda_3+2\mu_3)\left(k_{P3}^{(3)}\right)^2 A^t \exp\left[i\left(\xi x_1 - k_{P3}^{(3)}(x_3+h)-\omega t\right)\right] \\ +2\mu_3\xi k_{S3}^{(3)} B^t \exp\left[i\left(\xi x_1 - k_{S3}^{(3)}(x_3+h)-\omega t\right)\right], \end{aligned} \tag{4.6.29a}$$

$$\begin{aligned} \sigma_{31}'' = \mu_3\left[2\frac{\partial^2 \varphi^t}{\partial x_1 x_3} + \frac{\partial^2 \psi^t}{\partial x_1^2} - \frac{\partial^2 \psi^t}{\partial x_3^2}\right] &= \mu_3\Big\{2\xi k_{P3}^{(3)} A^t \exp\left[i\left(\xi x_1 - k_{P3}^{(3)}(x_3+h)-\omega t\right)\right] \\ -\xi^2 B^t \exp\left[i\left(\xi x_1 - k_{S3}^{(3)}(x_3+h)-\omega t\right)\right] \\ +\left(k_{S3}^{(3)}\right)^2 B^t \exp\left[i\left(\xi x_1 - k_{S3}^{(3)}(x_3+h)-\omega t\right)\right]\Big\}. \end{aligned} \tag{4.6.29b}$$

The governing equations can be written as

$$\rho \ddot{u}_i = \sigma_{ij,j}. \tag{4.6.30}$$

The first-order differential equations are obtained by reducing the order of the governing equations of the second-order differential equation

$$\begin{cases} \sigma_{11} = (\lambda+2\mu)u_{1,1} + \lambda u_{3,3} \\ \sigma_{13} = \mu u_{1,3} + \mu u_{3,1} \\ \sigma_{33} = \lambda u_{1,1} + (\lambda+2\mu)u_{3,3} \\ \rho u_{1,tt} = \sigma_{11,1} + \sigma_{13,3} \\ \rho u_{3,tt} = \sigma_{31,1} + \sigma_{33,3} \end{cases}. \tag{4.6.31}$$

Equation (4.6.31) are further simplified to

$$\begin{cases} \mu u_{1,3} = -i\xi\mu u_3 + \sigma_{13} \\ (\lambda+2\mu)u_{3,3} = -i\xi\lambda u_1 + \sigma_{33} \\ i\xi\lambda u_{3,3} + \sigma_{13,3} = \xi^2(\lambda+2\mu)u_1 - \rho\omega^2 u_1 \\ i\xi\mu u_{1,3} + \sigma_{33,3} = \xi^2\mu u_3 - \rho\omega^2 u_3 \end{cases}. \tag{4.6.32}$$

Define the displacement vector $\boldsymbol{U} = [u_1, u_3]^T$, the stress vector $\boldsymbol{T} = [\sigma_{13}, \sigma_{33}]^T$, and the state vector

$$\boldsymbol{\xi} = \begin{bmatrix} \boldsymbol{U} \\ \boldsymbol{T} \end{bmatrix} = \begin{bmatrix} u_1(x_3) \\ u_3(x_3) \\ \sigma_{13}(x_3) \\ \sigma_{33}(x_3) \end{bmatrix}. \tag{4.6.33}$$

According to Eq. (4.6.32), we can obtain

$$\begin{bmatrix} \mu & 0 & 0\,0 \\ 0 & \lambda+2\mu & 0\,0 \\ 0 & i\xi\lambda & 1\,0 \\ i\xi\mu & 0 & 0\,1 \end{bmatrix} \begin{bmatrix} u_{1,3} \\ u_{3,3} \\ \sigma_{13,3} \\ \sigma_{33,3} \end{bmatrix} = \begin{bmatrix} 0 & -i\xi\mu & 1\,0 \\ -i\xi\lambda & 0 & 0\,1 \\ \xi^2(\lambda+2\mu)-\rho\omega^2 & 0 & 0\,0 \\ 0 & \xi^2\mu-\rho\omega^2 & 0\,0 \end{bmatrix} \begin{bmatrix} u_1 \\ u_3 \\ \sigma_{13} \\ \sigma_{33} \end{bmatrix}. \tag{4.6.34}$$

By introducing the partition matrix, Eq. (4.6.34) can be rewritten as

$$\begin{bmatrix} \boldsymbol{\Gamma}_{33}(x_3) & \boldsymbol{0} \\ i\xi\boldsymbol{\Gamma}_{13}(x_3) & \boldsymbol{I} \end{bmatrix} \begin{bmatrix} \frac{\partial \boldsymbol{U}}{\partial x_3} \\ \frac{\partial \boldsymbol{T}}{\partial x_3} \end{bmatrix} = \begin{bmatrix} -i\xi\boldsymbol{\Gamma}_{31}(x_3) & \boldsymbol{I} \\ \xi^2\boldsymbol{\Gamma}_{11}(x_3)-\rho\omega^2\boldsymbol{I} & \boldsymbol{0} \end{bmatrix} \begin{bmatrix} \boldsymbol{U} \\ \boldsymbol{T} \end{bmatrix}, \tag{4.6.35}$$

where

$$\boldsymbol{\Gamma}_{33} = \begin{bmatrix} \mu & 0 \\ 0 & \lambda+2\mu \end{bmatrix}, \boldsymbol{\Gamma}_{13} = \begin{bmatrix} 0 & \lambda \\ \mu & 0 \end{bmatrix}, \boldsymbol{\Gamma}_{31} = \begin{bmatrix} 0 & \mu \\ \lambda & 0 \end{bmatrix}, \boldsymbol{\Gamma}_{11} = \begin{bmatrix} \lambda+2\mu & 0 \\ 0 & \mu \end{bmatrix}, \mathbf{I} = \begin{bmatrix} 1 & 0 \\ 0 & 1 \end{bmatrix}.$$

Write Eq. (4.6.35) as a matrix equation sets

$$\begin{cases} \boldsymbol{\Gamma}_{33}(x_3)\cdot\boldsymbol{U}' = -i\xi\boldsymbol{\Gamma}_{31}(x_3)\cdot\boldsymbol{U}+\boldsymbol{U} \\ i\xi\boldsymbol{\Gamma}_{13}(x_3)\cdot\boldsymbol{U}'+\boldsymbol{T}' = (\xi^2\boldsymbol{\Gamma}_{11}(x_3)-\rho\omega^2\boldsymbol{I})\cdot\boldsymbol{U} \end{cases}, \tag{4.6.36}$$

where $\begin{bmatrix} \boldsymbol{U}'(x_3) \\ \boldsymbol{T}'(x_3) \end{bmatrix} = \frac{\partial}{\partial x_3}\begin{bmatrix} \boldsymbol{U}(x_3) \\ \boldsymbol{T}(x_3) \end{bmatrix}$. Assuming $\boldsymbol{X}(x_3) = (\boldsymbol{\Gamma}_{33}(x_3))^{-1}$, then Eqs. (4.6.36) can be further written as

$$\begin{cases} \boldsymbol{U}'(x_3) = -i\xi\boldsymbol{X}(x_3)\cdot\boldsymbol{\Gamma}_{31}(x_3)\cdot\boldsymbol{U}(x_3)+\boldsymbol{X}(x_3)\cdot\boldsymbol{T}(x_3) \\ \boldsymbol{T}'(x_3) = \left[\xi^2(\boldsymbol{\Gamma}_{11}(x_3)-\boldsymbol{\Gamma}_{13}(x_3)\cdot\boldsymbol{X}(x_3)\cdot\boldsymbol{\Gamma}_{31}(x_3))-\rho(x_3)\omega^2\boldsymbol{I}\right]\cdot\boldsymbol{U}(x_3)\,. \\ \qquad -i\xi\boldsymbol{\Gamma}_{13}(x_3)\cdot\boldsymbol{X}(x_3)\cdot\boldsymbol{T}(x_3) \end{cases} \tag{4.6.37}$$

Thus, the state transfer differential equation can be obtained as

$$\frac{\mathrm{d}\boldsymbol{\xi}}{\mathrm{d}x_3} = i\boldsymbol{A}(x_3)\cdot\boldsymbol{\xi}, \tag{4.6.38}$$

where

$$\boldsymbol{A}(x_3) = \begin{bmatrix} -\xi \boldsymbol{X}(x_3) \cdot \boldsymbol{\Gamma}_{31}(x_3) & -i\boldsymbol{X}(x_3) \\ -i\xi^2(\boldsymbol{\Gamma}_{11}(x_3) - \boldsymbol{\Gamma}_{13}(x_3) \cdot \boldsymbol{X}(x_3) \cdot \boldsymbol{\Gamma}_{31}(x_3)) + i\rho(x_3)\omega^2 \boldsymbol{I} & -\xi \boldsymbol{\Gamma}_{13}(x_3) \cdot \boldsymbol{X}(x_3) \end{bmatrix}. \tag{4.6.39}$$

Assuming the transfer matrix **Q** satisfy

$$\boldsymbol{\xi}(x_3) = \boldsymbol{Q}\left(x_3, x_3^0\right) \cdot \boldsymbol{\xi}\left(x_3^0\right), \ \boldsymbol{Q}\left(x_3^0, x_3^0\right) = \boldsymbol{I}. \tag{4.6.40}$$

Combining it with the state transfer matrix equation, we can obtain

$$\frac{\mathrm{d}\boldsymbol{Q}\left(x_3, x_3^0\right)}{\mathrm{d}x_3} = i\boldsymbol{A}(x_3) \cdot \boldsymbol{Q}\left(x_3, x_3^0\right). \tag{4.6.41}$$

When the interlayer is homogeneous, $\boldsymbol{A}(x_3)$ is a constant matrix independent of coordinate, i.e. **A**. The solution of Eq. (4.6.41) can be expressed as

$$\boldsymbol{Q}(h) = \mathrm{e}^{i\boldsymbol{A}h}. \tag{4.6.42}$$

Assuming the eigenvalues of **A** are $\lambda_n (n = 1, \cdots, 4)$, and their corresponding eigenvectors constitute the matrixes **W**. The two eigenvalues less than zero are defined as λ_1^-, λ_2^-. Assume $\boldsymbol{\beta}_{x_3}^- = \mathrm{diag}(\lambda_1^-, \lambda_2^-)$. The eigenvector corresponding to the eigenvalue less than zero is $\boldsymbol{W}^- = \begin{bmatrix} \boldsymbol{P}^- \\ \boldsymbol{D}^- \end{bmatrix}$. The two eigenvalues greater than zero are defined as λ_3^+, λ_4^+. Assume $\boldsymbol{\beta}_{x_3}^+ = \mathrm{diag}(\lambda_3^+, \lambda_4^+)$. The eigenvector corresponding to the eigenvalue greater than zero is $\boldsymbol{W}^+ = \begin{bmatrix} \boldsymbol{P}^+ \\ \boldsymbol{D}^+ \end{bmatrix}$ (**P** and **D** represent the generalized displacement and the generalized stress, respectively). The coefficient matrix can be expressed as

$$\mathbf{A} = \begin{bmatrix} \boldsymbol{P}^- & \boldsymbol{P}^+ \\ \boldsymbol{D}^- & \boldsymbol{D}^+ \end{bmatrix} \begin{bmatrix} \boldsymbol{\beta}_{x_3}^- & 0 \\ 0 & \boldsymbol{\beta}_{x_3}^+ \end{bmatrix} \begin{bmatrix} \boldsymbol{P}^- & \boldsymbol{P}^+ \\ \boldsymbol{D}^- & \boldsymbol{D}^+ \end{bmatrix}^{-1} = \boldsymbol{W} \cdot \boldsymbol{\beta}_{x_3} \cdot \boldsymbol{W}^{-1}. \tag{4.6.43}$$

Thus, the transfer matrix can be expressed as

$$\begin{aligned} &\mathbf{Q}\left(x_3, x_3^0\right) = \mathrm{e}^{i\mathbf{A}\left(x_3 - x_3^0\right)} \\ &= \mathbf{W} \cdot \begin{bmatrix} \mathrm{e}^{i\lambda_1^-\left(x_3 - x_3^0\right)} & 0 & 0 & 0 \\ 0 & \mathrm{e}^{i\lambda_2^-\left(x_3 - x_3^0\right)} & 0 & 0 \\ 0 & 0 & \mathrm{e}^{i\lambda_3^+\left(x_3 - x_3^0\right)} & 0 \\ 0 & 0 & 0 & \mathrm{e}^{i\lambda_4^+\left(x_3 - x_3^0\right)} \end{bmatrix} \cdot \mathbf{W}^{-1}. \end{aligned} \tag{4.6.44}$$

Inserting Eq. (4.6.44) into Eq. (4.6.40), the state vector can be expressed as

$$\boldsymbol{\xi}(x_3) = \mathbf{Q}(x_3, x_3^0) \cdot \boldsymbol{\xi}(x_3^0)$$
$$= \mathbf{W} \cdot \begin{bmatrix} e^{i\lambda_1^-(x_3-x_3^0)} & 0 & 0 & 0 \\ 0 & e^{i\lambda_2^-(x_3-x_3^0)} & 0 & 0 \\ 0 & 0 & e^{i\lambda_3^+(x_3-x_3^0)} & 0 \\ 0 & 0 & 0 & e^{i\lambda_4^+(x_3-x_3^0)} \end{bmatrix} \cdot \mathbf{W}^{-1} \cdot \boldsymbol{\xi}(x_3^0). \tag{4.6.45}$$

According to Eqs. (4.1.7) and (4.1.9), we can obtain

$$\boldsymbol{X}_1 = \begin{bmatrix} i\xi & i\xi & ik_{s3}^{(1)} & -ik_{s3}^{(1)} \\ -ik_{p3}^{(1)} & ik_{p3}^{(1)} & i\xi & i\xi \\ 2\mu_1\xi k_{p3}^{(1)} & -2\mu_1\xi k_{p3}^{(1)} & \mu_1((k_{s3}^{(1)})^2 - \xi^2) & \mu_1((k_{s3}^{(1)})^2 - \xi^2) \\ -\lambda_1\xi^2 - (\lambda_1 + 2\mu_1)(k_{p3}^{(1)})^2 & -\lambda_1\xi^2 - (\lambda_1 + 2\mu_1)(k_{p3}^{(1)})^2 & 2\mu_1\xi k_{s3}^{(1)} & -2\mu_1\xi k_{s3}^{(1)} \end{bmatrix},$$

$$\boldsymbol{X}_3 = \begin{bmatrix} i\xi & ik_{s3}^{(3)} \\ -ik_{p3}^{(3)} & i\xi \\ 2\mu_3\xi k_{p3}^{(3)} & \mu_3[(k_{s3}^{(3)})^2 - \xi^2] \\ -\lambda_3\xi^2 - (\lambda_3 + 2\mu_3)(k_{p3}^{(3)})^2 & 2\mu_3\xi k_{p3}^{(3)} \end{bmatrix}.$$

The displacement and stress at interface 2 ($x_3 = 0^+$) can be written as

$$\boldsymbol{\xi}_{x_3=0^+} = \boldsymbol{X}_1 \cdot \begin{bmatrix} A^i \\ A^r \\ B^i \\ B^r \end{bmatrix} e^{i(\xi x_1 - \omega t)}. \tag{4.6.46}$$

The displacement and stress at interface 1 ($x_3 = -h^-$) can be written as

$$\boldsymbol{\xi}''_{x_3=-h^-} = \boldsymbol{X}_2 \cdot \begin{bmatrix} A^t \\ B^t \end{bmatrix} e^{i(\xi x_1 - \omega t)}. \tag{4.6.47}$$

For the perfect interface, the displacement and stress at interface 1 and 2 meet the continuity conditions

$$u_1|_{x_3=0} = u_1'\big|_{x_3=0}, \quad u_3|_{x_3=0} = u_3'\big|_{x_3=0}, \tag{4.6.48a}$$

$$\sigma_{33}|_{x_3=0} = \sigma_{33}'\big|_{x_3=0}, \quad \sigma_{31}|_{x_3=0} = \sigma_{31}'\big|_{x_3=0}, \tag{4.6.48b}$$

$$u_1'\big|_{x_3=-h} = u_1''\big|_{x_3=-h}, \quad u_3'\big|_{x_3=-h} = u_3''\big|_{x_3=-h}, \tag{4.6.49a}$$

$$\sigma_{33}'\big|_{x_3=-h} = \sigma_{33}''\big|_{x_3=-h}, \quad \sigma_{31}'\big|_{x_3=-h} = \sigma_{31}''\big|_{x_3=-h}, \tag{4.6.49b}$$

or equivalently

$$\boldsymbol{\xi}'_{x_3=-h^+} = \boldsymbol{\xi}''_{x_3=-h^-}, \quad \boldsymbol{\xi}_{x_3=0^+} = \boldsymbol{\xi}'_{x_3=0^-}, \tag{4.6.50}$$

where

$$\boldsymbol{\xi} = \begin{bmatrix} \boldsymbol{U} \\ \boldsymbol{T} \end{bmatrix} = \begin{bmatrix} u_1(x_3) \\ u_3(x_3) \\ \sigma_{13}(x_3) \\ \sigma_{33}(x_3) \end{bmatrix}, \boldsymbol{\xi}' = \begin{bmatrix} \boldsymbol{U}' \\ \boldsymbol{T}' \end{bmatrix} = \begin{bmatrix} u_1'(x_3) \\ u_3'(x_3) \\ \sigma_{13}'(x_3) \\ \sigma_{33}'(x_3) \end{bmatrix}, \boldsymbol{\xi}'' = \begin{bmatrix} \boldsymbol{U}'' \\ \boldsymbol{T}'' \end{bmatrix} = \begin{bmatrix} u_1''(x_3) \\ u_3''(x_3) \\ \sigma_{13}''(x_3) \\ \sigma_{33}''(x_3) \end{bmatrix}. \tag{4.6.51}$$

Thus, the state vectors in medium 1 and 3 satisfy

$$\boldsymbol{\xi}''\left(x_3 = -h^-\right) = \boldsymbol{Q}(h) \cdot \boldsymbol{\xi}\left(x_3 = 0^+\right). \tag{4.6.52}$$

According to Eqs. (4.6.46), (4.6.47), and (4.6.52), we can obtain

$$\boldsymbol{X}_3 \begin{bmatrix} A^t \\ B^t \end{bmatrix} = \boldsymbol{Q}(h) \cdot \boldsymbol{X}_1 \cdot \begin{bmatrix} A^i \\ A^r \\ B^i \\ B^r \end{bmatrix}. \tag{4.6.53}$$

By the partitioning of matrix, the amplitudes of reflection and transmission waves can be obtained from Eq. (4.6.53) as

$$\begin{Bmatrix} A^r \\ B^r \end{Bmatrix} = \begin{bmatrix} R_{AA} & R_{AB} \\ R_{BA} & R_{BB} \end{bmatrix} \begin{Bmatrix} A^i \\ B^i \end{Bmatrix}, \tag{4.6.54a}$$

$$\begin{Bmatrix} A^t \\ B^t \end{Bmatrix} = \begin{bmatrix} T_{AA} & T_{AB} \\ T_{BA} & T_{BB} \end{bmatrix} \begin{Bmatrix} A^i \\ B^i \end{Bmatrix}. \tag{4.6.54b}$$

The matrixes **R** and **T** in Eq. (4.6.54) are called reflection matrix and transmission matrix of the interlayer. If the amplitudes satisfy $A = 0$ and $B \neq 0$, the incident wave is SV wave. If the amplitudes satisfy $A \neq 0$, and $B = 0$, the incident wave is P wave. If the amplitudes satisfy $A \neq 0$ and $B \neq 0$, it means that P wave and SV wave are incident simultaneously. Thus, R_{AA} represents the reflection coefficient of P wave when the incident wave is only P wave. R_{AB} represents the reflection coefficient of P wave when the incident wave is only SV wave. T_{AA} represents the transmission coefficient of P wave when the incident wave is only P wave. T_{AB} represents the transmission coefficient of P wave when the incident wave is only SV wave. The physical meaning of other coefficients can be understood similarly.

Compared with other methods introduced in previous sections, the most obvious advantage of the state transfer equation is that the state transfer equation method is capable to deal with the inhomogeneous interlayer conveniently. For inhomogeneous interlayer, material parameters depend on the position. And the solution of Eq. (4.6.41) can be expressed as

$$\boldsymbol{Q}(h) = \boldsymbol{Q}\left(x_3^0 + h, x_3^0\right) = \mathrm{e}^{\boldsymbol{\Lambda}\left(x_3^0+h, x_3^0\right)}, \tag{4.6.55}$$

where $\boldsymbol{\Lambda}\left(x_3^0 + h, x_3^0\right)$ is an infinite series of the matrix $\boldsymbol{A}(x_3)$

$$\begin{aligned}\Lambda\left(x_3^0 + h, x_3^0\right) = &\int_{x_3^0}^{x_3^0+h} i\mathbf{A}(\tau)\mathrm{d}\tau + \frac{1}{2}\int_{x_3^0}^{x_3^0+h} \mathrm{d}\tau_1 \int_{x_3^0}^{\tau_1} \mathrm{d}\tau_2[i\mathbf{A}(\tau_1), i\mathbf{A}(\tau_2)] \\ &+ \frac{1}{12}\int_{x_3^0}^{x_3^0+h} \mathrm{d}\tau_1 \int_{x_3^0}^{\tau_1} \mathrm{d}\tau_2 \int_{x_3^0}^{\tau_1} \mathrm{d}\tau_3([i\mathbf{A}(\tau_3), [i\mathbf{A}(\tau_2), i\mathbf{A}(\tau_1)]]) \\ &+ \frac{1}{4}\int_{x_3^0}^{x_3^0+h} \mathrm{d}\tau_1 \int_{x_3^0}^{\tau_1} \mathrm{d}\tau_2 \int_{x_3^0}^{\tau_2} \mathrm{d}\tau_3([[i\mathbf{A}(\tau_3), i\mathbf{A}(\tau_2)], i\mathbf{A}(\tau_1)]) + \cdots,\end{aligned} \tag{4.6.56}$$

which is called the Magnus series. $[i\boldsymbol{A}(\tau_m), i\boldsymbol{A}(\tau_n)]$ is the commutator in Lie algebras and satisfies

$$[i\boldsymbol{A}(\tau_m), i\boldsymbol{A}(\tau_n)] = i\boldsymbol{A}(\tau_m)\cdot i\boldsymbol{A}(\tau_n) - i\boldsymbol{A}(\tau_n)\cdot i\boldsymbol{A}(\tau_m). \tag{4.6.57}$$

To calculate matrix $\boldsymbol{\Lambda}$ directly from Eq. (4.6.56) is very difficult. Usually, the following approximate calculation method is used.

(1) Calculate of $\boldsymbol{\Lambda}$

By truncating the Magnus series, i.e. Eq. (4.6.56), the approximate solution of $\boldsymbol{\Lambda}\left(x_3^0 + h, x_3^0\right)$ can be computed in the following simple and efficient way. By introducing the matrix $\boldsymbol{M}_j$,

$$\boldsymbol{M}_j = \frac{1}{h^j}\int_{x_3^0}^{x_3^0+h}\left(\tau - \left(x_3^0 + \frac{h}{2}\right)\right)^j i\boldsymbol{A}(\tau)d\tau,\ j = 0, 1, 2, \ldots. \tag{4.6.58}$$

It is noted that only single integral is included and the multiple integral is excluded. The truncated Magnus series $\boldsymbol{\Lambda}\left(x_3^0 + h, x_3^0\right)$ can be written as a linear combination of $\boldsymbol{M}_j$ and their commutators. The sixth-order approximation of the Magnus series can be written as

$$\boldsymbol{\Lambda}^{(6)} = \boldsymbol{M}_0 + \frac{1}{240}[-20\boldsymbol{b}_1 - \boldsymbol{b}_3 + \boldsymbol{s}_1, \boldsymbol{b}_1 + \boldsymbol{s}_2], \tag{4.6.59}$$

where

$$s_1 = [b_1, b_2], s_2 = -\frac{1}{60}[b_1, 2b_3 + s_1], \tag{4.6.60}$$

$$b_1 = \frac{3}{4}(3M_0 - 20M_2), b_2 = 12M_1, b_3 = -15(M_0 - 12M_2). \tag{4.6.61}$$

(2) Calculation of e^A or $Q(h)$

The approximate solution of e^A is obtained by Padé approximation. The 2 m order Pade approximation can be written as

$$Q(h) = (P_m(-A))^{-1} \cdot (P_m(A)), \tag{4.6.62}$$

where

$$P_0(A) = E, P_1(A) = 2E + A, \tag{4.6.63}$$

$$P_m(A) = 2(2m-1)P_{m-1}(A) + A^2 \cdot P_{m-2}(A). \tag{4.6.64}$$

By using the 8 order Padé approximation, the transfer matrix of inhomogeneous interlayer is

$$Q(h) = (P_4(-A))^{-1} \cdot (P_4(A)). \tag{4.6.65}$$

4.7 Bloch Waves in Periodic Layered Structures

The propagation of elastic waves in periodic layered structures has the characteristic of "band gap" [19–23]. The elastic waves with frequency falling into the forbidden band cannot propagate through periodic layered structures, which means that periodic layered structures have frequency selectivity for elastic waves propagation. Periodic layered structures are also known as one-dimensional phononic crystals. Due to the existence of periodic structure, elastic waves have Bragg reflection at the interface, and some elastic waves of certain frequencies will have destructive interference, so they cannot propagate in this periodic structure. The "band gap" property of phononic crystals has broad application prospects in the fields of vibration reduction, sound insulation, acoustic filtering and acoustic functional devices. By using the transfer matrix method (TMM) and the Bloch wave theory in the periodic structure, the transfer matrix of one typical cell and the dispersion equation are derived for the periodically laminated binary system with perfect interfaces. Then, it is derived that the transfer matrices of three types of imperfect interfaces, i.e. the displacement vector jumps, the traction vector jumps and the displacement and traction vectors jumps simultaneously, which are embedded into the transfer matrix of one typical single cell to study the dispersion relationship of the Bloch wave and the corresponding "band gap." Band gaps in the case of imperfect interfaces are compared with that in the case of perfect interfaces to discuss the influence of imperfect interfaces on band gaps.

Consider a one-dimensional periodic layered structure formed by periodic repetition of two different solids, as showed in Fig. 4.11. The constitutive constants of these two materials A and B are denoted by elasticity modulus $E_j\left(E_j = \frac{\mu_j(3\lambda_j+2\mu_j)}{(\lambda_j+\mu_j)}\right)$, mass density ρ_j and Lame constant λ_j and μ_j, respectively. The total thickness of one typical cell is $a = a_1 + a_2$. The thicknesses of two sublayers are a_1 and a_2, respectively. The x-axis is perpendicular to the planar interface of two sublayers. At the interface of each sublayer, due to the reflection and transmission of elastic waves, there are forward waves and backward waves with opposite propagation directions in the sublayer, as shown in Fig. 4.12. The forward and backward waves in material A are denoted as A_1 and A_2 respectively. The forward and backward waves in material B are denoted as B_1 and B_2 respectively.

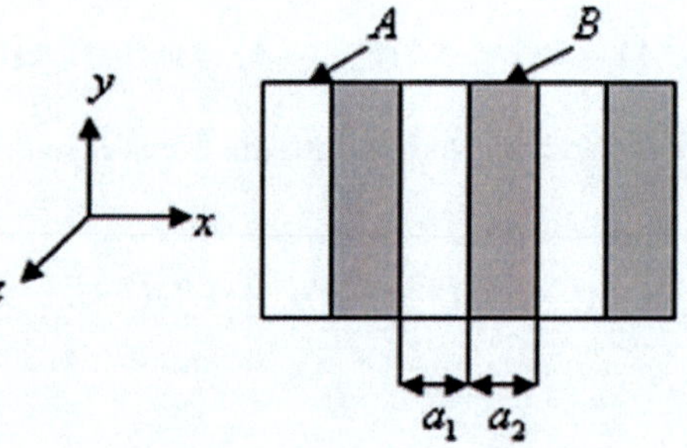

Fig. 4.11 Schematic diagram of a periodic layered structure (one-dimensional phononic crystal)

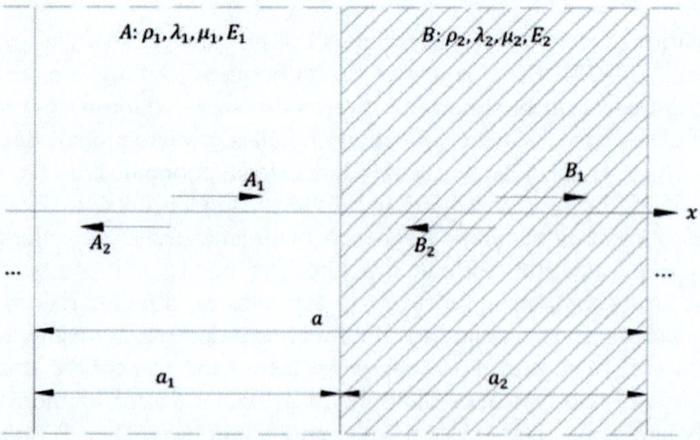

Fig. 4.12 Schematic diagram of plane waves in each sublayer of one typical single cell in the periodic layered structure

Let us first consider the SH waves propagating along the x-axis and the polarization direction is z-axis, namely

$$\boldsymbol{u}(x, y, z) = \boldsymbol{u}(x) = u_z(x)\boldsymbol{e}_z. \tag{4.7.1}$$

So, the motion equation degenerates into

$$\frac{\partial^2 u_z(x)}{\partial x^2} + k_t^2 u_z(x) = 0, \tag{4.7.2}$$

where $k_t = \frac{\omega}{c_t}$ and $c_t = \sqrt{\frac{\mu}{\rho}}$. The solution of above equation is

$$u_z(x) = A_1 e^{ik_t x} + A_2 e^{-ik_t x}. \tag{4.7.3}$$

The related stress component is

$$\tau = \mu \frac{\partial u_z}{\partial x} = i\mu k_t \left(A_1 e^{ik_t x} - A_2 e^{-ik_t x}\right). \tag{4.7.4}$$

Similarly, for the P wave propagating along the x-axis,

$$\boldsymbol{u}(x, y, z) = \boldsymbol{u}(x) = u_x(x)\boldsymbol{e}_x, \tag{4.7.5}$$

the motion equation degenerates into

$$\frac{\partial^2 u_x(x)}{\partial x^2} + k_l^2 u_x(x) = 0, \tag{4.7.6}$$

where $k_l = \frac{\omega}{c_l}$ and $c_l = \sqrt{\frac{(\lambda+2\mu)}{\rho}}$. The solution of above equation is

$$u_x(x) = A_1 e^{ik_l x} + A_2 e^{-ik_l x}. \tag{4.7.7}$$

The related stress component is

$$\sigma = (\lambda + 2\mu)\frac{\partial u_x}{\partial x} = i(\lambda + 2\mu)k_l\left(A_1 e^{ik_l x} - A_2 e^{-ik_l x}\right). \tag{4.7.8}$$

The displacement and stress of SH waves or P waves propagating along the normal direction of the interface can be uniformly written as

$$u_j(x_j) = A_1 e^{ik_j x_j} + A_2 e^{-ik_j x_j}, \tag{4.7.9}$$

$$\sigma_j = (\lambda_j + 2\mu_j)\frac{\partial u_j}{\partial x} = ik_j\rho_j c_{lj}^2\left(A_1 e^{ik_j x_j} - A_2 e^{-ik_j x_j}\right), \tag{4.7.10a}$$

$$\tau_j = \mu_j \frac{\partial u_j}{\partial x} = ik_j\rho_j c_{tj}^2 \left(A_1 e^{ik_j x_j} - A_2 e^{-ik_j x_j}\right), \tag{4.7.10b}$$

where $c_{tj} = \sqrt{\frac{\mu_j}{\rho_j}}$ and $c_{lj} = \sqrt{\frac{(\lambda_j + 2\mu_j)}{\rho_j}}$.

Consider the periodicity of layered structure

$$u(x) = u(x + na). \tag{4.7.11}$$

The Bloch waves propagating in the periodic layered structure can be expressed as

$$u(x,t) = A(q,x)e^{i(qx - \omega t)}, \tag{4.7.12a}$$

or

$$u(x) = A(q,x)e^{iqx}, \tag{4.7.12b}$$

where the amplitude $A(q,x)$ satisfies the periodicity condition

$$A(q,x) = A(q,x+a). \tag{4.7.13}$$

Based on Eqs. (4.7.3) and (4.7.11), it is obtained that

$$A(q,x_j) = e^{-iqx_j}\left(A_1 e^{ik_j x_j} + A_2 e^{-ik_j x_j}\right). \tag{4.7.14}$$

As $A(q,x)$ is a periodic function, replacing $x - na$ with x leads to

$$A(q,x_j) = e^{-iq(x_j - na)}\left[A_1 e^{ik_j(x_j - na)} + A_2 e^{-ik_j(x_j - na)}\right]. \tag{4.7.15}$$

When $na < x < na + a_1$, it is obtained that the expressions of displacement and stress components from Eqs. (4.7.11) and (4.7.14)

$$u_1(x) = e^{iqna}\left[A_1 e^{ik_1(x - na)} + A_2 e^{-ik_1(x - na)}\right], \tag{4.7.16a}$$

$$\sigma_1(x) = ik_1\rho_1 c_{l1}^2 e^{iqna}\left[A_1 e^{ik_1(x - na)} - A_2 e^{-ik_1(x - na)}\right], \tag{4.7.16b}$$

$$\tau_1(x) = ik_1\rho_1 c_{t1}^2 e^{iqna}\left[A_1 e^{ik_1(x - na)} - A_2 e^{-ik_1(x - na)}\right]. \tag{4.7.16c}$$

Similarly, when $na + a_1 < x < (n+1)a$, it is obtained that the expressions of displacement and stress components

$$u_2(x) = e^{iqna}\left[B_1 e^{ik_2(x - na - a_1)} + B_2 e^{-ik_2(x - na - a_1)}\right], \tag{4.7.17a}$$

$$\sigma_2(x) = ik_2\rho_2 c_{l2}^2 e^{iqna}\left[B_1 e^{ik_2(x-na-a_1)} - B_2 e^{-ik_2(x-na-a_1)}\right], \tag{4.7.17b}$$

$$\tau_2(x) = ik_2\rho_2 c_{t2}^2 e^{iqna}\left[B_1 e^{ik_2(x-na-a_1)} - B_2 e^{-ik_2(x-na-a_1)}\right]. \tag{4.7.17c}$$

For the SH waves, the expressions of displacement and stress components at the left and right side of material A (labeled L and R) are

$$u_{1L} = A_1 + A_2, \tag{4.7.18a}$$

$$\tau_{1L} = ik_1\rho_1 c_1^2 (A_1 - A_2), \tag{4.7.18b}$$

$$u_{1R} = A_1 e^{ik_1 a_1} + A_2 e^{-ik_1 a_1}, \tag{4.7.18c}$$

$$\tau_{1R} = ik_1\rho_1 c_1^2\left(A_1 e^{ik_1 a_1} - A_2 e^{-ik_1 a_1}\right), \tag{4.7.18d}$$

or in the matrix form

$$\begin{Bmatrix} u_{1,L}^{(i)} \\ \tau_{1,L}^{(i)} \end{Bmatrix} = \begin{pmatrix} 1 & 1 \\ ik_1\rho_1 c_1^2 & -ik_1\rho_1 c_1^2 \end{pmatrix}\begin{Bmatrix} A_1 \\ A_2 \end{Bmatrix} = \boldsymbol{D_L}\cdot\begin{Bmatrix} A_1 \\ A_2 \end{Bmatrix}, \tag{4.7.19b}$$

$$\begin{Bmatrix} u_{1,R}^{(i)} \\ \tau_{1,R}^{(i)} \end{Bmatrix} = \begin{pmatrix} e^{ik_1 a_1} & e^{-ik_1 a_1} \\ ik_1\rho_1 c_1^2 e^{ik_1 a_1} & -ik_1\rho_1 c_1^2 e^{-ik_1 a_1} \end{pmatrix}\begin{Bmatrix} A_1 \\ A_2 \end{Bmatrix} = \boldsymbol{D_R}\cdot\begin{Bmatrix} A_1 \\ A_2 \end{Bmatrix}. \tag{4.7.19b}$$

Then

$$\begin{Bmatrix} u_{1,R}^{(i)} \\ \tau_{1,R}^{(i)} \end{Bmatrix} = \boldsymbol{D_R}\cdot\begin{Bmatrix} A_1 \\ A_2 \end{Bmatrix} = \boldsymbol{D_R}\cdot\boldsymbol{D_L}^{-1}\begin{Bmatrix} u_{1,L}^{(i)} \\ \tau_{1,L}^{(i)} \end{Bmatrix} = \boldsymbol{T}_1'\cdot\begin{Bmatrix} u_{1,L}^{(i)} \\ \tau_{1,L}^{(i)} \end{Bmatrix}, \tag{4.7.20}$$

where the explicit expression of the elements in the transfer matrix of material A is

$$\boldsymbol{T}_1' = \begin{bmatrix} \cos k_1 a_1 & \frac{\sin k_1 a_1}{i\omega\rho_1 c_1} \\ -i\omega\rho_1 c_1 \sin k_1 a_1 & \cos k_1 a_1 \end{bmatrix}. \tag{4.7.21}$$

Similarly, the transfer matrix of material B is

$$T_2' = \begin{bmatrix} \cos k_2 a_2 & \frac{\sin k_2 a_2}{i\omega\rho_2 c_2} \\ -i\omega\rho_2 c_2 \sin k_2 a_2 & \cos k_2 a_2 \end{bmatrix}, \tag{4.7.22}$$

where $k_j = \frac{\omega}{c_j}$ and $c_j = c_{tj} = \sqrt{\frac{\mu_j}{\rho_j}}(j = 1, 2)$ is the wave velocity of SH waves in each sublayer. Furthermore, the transfer matrices of materials A and B for the P waves can be obtained through the variation of $c_j = c_{lj} = \sqrt{\frac{(\lambda_j + 2\mu_j)}{\rho_j}}$.

A composite layered structure constituted of solid A and solid B is called an element layer (or a typical unit) and is denoted by $i(i = 1, 2, \cdots, n)$. Solid A and solid B is called sublayer (denoted by $j = 1, 2$) of an element layer. The state vectors of the left or right surfaces of each sublayer are defined as

$$\boldsymbol{V}_{jL}^{(i)} = \left\{\tau_{zxjL}^{(i)}, u_{zjL}^{(i)}\right\}^T, \quad \boldsymbol{V}_{jR}^{(i)} = \left\{\tau_{zxjR}^{(i)}, u_{zjR}^{(i)}\right\}^T, \tag{4.7.23a}$$

for the propagation of SH waves, and

$$\boldsymbol{V}_{jL}^{(i)} = \left\{\sigma_{xjL}^{(i)}, u_{xjL}^{(i)}\right\}^T, \quad \boldsymbol{V}_{jR}^{(i)} = \left\{\sigma_{xjR}^{(i)}, u_{xjR}^{(i)}\right\}^T, \tag{4.7.23b}$$

for the propagation of P waves. State vectors of the left and right surfaces of the same sublayer are related by the transfer matrix of a sublayer, namely

$$\boldsymbol{V}_{jR}^{(i)} = \boldsymbol{T}'_j \boldsymbol{V}_{jL}^{(i)}. \tag{4.7.24}$$

State vectors of the left and fight surfaces of a typical single cell are related by the transfer matrix of a typical single cell

$$\boldsymbol{V}_{R}^{(i)} = \boldsymbol{T}_i \boldsymbol{V}_{L}^{(i)}. \tag{4.7.25}$$

For the perfect interface, the displacement and the traction vectors are continuous along the normal and tangent direction of the planar interface, namely

$$\boldsymbol{V}_{jR}^{(i)} = \boldsymbol{V}_{(j+1)L}^{(i)}, \ \boldsymbol{V}_{R}^{(i)} = \boldsymbol{V}_{L}^{(i+1)}. \tag{4.7.26}$$

So, the transfer matrix of a typical single cell is

$$\boldsymbol{T}_i = \boldsymbol{T}'_2 \boldsymbol{T}'_1. \tag{4.7.27}$$

Based on Eqs. (4.7.21), (4.7.22) and (4.7.27), it is derived that

$$\boldsymbol{T}_i = \begin{bmatrix} \cos\gamma_1 \cos\gamma_2 - \frac{w_2}{w_1}\sin\gamma_1 \sin\gamma_2 & \frac{1}{\omega}\left(\frac{1}{w_1}\sin\gamma_1\cos\gamma_2 + \frac{1}{w_2}\cos\gamma_1\sin\gamma_2\right) \\ -\omega(w_1\sin\gamma_1\cos\gamma_2 + w_2\cos\gamma_1\sin\gamma_2) & \cos\gamma_1\cos\gamma_2 - \frac{w_1}{w_2}\sin\gamma_1\sin\gamma_2 \end{bmatrix}, \tag{4.7.28}$$

where the dimensionless wave numbers $\gamma_1 = k_1 a_1 = \omega \frac{a_1}{c_1}$ and $\gamma_2 = k_2 a_2 = \omega \frac{a_2}{c_2}$, the dimensionless wave impedance $w_1 = i\rho_1 c_1$ and $w_2 = i\rho_2 c_2$.

For the one-dimensional periodic structure, Eqs. (4.7.12) and (4.7.13) lead to

$$\begin{Bmatrix} u(na) \\ \tau(na) \end{Bmatrix} = e^{ika} \begin{Bmatrix} u((n-1)a) \\ \tau((n-1)a) \end{Bmatrix}, \tag{4.7.29}$$

which is the well-known Bloch theorem and k is the Bloch wave vector. Combination of Eqs. (4.7.25) and (4.7.29) leads to

$$\left| \boldsymbol{T}_i - e^{ika} \boldsymbol{I} \right| = 0. \tag{4.7.30}$$

Inserting Eq. (4.7.28) into Eq. (4.7.30) leads to

$$\begin{vmatrix} \cos\gamma_1 \cos\gamma_2 - \frac{w_2}{w_1} \sin\gamma_1 \sin\gamma_2 - e^{ika} & \frac{1}{\omega}\left(\frac{1}{w_1} \sin\gamma_1 \cos\gamma_2 + \frac{1}{w_2} \cos\gamma_1 \sin\gamma_2\right) \\ -\omega(w_1 \sin\gamma_1 \cos\gamma_2 + w_2 \cos\gamma_1 \sin\gamma_2) & \cos\gamma_1 \cos\gamma_2 - \frac{w_1}{w_2} \sin\gamma_1 \sin\gamma_2 - e^{ika} \end{vmatrix} = 0. \tag{4.7.31}$$

The expansion form of above equation is

$$\left(e^{ika}\right)^2 - \left[2\cos\gamma_1 \cos\gamma_2 - \left(\frac{\rho_2 c_2}{\rho_1 c_1} + \frac{\rho_1 c_1}{\rho_2 c_2}\right) \sin\gamma_1 \sin\gamma_2\right] e^{ika} + 1 = 0. \tag{4.7.32}$$

The solution of Eq. (4.7.32) can be written as

$$\cos ka = \cos k_1 a_1 \cos k_2 a_2 - \frac{1}{2}\left(\frac{\rho_2 c_2}{\rho_1 c_1} + \frac{\rho_1 c_1}{\rho_2 c_2}\right) \sin k_1 a_1 \sin k_2 a_2. \tag{4.7.33}$$

This is the dispersion equation of Bloch waves in the one-dimension periodic laminated structure [23].

Considering the periodicity of layered structure, the dispersion equation can be discussed only in the first Brillouin zone, which means that the range of wave number k of Bloch wave is limited to $\left[-\frac{\pi}{a}, \frac{\pi}{a}\right]$. Based on Eq. (4.7.33), a family of dispersion curves can be obtained. According to these dispersion curves of Bloch wave, the band gaps structure of one-dimensional phononic crystal can be further analyzed.

When the direction of the elastic wave propagating in the sublayer has an angle with the normal direction of the interface, the elastic wave field in the sublayer is not only a function of the coordinate x but also a function of the coordinate y. Consider the oblique propagation of SH waves in each sublayer, as showed in Fig. 4.13. Assume SH waves propagate in the Oxy plane and the particle polarization is in the z direction,

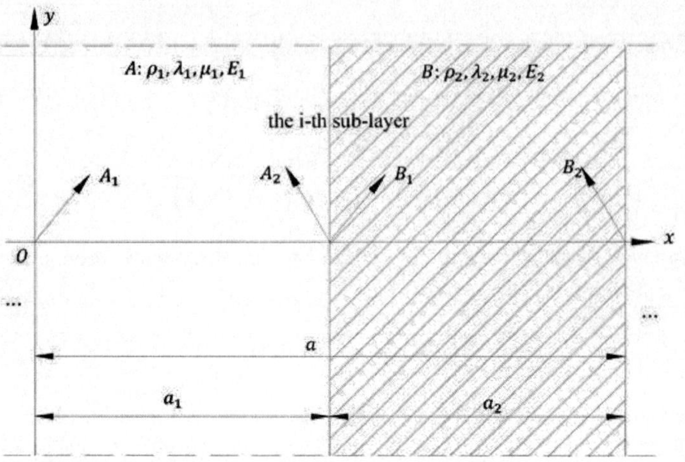

Fig. 4.13 Schematic diagram of SH waves propagation in a binary periodic layered structure with the propagation direction inconsistent with interface normal direction

the motion equation of oblique propagation of SH waves in the Oxy plane is

$$\frac{\partial^2 u_z(x, y)}{\partial x^2} + \frac{\partial^2 u_z(x, y)}{\partial y^2} + k_t^2 u_z(x, y) = 0, \quad (4.7.34)$$

where $k_t^2 = \frac{\omega}{c_t}$ and $c_t = \sqrt{\frac{\mu}{\rho}}$. The solution of above equation is

$$u_z(x, y) = \mathrm{e}^{ik_y y}\left(A_1\mathrm{e}^{iqx} + A_2\mathrm{e}^{-iqx}\right), \quad (4.7.35)$$

where $q = \sqrt{k_t^2 - k_y^2} = \sqrt{\left(\frac{\omega}{c_t}\right)^2 - k_y^2}$. k_y is the apparent waves number. Based on the Snell law, SH waves in each sublayer all have the same apparent wave number. So, the apparent wave number can be used to characterize the degree of inconsistency between SH waves and the interface normal direction. When $na < x < na + a_1$

$$u_z(x, y) = \mathrm{e}^{ikna}\left[A_+\mathrm{e}^{iq_1(x-na)} + A_-\mathrm{e}^{-iq_1(x-na)}\right]\mathrm{e}^{ik_y y}, \quad (4.7.36)$$

$$\tau_{zx}(x, y) = iq_1\mu\mathrm{e}^{ikna}\left[A_+\mathrm{e}^{iq_1(x-na)} - A_-\mathrm{e}^{-iq_1(x-na)}\right]\mathrm{e}^{ik_y y}. \quad (4.7.37)$$

When $na + a_1 < x < (n+1)a$, the expression of displacement and the stress can also be obtained

$$u_z(x, y) = \mathrm{e}^{ikna}\left[B_+\mathrm{e}^{iq_2(x-na-a_1)} - B_-\mathrm{e}^{-iq_2(x-na-a_1)}\right]\mathrm{e}^{ik_y y}, \quad (4.7.38)$$

$$\tau_{zx}(x, y) = iq_2\mu e^{ikna}\left[B_+ e^{iq_2(x-na-a_1)} + B_- e^{-iq_2(x-na-a_1)}\right]e^{ik_y y}, \tag{4.7.39}$$

where $q_1 = \sqrt{\left(\frac{\omega}{c_{t1}}\right)^2 - k_y^2}$ and $q_2 = \sqrt{\left(\frac{\omega}{c_{t2}}\right)^2 - k_y^2}$.

The transfer matrix of each sublayer is

$$T'_j = \begin{bmatrix} \cos q_{tj}a_j & \frac{\sin q_{tj}a_j}{iq_{t\,j}\mu_j} \\ iq_{tj}\mu_j \sin q_{tj}a_j & \cos q_{tj}a_j \end{bmatrix}, \tag{4.7.40}$$

where $q_{tj} = \sqrt{\left(\frac{\omega}{c_{tj}}\right)^2 - k_y^2}$, $j = 1, 2$. Based on $T_i = T'_1 T'_2$, the transfer matrix of one single typical cell is

$$T_i = \begin{bmatrix} \cos q_1 a_1 \cos q_2 a_2 - \frac{\mu_1 q_{t1}}{\mu_2 q_{t2}} \sin q_1 a_1 \sin q_2 a_2 & i\frac{1}{\mu_1 q_{t1}}\left(\sin q_1 a_1 \cos q_2 a_2 + \frac{\mu_1 q_{t1}}{\mu_2 q_{t2}} \cos q_1 a_1 \sin q_2 a_2\right) \\ i\mu_1 q_{t1}\left(\sin q_1 a_1 \cos q_2 a_2 + \frac{\mu_2 q_{t2}}{\mu_1 q_{t1}} \cos q_1 a_1 \sin q_2 a_2\right) & \cos q_1 a_1 \cos q_2 a_2 - \frac{\mu_2 q_{t2}}{\mu_1 q_{t1}} \sin q_1 a_1 \sin q_2 a_2 \end{bmatrix}. \tag{4.7.41}$$

Based on the Bloch theorem, the eigen equation (or secular equation) is obtained

$$\left|T_i - e^{ika} I\right| = 0. \tag{4.7.42}$$

The solution of Eq. (4.7.42) is

$$\cos ka = \cos q_1 a_1 \cos q_2 a_2 - \frac{1}{2}\left(\frac{q_1\mu_1}{q_2\mu_2} + \frac{q_2\mu_2}{q_1\mu_1}\right)\sin q_1 a_1 \sin q_2 a_2. \tag{4.7.43}$$

This is the dispersion equation with respect to oblique propagation SH waves, i.e. $\omega = \omega(k, k_y)$. When $k_y = 0$, Eq. (4.7.43) degenerates into the dispersion equation with respect to normal propagation SH waves, i.e. Eq. (4.7.33).

Consider the oblique propagation of *P* and SV waves now. When the propagation direction of the elastic wave in the sublayer has an angle with the normal direction of the interface, the coupling of *P* wave and SV wave exists when the reflection and transmission occur at the interface. Therefore, there are both *P* waves and SV waves in the forward and backward waves of each sublayer, and they all contribute to the displacement and stress.

Assume *P* and SV waves propagate in the Oxy plane, and the displacement is

$$\boldsymbol{u} = u_x(x, y)\boldsymbol{e}_x + u_y(x, y)\boldsymbol{e}_y. \tag{4.7.44}$$

The motion equation is

$$\begin{cases} (\lambda+2\mu)\left(\frac{\partial^2 u_x}{\partial x^2}+\frac{\partial^2 u_y}{\partial x\partial y}\right)+\mu\left(\frac{\partial^2 u_x}{\partial y^2}-\frac{\partial^2 u_y}{\partial x\partial y}\right)=-\rho\omega^2 u_x \\ (\lambda+2\mu)\left(\frac{\partial^2 u_x}{\partial x\partial y}+\frac{\partial^2 u_y}{\partial y^2}\right)+\mu\left(\frac{\partial^2 u_y}{\partial x^2}-\frac{\partial^2 u_x}{\partial y\partial x}\right)=-\rho\omega^2 u_y \end{cases}. \tag{4.7.45}$$

Introduce two potential functions $\phi(x, y)$ and $\psi(x, y)\boldsymbol{e}_z$ to express the displacement, namely

$$\boldsymbol{u}(x, y)=\nabla\phi(x, y)+\nabla\times\psi(x, y)\boldsymbol{e}_z. \tag{4.7.46}$$

Then, the motion equation is decoupled into

$$\nabla^2\phi(x, y)+k_l^2\phi(x, y)=0, \tag{4.7.47a}$$

$$\nabla^2\psi(x, y)+k_t^2\psi(x, y)=0, \tag{4.7.47b}$$

where $k_t=\frac{\omega}{c_t}$, $k_l=\frac{\omega}{c_l}$, $c_t=\sqrt{\frac{\mu}{\rho}}$ and $c_l=\sqrt{\frac{\lambda+2\mu}{\rho}}$. The solution of Eq. (4.7.47) is

$$\phi(x, y)=\mathrm{e}^{ik_y y}\left(A_1\mathrm{e}^{i\alpha x}+A_2\mathrm{e}^{-i\alpha x}\right), \tag{4.7.48a}$$

$$\psi(x, y)=\mathrm{e}^{ik_y y}\left(B_1\mathrm{e}^{i\beta x}+B_2\mathrm{e}^{-i\beta x}\right), \tag{4.7.48b}$$

where $\alpha=\sqrt{\left(\frac{\omega}{c_l}\right)^2-k_y^2}$ and $\beta=\sqrt{\left(\frac{\omega}{c_t}\right)^2-k_y^2}$. A_1, A_2, B_1 and B_2 are four unknown constants. Furthermore, the expressions of displacement and stress are obtained as

$$u_x=\frac{\partial\varphi}{\partial x}+\frac{\partial\psi}{\partial y}=\alpha\mathrm{e}^{ik_y y}\left(A_1\mathrm{e}^{\alpha x}-A_2\mathrm{e}^{-\alpha x}\right)+ik_y\mathrm{e}^{ik_y y}\left(B_1\mathrm{e}^{\beta x}+B_2\mathrm{e}^{-\beta x}\right),$$
$$u_y=\frac{\partial\varphi}{\partial y}-\frac{\partial\psi}{\partial x}=ik_y\mathrm{e}^{ik_y y}\left(A_1\mathrm{e}^{\alpha x}+A_2\mathrm{e}^{-\alpha x}\right)-\beta\mathrm{e}^{ik_y y}\left(B_1\mathrm{e}^{\beta x}-B_2\mathrm{e}^{-\beta x}\right),$$

$$\begin{gathered}\sigma_x=\lambda\left(\frac{\partial^2\varphi}{\partial x^2}+\frac{\partial^2\varphi}{\partial y^2}\right)+2\mu\left(\frac{\partial^2\varphi}{\partial x^2}+\frac{\partial^2\psi}{\partial x\partial y}\right)\\ =(\lambda+2\mu)\alpha^2\mathrm{e}^{ik_y y}\left(A_1\mathrm{e}^{\alpha x}+A_2\mathrm{e}^{-\alpha x}\right)+(-\lambda k_y+2\mu\beta i)k_y\mathrm{e}^{ik_y y}\left(B_1\mathrm{e}^{\beta x}+B_2\mathrm{e}^{-\beta x}\right),\end{gathered}$$

$$\begin{aligned}\tau_{yx}&=\mu\left(2\frac{\partial^2\varphi}{\partial x\partial y}+\frac{\partial^2\psi}{\partial y^2}-\frac{\partial^2\psi}{\partial x^2}\right)\\ &=\mu 2ik_y\alpha\mathrm{e}^{ik_y y}\left(A_1\mathrm{e}^{\alpha x}-A_2\mathrm{e}^{-\alpha x}\right)-\mu\alpha^2\mathrm{e}^{ik_y y}\left(A_1\mathrm{e}^{\alpha x}+A_2\mathrm{e}^{-\alpha x}\right)\\ &\quad-\mu k_y^2\mathrm{e}^{ik_y y}\left(B_1\mathrm{e}^{\beta x}+B_2\mathrm{e}^{-\beta x}\right).\end{aligned} \tag{4.7.49}$$

The state vector of the left and right surfaces of each sublayer, i.e. $\boldsymbol{V}_{jL}^{(i)}$ and $\boldsymbol{V}_{jR}^{(i)}(j=1,2)$, is defined as

$$\boldsymbol{V}_{jL}^{(i)}=\left\{\sigma_{xjL}^{(i)}, \tau_{yxjL}^{(i)}, u_{xjL}^{(i)}, u_{yjL}^{(i)}\right\}^T, \tag{4.7.50a}$$

$$\boldsymbol{V}_{jR}^{(i)} = \left\{ \sigma_{xjR}^{(i)}, \tau_{yxjR}^{(i)}, u_{xjR}^{(i)}, u_{yjR}^{(i)} \right\}^{T}. \tag{4.7.50b}$$

They are related by the transfer matrix of sublayer, namely

$$\boldsymbol{V}_{jR}^{(i)} = \boldsymbol{T}'_{j} \boldsymbol{V}_{jL}^{(i)}, \tag{4.7.51}$$

where $\boldsymbol{T}'_{j}$ is a transfer matrix of 4×4 order. Redefine four unknown constants, then the displacement can be expressed as (time factors are omitted)

$$u_y(x, y) = \left[A_{j,R}^{(i)} \cos q_{lj}x - i A_{j,L}^{(i)} \sin q_{lj}x \right.$$
$$\left. + B_{j,R}^{(i)} \cos q_{tj}x - i B_{j,L}^{(i)} \sin q_{tj}x \right] e^{ik_y y}, \tag{4.7.52a}$$

$$u_x(x, y) = \left[\frac{q_{lj}}{ik_y} \left(i A_{j,R}^{(i)} \sin q_{lj}x - A_{j,L}^{(i)} \cos q_{lj}x \right) \right.$$
$$\left. - \frac{k_y}{q_{tj}} \left(i B_{j,R}^{(i)} \sin q_{tj}x - B_{j,L}^{(i)} \cos q_{tj}x \right) \right] e^{ik_y y}, \tag{4.7.52b}$$

where $q_{lj} = \sqrt{\left(\frac{\omega}{c_{lj}}\right)^2 - k_y^2}$, $q_{tj} = \sqrt{\left(\frac{\omega}{c_{tj}}\right)^2 - k_y^2}$, $c_{lj} = \sqrt{\frac{(\lambda_j + 2\mu_j)}{\rho_j}}$ and $c_{tj} = \sqrt{\frac{\mu_j}{\rho_j}}$.

After obtaining the transfer matrix of each sublayer, the transfer matrix of the typical cell can be obtained by $\boldsymbol{T} = \boldsymbol{T}'_2 \boldsymbol{T}'_1$.

According to the Bloch theorem, the eigen equation is obtained as

$$\left| \boldsymbol{T} - e^{ika} \boldsymbol{I} \right| = 0. \tag{4.7.53}$$

The solution of Eq. (4.7.53) is

$$\cos ka = \cos(q_{p1}a_1)\cos(q_{p2}a_2)$$
$$- \frac{1}{2} \left(\frac{q_{p1}\rho_1 c_{p1}^2}{q_{p2}\rho_2 c_{p2}^2} + \frac{q_{p2}\rho_2 c_{p2}^2}{q_{p1}\rho_1 c_{p1}^2} \right) \sin(q_{p1}a_1)\sin(q_{p2}a_2), \tag{4.7.54}$$

where p denotes l and t, which means longitudinal and transverse modes. Equation (4.7.54) is the dispersion equation, i.e. $\omega = \omega(k, k_y)$, for oblique propagation P and SV waves. When $k_y = 0$, it degrade to the dispersion equation for the normal propagation of P and SV waves.

A perfect interface means the displacement and the traction vectors are continuous along the normal and tangent direction of the planar interface. The boundary condition can be written as

$$[\boldsymbol{t}] = 0, [\boldsymbol{u}] = 0, \tag{4.7.55}$$

where $[\cdot]$ denotes the jump across interface. $[\boldsymbol{t}] = \boldsymbol{t}^+ - \boldsymbol{t}^-$ denotes the traction vector jump. $[u] = \boldsymbol{u}^+ - \boldsymbol{u}^-$ denotes the displacement vector jump. $\boldsymbol{t}^+, \boldsymbol{t}^-, \boldsymbol{u}^+$ and $\boldsymbol{u}^-$ denote the traction and displacement vectors on both sides of the interface, respectively.

For the imperfect interface, the displacement and the traction vectors do not satisfy the continuous condition. Three kinds of imperfect interfaces are discussed here. For simplicity, only the case of P wave or SV propagating along the normal of the interface is discussed, and the case of oblique propagation can be treated similarly. For the spring interface model, the traction vector is continuous but the displacement vector jumps. The boundary condition in the spring interface model can be expressed as

$$\sigma_{xs}^{(i+1)} = \sigma_{xs}^{(i)}, \quad u_s^{(i+1)} - u_s^{(i)} = F_s \sigma_s^{(i)} \, (s = x, y), \tag{4.7.56}$$

where F_s is compliance coefficient. For the interface of materials A and material B in the same cell and in the different cells, the boundary conditions can be expressed in terms of state vector as

$$\boldsymbol{V}_{jL'}^{(i)} = \boldsymbol{T}_3' \boldsymbol{V}_{(j-1)\boldsymbol{R}'}^{(i)} \quad \boldsymbol{V}_{jL}^{(i+1)} = \boldsymbol{T}_4' \boldsymbol{V}_{(j+1)\boldsymbol{R}}^{(i)} \; (j = 1, 2), \tag{4.7.57}$$

where $\boldsymbol{T}_3'$ and $\boldsymbol{T}_4'$ show the effect of imperfect interface. Their expressions are

$$\boldsymbol{T}_3' = \boldsymbol{T}_4' = \begin{bmatrix} 1 & F_s \\ 0 & 1 \end{bmatrix}. \tag{4.7.58}$$

For a phononic crystal with periodic layered structure of two components, the transfer matrix of a typical cell with an imperfect interface should be modified to

$$\boldsymbol{T} = \boldsymbol{T}_1' \boldsymbol{T}_3' \boldsymbol{T}_2' \boldsymbol{T}_4' = \begin{bmatrix} T_{11} & T_{12} \\ T_{21} & T_{22} \end{bmatrix}, \tag{4.7.59}$$

where

$$\begin{aligned}
T_{11} &= \cos k_1 a_1 \cos k_2 a_2 - \tfrac{w_2}{w_1} \sin k_1 a_1 \sin k_2 a_2 - F_s w_2 \cos k_1 a_1 \sin k_2 a_2, \\
T_{12} &= F_s \left(2 \cos k_1 a_1 \cos k_2 a_2 - \tfrac{w_2}{w_1} \sin k_1 a_1 \sin k_2 a_2\right) - F_s^2 w_2 \cos k_1 a_1 \sin k_2 a_2 \\
&\quad + \tfrac{1}{w_2} \cos k_1 a_1 \sin k_2 a_2 + \tfrac{1}{w_1} \sin k_1 a_1 \cos k_2 a_2, \\
T_{21} &= -w_1 \sin k_1 a_1 \cos k_2 a_2 - w_2 \cos k_1 a_1 \sin k_2 a_2 \\
&\quad + F_s w_1 w_2 \sin k_1 a_1 \sin k_2 a_2, \\
T_{22} &= -2F_s (w_1 \sin k_1 a_1 \cos k_2 a_2 + w_2 \cos k_1 a_1 \sin k_2 a_2) \\
&\quad + F_s^2 w_1 w_2 \sin k_1 a_1 \sin k_2 a_2 + \cos k_1 a_1 \cos k_2 a_2 - \tfrac{w_1}{w_2} \sin k_1 a_1 \sin k_2 a_2, \\
w_j &= \omega \rho_j c_j, \; j = 1, 2.
\end{aligned}$$

For the mass interface model, the displacement vector is continuous but the traction vector jumps. The boundary condition in the case of the mass interface model can be written as

$$u_s^{(i+1)} = u_s^{(i)}, \quad \sigma_{xs}^{(i+1)} - \sigma_{xs}^{(i)} = G_s u_s^{(i)}. \tag{4.7.60}$$

Similarly, the explicit expression of $\boldsymbol{T}'_3$ and $\boldsymbol{T}'_4$ can be written as

$$\boldsymbol{T}'_3 = \boldsymbol{T}'_4 = \begin{bmatrix} 1 & 0 \\ G_s & 1 \end{bmatrix}. \tag{4.7.61}$$

The transfer matrix of a typical cell with imperfect interfaces should be modified to $\boldsymbol{T} = \boldsymbol{T}'_1 \boldsymbol{T}'_3 \boldsymbol{T}'_2 \boldsymbol{T}'_4$.

For the spring–mass interface model, both displacement and traction vectors jump. The boundary condition in the case of the spring–mass interface model can be written as

$$u_s^{(i+1)} - u_s^{(i)} = F_s \frac{\sigma_{xs}^{(i+1)} + \sigma_{xs}^{(i)}}{2}, \quad \sigma_{xs}^{(i+1)} + \sigma_{xs}^{(i)} = G_s \frac{u_s^{(i+1)} + u_s^{(i)}}{2}. \tag{4.7.62}$$

The explicit expression of $\boldsymbol{T}'_3$ and $\boldsymbol{T}'_4$ can be written as

$$\boldsymbol{T}'_3 = \boldsymbol{T}'_4 = \begin{bmatrix} \frac{4+F_sG_s}{4-F_sG_s} & \frac{4F_s}{4+F_sG_s} \\ \frac{4G_s}{4-F_sG_s} & \frac{4+F_sG_s}{4-F_sG_s} \end{bmatrix}. \tag{4.7.63}$$

The transfer matrix of a typical cell with imperfect interfaces should be modified to $\boldsymbol{T} = \boldsymbol{T}'_1 \boldsymbol{T}'_3 \boldsymbol{T}'_2 \boldsymbol{T}'_4$. If the mass coefficient $G_s = 0$ and the compliance coefficient $F_s = 0$, Eq. (4.7.63) reduces to one unit matrix, namely the imperfect interface reduces to a perfect interface. If the mass coefficient $G_s = 0$ and the compliance coefficient $F_s \neq 0$, then the imperfect interface reduces to the spring interface model. If the mass coefficient $G_s \neq 0$ and the compliance coefficient $F_s = 0$, then the imperfect interface reduces to the mass interface model. The interface transfer matrix is not the unit matrix as long as the mass coefficient G_s and the compliance coefficient F_s are not zero simultaneously. They reflect the influence of imperfect interface on wave propagation.

The Bloch theorem is also true for imperfect interfaces. The dispersion equation of Bloch waves in the case of the spring interface model is

$$\begin{aligned} \cos ka = {} & \cos k_1 a_1 \cos k_2 a_2 - \frac{1}{2}\left(\frac{\rho_2 c_2}{\rho_1 c_1} + \frac{\rho_1 c_1}{\rho_2 c_2}\right) \sin k_1 a_1 \sin k_2 a_2 \\ & - F_s (w_1 \sin k_1 a_1 \cos k_2 a_2 + w_2 \cos k_1 a_1 \sin k_2 a_2) \\ & + \frac{1}{2} F_s^2 w_1 w_2 \sin k_1 a_1 \sin k_2 a_2, \end{aligned} \tag{4.7.64}$$

where F_s is the compliance coefficient. $\frac{\rho_1 c_1}{\rho_2 c_2}$ is impedance ratio, $w_1 = i\omega\rho_1 c_1$ and $w_2 = i\omega\rho_2 c_2$.

Table 4.1 Material density and elastic coefficient

Material	Mass density (kgm^{-3})	Lame constants Pa($kgm^{-1}s^{-2}$)
Pb	$\rho_{Pb} = 11600$	$\lambda_{Pb} = 4.23 \times 10^{10}$, $\mu_{Pb} = 1.49 \times 10^{10}$
Epoxy	$\rho_{epoxy} = 1180$	$\lambda_{epoxy} = 4.43 \times 10^{9}$, $\mu_{epoxy} = 1.59 \times 10^{9}$

Table 4.2 Compliance coefficients $\boldsymbol{F_s}$ and mass coefficients $\boldsymbol{G_s}$ in the case of imperfect interface

F_s	F_s	F_s/F_0	G_s	G_s/G_0
$F_{t1} = 2.5e{-}13$	$F_{n1} = 1.47e{-}13$	0.2	$G_1 = 1.278e11$	0.2
$F_{t2} = 1.25e{-}12$	$F_{n2} = 7.35e{-}3$	1	$G_2 = 6.39e11$	1

The dispersion equation of Bloch waves in the case of the mass interface model is

$$
\begin{aligned}
\cos ka = {} & \cos k_1 a_1 \cos k_2 a_2 - \frac{1}{2}\left(\frac{\rho_2 c_2}{\rho_1 c_1} + \frac{\rho_1 c_1}{\rho_2 c_2}\right) \sin k_1 a_1 \sin k_2 a_2 \\
& + G_s\left(\frac{1}{w_1} \sin k_1 a_1 \cos k_2 a_2 + \frac{1}{w_2} \cos k_1 a_1 \sin k_2 a_2\right) \\
& + \frac{G_s^2}{2 w_1 w_2} \sin k_1 a_1 \sin k_2 a_2,
\end{aligned}
\tag{4.7.65}
$$

where G_s is the mass coefficients.

The dispersion equation of Bloch waves in the case of the spring–mass interface model is

$$
\begin{aligned}
\cos ka = {} & \frac{1}{(4 - F_s G_s)^2}\left[(4 + F_s G_s)^2 + 16 F_s G_s\right] \cos k_1 a_1 \cos k_2 a_2 \\
& - \frac{1}{2(4 - F_s G_s)^2}\left[(4 + F_s G_s)^2\left(\frac{\rho_2 c_2}{\rho_1 c_1} + \frac{\rho_1 c_1}{\rho_2 c_2}\right)\right. \\
& \left. - 16\left(F_s^2 w_1 w_2 + \frac{G_s^2}{w_1 w_2}\right)\right] \sin k_1 a_1 \sin k_2 a_2 \\
& - \frac{4(4 + F_s G_s)}{(4 - F_s G_s)^2}\left[\left(F_s w_1 - \frac{G_s}{w_1}\right) \sin k_1 a_1 \cos k_2 a_2\right. \\
& \left. + \left(F_s w_2 - \frac{G_s}{w_2}\right) \cos k_1 a_1 \sin k_2 a_2\right].
\end{aligned}
\tag{4.7.66}
$$

A one-dimensional binary phononic crystal composed of lead (Pb) and epoxy was used as an example for numerical simulation. The material density and elastic coefficient of Pb and epoxy are given in Table 4.1. In the case of imperfect interfaces,

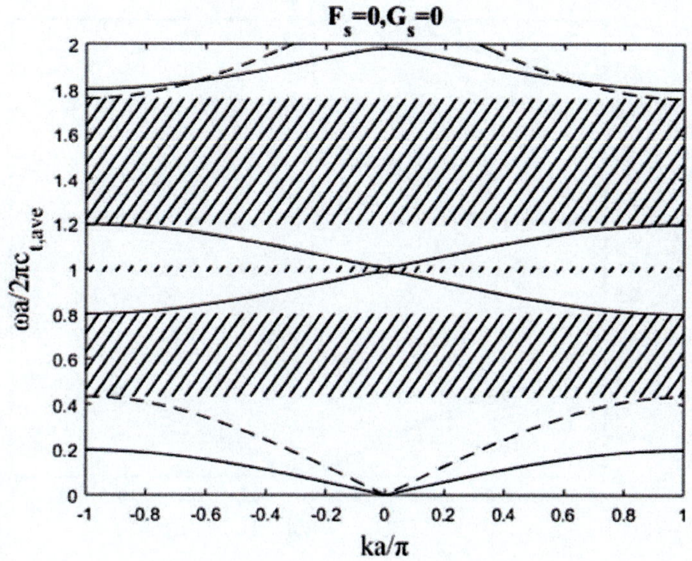

Fig. 4.14 Band gaps in the case of perfect interfaces (solid lines denote SH waves and dashed lines denote *P* waves)

the given compliance coefficient and the mass coefficient are listed in Table 4.2. It is assumed that both *P* and SH waves propagate in the normal direction of the interface. The filling fraction $f = \frac{a_{Pb}}{a} = 0.5$, where a_{Pb} is the thickness of the Pb sublayer. $a = 20$ mm is the lattice constant. The average velocity of the SH wave is $c_{t,ave} = \frac{(a_1 c_{t1} + a_2 c_{t2})}{a}$.

In the case of perfect interfaces, band gaps of one-dimensional phononic crystal considered are shown in Fig. 4.14. The solid line and the dashed line denote the dispersion curves of SH and *P* waves, respectively. Shadowed zones denote band gaps or stop bands. Figures 4.15, 4.16 and 4.17 show the dispersion curves of *P* wave and SH wave propagating along the normal line of the interface. It is observed that the dispersion curve moves toward the low frequency as the interface compliance coefficient increases. This is due to the increase of interface flexibility, which leads to the decrease of the stiffness of the entire laminar structure. But the effect of interface mass coefficient on dispersion curve is opposite to that of interface compliance coefficient. The high frequency mode is sensitive to the compliance coefficient of interface. The low-frequency modes are more sensitive to the mass coefficient of interface. The imperfect interface has a significant effect on the dispersion and "band gap" distribution of Bloch waves, which makes the number of "band gaps" increase. This phenomenon can be explained as following. On the one hand, the imperfect interface reduces the stiffness of the whole laminated structure, which makes the dispersion curves of *P* and SH waves shift toward low frequency. On the other hand, the imperfect interface induces of local resonance mode, which makes the dispersion curve more flat and thus helpful for the formation of band gap.

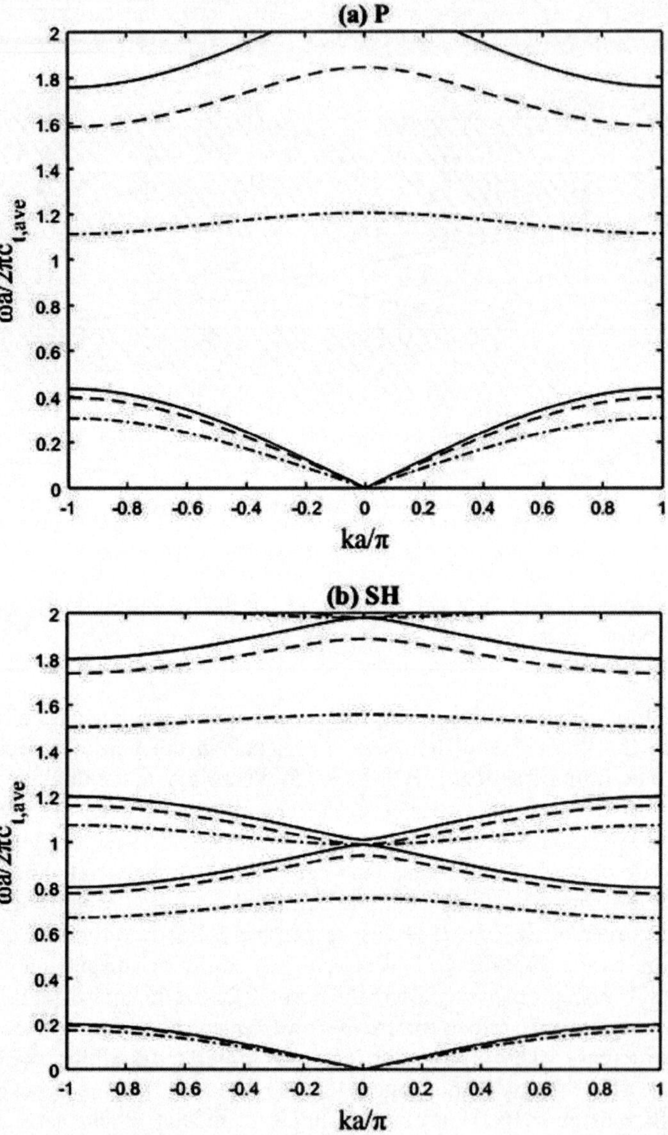

Fig. 4.15 Band gaps in the case of the spring interface model. **a** *P* mode; **b** SH mode. —$F_s = 0$, −−$F_s = 0.2F_0$, −.−$F_s = F_0$

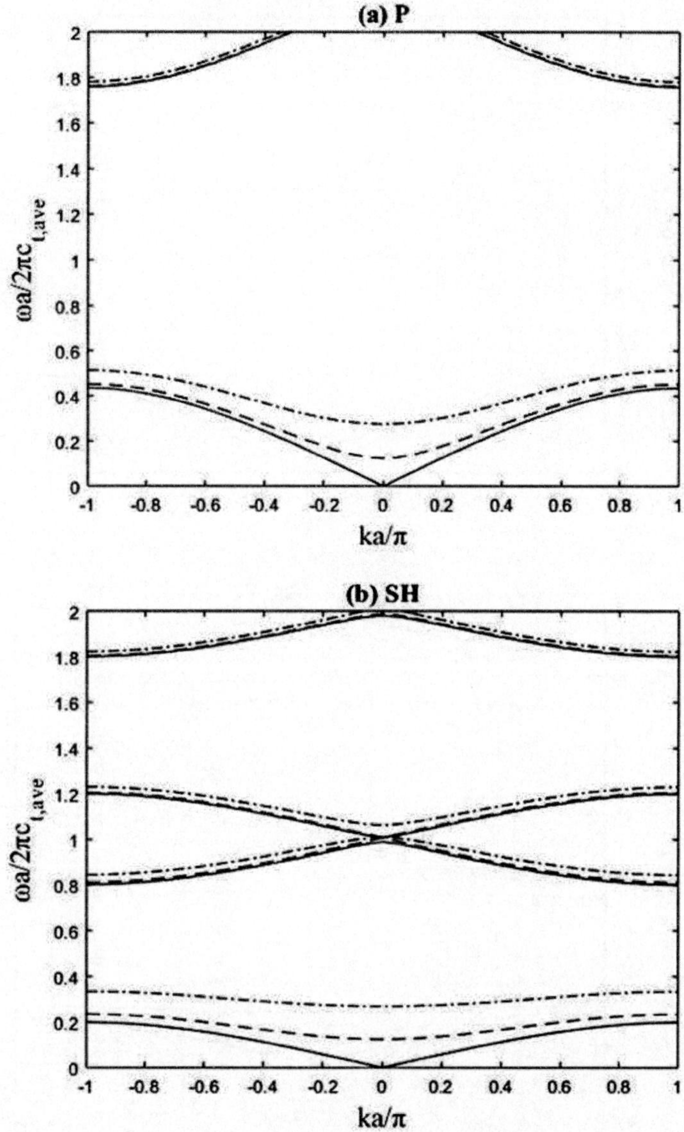

Fig. 4.16 Band gaps in the case of the mass interface model. **a** *P* mode; **b** SH mode. —$\boldsymbol{G_s = 0}$, −−$\boldsymbol{G_s = 0.2G_0}$, −.−$\boldsymbol{G_s = G_0}$

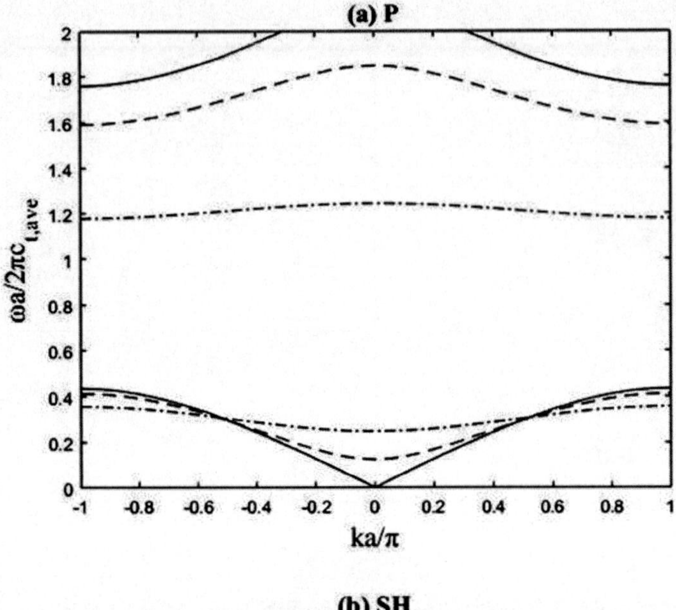

Fig. 4.17 Band gaps in the case of the spring–mass interface model. **a** *P* mode; **b** SH mode. —$F_s = 0, G_s = 0$, −−$F_s = 0.2F_0, G_s = 0.2G_0$, −.−$F_s = F_0, G_s = G_0$

Chapter 5
Surface Wave and Interface Wave

The reflected waves and transmitted waves formed on the interface will propagate into the interior of elastic medium with the increase of time, so these waves are called bulk waves or body waves. Compared with the bulk waves, there is another type of waves near the interface of elastic media. Its amplitude only has a certain value near the interface. As the distance from the interface or surface increases, the amplitude decreases rapidly. In terms of energy, their energy is concentrated near the surface or interface, so these waves are called surface waves and interface waves. The velocity of the surface/interface waves are smaller than that of the bulk waves.

In this chapter, we will mainly discuss the P-type surface waves, SV-type surface waves and resultant wave by their superposition, namely Rayleigh wave. We also discuss SH-type surface waves which existing inside of the cover layer on the half-space surface, namely Love waves. Moreover, the interface wave, which is similar to the Rayleigh wave but exists on both sides of interface with the energy concentrated near the interface of two semi-infinite homogeneous elastic media, namely Stanley wave is discussed. Finally, the rotational surface waves, which is the counterpart in the rotational coordinate system of surface waves discussed previously in the rectangular coordinate system, are discussed referenced to the rotational coordinate system.

5.1 P-type Surface Waves and SV-Type Surface Waves

In Chap. 3, we have discussed the SV wave incident on a free surface alone. When the incident angle of the SV wave is equal to the critical angle β_{cr}, the reflection angle of the reflected P wave reaches 90° and its propagation direction becomes along the x_1 direction, as shown in Fig. 5.1. When the incidence angle continues to increase, namely $\beta > \beta_{cr}$, the expression of the reflected P wave propagating along the x_1 direction can be written as

© Science Press 2022

P. Wei, *Theory of Elastic Waves*,
https://doi.org/10.1007/978-981-19-5662-1_5

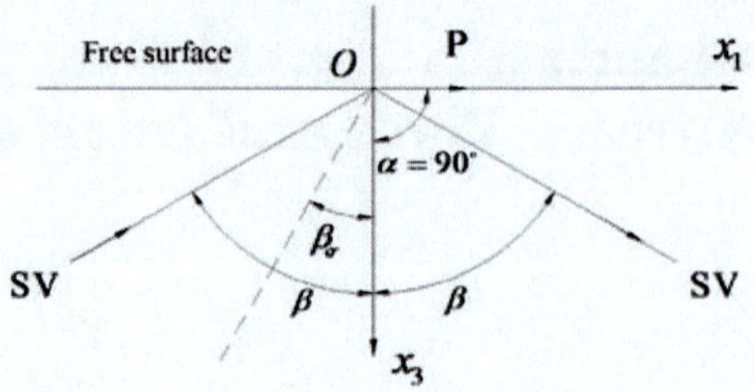

Fig. 5.1 P-type surface waves

$$\varphi = A\mathrm{e}^{-\xi x_3}\mathrm{e}^{i(k_{p1}x_1-\omega t)}. \tag{5.1.1}$$

The equiphase plane of the wave is perpendicular to the x_1 direction, namely

$$x_1 = \text{constant}. \tag{5.1.2}$$

The amplitudes on the wavefront or equiphase surface are no longer constant but decay exponentially with x_3. As shown in Fig. 5.2, the constant amplitude surface of this wave can be expressed as

$$x_3 = \text{constant}. \tag{5.1.3}$$

That is, the equi-amplitude surface is perpendicular to the x_3 axis. The plane wave in which the equal-phase plane and the equal-amplitude plane do not coincide is usually called **inhomogeneous wave** because the displacement distribution on the wavefront or equiphase surface is not uniform. The plane wave in which the equal-phase surface and the equal-amplitude surface overlap are called **homogeneous waves.** For the surface wave propagating along the x_1 direction, when the equal-amplitude surface (its normal direction is the attenuation direction of the wave) is perpendicular to the equal-phase surface (its normal direction is the propagation direction of the wave), the amplitude only exists in the thin layer near the free surface

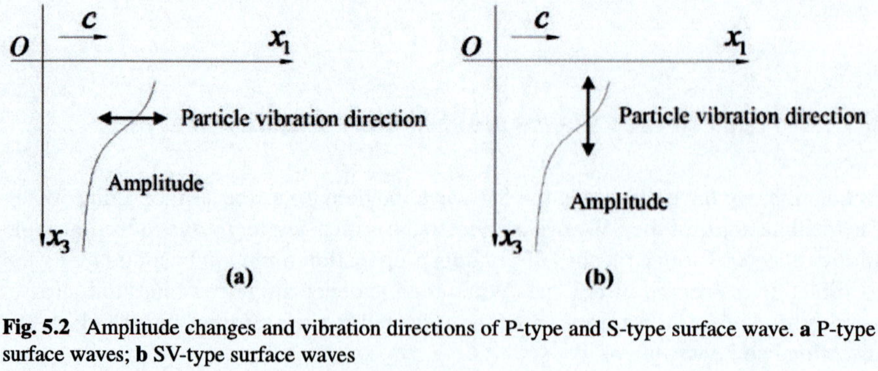

Fig. 5.2 Amplitude changes and vibration directions of P-type and S-type surface wave. **a** P-type surface waves; **b** SV-type surface waves

and will decays exponentially with the increasing distance away from free surface. Such inhomogeneous waves are usually called surface waves, and the surface waves represented by Eq. (5.1.1) are called P-type surface waves.

When the SV wave is incident from medium 1 to the interface between medium 1 and medium 2, as shown in Fig. 5.3, if

$$c_s^{(1)} < c_p^{(1)} < c_s^{(2)} < c_p^{(2)}, \tag{5.1.4}$$

is satisfied, there are 3 critical angles for the incident wave, namely when

$$\beta_1 = \beta_{cr_1}, \tag{5.1.5}$$

is satisfied, the transmitted P wave propagates along the x_1 direction; when

Fig. 5.3 Three critical angles for incident SV waves. **a** The first critical angle; **b** The second critical angle; **c** The third critical angle

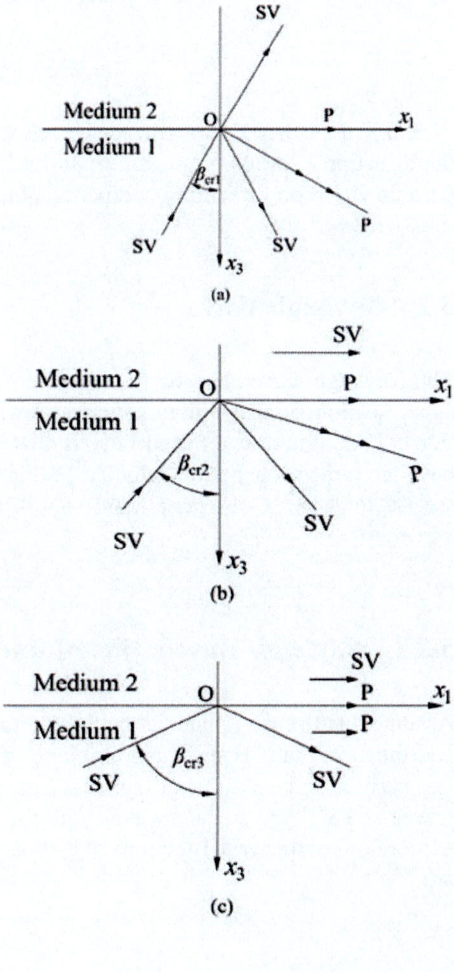

$$\beta_2 = \beta_{cr_2}, \tag{5.1.6}$$

is satisfied, the transmitted SV wave propagates along the x_1 direction; when

$$\beta_3 = \beta_{cr_3}, \tag{5.1.7}$$

is satisfied, the reflected P wave propagates along the x_1 direction.

When the incident angle β is greater than β_{cr3}, the reflected P wave, transmitted P wave and transmitted SV wave all become surface waves, and their expressions can be written as

$$\varphi^r = A^r \mathrm{e}^{-\xi_1 x_3} \mathrm{e}^{i\left(k_{p1}^{(1)} x_1 - \omega t\right)}, \tag{5.1.8}$$

$$\varphi^t = A^t \mathrm{e}^{\xi_2 x_3} \mathrm{e}^{i\left(k_{p1}^{(2)} x_1 - \omega t\right)}, \tag{5.1.9}$$

$$\psi^t = B^t \mathrm{e}^{\xi_3 x_3} \mathrm{e}^{i\left(k_{p2}^{(2)} x_1 - \omega t\right)}. \tag{5.1.10}$$

φ^r and φ^t are called P-type surface waves, whose particle vibration direction is along the direction of wave propagation; and ψ^t is called SV-type surface waves, whose particle vibration direction is perpendicular to the direction of wave propagation.

5.2 Rayleigh Wave

The Rayleigh wave near the free surface of the elastic half-space is formed by the superposition of the inhomogeneous P wave and SV wave under certain conditions. The condition referred here is that these two inhomogeneous plane waves have the same propagation velocity. In this section, we discuss the wave function of the Rayleigh wave, the generation conditions and some properties of the Rayleigh wave.

5.2.1 *Rayleigh Wave's Wave Function*

Assume that the x_1x_2 plane is the free surface of a homogeneous elastic half-space, and the x_1x_3 plane is the incident plane of the wave. The Rayleigh wave near the free surface of the elastic half-space is formed by the superposition of P-type surface waves and SV-type surface waves with the same phase velocity. Assuming that the expressions of the wave functions of P-type surface waves and SV-type surface waves are

$$\varphi(x_1, x_3, t) = \Phi(x_3)e^{ik(x_1-ct)}, \tag{5.2.1}$$

$$\psi(x_1, x_3, t) = \Psi(x_3)e^{ik(x_1-ct)}. \tag{5.2.2}$$

Substituting Eqs. (5.2.1) and (5.2.2) into the wave equation of the potential function, we have

$$\Phi''(x_3) + k^2\left(\frac{c^2}{c_P^2} - 1\right)\Phi = 0, \tag{5.2.3}$$

$$\Psi''(x_3) + k^2\left(\frac{c^2}{c_S^2} - 1\right)\Psi = 0. \tag{5.2.4}$$

The solutions of Eqs. (5.2.3) and (5.2.4) depend on the values of $p = \frac{c^2}{c_P^2} - 1$ and $q = \frac{c^2}{c_S^2} - 1$.

Let's analyze the following five situations:

(1) When $c > c_P > c_S$, we have $p > 0$ and $q > 0$, then the solutions of Eqs. (5.2.3) and (5.2.4) can be written as

$$\Phi(x_3) = A_1e^{-ik\sqrt{p}x_3} + A_2e^{ik\sqrt{p}x_3}, \tag{5.2.5}$$

$$\Psi(x_3) = B_1e^{-ik\sqrt{q}x_3} + B_2e^{ik\sqrt{q}x_3}. \tag{5.2.6}$$

Then, P wave and SV wave propagate into the interior of medium and thus both are body waves, which corresponds to the situation where the SV wave is incident on the free surface at an angle smaller than the critical angle.

(2) When $c = c_P > c_S$, we have $p = 0$ and $q > 0$, then the solutions of Eqs. (5.2.3) and (5.2.4) can be written as

$$\Phi(x_3) = A_1x_3 + A_2, \tag{5.2.7}$$

$$\Psi(x_3) = B_1e^{-ik\sqrt{q}x_3} + B_2e^{ik\sqrt{q}x_3}, \tag{5.2.8}$$

wherein, $A_1 = 0$ is required, which ensures that the amplitude is bounded at $x_3 \to \infty$. At this time, the P wave and SV wave are still homogeneous plane waves, but the P wave propagates in the x_1 direction and the SV wave propagates into the interior of medium, which corresponds to the situation where the SV wave is incident on the free surface at a critical angle.

(3) When $c_S < c < c_P$, we have $p < 0$ and $q > 0$, then the solutions of Eqs. (5.2.3) and (5.2.4) can be written as

$$\Phi(x_3) = A_1e^{-k\sqrt{-p}x_3} + A_2e^{k\sqrt{-p}x_3}, \tag{5.2.9}$$

$$\Psi(x_3) = B_1 \mathrm{e}^{-ik\sqrt{q}x_3} + B_2 \mathrm{e}^{ik\sqrt{q}x_3} \tag{5.2.10}$$

wherein, $A_2 = 0$ is required, which ensures that the amplitude is bounded at $x_3 \to \infty$. At this time, the SV wave is still a homogeneous wave and propagates into the interior of medium. But the P wave has become a P-type surface wave. This case corresponds to the situation where the SV wave is incident beyond the critical angle.

(4) When $c = c_S < c_P$, we have p<0 and $q = 0$, then the solutions of Eqs. (5.2.3) and (5.2.4) can be written as

$$\Phi(x_3) = A_1 \mathrm{e}^{-k\sqrt{-p}x_3} + A_2 \mathrm{e}^{k\sqrt{-p}x_3}, \tag{5.2.11}$$

$$\Psi(x_3) = B_1 x_3 + B_2 \tag{5.2.12}$$

wherein, $A_2 = 0$ and $B_1 = 0$ are required, which ensures that the amplitude is bounded at $x_3 \to \infty$. At this time, both the P wave and the SV wave propagating along the x_1 direction, which corresponds to the grazing incidence of the SV wave.

(5) When $c < c_S < c_P$, we have $p < 0$ and $q < 0$, then the solutions of Eqs. (5.2.3) and (5.2.4) can be written as

$$\Phi(x_3) = A_1 \mathrm{e}^{-k\sqrt{-p}x_3} + A_2 \mathrm{e}^{k\sqrt{-p}x_3}, \tag{5.2.13}$$

$$\Psi(x_3) = B_1 \mathrm{e}^{-k\sqrt{-q}x_3} + B_2 \mathrm{e}^{k\sqrt{-q}x_3}, \tag{5.2.14}$$

wherein, $A_2 = B_2 = 0$ is required, which ensures that the amplitudes of P wave and SV wave are both bounded at $x_3 \to \infty$. At this time, both P wave and the SV wave become inhomogeneous plane wave propagating along x_1 direction, which is just the situation we are concerned about.

Substituting Eqs. (5.2.13) and (5.2.14) into Eqs. (5.2.1) and (5.2.2), the wave functions of the Rayleigh wave are obtained as

$$\varphi(x_1, x_3, t) = A_1 \mathrm{e}^{-k\sqrt{-p}x_3} \mathrm{e}^{ik(x_1 - ct)}, \tag{5.2.15}$$

$$\psi(x_1, x_3, t) = B_1 \mathrm{e}^{-k\sqrt{-q}x_3} \mathrm{e}^{ik(x_1 - ct)}. \tag{5.2.16}$$

It is obtained under the condition $c < c_s < c_p$; that is, the wave speed of the Rayleigh wave is less than that of transverse wave.

5.2.2 Rayleigh Equation

Rayleigh waves propagate on the surface of half-space; therefore, the free-stress surface condition must be satisfied. Next, we will study the conditions for the generation of Rayleigh waves and the propagation speed of Rayleigh waves. Let

$$r = \sqrt{-p} = \left(1 - (c/c_P)^2\right)^{\frac{1}{2}}, \tag{5.2.17a}$$

$$s = \sqrt{-q} = \left(1 - (c/c_s)^2\right)^{\frac{1}{2}}. \tag{5.2.17b}$$

Equations (5.2.15) and (5.2.16) can be rewritten as

$$\varphi(x_1, x_3, t) = A_1 \mathrm{e}^{-krx_3} \mathrm{e}^{ik(x_1 - ct)}, \tag{5.2.18a}$$

$$\psi(x_1, x_3, t) = B_1 \mathrm{e}^{-ksx_3} \mathrm{e}^{ik(x_1 - ct)}. \tag{5.2.18b}$$

Substituting Eq. (5.2.18) into free surface conditions

$$\sigma_{33}\big|_{x_3=0} = 0, \tag{5.2.19}$$

$$\sigma_{31}\big|_{x_3=0} = 0. \tag{5.2.20}$$

We have

$$\left[c_P^2 \nabla^2 \varphi + 2c_S^2 \left(\frac{\partial^2 \psi}{\partial x_1 \partial x_3} - \frac{\partial^2 \varphi}{\partial x_1^2}\right)\right]\bigg|_{x_3=0} = 0, \tag{5.2.21a}$$

$$\left[2\frac{\partial^2 \varphi}{\partial x_1 \partial x_3} + \frac{\partial^2 \psi}{\partial x_1^2} - \frac{\partial^2 \psi}{\partial x_3^2}\right]\bigg|_{x_3=0} = 0. \tag{5.2.21b}$$

Substituting Eqs. (5.2.18) and (5.2.19) into Eq. (5.2.21), we obtain

$$\left(2c_S^2 - c^2\right)A_1 - 2isc_S^2 B_1 = 0, \tag{5.2.22a}$$

$$2irc_S^2 A_1 + \left(2c_S^2 - c^2\right)B_1 = 0. \tag{5.2.22b}$$

A nontrivial solution of Eq. (5.2.22) requires

$$\begin{vmatrix} 2c_S^2 - c^2 & -2isc_S^2 \\ 2irc_S^2 & 2c_S^2 - c^2 \end{vmatrix} = 0, \tag{5.2.23}$$

namely

$$\left(2c_S^2 - c^2\right)^2 - 4rsc_S^4 = 0. \tag{5.2.24}$$

Substituting Eq. (5.2.17) into Eq. (5.2.24), we have

$$\left(2 - \frac{c^2}{c_S^2}\right)^2 - 4\sqrt{1 - \frac{c^2}{c_P^2}} \cdot \sqrt{1 - \frac{c^2}{c_S^2}} = 0. \tag{5.2.25}$$

Equation (5.2.25) is so-called Rayleigh equation. When the material parameters of the medium are given, the longitudinal and transverse wave speeds in the medium are determined. The Rayleigh wave velocity c can be solved from the Rayleigh equation. However, the Rayleigh equation is a nonlinear equation about the wave velocity c, which generally needs to be solved by numerical methods (such as dichotomy).

Introducing

$$x = \left(\frac{c}{c_S}\right)^2, \quad m^2 = \left(\frac{c_S}{c_P}\right)^2,$$

then, the Rayleigh equation can be reduced to the dimensionless form, namely

$$x[x^3 - 8x^2 + 8(3 - 2m^2)x + 16(m^2 - 1)] = 0. \tag{5.2.26}$$

Obviously, $x = 0$ is a root of this equation, but it leads to $c = 0$, which is not a nonzero solution. So, the problem boils down to finding the roots of the following equation

$$x^3 - 8x^2 + 8(3 - 2m^2)x + 16(m^2 - 1) = 0. \tag{5.2.27}$$

It is easy to verify that the left side of Eq. (5.2.27) is a negative number when $x = 0$ while a positive number when $x = 1$. Therefore, in the interval (0, 1), there should be at least one root. Now, we have known

$$c < c_S. \tag{5.2.28}$$

Let the speed of Rayleigh wave be $c = c_R$, namely

$$c < c_S < c_P. \tag{5.2.29}$$

Equation (5.2.29) is consistent with the conclusion of case (5); that is, the wave speed of the Rayleigh wave is less than the wave speed of shear wave.

Considering

$$m^2 = \left(\frac{c_S}{c_P}\right)^2 = \frac{1 - 2\nu}{2(1 - \nu)}, \tag{5.2.30}$$

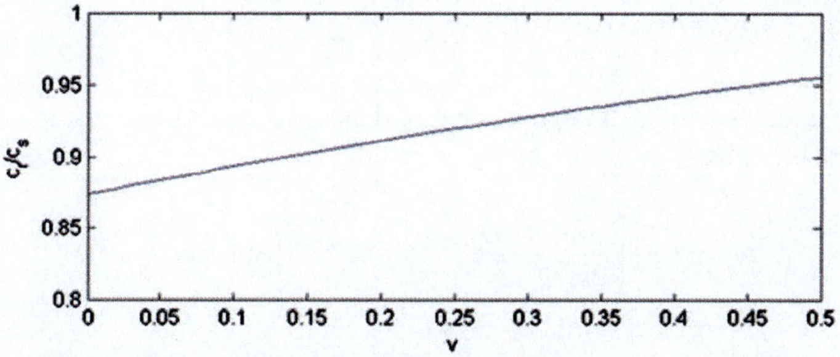

Fig. 5.4 Rayleigh wave speed c_R versus Poisson's ratio ν

we can see that the Rayleigh wave speed only depends on Poisson's ratio of the material, and Poisson's ratio of the real material satisfies

$$0 \leq \nu \leq \frac{1}{2}. \tag{5.2.31}$$

Therefore, the real root of Eq. (5.2.27) in the interval (0,1) is unique. The dependence of the Ryleigh wave velocity on Poisson's ratio is shown $\nu = \frac{1}{4}$ in Fig. 5.4. For Poisson solids, we have $\lambda = \mu$ or, namely $m^3 = \frac{1}{3}$. Further, we get the Rayleigh wave velocity

$$c_R = 0.9194 c_S. \tag{5.2.32}$$

Namely, the Rayleigh wave is slightly slower than the shear wave. It should be pointed out that the propagation speed of Rayleigh waves does not depend on frequency, So Rayleigh waves are not dispersive in their propagation process, which is an important feature of Rayleigh waves.

5.2.3 *The Displacement Field of the Ryleigh Wave*

Let's discuss the motion trajectory of the particle of the Rayleigh wave. From the wave function expression, i.e. Eq. (5.2.18), we have

$$\begin{aligned} u_1 &= \frac{\partial \varphi}{\partial x_1} - \frac{\partial \psi}{\partial x_3} \\ &= ikA_1 \left[\mathrm{e}^{-krx_3} - \frac{2rsc_S^2}{2c_S^2 - c_R^2} \mathrm{e}^{-ksx_3} \right] \mathrm{e}^{i(kx_1 - \omega t)}, \end{aligned} \tag{5.2.33}$$

$$u_3 = \frac{\partial \varphi}{\partial x_3} + \frac{\partial \psi}{\partial x_1}$$

$$= krA_1\left[-\mathrm{e}^{-krx_3} + \frac{2c_S^2}{2c_S^2 - c_R^2}\mathrm{e}^{-ksx_3}\right]\mathrm{e}^{i(kx_1-\omega t)}. \tag{5.2.34}$$

The real displacement can take the real part of the above complex number expression, namely

$$\begin{aligned} u_1 &= kA_1\left[\frac{2rsc_S^2}{2c_S^2 - c_R^2}\mathrm{e}^{-ksx_3} - \mathrm{e}^{-krx_3}\right]\sin(kx_1 - \omega t) \\ &= a(x_3)\sin(kx_1 - \omega t), \end{aligned} \tag{5.2.35}$$

$$\begin{aligned} u_3 &= krA_1\left[\frac{2c_S^2}{2c_S^2 - c_R^2}\mathrm{e}^{-ksx_3} - \mathrm{e}^{-krx_3}\right]\cos(kx_1 - \omega t) \\ &= b(x_3)\cos(kx_1 - \omega t). \end{aligned} \tag{5.2.36}$$

Then, we have

$$\frac{u_1^2(x_1, x_3, t)}{a^2(x_3)} + \frac{u_3^2(x_1, x_3, t)}{b^2(x_3)} = 1. \tag{5.2.37}$$

Equation (5.2.37) indicates that the particle motion trajectory of the Rayleigh wave is an ellipse, where $a(x_3)$ and $b(x_3)$ are the long and short semi-axes of the ellipse. The long semi-axes $a(x_3)$ is perpendicular to the free surface, and the short semi-axes $b(x_3)$ is parallel to the free surface, as shown in Fig. 5.5. Note that the long and short semi-axes $a(x_3)$ and $b(x_3)$ are exponential functions of the coordinate x_3, and the exponentially decreases with the increase of x_3. Therefore, the amplitude of the particle polarization gradually decreases as the depth increases, namely the vibration energy or vibration intensity gradually decreases, as shown in Fig. 5.6. This is the main reason why Rayleigh waves are called surface waves. For a Poisson solid, at the free surface ($x_3 = 0$), the displacement can be written as

$$u_1 = 0.426kA_1\sin(kx_1 - \omega t),$$

$$u_3 = 0.622kA_1\cos(kx_1 - \omega t).$$

The ratio of the long axis to the short axis is about 3:2.

According to the discussion in the previous three subsections, we can summarize the main properties of Rayleigh surface waves as follows:

(1) The Rayleigh wave is formed by the superposition of an inhomogeneous P wave and an inhomogeneous SV wave. It propagates along the surface and attenuates in the vertical direction. The higher the frequency, the faster the attenuation.
(2) During the propagation of the Rayleigh wave, the trajectory of the particle motion is an inverted ellipse. The long axis is perpendicular to the free surface,

Fig. 5.5 Particle vibration mode of Rayleigh wave

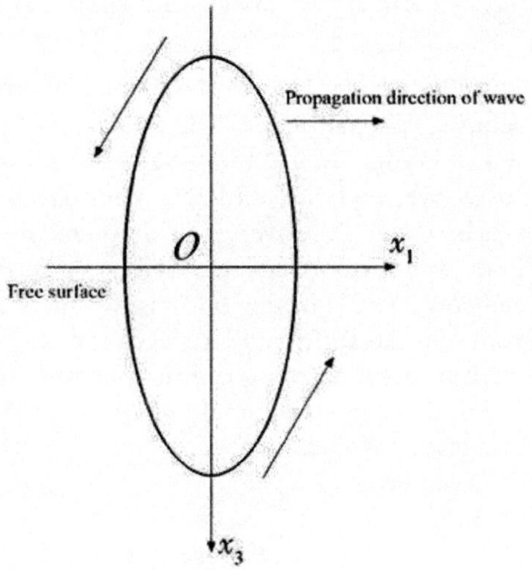

Fig. 5.6 Motion trajectory of the particle at different positions along x_3 direction

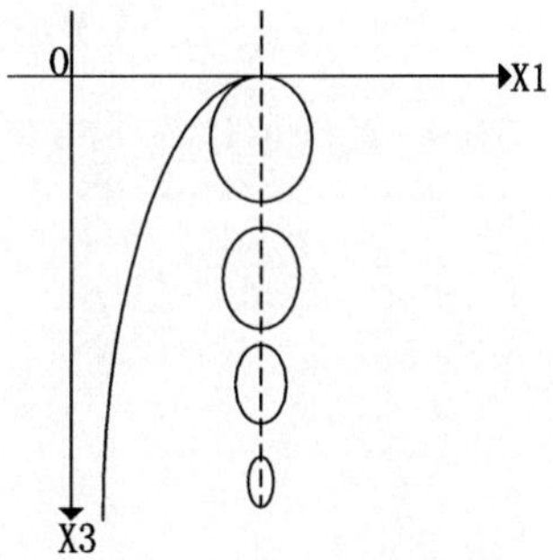

and the short axis is parallel to the free surface; that is, the vertical displacement component is greater than the horizontal displacement component.

(3) The Rayleigh wave speed is slightly smaller than the shear wave speed, and independent of frequency, that is, non-dispersive.

(4) In theory, the propagation of Rayleigh waves along a free surface is non-attenuated. But, in fact, the material is always viscoelastic, and there is inevitably viscous dissipation. Therefore, Rayleigh waves will also attenuate during propagation. Nevertheless, the Rayleigh wave propagates in the thin surface layer, the attenuation of its energy along the horizontal direction is much smaller than the body wave; thus, its propagation distance is longer. The farther away from the wave source, the more obvious the influence of surface waves which makes Rayleigh wave plays a greater role than body waves in the destruction of ground buildings caused by earthquakes.

5.2.4 *Rayleigh Wave in Elastic Half-Space with Cover Layer*

Consider a cover layer with completely different elastic properties over a homogeneous elastic half-space. The thickness of cover layer is h, and the configuration is shown in Fig. 5.7. Let the rectangular coordinate plane ox_1x_2 be the free surface of cover layer, and the coordinate plane ox_1x_3 be the plane where Rayleigh wave propagates. P and SV waves exist simultaneously in the cover layer and the half-space. Body waves are reflected from the two surfaces of the cover layer and form upward and downward traveling body waves in the cover layer, which further interfere with each other to form guided waves propagating along the cover layer. Only surface wave propagating along x_1 direction exists in the lower half-space for Rayleigh wave. The displacement potential functions in the cover layer and the lower half-space are expressed as follows.

Cover layer

$$\text{P Wave: } \varphi'(x_1, x_3, t) = \Phi'(x_3)e^{ik(x_1-ct)}, \tag{5.2.38a}$$

$$\text{SV Wave: } \psi'(x_1, x_3, t) = \Psi'(x_3)e^{ik(x_1-ct)}. \tag{5.2.38b}$$

Inserting Eq. (5.2.38) into the governing equation of motion leads to

$$\Phi'_{,x_3x_3} + k^2p'\Phi' = 0, \tag{5.2.39a}$$

$$\Psi'_{,x_3x_3} + k^2q'\Psi' = 0, \tag{5.2.39a}$$

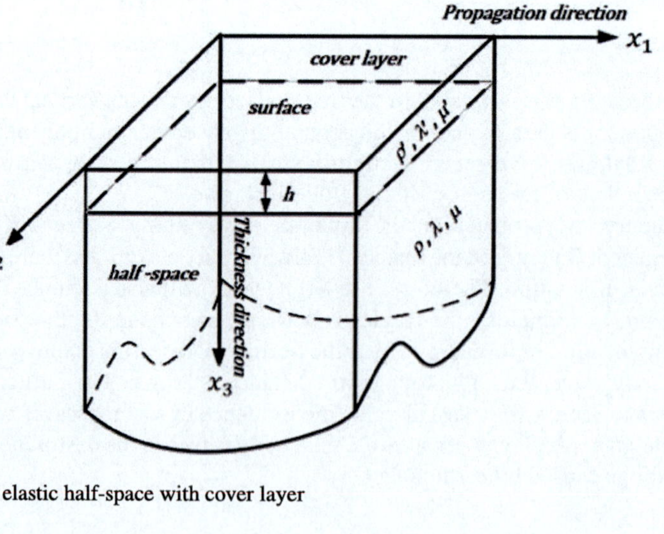

Fig. 5.7 The elastic half-space with cover layer

where

$$p' = \frac{c^2}{c_p'^2} - 1, \quad q' = \frac{c^2}{c_s'^2} - 1. \tag{5.2.40}$$

The solutions of Eq. (5.2.39) are

$$\Phi' = A'e^{ik\sqrt{p'}x_3} + B'e^{-ik\sqrt{p'}x_3}, \tag{5.2.41a}$$

$$\Psi' = C'e^{ik\sqrt{q'}x_3} + D'e^{-ik\sqrt{q'}x_3}. \tag{5.2.41b}$$

Therefore, the potential functions can be expressed as

$$\begin{aligned} \varphi'(x_1, x_3, t) &= \Phi'(x_3)e^{ik(x_1-ct)} \\ &= \left[A'e^{ik\sqrt{p'}x_3} + B'e^{-ik\sqrt{p'}x_3}\right]e^{ik(x_1-ct)}, \end{aligned} \tag{5.2.42a}$$

$$\begin{aligned} \psi'(x_1, x_3, t) &= \Psi'(x_3)e^{ik(x_1-ct)} \\ &= \left[C'e^{ik\sqrt{q'}x_3} + D'e^{-ik\sqrt{q'}x_3}\right]e^{ik(x_1-ct)}. \end{aligned} \tag{5.2.42b}$$

Based on Eq. (5.2.42), the displacements and the stresses can be obtained as

$$\begin{aligned} u_1' &= \frac{\partial \varphi'}{\partial x_1} - \frac{\partial \psi'}{\partial x_3} = \left[ik\Phi'(x_3) - \Psi'_{,x_3}(x_3)\right]e^{ik(x_1-ct)} \\ &= \left[\left(A'e^{ik\sqrt{p'}x_3} + B'e^{-ik\sqrt{p'}x_3}\right) - \sqrt{q'}\left(C'e^{ik\sqrt{q'}x_3} - D'e^{-ik\sqrt{q'}x_3}\right)\right]ike^{ik(x_1-ct)}, \end{aligned} \tag{5.2.43a}$$

$$\begin{aligned} u_3' &= \frac{\partial \varphi'}{\partial x_3} + \frac{\partial \psi'}{\partial x_1} = \left[\Phi'_{,x_3}(x_3) + ik\Psi'(x_3)\right]e^{ik(x_1-ct)} \\ &= \left[\sqrt{p'}\left(A'e^{ik\sqrt{p'}x_3} - B'e^{-ik\sqrt{p'}x_3}\right) + \left(C'e^{ik\sqrt{q'}x_3} + D'e^{-ik\sqrt{q'}x_3}\right)\right]ike^{ik(x_1-ct)}, \end{aligned} \tag{5.2.43b}$$

$$\begin{aligned} \sigma_{33}' &= \lambda'\left(\frac{\partial u_1'}{\partial x_1} + \frac{\partial u_3'}{\partial x_3}\right) + 2\mu'\left(\frac{\partial u_3'}{\partial x_3}\right) \\ &= \lambda'\left[-k^2\Phi'(x_3) - ik\Psi'_{,x_3}(x_3) + \Phi'_{,x_3x_3}(x_3) + ik\Psi'_{,x_3}(x_3)\right]e^{ik(x_1-ct)} \\ &\quad + 2\mu'\left[\Phi'_{,x_3x_3}(x_3) + ik\Psi'_{,x_3}(x_3)\right]e^{ik(x_1-ct)} \\ &= \left\{\lambda'\left[-k^2\Phi'(x_3) - k^2p'\Phi'(x_3)\right] + 2\mu'\left[-k^2p'\Phi'(x_3) + ik\Psi'_{,x_3}(x_3)\right]\right\}e^{ik(x_1-ct)} \\ &= \left\{\left[\lambda'(-1-p') - 2\mu'p'\right]\left(A'e^{ik\sqrt{p'}x_3} + B'e^{-ik\sqrt{p'}x_3}\right)\right. \\ &\quad \left. + 2\mu'\sqrt{q'}\left(-C'e^{ik\sqrt{q'}x_3} + D'e^{-ik\sqrt{q'}x_3}\right)\right\}k^2e^{ik(x_1-ct)}, \end{aligned} \tag{5.2.44a}$$

$$\begin{aligned}\sigma'_{31} &= \mu'\left(\frac{\partial u'_1}{\partial x_3} + \frac{\partial u'_3}{\partial x_1}\right)\\ &= \mu'\left[ik\Phi'_{,x_3}(x_3) - \Psi'_{,x_3x_3}(x_3)\right]e^{ik(x_1-ct)} + \mu'\left[ik\Phi'_{,x_3}(x_3) - k^2\Psi'(x_3)\right]e^{ik(x_1-ct)}\\ &= \mu'\Big[2\sqrt{p'}\left(-A'e^{ik\sqrt{p'}x_3} + B'e^{-ik\sqrt{p'}x_3}\right)\\ &\quad+(q'-1)\left(C'e^{ik\sqrt{q'}x_3} + D'e^{-ik\sqrt{q'}x_3}\right)\Big]k^2e^{ik(x_1-ct)}.\end{aligned} \tag{5.2.44b}$$

Inserting Eq. (5.2.44) into the boundary conditions

$$\sigma'_{33}|_{x_3=0} = \sigma'_{31}|_{x_3=0} = 0, \tag{5.2.45}$$

we get

$$\left[\lambda'(-1-p') - 2\mu' p'\right](A' + B') + 2\mu'\sqrt{q'}(-C' + D') = 0, \tag{5.2.46a}$$

$$2\sqrt{p'}(-A' + B') + (q' - 1)(C' + D') = 0. \tag{5.2.46b}$$

On the other hand, the potential functions in the lower half-space can also be expressed as

$$\text{P wave: } \varphi(x_1, x_3, t) = \Phi(x_3)e^{ik(x_1-ct)}, \tag{5.2.47a}$$

$$\text{SV wave: } \psi(x_1, x_3, t) = \Psi(x_3)e^{ik(x_1-ct)}. \tag{5.2.47b}$$

After instituting Eq. (5.2.47) into the governing equations of motion, we get

$$\Phi_{,x_3x_3} + k^2 p\Phi = 0, \tag{5.2.48a}$$

$$\Psi_{,x_3x_3} + k^2 q\Psi = 0, \tag{5.2.48b}$$

where

$$p = \frac{c^2}{c_p^2} - 1, \quad q = \frac{c^2}{c_s^2} - 1. \tag{5.2.49}$$

The solution of Eq. (5.2.48) can be one of the following five cases dependent upon the value of p and q:

(1) When $p > 0$ and $q > 0, c > c_p > c_s$. The propagation directions of P wave and SV wave both points into the interior of half-space, which means that P wave and SV wave are both bulk waves.

(2) When $p = 0$ and $q > 0, c = c_p > c_s$. P wave becomes the wave propagating along x_1 direction while SV wave still propagates toward the interior of half-space.
(3) When $p < 0$ and $q > 0, c_p > c > c_s$. P wave has become into the P-type surface wave while SV wave remains bulk wave and propagates in the interior of half-space.
(4) When $p < 0$ and $q = 0, c_p > c = c_s$. P wave and SV wave both propagate along the surface of half-space.
(5) When $p < 0$ and $q < 0, c < c_S < c_P$. Both of P wave and SV wave have become into the inhomgeneous plane waves which propagate along the surface of half-space, i.e. x_1 direction. The interference of them results in the Rayleigh wave propagating along x_1 direction but decays along the depth.

For the Rayleigh wave, the potential function in the lower half-space is

$$\text{P wave: } \varphi(x_1, x_3, t) = \Phi(x_3)\mathrm{e}^{ik(x_1-ct)} = A\mathrm{e}^{-k\sqrt{-p}x_3}\mathrm{e}^{ik(x_1-ct)}, \tag{5.2.50a}$$

$$\text{SV wave: } \psi(x_1, x_3, t) = \Psi(x_3)\mathrm{e}^{ik(x_1-ct)} = C\mathrm{e}^{-k\sqrt{-q}x_3}\mathrm{e}^{ik(x_1-ct)}. \tag{5.2.50b}$$

The displacement and stress components are

$$u_1 = \frac{\partial\varphi}{\partial x_1} - \frac{\partial\psi}{\partial x_3} = \left(iA\mathrm{e}^{-k\sqrt{-p}x_3} + \sqrt{-q}C\mathrm{e}^{-k\sqrt{-q}x_3}\right)k\mathrm{e}^{ik(x_1-ct)}, \tag{5.2.51a}$$

$$u_3 = \frac{\partial\varphi}{\partial x_3} + \frac{\partial\psi}{\partial x_1} = \left(-\sqrt{-p}A\mathrm{e}^{-k\sqrt{-p}x_3} + iC\mathrm{e}^{-k\sqrt{-q}x_3}\right)k\mathrm{e}^{ik(x_1-ct)}, \tag{5.2.51b}$$

$$\begin{aligned}\sigma_{33} &= \lambda\left(\frac{\partial u_1}{\partial x_1} + \frac{\partial u_3}{\partial x_3}\right) + 2\mu\left(\frac{\partial u_3}{\partial x_3}\right) \\ &= \left[-\lambda(1+p)A\mathrm{e}^{-k\sqrt{-p}x_3} + 2\mu\left(-pA\mathrm{e}^{-k\sqrt{-p}x_3} - i\sqrt{-q}C\mathrm{e}^{-k\sqrt{-q}x_3}\right)\right]k^2\mathrm{e}^{ik(x_1-ct)},\end{aligned} \tag{5.2.52a}$$

$$\sigma_{31} = \mu\left(\frac{\partial u_1}{\partial x_3} + \frac{\partial u_3}{\partial x_1}\right) = \mu\left[-i2\sqrt{-p}A\mathrm{e}^{-k\sqrt{-p}x_3} + (q-1)C\mathrm{e}^{-k\sqrt{-q}x_3}\right]k^2\mathrm{e}^{ik(x_1-ct)}. \tag{5.2.52b}$$

The interface continuity conditions between the cover layer and the lower half-space are

$$\sigma'_{33}|_{x_3=h} = \sigma_{33}|_{x_3=h}, \tag{5.2.53a}$$

$$\sigma'_{31}|_{x_3=h} = \sigma_{31}|_{x_3=h}, \tag{5.2.53b}$$

$$u'_3|_{x_3=h} = u_3|_{x_3=h}, \tag{5.2.53c}$$

$$u_1'|_{x_3=h} = u_1|_{x_3=h}. \tag{5.2.53d}$$

Substituting the displacement and stress expressions of the cover layer and the lower half-space into the interface continuity conditions leads to

$$\begin{aligned} &-[\lambda'(1+p')+2\mu'p']\mathrm{e}^{ik\sqrt{p'}h}A' - [\lambda'(1+p')+2\mu'p']\mathrm{e}^{-ik\sqrt{p'}h}B' \\ &-2\mu'\sqrt{q'}\mathrm{e}^{ik\sqrt{q'}h}C' + 2\mu'\sqrt{q'}\mathrm{e}^{-ik\sqrt{q'}h}D' \\ &+[\lambda(1+p)+2\mu p]\mathrm{e}^{-k\sqrt{-p}h}A + 2i\mu\sqrt{-q}\mathrm{e}^{-k\sqrt{-q}h}C = 0, \end{aligned} \tag{5.2.54a}$$

$$\begin{aligned} &-2\mu'\sqrt{p'}\mathrm{e}^{ik\sqrt{p'}h}A' + 2\mu'\sqrt{p'}\mathrm{e}^{-ik\sqrt{p'}h}B' \\ &+\mu'(q'-1)\mathrm{e}^{ik\sqrt{q'}h}C' + \mu'(q'-1)\mathrm{e}^{-ik\sqrt{q'}h}D' \\ &+i2\mu\sqrt{-p}\mathrm{e}^{-k\sqrt{-p}h}A + (1-q)\mu\mathrm{e}^{-k\sqrt{-q}h}C = 0, \end{aligned} \tag{5.2.54b}$$

$$\begin{aligned} &i\mathrm{e}^{ik\sqrt{p'}h}A' + i\mathrm{e}^{-ik\sqrt{p'}h}B' - i\sqrt{q'}\mathrm{e}^{ik\sqrt{q'}h}C' + i\sqrt{q'}\mathrm{e}^{-ik\sqrt{q'}h}D' \\ &-i\mathrm{e}^{-k\sqrt{-p}h}A - \sqrt{-q}\mathrm{e}^{-k\sqrt{-q}h}C = 0, \end{aligned} \tag{5.2.54c}$$

$$\begin{aligned} &i\sqrt{p'}\mathrm{e}^{ik\sqrt{p'}h}A' - i\sqrt{p'}\mathrm{e}^{-ik\sqrt{p'}h}B' + i\mathrm{e}^{ik\sqrt{q'}h}C' + i\mathrm{e}^{-ik\sqrt{q'}h}D' \\ &+\sqrt{-p}\mathrm{e}^{-k\sqrt{-p}h}A - i\mathrm{e}^{-k\sqrt{-q}h}C = 0. \end{aligned} \tag{5.2.54d}$$

Combining the traction-free surface condition on the upper surface of the cover layer, i.e. Eq. (5.2.46), with the interface continuity conditions on the lower surface of the cover layer, i.e. Eq. (5.2.54), and expressing them in the form of matrix, lead to

$$MX = 0, \tag{5.2.54d}$$

where $X = (A', B', C', D', A, C)^{\mathrm{T}}$. The transpose of coefficient matrix, i.e. M^{T}, is

$$\begin{vmatrix} \overline{p'}\mathrm{e}^{ik\sqrt{p'}h} & -2\mu'\sqrt{p'}\mathrm{e}^{ik\sqrt{p'}h} & i\mathrm{e}^{ik\sqrt{p'}h} & i\sqrt{p'}\mathrm{e}^{ik\sqrt{p'}h} & \overline{p'} & -2\sqrt{p'} \\ \overline{p'}\mathrm{e}^{-ik\sqrt{p'}h} & 2\mu'\sqrt{p'}\mathrm{e}^{-ik\sqrt{p'}h} & i\mathrm{e}^{-ik\sqrt{p'}h} & -i\sqrt{p'}\mathrm{e}^{-ik\sqrt{p'}h} & \overline{p'} & 2\sqrt{p'} \\ -\overline{q'}\mathrm{e}^{ik\sqrt{q'}h} & \mu'(q'-1)\mathrm{e}^{ik\sqrt{q'}h} & -i\sqrt{q'}\mathrm{e}^{ik\sqrt{q'}h} & i\mathrm{e}^{ik\sqrt{q'}h} & -\overline{q'} & q'-1 \\ 2\overline{q'}\mathrm{e}^{-ik\sqrt{q'}h} & \mu'(q'-1)\mathrm{e}^{-ik\sqrt{q'}h} & i\sqrt{q'}\mathrm{e}^{-ik\sqrt{q'}h} & i\mathrm{e}^{-ik\sqrt{q'}h} & 2\overline{q'} & q'-1 \\ -\overline{p}\mathrm{e}^{-k\sqrt{-p}h} & i2\mu\sqrt{-p}\mathrm{e}^{-k\sqrt{-p}h} & -i\mathrm{e}^{-k\sqrt{-p}h} & \sqrt{-p}\mathrm{e}^{-k\sqrt{-p}h} & 0 & 0 \\ 2i\mu\sqrt{-q}\mathrm{e}^{-k\sqrt{-q}h} & (1-q)\mu\mathrm{e}^{-k\sqrt{-q}h} & -\sqrt{-q}\mathrm{e}^{-k\sqrt{-q}h} & -i\mathrm{e}^{-k\sqrt{-q}h} & 0 & 0 \end{vmatrix}, \tag{5.2.55}$$

where

$$c_p^2 = \frac{\lambda + 2\mu}{\rho}, c_s^2 = \frac{\mu}{\rho}, \overline{p\prime} = -[\lambda\prime(1 + p') + 2\mu\prime p\prime],$$
$$\bar{p} = -[\lambda(1 + p) + 2\mu p], 2\mu\prime\sqrt{q\prime} = \overline{q\prime}.$$

The nonzero solution of Eq. (5.2.55) exists requires that the determinant of the coefficients matrix must be zero, i.e.

$$det(M) = det(M^{\mathrm{T}}) = f(c, kh) = 0. \tag{5.2.56}$$

Different from Rayleigh waves in elastic half-space, the existence of cover layer makes Rayleigh wave in the configuration of elastic half-space with cover layer showing dispersive feature. The body waves in cover layer are reflected by multiple times on the upper and lower surfaces of cover layer and generate upward and downward bulk waves. These upward and downward bulk waves interfere with each other to resultantly form guided waves propagating along the cover layer. In the lower half-space, there are only inhomogeneous waves propagating along the interface, namely Rayleigh waves. Guided waves in the cover layer and inhomogeneous waves in the lower half-space are coupled together by the interface continuous conditions to form a unified propagation velocity, namely Rayleigh wave velocity. Therefore, Rayleigh waves in elastic half-space with cover layer are strongly affected by cover layer and behave as dispersive surface waves. However, this dispersion characteristic belongs to geometrical dispersion, which is essentially different from the dispersion characteristic of wave propagating in viscoelastic materials where the dispersion is caused by viscoelasticity and thus belongs to physical dispersion. In order to show the dispersion feature of Rayleigh wave, the dispersion curves of Rayleigh wave under different material combinations are given below.

(1) Combination case 1 (Figs. 5.8 and 5.9; Table 5.1)

$$c_{p(Steel)} = 5911.89\,\text{m/s}, c_{s(Steel)} = 3229.04\,\text{m/s}, c_{p(Al)} = 6353.77\,\text{m/s},$$
$$c_{s(Al)} = 3132.86\,\text{m/s}, c_{R(Al)} = 2924.324\,\text{m/s}, c_{R(Steel)} = 2988.28/\text{s}.$$
$$c_{p(Steel)} < c_{p(Al)}, c_{s(Al)} < c_{s(Steel)}.$$

(2) Combination case 2 (Figs. 5.10, 5.11 and 5.12; Table 5.2)

$$c_{p1} = 4698.96\,\text{m/s}, c_{s1} = 3000.02\,\text{m/s}, c_{R1} = 2709.87\,\text{m/s},$$
$$c_{p2} = 8000\,\text{m/s}, c_{s2} = 5000\,\text{m/s}, c_{R2} = 4537.2\,\text{m/s}.$$

(3) Combination case 3 (Figs. 5.13, 5.14 and 5.15; Table 5.3)

$$c_{p1} = 6300\,\text{m/s}, c_{s1} = 3099\,\text{m/s}, c_{R1} = 2894\,\text{m/s},$$
$$c_{p2} = 4000\,\text{m/s}, c_{s2} = 2500\,\text{m/s}, c_{R2} = 2268\,\text{m/s}.$$

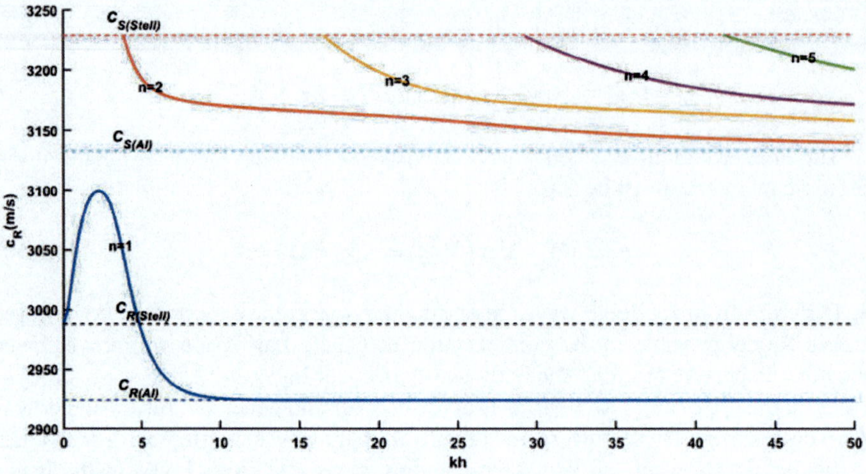

Fig. 5.8 Rayleigh wave speed versus dimensionless wavenumber kh under combination of Al as covered layer and steel as substrate

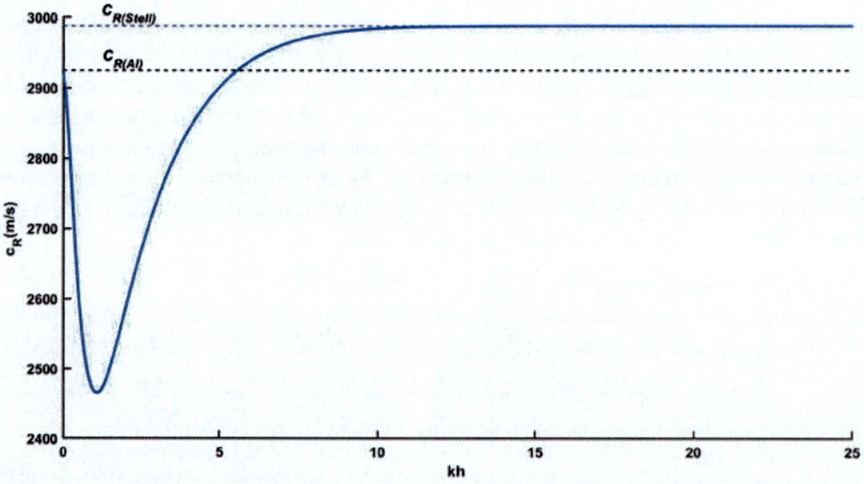

Fig. 5.9 Rayleigh wave speed versus dimensionless wavenumber kh under combination of steel as covered layer and Al as substrate

Table 5.1 The material parameters of cover layer and substrate

Material	$\rho(\text{kg/m}^3)$	λ(Gpa)	μ(Gpa)
Steel	7874	111	82.1
Al	2700	56	26.5

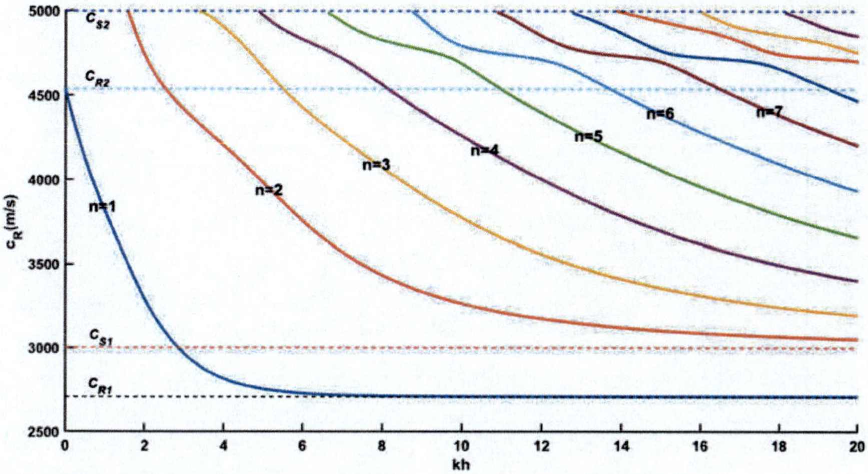

Fig. 5.10 Rayleigh wave speed versus dimensionless wavenber kh under combination of material 1 as cover layer and material 2 as substrate

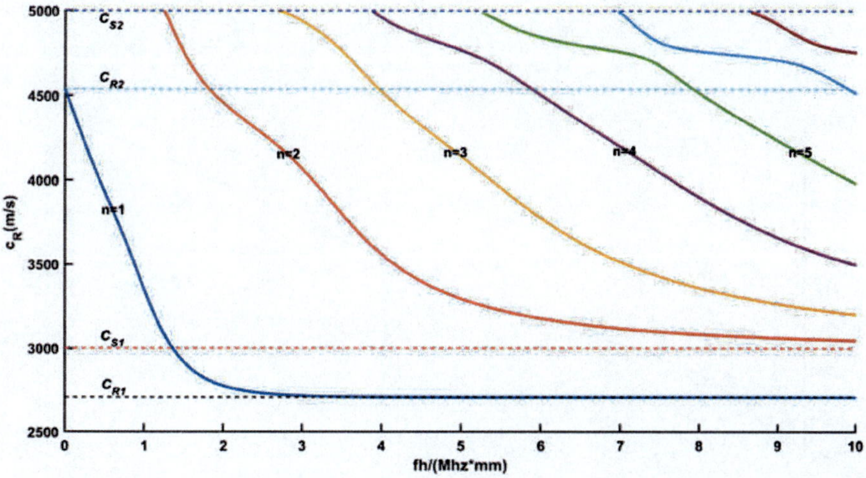

Fig. 5.11 Rayleigh wave speed versus the frequency-thickness production fh under combination of material 1 as cover layer and material 2 as substrate

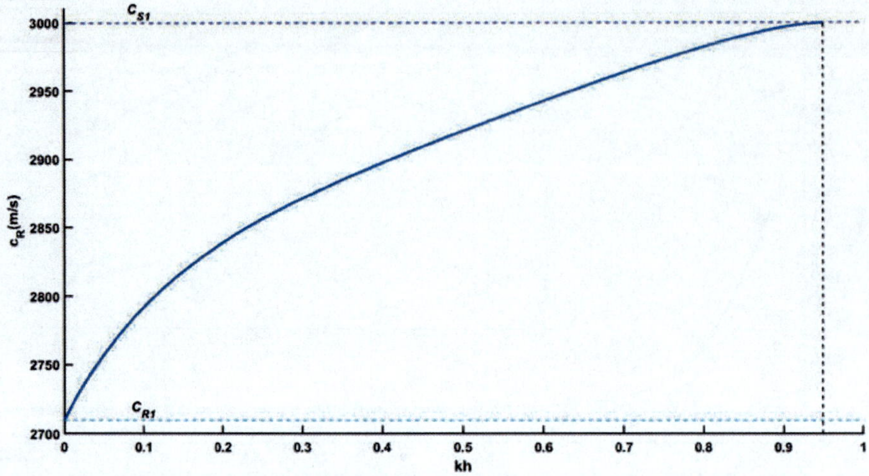

Fig. 5.12 Rayleigh wave speed versus dimensionless wavenumber kh under combination of material 2 as cover layer and material 1 as substrate

Table 5.2 Material parameters of cover layer and substrate

Material	$\rho\left(\text{kg/m}^3\right)$	λ(Gpa)	μ(Gpa)
Material 1	333.33	1.36	3
Material 2	280	3.92	7

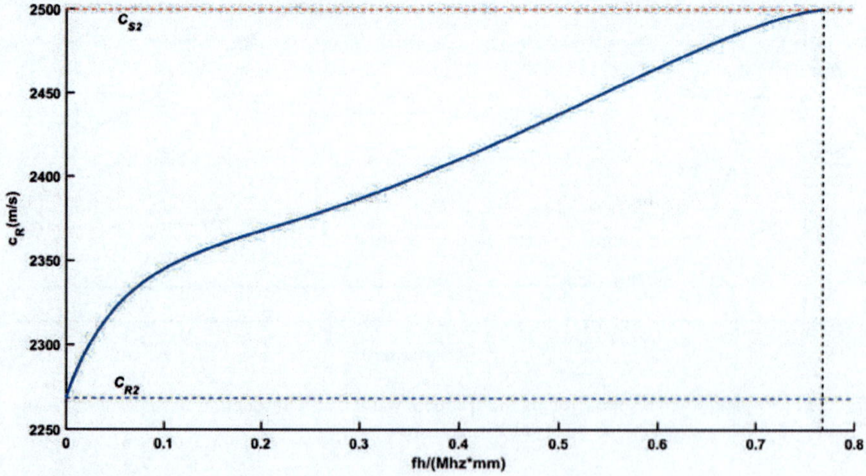

Fig. 5.13 Rayleigh wave speed versus the frequency-thickness production fh under combination of material 1 as cover layer and material 2 as substrate

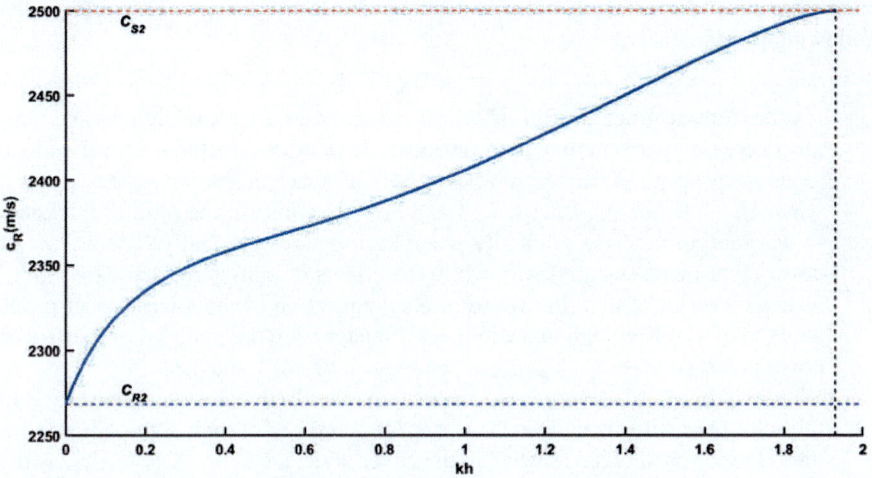

Fig. 5.14 Rayleigh wave speed versus the dimensionless wavenumber kh under combination of material 1 as cover layer and material 2 as substrate

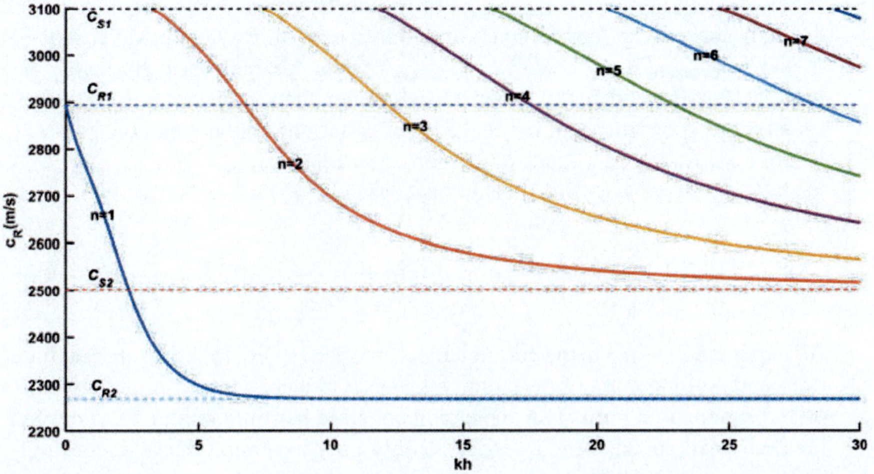

Fig. 5.15 Rayleigh wave speed versus the dimensionless wavenumber kh under combination of material 2 as cover layer and material 1 as substrate

Table 5.3 Material parameters of cover layer and substrate

Material	$\rho(\mathrm{kg/m^3})$	λ(Gpa)	μ(Gpa)
Material 1	271.6	5.56	2.61
Material 2	240	0.84	1.5

Based on the dispersion curves for three combination cases, we can observe the following phenomenon:

(1) When kh gradually decreases, which means that the ratio of thickness to wavelength decrease gradually, the effects of cover layer on the Rayleigh wave gradually decrease. As result of that, the speed of Rayleigh wave of lowest-order mode tends to the speed of Rayleigh wave in the half-space, namely $c \to c_{R(halfspace)}$ when $kh \to 0$; see Figs. 5.10, 5.11 and 5.15. In contrast, the ratio of thickness to wavelength increase gradually when kh increases gradually. The effects of cover layer increases gradually while the effects of half-space decrease gradually. As a result of that, the speed of Rayleigh wave of the lowest-order mode tends to that of Rayleigh wave in the half-space with the material properties of cover layer, i.e. $c \to c_{R(coverlayer)}$; see Figs. 5.10, 5.11 and 5.15.

(2) When $c_{s(coverlayer)} < c_{s(halfspace)}$, more than one Rayleigh wave modes exist in the half-space with cover layer. The propagation speed of each high-order modes tends to the speed of shear wave in the cover layer, i.e. $c \to c_{S(coverlayer)}$, as the dimensionless wavenumber kh increases. Besides, there are cut-off wavenumber (or cut-off frequency) for all higher-order modes when kh decreases gradually; see Figs. 5.10, 5.11 and 5.15.

(3) When $c_{s(coverlayer)} > c_{s(halfspace)}$, only one Rayleigh wave mode exists in the half-space with cover layer and the speed of Rayleigh wave satisfies $c \to c_{R(coverlayer)}$ or $c \to c_{s(halfspace)}$ when kh increases gradually; see Figs. 5.12, 5.13 and 5.14. When kh is greater than a special value, the real speed (corresponding with the real root of equation) may not exists.

(4) When $c_{s(halfspace)} < c_{s(coverlayer)}$, then, $c < c_{s(halfspace)} < c_{s(coverlayer)}$, $c < c_{P(halfspace)}$,$c < c_{P(coverlayer)}$. This further leads to

$$p' = \frac{c^2}{c_P'^2} - 1 < 0, q' = \frac{c^2}{c_s'^2} - 1 < 0, p = \frac{c^2}{c_P^2} - 1 < 0, q = \frac{c^2}{c_s^2} - 1 < 0.$$

All exponential terms in the coefficients matrix M, i.e. Eq. (5.2.55), are real functions and no longer are of periodicity. kh as an independent variable only exists on the exponential term. The dispersion equation has only unique real solution, i.e. real wave speed, which generates only one dispersion curve accordingly. This explains why there is no higher-order mode existed.

(5) When $c_{s(coverlayer)} < c_{s(halfspace)}$, there are two cases as follows.

(a) When $c < c_{s(coverlayer)}$, $c < c_{s(coverlayer)} < c_{s(halfspace)}$. The dispersion equation has only unique real solution. In other words, there is only one dispersive curve.

(b) When $c > c_{s(coverlayer)}$, $then$, $c_{s(coverlayer)} < c < c_{s(halfspace)}$ and

$$p' = \frac{c^2}{c_P'^2} - 1, q' = \frac{c^2}{c_s'^2} - 1 > 0, p = \frac{c^2}{c_P^2} - 1 < 0, q = \frac{c^2}{c_s^2} - 1 < 0.$$

The exponential terms containing $e^{\pm ik\sqrt{q'}h}$ in the coefficient matrix M are of periodicity. The dispersion equation has multiple solutions. This explains why there are higher-order modes appeared.

(6) It is observed that the speed of Rayleigh wave of lowest-order mode does not change monotonously as the dimensionless wavenumber kh increases gradually from zero; see Figs. 5.8 and 5.9. This is because that the shear wave speed of Al is smaller than that of substrate, i.e. $c_{s(Al)} < c_{s(Steel)}$, but the speed of P wave in Al is greater than that of steel, i.e. $c_{p(Steel)} < c_{p(Al)}$. Therefore, the dispersion feature of Rayleigh wave is not only dependent on the speed of shear waves but also the speeds of compression waves in the cover layer and substrate.

In order to better study the influence of different combinations of cover layer and lower half-space on the dispersion curves. Let's introduces one parameter, i.e. $\alpha = c_{S(\text{cover layer})}/c_{S(\text{half-space})} = c_{P(\text{cover layer})}/c_{P(half-space)}$. Figure 5.16 provides the dispersion curves under different combination. It can be seen that there are more than one dispersion curves existed when $\alpha \leq 1$. With the increase of α, each curve becomes more flat. When $\alpha = 0.99$, the dispersive curve of the lowest-order mode tends to a straight line and the dispersive curve of second-order mode is also very flat. The reason is that $c_{S(coverlayer)}$ and $c_{S(halfspace)}$ are very close and $c_{R(coverlayer)}$ and $c_{R(halfspace)}$ are also very close each other. When $\alpha = 1$, only the lowest-order mode retain while the higher-order modes disappear. It reduces to the situation of Rayleigh wave in elastic half-space. When $\alpha \geq 1$, there is only one dispersion curve and the dispersive curve becomes more steep with α increases gradually; see Fig. 5.18. The cut-off wavenumber where the phase velocity reached $c_{S(halfspace)}$ also decreases gradually. Figures 5.17 and 5.19 show the evolutionary process of dispersion curves of each order modes for different modulus ratio α at the region $(\alpha \leq 1)$ and $(0.5 \leq \alpha \leq 1.8)$.

When the cover layer or the substrate is viscoelastic medium, the elastic modulus of material becomes the frequency-dependent complex elastic modulus due to the viscous dissipation. Accordingly, the wavenumber of Rayleigh wave becomes complex wavenumber. The dispersion equation becomes the nonlinear equation with complex coefficients, and the roots of the dispersion equation should be searched in the complex domain. The dichotomizing search method which is usually used for root-finding of real functions with single variable will no longer be applicable, then Muller method can be considered as an alternative approach [32]. Three initial values are firstly selected in the Muller method, then performing iteration according to certain rules and replacing the old root with the new root. After the initial value is updated, the root-searching is performed for the next root. The process is continuously repeated until the root that meets the accuracy requirement is obtained. Many scholars have adopted the Muller method to study the dispersion characteristics of Rayleigh waves, and the Muller method has also been modified and improved in the practical application [33–35].

An alternative to root-finding is to solve the underlying differential equations directly by using a numerical scheme based on spectral methods. Unlike root-finding method, it is not based on the exact dispersion equation but a numerical interpolation

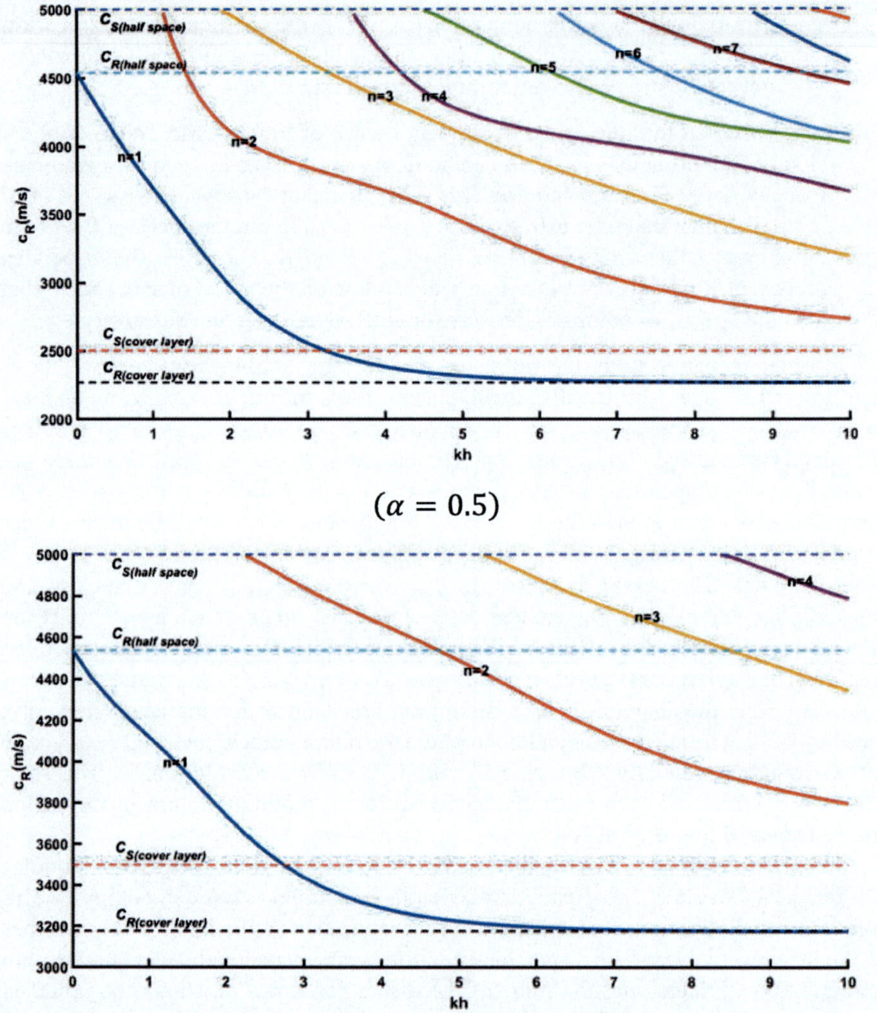

Fig. 5.16 Rayleigh wave speed versus the dimension wavenumber kh for different ratio of modulus α

of unknown function and the governing differential equation as well as the boundary conditions. For introductory material on spectral methods, the reader is referred to the books by Fornberg [36], Trefethen [37], Boyd [38], Shen [39] and the software suite by Weideman and Reddy [40]. The basic idea is to use spectral differentiation matrices to discretize the underlying equations and solves the corresponding equations as a generalized matrix eigenvalue problem. The eigenvalues are corresponding

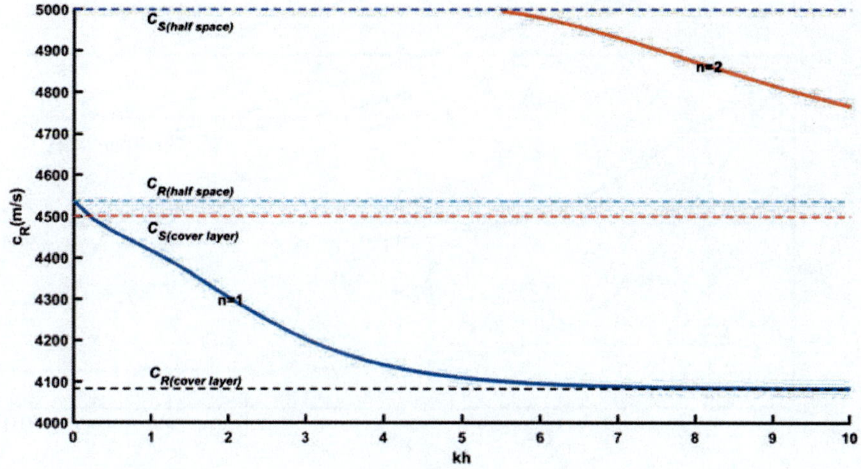

$(\alpha = 0.9)$

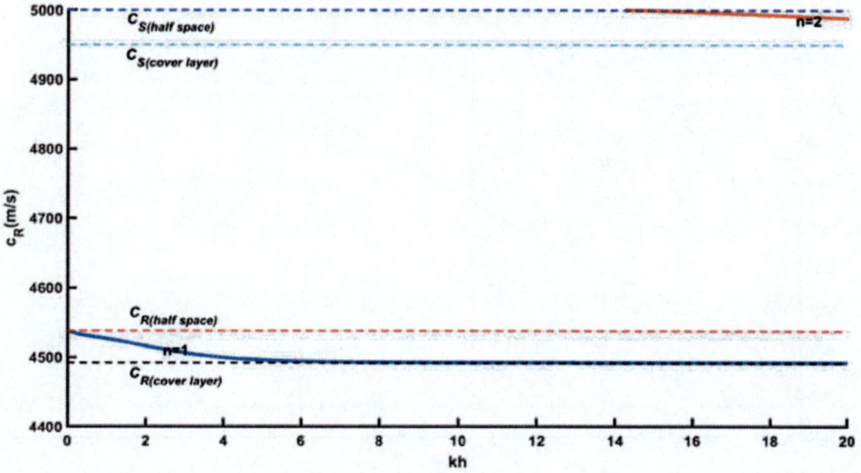

$(\alpha = 0.99)$

Fig. 5.16 (continued)

to the frequency for a given wave number or on the contrary. This is then solved using a standard algorithm MATLAB's eig routine. Dispersion curves are generated by varying the wave number and repeating the procedure. Spectral methods are particularly well suited to solve eigenvalue problems because the number of interpolation points, N, required for a given accuracy is considerably smaller than for finite difference or element schemes (where the shape function is local in character with

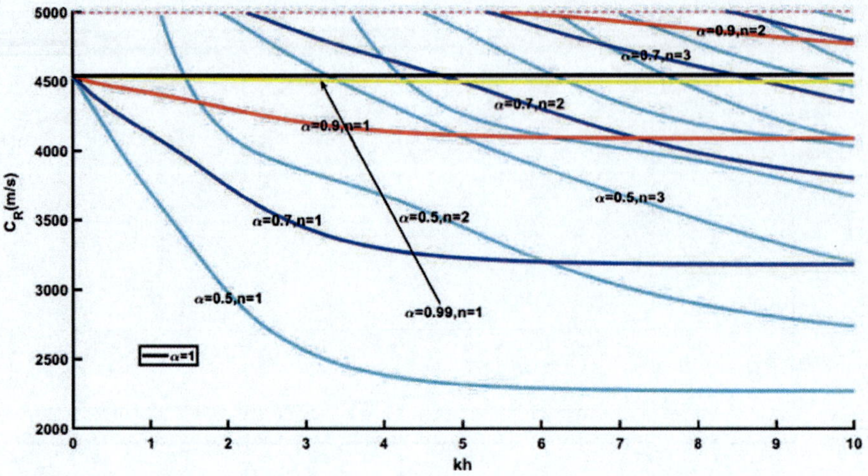

Fig. 5.17 Rayleigh wave speed versus the dimension wavenumber for different ratio of modulus α ($\alpha \leq 1$)

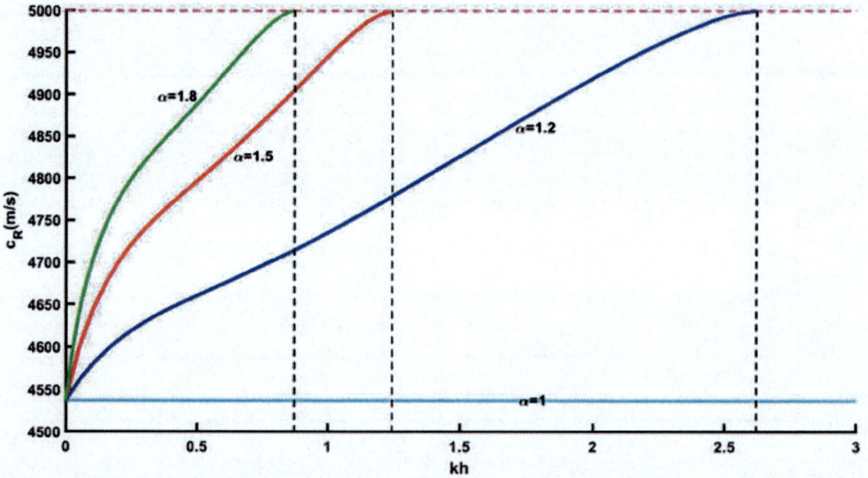

Fig. 5.18 Rayleigh wave speed versus the dimension wavenumber for different ratio of modulus α ($\alpha \geq 1$)

finite regularity, in contrast, the spectral method employs globally smooth function as shape function or trial function). For certain scenarios, where root-finding method does not work so well, such as curved, damped, inhomogeneous and anisotropic media, where the spectral methods represent a marked improvement and have some advantages. (1) The eigenvectors which represent the mode shapes are automatically generated. However, it needs a separate calculation for the root-finding approach. (2)

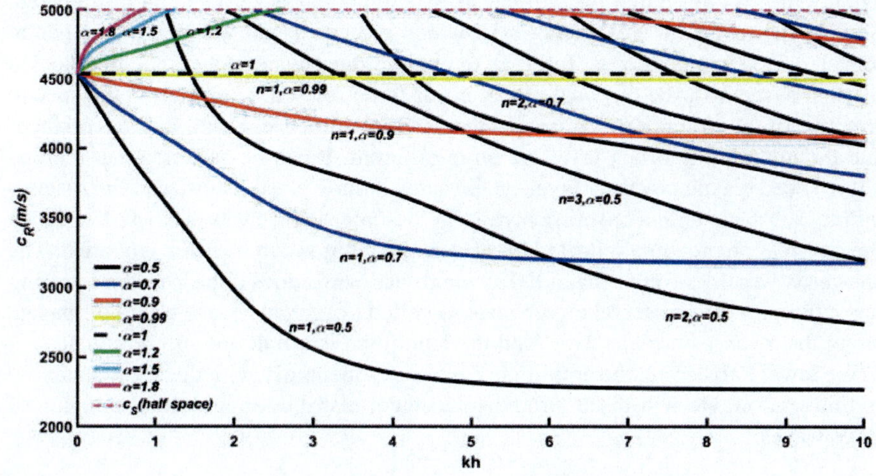

Fig. 5.19 Rayleigh wave speed versus the dimension wavenumber kh for different ratio of modulus α $(0.5 \le \alpha \le 1.8)$

The lowest eigenvalues are guaranteed to be found without any being missed. One of the pitfalls of root-finding method is that roots can be missed, particularly when the characteristic function varies rapidly [41,42]. (3) The approach is flexible and can easily be extended to handle a wide variety of scenarios, e.g. viscoelastic, inhomogeneous and anisotropic media where the traditional root-finding method is limited or hard to realize. The readers who are interesting in related works are referred to references [43–48].

5.3 Love Wave

In Sect. 5.2, we have discussed the Rayleigh wave whose displacement components are in the x_1x_3 plane and without horizontal lateral displacement. Its phase velocity is independent of frequency, namely non-dispersive. However, seismograph records show that there is a dispersive wave with a large lateral displacement on the ground surface, which indicates that there is another type of wave. In this section, we will discuss whether there are SH-type surface waves existing. The existence of SH-type surface waves requires

$$u_2(x_1, x_3, t) = A(x_3)\mathrm{e}^{ik(x_1-ct)},$$

then we have

$$\sigma_{32}(x_1, x_3, t) = \mu\frac{\partial u_2}{\partial x_3} = \mu A'(x_3)\mathrm{e}^{ik(x_1-ct)} \neq 0,$$

namely the condition that the surface stress component is zero cannot be satisfied. So SH-type surface waves cannot exist which is inconsistent with the phenomenon reflected in seismic records. In order to explain the physical phenomena reflected in seismic records, the physical model needs to be revised. Considering the special structure of the stratum, we notice that the material properties of the surface soil layer and the underground rock layer are quite different. It can be assumed that there is a low-speed elastic covering layer on the semi-infinite elastic substrate. For a semi-infinite substrate with a covering layer, can SH-type surface waves exist? If so, then the physical phenomena reflected in the seismic records can be fully explained. The answer is yes. It is found that a SH-type surface wave does appear in the covering layer after intensive researches, which is so-called Love wave. Love waves propagate along the x_1-axis on plane x_1x_3, and the amplitude is a function of coordinates x_3. In the lower half-space, the amplitude decays exponentially with the coordinates x_3. In this section, we will discuss the displacement distribution and the dispersion of Love waves.

5.3.1 *The Displacement Distribution of Love Wave*

Consider a homogeneous elastic half-space with a covering layer whose thickness is *h*, as shown in Fig. 5.20. Assuming that SH-type surface waves propagating along the x_1 direction and displacement amplitude changes with x_3. Thus, the displacement fields in medium 1 and medium 2 can be written as

$$u_2 = f(x_3)\mathrm{e}^{ik(x_1-ct)}, \tag{5.3.1a}$$

$$u_2' = g(x_3)\mathrm{e}^{ik(x_1-ct)}, \tag{5.3.1b}$$

where k and c are the wave number and phase velocity of the surface wave, respectively.

The displacement fields u and u' in medium 1 and medium 2 satisfy the governing equation

$$\frac{\partial^2 u_2}{\partial x_1^2} + \frac{\partial^2 u_2}{\partial x_3^2} = \frac{1}{c_S^2}\frac{\partial^2 u_2}{\partial t^2} \tag{5.3.2a}$$

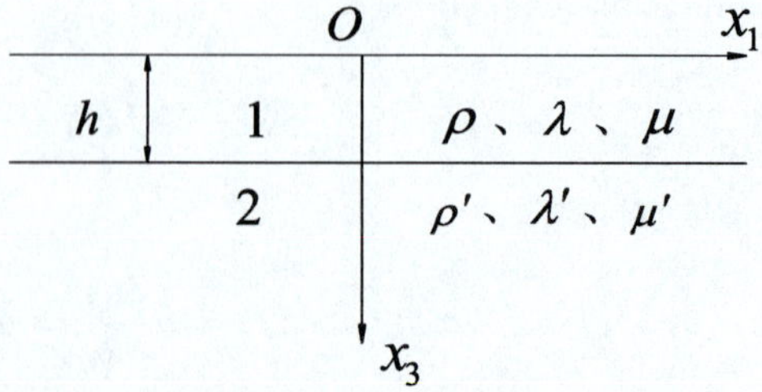

Fig. 5.20 The elastic half-space with covering layer

$$\frac{\partial^2 u_2'}{\partial x_1^2} + \frac{\partial^2 u_2'}{\partial x_3^2} = \frac{1}{c_S'^2}\frac{\partial^2 u_2'}{\partial t^2}. \tag{5.3.2b}$$

Substituting Eq. (5.3.1) into Eq. (5.3.2), we have

$$f''(x_3) + k^2(\frac{c^2}{c_S^2} - 1)f(x_3) = 0, \tag{5.3.3a}$$

$$g''(x_3) + k^2(\frac{c^2}{c_S'^2} - 1)g(x_3) = 0. \tag{5.3.3b}$$

Further, introduce two new quantities

$$p^2 = \frac{c^2}{c_S^2} - 1, \tag{5.3.4a}$$

$$q^2 = \frac{c^2}{c_S'^2} - 1, \tag{5.3.4b}$$

the solution of Eq. (5.3.3) can be written as

$$f(x_3) = C_1 \mathrm{e}^{ikpx_3} + C_2 \mathrm{e}^{-ikpx_3}, \tag{5.3.5a}$$

$$g(x_3) = C_1' \mathrm{e}^{ikqx_3} + C_2' \mathrm{e}^{-ikqx_3}. \tag{5.3.5b}$$

Further, the expression of displacement can be written as

$$u_2 = (C_1 \mathrm{e}^{ikpx_3} + C_2 \mathrm{e}^{-ikpx_3})\mathrm{e}^{ik(x_1 - ct)}, \tag{5.3.6a}$$

$$u_2' = (C_1' \mathrm{e}^{ikqx_3} + C_2' \mathrm{e}^{-ikqx_3})\mathrm{e}^{ik(x_1 - ct)}. \tag{5.3.6b}$$

It is required $u_2' \to 0$ when $x_3 \to \infty$. Thus, q should be a pure imaginary number and $C_2' = 0$. Let

$$q = i\xi, \xi = \sqrt{1 - \frac{c^2}{c_s'^2}}, \tag{5.3.7}$$

and using Eq. (5.3.7), we have

$$1 - \frac{c^2}{c_s'^2} > 0, \tag{5.3.8}$$

namely

$$c < c_s'. \tag{5.3.9}$$

Further, using the boundary conditions

$$\sigma_{23}\big|_{x_3=0} = 0, \tag{5.3.10}$$

$$u_2\big|_{x_3=h} = u_2'\big|_{x_3=h}, \tag{5.3.11a}$$

$$\sigma_{23}\big|_{x_3=h} = \sigma_{23}'\big|_{x_3=h}, \tag{5.3.11b}$$

we have

$$C_1 = C_2, \tag{5.3.12a}$$

$$C_1 e^{ikph} + C_2 e^{-ikph} = C_1' e^{-k\xi h}, \tag{5.3.12b}$$

$$i\mu p(C_1 e^{ikph} - C_2 e^{-ikph}) = -\mu'\xi C_1' e^{-k\xi h}. \tag{5.3.12c}$$

Substituting C_1=C_2 into the other two formulas and eliminate C_2, we have

$$\left(e^{ikph} + e^{-ikph}\right)C_1 - e^{-k\xi h} C_1' = 0, \tag{5.3.13a}$$

$$i\mu p\left(e^{ikph} - e^{-ikph}\right)C_1 + \mu'\xi e^{-k\xi h} C_1' = 0. \tag{5.3.13b}$$

From Eq. (5.3.13), we obtain

$$C_1' = 2C_1 \cos(kph) e^{k\xi h}. \tag{5.3.14}$$

Take the real part of Eq. (5.3.6), and set $C = 2C_1$, then the displacements in the two media can be rewritten as

$$u_2 = C \cos(kpx_3) e^{ik(x_1 - ct)}, \tag{5.3.15a}$$

$$u_2' = C \cos(kph) e^{-k\xi(x_3 - h)} e^{ik(x_1 - ct)}. \tag{5.3.15b}$$

Equation (5.3.15) shows that **the amplitude of the wave in medium 1 changes periodically in the direction x_3 which is of the characteristic of standing waves. In medium 2, the amplitude decays exponentially with the increase of x_3.** But they all propagate along the x_1 direction with the phase velocity c, and together form the Love wave.

5.3.2 The Dispersion Equation of Love Wave

A nontrivial solution of Eq. (5.3.13) requires

$$\begin{vmatrix} \mathrm{e}^{ikph} + \mathrm{e}^{-ikph} & -\mathrm{e}^{-k\xi h} \\ i\left(\mathrm{e}^{ikph} - \mathrm{e}^{-ikph}\right)\mu p & \mu' \xi \mathrm{e}^{-k\xi h} \end{vmatrix} = 0. \tag{5.3.16}$$

Further, we have

$$\tan khp = \frac{\mu' \xi}{\mu p} = a\left(\frac{1 - c^2/c_S'^2}{c^2/c_S^2 - 1}\right)^{\frac{1}{2}} = aF(c^2), \tag{5.3.17}$$

where $a = \mu'/\mu$, Eq. (5.3.17) is called the dispersion equation of Love waves. It can be seen that the speed of the Love wave is related to the wave number and frequency, namely the Love wave is a dispersive wave. In addition, when the right end of Eq. (5.3.17) is given, due to the periodicity of the tangent function, there will be many roots corresponding to it, as shown in Fig. 5.21. Therefore, Love waves have many dispersion curves. Each dispersion curve represents Love waves with a specific dispersion property.

The existence of Love waves require that the right side of Eq. (5.3.17) is real number. Then, we have

$$c_S < c < c_S'. \tag{5.3.18}$$

That is, the wave speed of Love wave is greater than that of the transverse body wave of the covering layer but less than that of the elastic half-space of the bottom

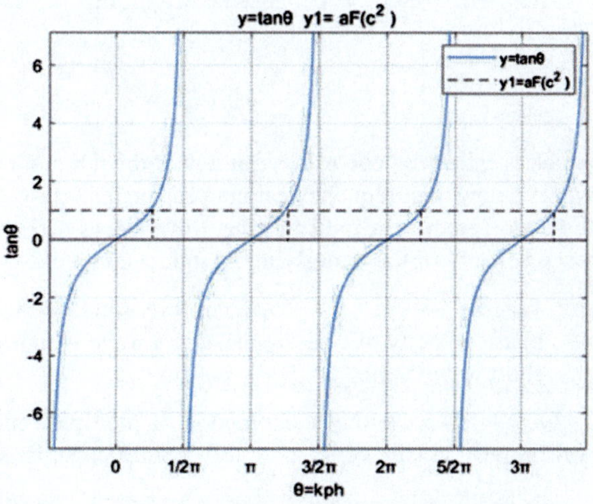

Fig. 5.21 Root distribution of the dispersion equation

layer. Therefore, Love waves exist under the condition that the softer covering layer covers the harder half-space, and the upper layer and lower layer are called the low-speed layer and high-speed layer, respectively.

Next, let's find the real roots of Eq. (5.3.17) that satisfy the condition $c_S < c < c'_S$. For this purpose, we will discuss it as follows:

(1) When $c \to c'_S$, $F(c^2) \to 0$; and when $c \to c_S$, $F(c^2) \to +\infty$. Obviously, for the solution $c = c_L$ in the range of $c_S < c < c'_S$, one can always find a wave number k corresponding to it, namely Formula (5.3.17) is always solvable.
(2) Equation (5.3.17) can be rewritten as

$$kh = \left(\frac{c^2}{c_S^2} - 1\right)^{-\frac{1}{2}} \left[\arctan a\left(\frac{1 - c^2/c_S'^2}{c^2/c_S^2 - 1}\right)^{\frac{1}{2}} + n\pi\right]. \quad n = 0, 1, 2, 3 \ldots. \tag{5.3.19}$$

From Eq. (5.3.19), the following conclusions can be made:

(a) When $n = 0$, for any positive wave number k, a wave velocity can be found from the right side of Eq. (5.3.19) to correspond to it. By this way, a dispersion curve can be obtained, namely $c_L^{(0)}(k) \sim k$, which is called the lowest-order dispersion curve of Love wave. Let $\lambda = 2\pi/k$ be the corresponding wavelength. It means that any wavelength between zero and infinity can exist for the lowest-order Love wave ($n = 0$).
(b) When $n \neq 0$, the minimum wave number is

$$k_{cr}^{(n)} h = n\pi \left(\frac{c_S'^2}{c_S^2} - 1\right)^{-\frac{1}{2}}. \tag{5.3.20}$$

Then, we have $c_L^{(n)} = c'_S$. When $k < k_{cr}^{(n)}$, Eq. (5.3.19) has no real number solution. Thus

$$\omega_{cr}^{(n)} = k_{cr}^{(n)} c'_S, \tag{5.3.21}$$

which is called the **cut-off frequency** of the nth mode of Love wave; in other word, waves below this frequency cannot propagate in the covering layer. The existence of the cut-off frequency indicates that Love waves with the speed higher than the shear wave velocity of the underlying medium cannot exist.

(c) When $c \to c_S$, $k \to \infty$. Therefore, the Love wave that can propagate at the phase velocity of $c \to c_S$ must be a wave with a very high frequency or a very short wavelength.

We can also use the phenomenon of multiple reflections to explain why Love waves must satisfy $c_S < c < c'_S$ condition. Suppose there is a SH wave in the covering layer propagates toward the interface between the covering layer and the substrate. When the incident angle satisfies

$$\gamma < \gamma_{cr} = \sin^{-1}\left(c_S/c_S'\right). \tag{5.3.22}$$

There are waves reflected back into the covering layer, and also, there are waves transmitted into the substrate. After such multiple reflections, the wave remaining in the covering layer decays quickly, and the energy is basically transferred to waves that propagate in the half-space, as shown in Fig. 5.22. The waves in the covering layer that have energy leakage during the propagation are usually called **leaky waves.** However, when the incident angle $\gamma > \gamma_{cr}$, total reflection will occur and forming resultant SH-type waves in the covering layer, and the energy is completely trapped in the covering layer, as shown in Fig. 5.23. Therefore, Love waves, in essence, are produced by the interference of plane waves that generated from multiple total reflections at the free surface and the interface and finally are captured in the covering layer. In fact, **the dispersion equation exactly expresses the interference condition of these plane waves.**

Finally, we will discuss the displacement fields of each order Love wave which propagates in the covering layer and the underlying half-space along the x_3 direction. From Eq. (5.3.15a), we have

$$|u_2| = C\cos(k^{(n)}px_3), \tag{5.3.23a}$$

or

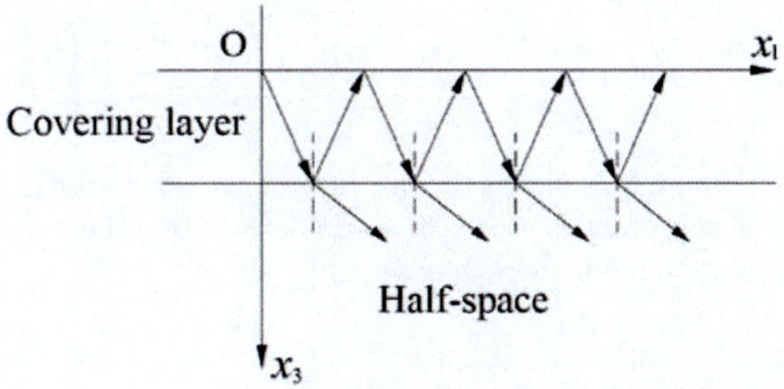

Fig. 5.22 A wave (leakage wave) which propagates along x_1 direction in the covering layer but with energy leaking into the lower half-space

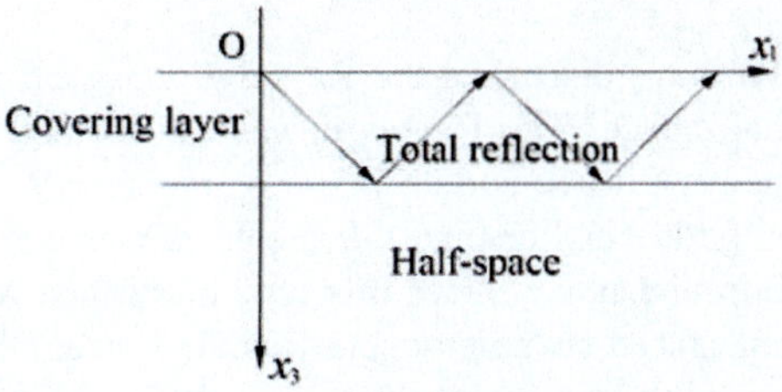

Fig. 5.23 A wave (love wave) which propagates along x_1 direction in the covering layer but without energy leakage

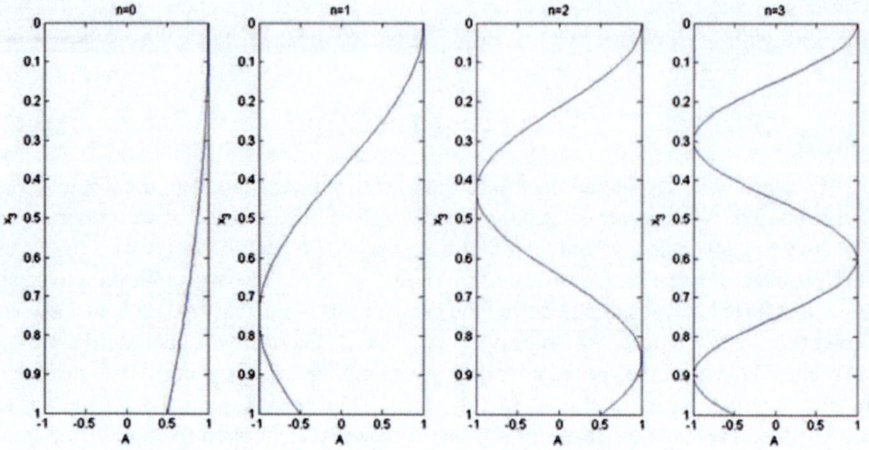

Fig. 5.24 Sketch of vibration mode of each order Love wave ($\theta_0 = 60°$)

$$A = \frac{|u_2|}{C} = \cos(k^{(n)} ph\overline{x}_3),\ (\overline{x}_3 = x_3/h). \tag{5.3.23b}$$

When $x_3 = h$,

$$|u_2| = C\cos(k^{(n)} ph) = C\cos(\theta_0 + n\pi), \tag{5.3.24}$$

where

$$\theta_0 = \arctan a\left(\frac{1 - c^2/c_s'^2}{c^2/c_s^2 - 1}\right)^{\frac{1}{2}} \in (0, \pi). \tag{5.3.25}$$

Figure 5.24 shows the displacement distribution along the x_3 direction generated by each order Love wave when $\theta_0 = 60°$. These curves are generally called the vibration mode of each order Love wave.

5.4 Stoneley Wave

We have discussed the Rayleigh wave propagating along the surface of an elastic half-space. The Rayleigh wave is formed by the superposition of an inhomogeneous P wave and an inhomogeneous SV wave. The velocity of Rayleigh wave is slightly smaller than the shear wave speed, and independent of frequency, that is, non-dispersive. Does this type of surface wave also exist near the two sides of the interface between two elastic half-spaces? This wave propagates along the interface with the same propagation velocity, and the amplitudes on both sides of the interface decay exponentially as the distance from the interface increases. This chapter will discuss this issue. The answer to the question is yes, and such surface waves are usually called Stoneley waves.

5.4.1 Wave Function of Stoneley Wave

Consider two semi-infinite elastic media, the x_1x_2 plane is the interface, and the x_1x_3 plane is the plane where the wave propagates. From the discussion of Rayleigh waves, it is known that the superposition of the P-type and SV-type surface waves will form an elliptically polarized Rayleigh wave on both sides of the interface. If the Rayleigh waves formed on both sides of the interface interact (required by the continuity condition of the interface) and eventually cause the propagation along the interface at the same speed, a new wave will be formed. This kind of wave propagates along x_1-axis direction, and its amplitude decays exponentially with the increase of $|x_3|$ along the x_3-axis direction. At $|x_3| \to \infty$, the amplitude of the wave function decays to zero, as shown in Fig. 5.25, the wave function expressions in the two media can be written as:

$$\varphi_1(x_1, x_3, t) = \Phi_1(x_3)\mathrm{e}^{ik_1(x_1-c_1t)}, \quad x_3 \geq 0 \tag{5.4.1}$$

$$\psi_1(x_1, x_3, t) = \Psi_1(x_3)\mathrm{e}^{ik_1(x_1-c_1t)}, \quad x_3 \geq 0 \tag{5.4.2}$$

$$\varphi_2(x_1, x_3, t) = \Phi_2(x_3)\mathrm{e}^{ik_2(x_1-c_2t)}, \quad x_3 \leq 0 \tag{5.4.3}$$

$$\psi_2(x_1, x_3, t) = \Psi_2(x_3)\mathrm{e}^{ik_2(x_1-c_2t)}, \quad x_3 \leq 0 \tag{5.4.4}$$

The continuous conditions at the interface $x_3 = 0$ require

$$k_1 = k_2 = k, \tag{5.4.5}$$

$$c_1 = c_2 = c, \tag{5.4.6}$$

where, $c = \frac{\omega}{k}$ is the phase velocity, $\Phi_1(x_3)$, $\Psi_1(x_3)$, $\Phi_2(x_3)$, $\Psi_2(x_3)$ are the amplitude of P wave and SV wave, respectively. The above wave functions satisfy the following the longitudinal wave and transverse wave equations

$$\nabla^2\varphi = \frac{1}{c_P^2}\frac{\partial^2\varphi}{\partial t^2}, \tag{5.4.7}$$

$$\nabla^2\psi = \frac{1}{c_S^2}\frac{\partial^2\psi}{\partial t^2}. \tag{5.4.8}$$

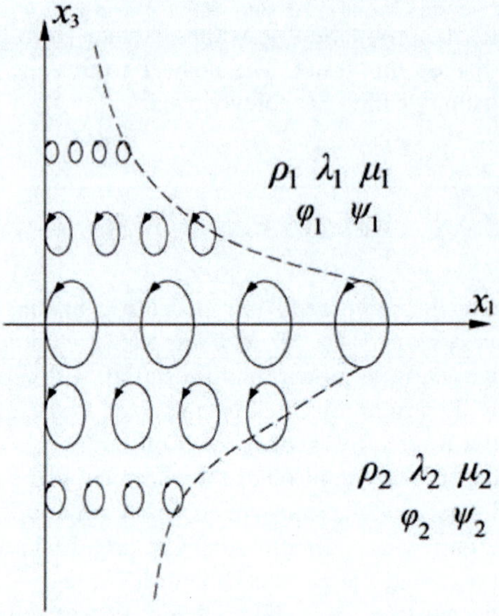

Fig. 5.25 Amplitude attenuation and particle polarization of Stoneley waves

Inserting φ_1, ψ_1 and φ_2, ψ_2 into Eqs. (5.4.7) and (5.4.8), we have

$$\Phi_1''(x_3) + k^2\left(\frac{c^2}{c_{P1}^2} - 1\right)\Phi_1(x_3) = 0, \quad x_3 \geq 0 \tag{5.4.9}$$

$$\Psi_1''(x_3) + k^2\left(\frac{c^2}{c_{S1}^2} - 1\right)\Psi_1(x_3) = 0, \quad x_3 \geq 0 \tag{5.4.10}$$

$$\Phi_2''(x_3) + k^2\left(\frac{c^2}{c_{P2}^2} - 1\right)\Phi_2(x_3) = 0, x_3 \leq 0 \tag{5.4.11}$$

$$\Psi_2''(x_3) + k^2\left(\frac{c^2}{c_{S2}^2} - 1\right)\Psi_2(x_3) = 0, \quad x_3 \leq 0, \tag{5.4.12}$$

where

$$c_{P1} = \sqrt{\frac{\lambda_1 + 2\mu_1}{\rho_1}}, \quad c_{S1} = \sqrt{\frac{\mu_1}{\rho_1}},$$

$$c_{P2} = \sqrt{\frac{\lambda_2 + 2\mu_2}{\rho_2}}, \quad c_{S2} = \sqrt{\frac{\mu_2}{\rho_2}}.$$

Let

$$p_1^2 = 1 - \frac{c^2}{c_{P1}^2}, \quad q_1^2 = 1 - \frac{c^2}{c_{S1}^2},$$

$$p_2^2 = 1 - \frac{c^2}{c_{P2}^2}, \quad q_2^2 = 1 - \frac{c^2}{c_{S2}^2}.$$

The solution of Eqs. (5.4.9)–(5.4.12) depends on the values of p_1, p_2, q_1 and q_2. Similar to Rayleigh wave, only when

$$c < \min(c_{S1}, c_{S2}), \tag{5.4.13}$$

the amplitude on both sides of the interface decays exponentially with increasing $|x_3|$.

The solution of Eqs. (5.4.9)–(5.4.12) can be expressed as

$$\Phi_1(x_3) = A_1 \mathrm{e}^{-kp_1x_3} + A_1' \mathrm{e}^{kp_1x_3}, \tag{5.4.14}$$

$$\Psi_1(x_3) = B_1 \mathrm{e}^{-kq_1x_3} + B_1' \mathrm{e}^{kq_1x_3}, \tag{5.4.15}$$

$$\Phi_2(x_3) = A_2 \mathrm{e}^{kp_2x_3} + A_2' \mathrm{e}^{-kp_2x_3}. \tag{5.4.16}$$

$$\Psi_2(x_3) = B_2 \mathrm{e}^{kq_2x_3} + B_2' \mathrm{e}^{-kq_2x_3}. \tag{5.4.17}$$

According to the radiation condition that the amplitude is zero at $|x_3| \to \infty$, we have

$$A_1' = B_1' = A_2' = B_2' = 0. \tag{5.4.18}$$

Then, the wave functions Eqs. (5.4.1)–(5.4.4) can be expressed as

$$\varphi_1 = A_1 \mathrm{e}^{-kp_1x_3} \mathrm{e}^{ik(x_1-ct)}, \quad x_3 \geq 0 \tag{5.4.19}$$

$$\psi_1 = B_1 \mathrm{e}^{-kq_1x_3} \mathrm{e}^{ik(x_1-ct)}, \quad x_3 \geq 0 \tag{5.4.20}$$

$$\varphi_2 = A_2 \mathrm{e}^{kp_2x_3} \mathrm{e}^{ik(x_1-ct)}, \quad x_3 \leq 0 \tag{5.4.21}$$

$$\psi_2 = B_2 \mathrm{e}^{kq_2x_3} \mathrm{e}^{ik(x_1-ct)}, \quad x_3 \leq 0. \tag{5.4.22}$$

5.4.2 Stoneley Equation

Considering the interface continuity conditions

$$u_1^{(1)}\big|_{x_3=0} = u_1^{(2)}\big|_{x_3=0}, \tag{5.4.23a}$$

$$u_3^{(1)}\big|_{x_3=0} = u_3^{(2)}\big|_{x_3=0}, \tag{5.4.23b}$$

$$\sigma_{33}^{(1)}\big|_{x_3=0} = \sigma_{33}^{(2)}\big|_{x_3=0}, \tag{5.4.23c}$$

$$\sigma_{31}^{(1)}\big|_{x_3=0} = \sigma_{31}^{(2)}\big|_{x_3=0}, \tag{5.4.23d}$$

we have

$$A_1 - iq_1B_1 - A_2 - iq_2B_2 = 0, \tag{5.4.24a}$$

$$p_1A_1 - iB_1 + p_2A_2 + iB_2 = 0, \tag{5.4.24b}$$

$$-\mu_1(q_1^2+1)A_1 + 2i\mu_1q_1B_1 + \mu_2(q_2^2+1)A_2 + 2i\mu_2q_2B_2 = 0, \tag{5.4.24c}$$

$$2i\mu_1p_1A_1 + \mu_1(q_1^2+1)B_1 + 2i\mu_2p_2A_2 - \mu_2(q_2^2+1)B_2 = 0. \tag{5.4.24d}$$

Equation (5.4.24) is written in matrix form as

$$\begin{pmatrix} 1 & -q_1 & -1 & -q_2 \\ p_1 & -1 & p_2 & 1 \\ -\mu_1(q_1^2+1) & 2\mu_1q_1 & \mu_2(q_2^2+1) & 2\mu_2q_2 \\ 2\mu_1p_1 & \mu_1(q_1^2+1) & 2\mu_2p_2 & \mu_2(q_2^2+1) \end{pmatrix} \begin{pmatrix} A_1 \\ iB_1 \\ A_2 \\ iB_2 \end{pmatrix} = \begin{pmatrix} 0 \\ 0 \\ 0 \\ 0 \end{pmatrix}. \tag{5.4.25}$$

A nontrivial solution of Eq. (5.5.25) requires

$$\begin{vmatrix} 1 & -q_1 & -1 & -q_2 \\ p_1 & -1 & p_2 & 1 \\ -\mu_1(q_1^2+1) & 2\mu_1q_1 & \mu_2(q_2^2+1) & 2\mu_2q_2 \\ 2\mu_1p_1 & \mu_1(q_1^2+1) & 2\mu_2p_2 & \mu_2(q_2^2+1) \end{vmatrix} = 0. \tag{5.4.26}$$

Let $c = c_{st}$, Eq. (5.4.26) is a sixth-order equation with respect to c_{st},

$$\begin{aligned} &c_{st}^4\left[(\rho_2-\rho_1)^2 - (\rho_1p_2+\rho_2p_1)(\rho_1q_2+\rho_2q_1)\right] \\ &+4c_{st}^2[(\mu_1-\mu_2)(\rho_2-\rho_1) + (\rho_1p_1+\rho_2p_1)(\mu_1q_2+\mu_2q_1) \end{aligned}$$

$$4(\mu_1 - \mu_2)[(\mu_1(1 + p_1q_1)(1 - p_2q_2) - \mu_2(1 - p_1q_1)(1 + p_2q_2)] = 0. \quad (5.4.27)$$

This is the **Stoneley equation**. Because p_1,p_2,q_1 and q_2 are all function of c_{st}, Stoneley equation is actually a sixth-order equation with respect to c_{st}, and usually very difficult to obtain the analytic solution. Consider that traveling waves correspond to real number roots, if we ignoring complex number roots, the real number roots of the equation can generally be obtained in the real number domain with the help of dichotomy.

Because c_{st} represents the wave velocity along the x_1 direction, it can be seen from Eqs. (5.4.19)–(5.4.22) that this wave is a inhomogeneous wave, and its amplitude decays exponentially with $|x_3| \to +\infty$. Because the energy of this wave is mainly concentrated near the interface, this wave is usually called interface wave. The interface wave was first discovered by Stoneley (1924), so it is also called Stoneley wave. Stoneley waves can also be regarded as a generalized Rayleigh wave; that is, two Rayleigh waves on both sides of the interface propagate along the interface but with the same phase velocity.

Equation (5.4.27) is the equation that Stoneley wave velocity c_{st} must satisfy. If the real number roots can be obtained from this equation, it indicates that Stoneley waves exist indeed; otherwise, not exist. There are two problems here. The first question is whether there must be real roots for any combination of two media? The answer is no; that is, Stoneley waves do not necessarily exist for any combination of two media. Cagniand studied the existence of Stoneley waves under different combinations of material parameters, made calculated for Poisson solids ($\lambda = \mu$) and gave the material combination area where Stoneley waves exist, as shown in Fig. 5.26. The shaded area is the area where the Stoneley wave exists. Another question is, which is faster, Stoneley wave or Rayleigh wave? Koppe (1948) studied this question and gave a clear answer; that is, the wave speed of Stoneley wave is higher than that of Rayleigh wave while smaller than that of shear wave in the two media, namely

$$\max(c_{R1}, c_{R2}) < c_{st} < \min(c_{S1}, c_{S2}).$$

where c_{R1} and c_{R2} are the Rayleigh wave velocities of the two media, c_{S1} and c_{S2} are the transversal wave velocities of the two media. In addition, it can be seen from the Stoneley equation Eq. (5.4.27) that the speed of the Stoneley wave is independent of frequency. That is to say, similar to the Rayleigh wave, the Stoneley wave is also non-dispersive.

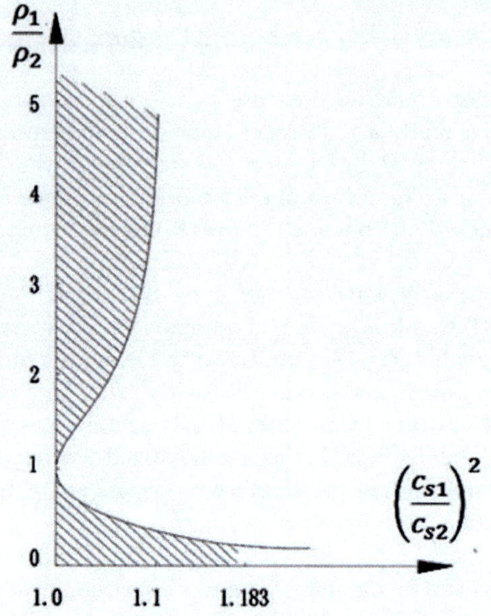

Fig. 5.26 Existence region of Stoneley wave at the Poisson solid interface

5.5 Torsional Surface Wave

The surface and interface waves are studied in the rectangular coordinate system in the previous several sections. Now, let's discuss the surface waves in the circular cylindrical coordinate system, as shown in Fig. 5.27. Consider a system of cylindrical coordinates (r, θ, z) having unit base vectors $(\boldsymbol{e}_r, \boldsymbol{e}_\theta, \boldsymbol{e}_z)$. The displacement field of torsional surface waves can be expressed as

$$u_r(r, z, t) = 0, u_\theta(r, z, t) \neq 0, u_z(r, z, t) = 0. \tag{5.5.1}$$

The above-mentioned displacement fields lead to nontrivial shear stresses in an elastic underlying half-space, which is not consistent with the free-stress surface condition. Therefore, torsional surface waves do not exist in an elastic half-space. However, the situation is different for the half-space with covering layer. The torsional surface waves can exist in an elastic half-space with a covering layer, which is like that Love waves do not exist in an elastic half-space but can exist in an elastic half-space with a covering layer.

According to (5.5.1), the displacement field of torsional surface waves in the covering layer can be expressed as

$$\boldsymbol{u} = u_\theta(r, z, t)\boldsymbol{e}_\theta = u(r, z)\mathrm{e}^{-i\omega t}\boldsymbol{e}_\theta. \tag{5.5.2}$$

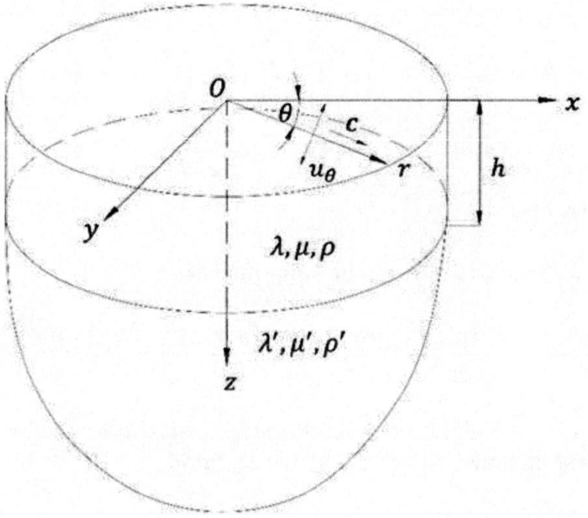

Fig. 5.27 Torsional surface waves in an elastic half-space with a covering layer

In the circular cylindrical coordinate system, the gradient operator is

$$\nabla = \frac{\partial}{\partial r}\boldsymbol{e}_r + \frac{\partial}{r\partial\theta}\boldsymbol{e}_\theta + \frac{\partial}{\partial z}\boldsymbol{e}_z. \tag{5.5.3}$$

The partial differential of unit base vectors has the following properties

$$\frac{\partial \boldsymbol{e}_r}{\partial r} = 0,\ \frac{\partial \boldsymbol{e}_r}{\partial \theta} = \boldsymbol{e}_\theta,\ \frac{\partial \boldsymbol{e}_r}{\partial z} = 0, \tag{5.5.4a}$$

$$\frac{\partial \boldsymbol{e}_\theta}{\partial r} = 0,\ \frac{\partial \boldsymbol{e}_\theta}{\partial \theta} = -\boldsymbol{e}_r,\ \frac{\partial \boldsymbol{e}_\theta}{\partial z} = 0, \tag{5.5.4b}$$

$$\frac{\partial \boldsymbol{e}_z}{\partial r} = \frac{\partial \boldsymbol{e}_z}{\partial \theta} = \frac{\partial \boldsymbol{e}_z}{\partial z} = 0. \tag{5.5.4c}$$

Inserting the displacement field of torsional surface waves into the governing equation of motion in terms of the displacement components, i.e.

$$(\lambda + 2\mu)\nabla(\nabla \cdot \boldsymbol{u}) - \mu\nabla \times \nabla \times \boldsymbol{u} = \rho\ddot{\boldsymbol{u}}, \tag{5.5.5}$$

leads to

$$
\begin{aligned}
&(\lambda+2\mu)\left(\frac{\partial}{\partial r}\boldsymbol{e}_r+\frac{\partial}{r\partial\theta}\boldsymbol{e}_\theta+\frac{\partial}{\partial z}\boldsymbol{e}_z\right)\left[\left(\frac{\partial}{\partial r}\boldsymbol{e}_r+\frac{\partial}{r\partial\theta}\boldsymbol{e}_\theta+\frac{\partial}{\partial z}\boldsymbol{e}_z\right)\cdot u(r,z)\boldsymbol{e}_\theta\right]\\
&-\mu\left(\frac{\partial}{\partial r}\boldsymbol{e}_r+\frac{\partial}{r\partial\theta}\boldsymbol{e}_\theta+\frac{\partial}{\partial z}\boldsymbol{e}_z\right)\times\left(\frac{\partial}{\partial r}\boldsymbol{e}_r+\frac{\partial}{r\partial\theta}\boldsymbol{e}_\theta+\frac{\partial}{\partial z}\boldsymbol{e}_z\right)\times u(r,z)\boldsymbol{e}_\theta\\
&=-\rho\omega^2u(r,z)\boldsymbol{e}_\theta.
\end{aligned}
\tag{5.5.6}
$$

According to (5.5.4), (5.5.6) can be simplified as

$$
\mu\left(\frac{\partial^2u(r,z)}{\partial r^2}+\frac{\partial u(r,z)}{r\partial r}-\frac{u(r,z)}{r^2}+\frac{\partial^2u(r,z)}{\partial z^2}\right)=-\rho\omega^2u(r,z).
\tag{5.5.7}
$$

In order to solve the above equation through the Hankel transform, perform the weighted integral on both sides of the above equation

$$
\begin{aligned}
&\int_0^\infty\left(\frac{\partial^2}{\partial r^2}+\frac{\partial}{r\partial r}-\frac{1}{r^2}+\frac{\partial^2}{\partial z^2}\right)u(r,z)\cdot J_1(\xi r)r\mathrm{d}r\\
&=-\int_0^\infty\frac{\rho\omega^2}{\mu}u(r,z)J_1(\xi r)r\mathrm{d}r,
\end{aligned}
\tag{5.5.8}
$$

where $J_1(\xi r)$ is the first kind of Bessel function of order one. Then,

$$
\begin{aligned}
&\int_0^\infty\left(\frac{\partial^2}{\partial r^2}+\frac{\partial}{r\partial r}-\frac{1}{r^2}\right)u(r,z)\cdot J_1(\xi r)r\mathrm{d}r\\
&=-\left(\frac{\partial^2}{\partial z^2}+\frac{\rho\omega^2}{\mu}\right)\int_0^\infty u(r,z)\cdot J_1(\xi r)r\mathrm{d}r.
\end{aligned}
\tag{5.5.9}
$$

The definite integral on the left-hand side of the equation is as follows

$$\int_0^\infty \left(\frac{\partial^2}{\partial r^2}+\frac{\partial}{r\partial r}-\frac{1}{r^2}\right)u(r,z)\cdot J_1(\xi r)r\mathrm{d}r.$$

$$=\int_0^\infty J_1(\xi r)r\frac{\partial^2 u(r,z)}{\partial r^2}\mathrm{d}r+\int_0^\infty \frac{\partial u(r,z)}{\partial r}J_1(\xi r)\mathrm{d}r-\int_0^\infty u(r,z)\frac{J_1(\xi r)}{r}\mathrm{d}r$$

$$=\frac{\partial u(r,z)}{\partial r}J_1(\xi r)r\Big|_0^\infty-\int_0^\infty \frac{\partial u(r,z)}{\partial r}\left(J_1(\xi r)+\xi r J_1'(\xi r)\right)\mathrm{d}r$$

$$+\int_0^\infty \frac{\partial u(r,z)}{\partial r}J_1(\xi r)\mathrm{d}r-\int_0^\infty u(r,z)\frac{J_1(\xi r)}{r}\mathrm{d}r$$

$$=-\int_0^\infty \xi r J_1'(\xi r)\mathrm{d}u(r,z)-\int_0^\infty u(r,z)\frac{J_1(\xi r)}{r}\mathrm{d}r$$

$$=-\xi r J_1'(\xi r)u(r,z)\Big|_0^\infty+\int_0^\infty u(r,z)\left(\xi J_1'(\xi r)+r\xi^2 J_1''(\xi r)\right)\mathrm{d}r$$

$$-\int_0^\infty u(r,z)\frac{J_1(\xi r)}{r}\mathrm{d}r$$

$$=0+\int_0^\infty u(r,z)\left(\xi J_1'(\xi r)+r\xi^2 J_1''(\xi r)-\frac{J_1(\xi r)}{r}\right)\mathrm{d}r.$$

Consider the recursion formula of Bessel function

$$J_0(\xi r)-J_2(\xi r)=2J_1'(\xi r),$$
$$J_2(\xi r)=\frac{2}{\xi r}J_1(\xi r)-J_0(\xi r),$$
$$J_1'(\xi r)=J_0(\xi r)-\frac{J_1(\xi r)}{\xi r}.$$

The integrand can be simplified as

$$\xi J_1'(\xi r)+r\xi^2 J_1''(\xi r)-\frac{J_1(\xi r)}{r}$$
$$=\xi\left(J_0(\xi r)-\frac{J_1(\xi r)}{\xi r}\right)+r\xi^2 J_1''(\xi r)-\frac{J_1(\xi r)}{r}$$
$$=\xi\left(J_0(\xi r)-\frac{J_1(\xi r)}{\xi r}\right)+r\xi^2\left(-J_1(\xi r)-\frac{J_0(\xi r)}{\xi r}+\frac{2J_1(\xi r)}{(\xi r)^2}\right)-\frac{J_1(\xi r)}{r}$$
$$=-r\xi^2 J_1(\xi r).$$

Therefore, (5.5.8) can be simplified as

$$\int_0^\infty r\xi^2 J_1(\xi r)u(r,z)\mathrm{d}r = \left(\frac{\partial^2}{\partial z^2} + \frac{\rho\omega^2}{\mu}\right)\int_0^\infty u(r,z)J_1(\xi r)r\mathrm{d}r, \tag{5.5.10a}$$

or

$$\frac{\partial^2}{\partial z^2}\int_0^\infty u(r,z)J_1(\xi r)r\mathrm{d}r - \left(\xi^2 - \frac{\rho\omega^2}{\mu}\right)\int_0^\infty u(r,z)J_1(\xi r)r\mathrm{d}r = 0. \tag{5.5.10b}$$

In view of the axisymmetry of the problem and in order to suppress the r-dependence in the governing equations, the Hankel transform of order one is employed, i.e.,

$$f^*(\xi,z) \overset{\mathrm{def}}{=} \int_0^\infty f(r,z)J_1(\xi r)r\mathrm{d}r,\ f(r,z) = \int_0^\infty f^*(\xi,z)J_1(\xi r)\xi\mathrm{d}\xi. \tag{5.5.11}$$

Now, the partial differential governing equation has been transformed into the ordinary differential equation

$$\frac{\mathrm{d}^2 u^*(\xi,z)}{\mathrm{d}z^2} - \left(\xi^2 - \frac{\rho\omega^2}{\mu}\right)u^*(\xi,z) = 0. \tag{5.5.12}$$

The solution of the above equation is

$$u^*(\xi,z) = B\mathrm{e}^{-kz} + C\mathrm{e}^{kz}, \tag{5.5.13}$$

where $k = \sqrt{\xi^2 - \frac{\rho\omega^2}{\mu}}$. The displacement field can be obtained through the inverse Hankel transform on Eq. (5.5.13), i.e.,

$$u(r,z) = \int_0^\infty u^*(\xi,z)J_1(\xi r)\xi\mathrm{d}\xi. \tag{5.5.14}$$

Similarly, the displacement field in the half space can also be obtained. Considering the bounded conditions of displacement field at infinity $z \to \infty$., the displacement field of torsional surface waves in the elastic half-space is assumed as

$$u^{*'}(\xi',z) = D\mathrm{e}^{-k'z}, \tag{5.5.15}$$

where $k' = \sqrt{\xi'^2 - \frac{\rho'\omega'^2}{\mu'}}$. The displacement field can be obtained through the inverse Hankel transform with on Eq. (5.5.15)

$$u'(r, z) = \int_0^{\infty} u^{*\prime}(\xi, z) J_1(\xi r)\xi d\xi. \tag{5.5.16}$$

Inserting the displacement field of torsional surface waves $\boldsymbol{u} = u_\theta(r, z, t)\boldsymbol{e_\theta}$ into the strain tensor

$$\boldsymbol{\varepsilon} = \frac{1}{2}(\nabla \boldsymbol{u} + \boldsymbol{u}\nabla), \tag{5.5.17}$$

leads to

$$\begin{aligned}
\boldsymbol{\varepsilon} &= \frac{1}{2}\left(\frac{\partial}{\partial r}\boldsymbol{e_r} + \frac{\partial}{r\partial\theta}\boldsymbol{e_\theta} + \frac{\partial}{\partial z}\boldsymbol{e_z}\right)u_\theta(r, z, t)\boldsymbol{e_\theta} \\
&+ \frac{1}{2}u_\theta(r, z, t)\boldsymbol{e_\theta}\left(\frac{\partial}{\partial r}\boldsymbol{e_r} + \frac{\partial}{r\partial\theta}\boldsymbol{e_\theta} + \frac{\partial}{\partial z}\boldsymbol{e_z}\right) \\
&= \frac{1}{2}\left(\frac{\partial u_\theta(r, z, t)}{\partial r}\boldsymbol{e_r e_\theta} - \frac{u_\theta(r, z, t)}{r}\boldsymbol{e_\theta e_r} + \frac{\partial u_\theta(r, z, t)}{\partial z}\boldsymbol{e_z e_\theta}\right) \\
&+ \frac{1}{2}\left(\frac{\partial u_\theta(r, z, t)}{\partial r}\boldsymbol{e_\theta e_r} - \frac{u_\theta(r, z, t)}{r}\boldsymbol{e_r e_\theta} + \frac{\partial u_\theta(r, z, t)}{\partial z}\boldsymbol{e_\theta e_z}\right).
\end{aligned} \tag{5.5.18}$$

The non-vanishing components of the strain tensors are

$$\varepsilon_{r\theta} = \frac{1}{2}\left(\frac{\partial u_\theta(r, z, t)}{\partial r} - \frac{u_\theta(r, z, t)}{r}\right), \quad \varepsilon_{\theta z} = \frac{1}{2}\frac{\partial u_\theta(r, z, t)}{\partial z}. \tag{5.5.19}$$

Inserting non-vanishing components of the strain tensors into the constitutive equations

$$\tau_{pq} = \lambda\delta_{pq}\varepsilon_{jj} + 2\mu\varepsilon_{pq}, \tag{5.5.20}$$

leads to the non-vanishing components of the stress tensors

$$\tau_{r\theta} = 2\mu\varepsilon_{r\theta} = \mu\left(\frac{\partial u_\theta(r, z, t)}{\partial r} - \frac{u_\theta(r, z, t)}{r}\right), \tag{5.5.21a}$$

$$\tau_{\theta z} = 2\mu\varepsilon_{\theta z} = \mu\frac{\partial u_\theta(r, z, t)}{\partial z}. \tag{5.5.21b}$$

For the elastic half-space with a covering layer, the free-stress conditions are satisfied at the surface of the covering layer and the continuity conditions of displacements and stresses are satisfied at the interface between the elastic half-space and the covering layer. The boundary conditions at the surface $z = 0$ and the interface $z = h$ are

$$\tau_{z\theta}\big|_{z=0^+}=0,\ u_\theta\big|_{z=h^-}=u'_\theta\big|_{z=h^+},\ \tau_{z\theta}\big|_{z=h^-}=\tau'_{z\theta}\big|_{z=h^+}. \tag{5.5.22}$$

Inserting the solution of displacement and the stress tensor components in the elastic half-space and the covering layer into above boundary conditions leads to

$$\int_0^\infty \left(-\mu k B\mathrm{e}^{-kz}+\mu k C\mathrm{e}^{kz}\right)J_1(\xi r)\xi \mathrm{d}\xi\Bigg|_{z=0^+}=0, \tag{5.5.23a}$$

$$\int_0^\infty \left(B\mathrm{e}^{-kz}+C\mathrm{e}^{kz}\right)J_1(\xi r)\xi \mathrm{d}\xi\Bigg|_{z=h^-}=\int_0^\infty D\mathrm{e}^{-k'z}J_1(\xi' r)\xi' \mathrm{d}\xi'\Bigg|_{z=h^+}, \tag{5.5.23b}$$

$$\int_0^\infty \left(-\mu k B\mathrm{e}^{-kz}+\mu k C\mathrm{e}^{kz}\right)J_1(\xi r)\xi \mathrm{d}\xi\Bigg|_{z=h^-}=\int_0^\infty -\mu' k' D\mathrm{e}^{-k'z}J_1(\xi' r)\xi' \mathrm{d}\xi'\Bigg|_{z=h^+}. \tag{5.5.23c}$$

Considering the region $0 \le r < +\infty$, the angular wavenumber and angular frequency have these relationships

$$\xi=\xi',\ \omega=\omega'. \tag{5.5.24}$$

The boundary conditions at the surface $z = 0$ and the interface $z = h$ can be simplified as

$$-\mu k B\mathrm{e}^{-kz}+\mu k C\mathrm{e}^{kz}\big|_{z=0^+}=0, \tag{5.5.25a}$$

$$B\mathrm{e}^{-kz}+C\mathrm{e}^{kz}\big|_{z=h^-}=D\mathrm{e}^{-k'z}\big|_{z=h^+}, \tag{5.5.25b}$$

$$-\mu k B\mathrm{e}^{-kz}+\mu k C\mathrm{e}^{kz}\big|_{z=h^-}=-\mu' k' D\mathrm{e}^{-k'z}\big|_{z=h^+}. \tag{5.5.25c}$$

Above equations can be written as the following linear homogeneous equation system

$$\begin{bmatrix} -\mu k & \mu k & 0 \\ \mathrm{e}^{-kh} & \mathrm{e}^{kh} & -\mathrm{e}^{-k'h} \\ -\mu k\mathrm{e}^{-kh} & \mu k\mathrm{e}^{kh} & \mu' k'\mathrm{e}^{-k'h} \end{bmatrix}\begin{Bmatrix} B \\ C \\ D \end{Bmatrix}=\begin{Bmatrix} 0 \\ 0 \\ 0 \end{Bmatrix}, \tag{5.5.26}$$

which has a nontrivial solution if and only if the determinant of the matrix is zero

$$\begin{vmatrix} -\mu k & \mu k & 0 \\ \mathrm{e}^{-kh} & \mathrm{e}^{kh} & -\mathrm{e}^{-k'h} \\ -\mu k\mathrm{e}^{-kh} & \mu k\mathrm{e}^{kh} & \mu' k'\mathrm{e}^{-k'h} \end{vmatrix}=0. \tag{5.5.27}$$

The dispersion relation of torsional surface waves in an elastic half-space with a covering layer can be explicitly expressed as

$$e^{2h\sqrt{\xi^2-\frac{\rho}{\mu}\omega^2}} = \frac{\mu\sqrt{\xi^2-\frac{\rho}{\mu}\omega^2}-\mu'\sqrt{\xi^2-\frac{\rho'}{\mu'}\omega^2}}{\mu\sqrt{\xi^2-\frac{\rho}{\mu}\omega^2}+\mu'\sqrt{\xi^2-\frac{\rho'}{\mu'}\omega^2}}. \tag{5.5.28}$$

Define the wave speeds of torsional surface waves and shear waves in the elastic half-space and the covering layer

$$c=\frac{\omega}{\xi},\ c_t=\sqrt{\frac{\mu}{\rho}},\ c_t'=\sqrt{\frac{\mu'}{\rho'}}. \tag{5.5.29}$$

(1) In the case of $c < c_t < c_t'$, inserting (5.5.29) into (5.5.28) leads to

$$e^{2\xi h\sqrt{1-\left(\frac{c}{c_t}\right)^2}} = \frac{\mu\sqrt{1-\left(\frac{c}{c_t}\right)^2}-\mu'\sqrt{1-\left(\frac{c}{c_t'}\right)^2}}{\mu\sqrt{1-\left(\frac{c}{c_t}\right)^2}+\mu'\sqrt{1-\left(\frac{c}{c_t'}\right)^2}}. \tag{5.5.30}$$

As the left and right sides of (5.5.30) are positive and negative, respectively, (5.5.30) can't be true.

(2) In the case of $c_t < c < c_t'$, inserting (5.5.29) into (5.5.28) leads to

$$\cos\left[2\xi h\sqrt{\left(\frac{c}{c_t}\right)^2-1}\right]+i\sin\left[2\xi h\sqrt{\left(\frac{c}{c_t}\right)^2-1}\right] = \frac{i\mu\sqrt{\left(\frac{c}{c_t}\right)^2-1}-\mu'\sqrt{1-\left(\frac{c}{c_t'}\right)^2}}{i\mu\sqrt{\left(\frac{c}{c_t}\right)^2-1}+\mu'\sqrt{1-\left(\frac{c}{c_t'}\right)^2}}. \tag{5.5.31}$$

Separating the real part and the imaginary part of (5.5.31) leads to

$$\cos\left[2\xi h\sqrt{\left(\frac{c}{c_t}\right)^2-1}\right] = \frac{\mu^2\left[\left(\frac{c}{c_t}\right)^2-1\right]-\mu'^2\left[1-\left(\frac{c}{c_t'}\right)^2\right]}{\mu^2\left[\left(\frac{c}{c_t}\right)^2-1\right]+\mu'^2\left[1-\left(\frac{c}{c_t'}\right)^2\right]}, \tag{5.5.32a}$$

$$\sin\left[2\xi h\sqrt{\left(\frac{c}{c_t}\right)^2-1}\right]=\frac{2\mu\mu'\left[\left(\frac{c}{c_t}\right)^2-1\right]\left[1-\left(\frac{c}{c_t'}\right)^2\right]}{\mu^2\left[\left(\frac{c}{c_t}\right)^2-1\right]+\mu'^2\left[1-\left(\frac{c}{c_t'}\right)^2\right]}. \tag{5.5.32b}$$

Combination of (5.5.32a) and (5.5.32b) leads to

$$\tan\left[\xi h\sqrt{\left(\frac{c}{c_t}\right)^2-1}\right]=\frac{\mu'}{\mu}\sqrt{\frac{1-\left(\frac{c}{c_t'}\right)^2}{\left(\frac{c}{c_t}\right)^2-1}}. \tag{5.5.33}$$

(3) In the case of $c_t' \le c \le c_t$, inserting (5.5.29) into (5.5.28) leads to

$$e^{2\xi h\sqrt{1-\left(\frac{c}{c_t}\right)^2}}=\frac{\mu\sqrt{1-\left(\frac{c}{c_t}\right)^2}-i\mu'\sqrt{\left(\frac{c}{c_t'}\right)^2-1}}{\mu\sqrt{1-\left(\frac{c}{c_t}\right)^2}+i\mu'\sqrt{\left(\frac{c}{c_t'}\right)^2-1}}. \tag{5.5.34}$$

Separating the real part and the imaginary part of (5.5.34) leads to

$$e^{2\xi h\sqrt{1-\left(\frac{c}{c_t}\right)^2}}=\frac{\mu^2\left[1-\left(\frac{c}{c_t}\right)^2\right]-\mu'^2\left[\left(\frac{c}{c_t'}\right)^2-1\right]}{\mu^2\left[1-\left(\frac{c}{c_t}\right)^2\right]+\mu'^2\left[\left(\frac{c}{c_t'}\right)^2-1\right]}, \tag{5.5.35a}$$

$$0=\frac{2\mu\mu'\sqrt{1-\left(\frac{c}{c_t}\right)^2}\sqrt{\left(\frac{c}{c_t'}\right)^2-1}}{\mu^2\left[1-\left(\frac{c}{c_t}\right)^2\right]+\mu'^2\left[\left(\frac{c}{c_t'}\right)^2-1\right]}. \tag{5.5.35b}$$

From Eq. (5.5.35b), we get

$$c_t' = c \le c_t, \tag{5.5.36a}$$

or

$$c_t' \le c = c_t. \tag{5.5.36b}$$

Inserting (5.5.36a) and (5.5.36b) into (5.5.35a) leads to

$$c_t' = c = c_t. \tag{5.5.37}$$

Equation (5.5.37) shows that the shear wave speeds in the covering layer and the elastic half-space are identical. This is contradiction with the hypothesis, and thus, torsional surface waves cannot exist. Therefore, $c_t < c_t'$ is necessary, which means that torsional surface waves can only exist when the covering layer is wave-thinner medium and the elastic half-space is wave-denser medium.

(4) In the case of $c_t < c_t' < c$, inserting (5.5.29) into (5.5.28) leads to

$$\cos\left[2\xi h\sqrt{\left(\frac{c}{c_t}\right)^2-1}\right]+i\sin\left[2\xi h\sqrt{\left(\frac{c}{c_t}\right)^2-1}\right]$$
$$=\frac{i\mu\sqrt{\left(\frac{c}{c_t}\right)^2-1}-i\mu'\sqrt{\left(\frac{c}{c_t'}\right)^2-1}}{i\mu\sqrt{\left(\frac{c}{c_t}\right)^2-1}+i\mu'\sqrt{\left(\frac{c}{c_t'}\right)^2-1}}. \tag{5.5.38}$$

Separating the real part and the imaginary part of (5.5.38) leads to

$$\cos\left[2\xi h\sqrt{\left(\frac{c}{c_t}\right)^2-1}\right]=\frac{\mu\sqrt{\left(\frac{c}{c_t}\right)^2-1}-\mu'\sqrt{\left(\frac{c}{c_t'}\right)^2-1}}{\mu\sqrt{\left(\frac{c}{c_t}\right)^2-1}+\mu'\sqrt{\left(\frac{c}{c_t'}\right)^2-1}}, \tag{5.5.39a}$$

$$\sin\left[2\xi h\sqrt{\left(\frac{c}{c_t}\right)^2-1}\right]=0. \tag{5.5.39b}$$

Combination of (5.5.39b) and (5.5.39a) leads to

$$c_t' = c_t = c. \tag{5.5.40}$$

Equation (5.5.40) shows that the shear wave speeds in the covering layer and the elastic half-space are identical, which is also contradiction with the hypothesis. Therefore, $c_t < c < c_t'$ is necessary, which means that the velocity of the torsional surface wave is between the velocity of the covering layer and that of elastic half-space.

Choose $\frac{\mu}{\mu'} = 0.3$ and $\frac{\rho}{\rho'} = 0.6$. Based on (5.5.30), (5.5.33). and (5.5.40), the dispersion curves of torsional surface waves are plotted in Fig. 5.28, where the dimensionless velocity $\frac{c}{c_t}$ and the dimensionless wave number ξh are used.

After the investigation of torsional surface waves in homogeneous elastic half-space, we can further discuss the torsional surface waves in an inhomogeneous elastic half-space with a covering layer. Material parameters that influence the propagation properties of torsional surface waves mainly are the shear modulus μ and mass density ρ. A variety of inhomogeneities of material parameters exist in the actual problem. Here, the linear distribution of material parameters is discussed as example. Consider

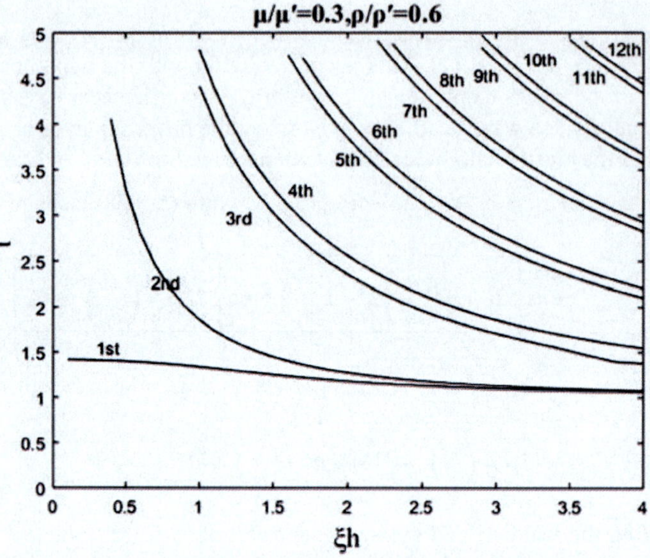

Fig. 5.28 Dispersion curves of torsional surface wave in the elastic half-space with a covering layer

a homogeneous covering layer of thickness h, which locates over a vertically inhomogeneous elastic half-space. The origin of the cylindrical coordinate system locates at the interface, and the positive z-axis is downward, as shown in Fig. 5.29. Both mass density and shear modulus are considered to be inhomogeneous. The variation of shear modulus and mass density is assumed as follows:

$\mu = \mu_0$, $\rho = \rho_0$, for the covering layer;
$\mu = \mu_1(1 + az)$, $\rho = \rho_1(1 + bz)$, for the inhomogeneous elastic half-space.

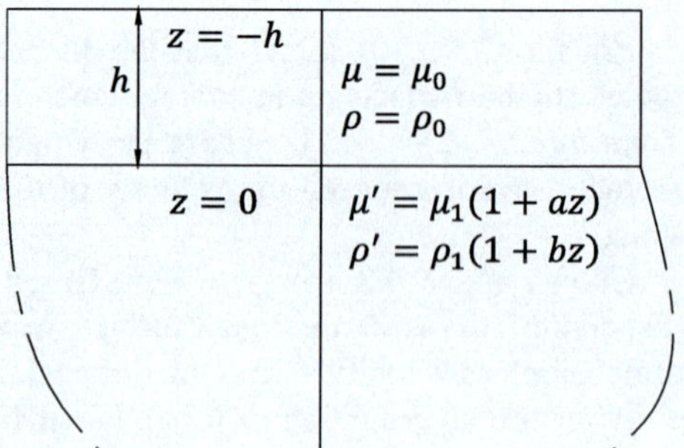

Fig. 5.29 A vertically inhomogeneous elastic half-space with a homogeneous covering layer of thickness ***h***

In the above equations, μ and ρ are the shear modulus and the mass density of the media, respectively. a and b are constants having dimensions of inverse of length.

For a torsional surface wave which propagates in the radial direction, the equation of motion is

$$\frac{\partial}{\partial r}\sigma_{r\theta} + \frac{\partial}{\partial r}\sigma_{z\theta} + \frac{2}{r}\sigma_{r\theta} = \rho(z)\frac{\partial^2 v}{\partial t^2}, \tag{5.5.41}$$

where $v(r, z, t)$ is the displacement along the θ direction. For an elastic medium, the stresses are related to the displacement component $v(r, z, t)$ by

$$\sigma_{r\theta} = \mu(z)\left(\frac{\partial v}{\partial r} - \frac{v}{r}\right), \sigma_{z\theta} = \mu(z)\frac{\partial v}{\partial z}. \tag{5.5.42}$$

Inserting (5.5.41) into (5.5.42) leads to

$$\mu(z)\left(\frac{\partial^2}{\partial r^2} + \frac{1}{r}\frac{\partial}{\partial r} - \frac{1}{r^2}\right)v + \frac{\partial}{\partial z}\left(\mu(z)\frac{\partial v}{\partial z}\right) = \rho(z)\frac{\partial^2 v}{\partial t^2}. \tag{5.5.43}$$

Assume a solution of (5.5.43) of the form

$$v = V(z)J_1(\xi r)e^{i\omega t}. \tag{5.5.44}$$

Then, $V(z)$. satisfies

$$V''(z) + \frac{\mu'(z)}{\mu(z)}V'(z) - \xi^2\left(1 - \frac{c^2}{c_s^2}\right)V(z) = 0, \tag{5.5.45}$$

where $c = \omega/\xi$ and $c_s = (\mu/\rho)^{1/2}$. $J_1(\xi r)$ is the Bessel function of first kind.

For the homogeneous covering layer

$$\mu = \mu_0, \rho = \rho_0. \tag{5.5.46}$$

According to Eq. (5.5.45), the displacement equation for the upper covering layer takes the form

$$V''(z) - \xi^2\left(1 - \frac{c^2}{c_0^2}\right)V(z) = 0, \tag{5.5.47}$$

which $c_0 = (\mu_0/\rho_0)^{1/2}$. The solution of (5.5.47) is

$$V(z) = A_1 \mathrm{e}^{\left(1-\frac{c^2}{c_0^2}\right)^{1/2}\xi z} + A_2 \mathrm{e}^{-\left(1-\frac{c^2}{c_0^2}\right)^{1/2}\xi z}. \tag{5.5.48}$$

Hence, the displacement in the homogeneous covering layer is

$$v = \left[A_1 e^{\left(1-\frac{c^2}{c_0^2}\right)^{1/2} \xi z} + A_2 e^{-\left(1-\frac{c^2}{c_0^2}\right)^{1/2} \xi z} \right] J_1(\xi r) e^{i\omega t}. \tag{5.5.49}$$

For the inhomogeneous elastic half-space

$$\mu = \mu_1(1 + az), \rho = \rho_1(1 + bz), \tag{5.5.50}$$

and Eq. (5.5.45) takes the form

$$V''(z) + \frac{a}{1 + az} V'(z) - \xi^2 \left[1 - \frac{c^2(1 + az)}{c_1^2(1 + bz)} \right] V(z) = 0, \tag{5.5.51}$$

where $c_1 = (\mu_1/\rho_1)^{1/2}$.

Inserting $V(z) = \phi(z)/(1 + az)^{1/2}$ into (5.5.51) leads to

$$\phi''(z) + \left\{ \frac{a^2}{4(1 + az)^2} - \xi^2 \left[1 - \frac{c^2(1 + az)}{c_1^2(1 + bz)} \right] \right\} \phi(z) = 0. \tag{5.5.52}$$

Inserting the dimensionless quantities $\gamma = \left[1 - c^2 b/c_1^2 a\right]^{1/2}$ and $\eta = 2\gamma\xi(1 + az)/a$ into (5.5.52) leads to the Whittaker equation

$$\phi''(\eta) + \left(-\frac{1}{4} + \frac{\frac{P}{2}}{\eta} + \frac{\frac{1}{4} - 0}{\eta^2} \right) \phi(\eta) = 0, \tag{5.5.53}$$

where

$$P = \frac{\omega^2(a - b)}{c_1^2 a^2 \gamma \xi}.$$

The solution of Eq. (5.5.53) satisfying the condition $\lim\limits_{z\to\infty} V(z) \to 0$ and $\lim\limits_{\eta\to\infty} \phi(\eta) \to 0$ is

$$\phi(\eta) = D_1 W_{P/2,0}(\eta), \tag{5.5.54}$$

where $W_{P/2,0}(\eta)$ is the Whittaker function.

$$W_{P/2,0}(\eta) = \frac{e^{-\frac{\eta}{2}} \eta^{\frac{P}{2}}}{\Gamma\left(\frac{1}{2} - \frac{\eta}{2}\right)} \int\limits_0^\infty e^{-\varepsilon} \varepsilon^{-\frac{\eta}{2}-\frac{1}{2}} \left(1 + \frac{\varepsilon}{\eta} \right)^{\frac{P}{2}-\frac{1}{2}} d\varepsilon. \tag{5.5.55}$$

Hence, the displacement component v_1 in the inhomogeneous elastic half-space is

$$v_1 = \frac{D_1 W_{P/2,0}\left[2\gamma\xi(1+az)/a\right]}{(1+az)^{1/2}} J_1(\xi r)e^{i\omega t}. \tag{5.5.56}$$

Assume the interface between homogeneous covering layer and the inhomogeneous elastic half-space is perfect interface. The following surface and interface conditions must be satisfied

$$v|_{z=0^-} = v_1|_{z=0^+}, \tag{5.5.57a}$$

$$\sigma_{z\theta}|_{z=0^-} = \sigma_{z\theta}|_{z=0^+}, \tag{5.5.57b}$$

$$\sigma_{z\theta}|_{z=-h} = 0. \tag{5.5.57c}$$

Expansion of the Whittaker function up to the linear terms leads to

$$W_{P/2,0}(\eta) \sim e^{-\frac{\eta}{2}}\eta^{\frac{P}{2}}(1+\eta A), \tag{5.5.58}$$

where $A = \frac{1-P}{2}$. Inserting this expression into above surface and interface conditions leads to

$$A_1 + A_2 = D_1\left(\frac{2\gamma\xi}{a}\right)^{1/2} e^{-\gamma\xi/a}\left(1 + \frac{2\gamma\xi A}{a}\right), \tag{5.5.59a}$$

$$\begin{aligned} &\mu_0\xi\left(1-\frac{c^2}{c_0^2}\right)^{1/2}(A_1 - A_2) \\ &= D_1\mu_1\left[2\gamma\xi A - \gamma\xi\left(1+\frac{2\gamma\xi A}{a}\right)\right]\left(\frac{2\gamma\xi}{a}\right)^{1/2} e^{-\gamma\xi/a}, \end{aligned} \tag{5.5.59b}$$

$$A_1 e^{-\left(1-\frac{c^2}{c_0^2}\right)^{1/2}\xi h} + A_2 e^{\left(1-\frac{c^2}{c_0^2}\right)^{1/2}\xi h} = 0. \tag{5.5.59c}$$

Equation (5.5.59) is a linear algebraic equation with respect to A_1, A_2 and D_1. The existence of nonzero solutions requires that the determinant of the coefficients be zero

$$\frac{\mu_1\gamma}{\mu_0}\left[1 - \frac{2A}{1+2\gamma\xi A/a}\right]\frac{1}{(c^2/c_0^2-1)^{1/2}} = \tan\left[\xi h(c^2/c_0^2-1)^{1/2}\right]. \tag{5.5.60}$$

Equation (5.5.60) gives the dispersion equation of torsional surface waves in a homogeneous covering layer over a vertically inhomogeneous half-space. This equation is a transcendental equation. Because of the periodicity of the tangent function, the equation has an infinite number of solutions, which means that torsional surface waves have an infinite number of modes.

Case 1: When $b \to 0$, the half-space has constant density. Then, $\gamma = 1$ and $P = c^2\xi/c_0^2 a$. Based on these results, Eq. (5.5.60) reduces to

$$\frac{\mu_1}{\mu_0}\left[1 - \frac{2A}{1 + 2\gamma\xi A/a}\right]\frac{1}{\left(c^2/c_0^2 - 1\right)^{1/2}} = \tan\left[\xi h\left(c^2/c_0^2 - 1\right)^{1/2}\right]. \tag{5.5.61}$$

Case 2: When $a \to 0$ and $b \to 0$, the layer and the half-space have both constant density and rigidity. Then, Eq. (5.5.60) reduces to

$$\frac{\mu_1}{\mu_0}\frac{\left(1 - c^2/c_1^2\right)^{1/2}}{\left(c^2/c_0^2 - 1\right)^{1/2}} = \tan\left[\xi h\left(c^2/c_0^2 - 1\right)^{1/2}\right]. \tag{5.5.62}$$

Equation (5.5.62) is exactly the dispersion equation of torsional surface waves in a homogeneous covering layer over a homogeneous half-space, i.e. Eq. (5.5.33). Further, it is noted that Eq. (5.5.62) shares the same mathematical form with the well-known equation for Love waves in a homogeneous layer over a homogeneous half-space, i.e. Eq. (5.3.17). This shows that in a layered isotropic homogeneous medium, torsional surface waves reduce gradually to Love waves as the distance from the origin increases.

Case 3: In the absence of the upper layer ($h \to 0$), Eq. (5.5.60) reduces to

$$\gamma\left[1 - \frac{2A}{1 + 2\gamma\xi A/a}\right] = 0. \tag{5.5.63}$$

This gives either

$$\frac{c}{c_1} = \left(\frac{a}{b}\right)^{1/2}, \tag{5.5.64}$$

or

$$\left[1 - \frac{2A}{1 + 2\gamma\xi A/a}\right] = 0. \tag{5.5.65}$$

Expansion of Eq. (5.5.65) leads to

$$X_1 C^3 + X_2 C^2 + X_3 C + 0, \tag{5.5.66}$$

where

$$C = \left(\frac{c}{c_1}\right)^{1/2},$$

$$X_1 = \frac{\xi^2 b}{a^2}\left(1 - \frac{b}{a}\right)^2,$$

$$X_2 = \left(\frac{2b}{a} - 1\right)^2 - \frac{\xi^2}{a^2}\left(1 - \frac{b}{a}\right)^2,$$
$$X_3 = 2\left(1 - \frac{2b}{a}\right).$$

Let

$$G = X_1^2 - \frac{X_1X_2X_3}{3} + 2\left(\frac{X_2}{3}\right)^3, \quad H_1 = \frac{X_1X_3}{3} - \left(\frac{X_2}{3}\right)^3. \tag{5.5.67}$$

Based on Cardano's formula, there are several situations as follows.

When $G^2 + 4H_1^3 > 0$, Eq. (5.5.66) leads to no positive real value of C. Hence, the wave velocity can be obtained from Eq. (5.5.64). i.e.

$$c = \left(\frac{a}{b}\right)^{1/2} c_1. \tag{5.5.68}$$

When $G^2 + 4H_1^3 < 0$, Eq. (5.5.66) leads to two different positive real values of C. Moreover, Eq. (5.5.64) also leads to one. Hence, there are totally three torsional surface waves. When the shear modulus μ and the mass density ρ are of same change law, i.e. $a = b$, and there is no covering layer at the same time; then, the torsional surface wave velocity $c = c_1$. This implies that torsional surface waves degenerate into shear waves.

When $b = 0$ and $a \neq 0$, the elastic half-space degenerates into the Gibson half-space. The wave velocity of torsional surface wave is

$$c = c_1\left(\frac{1}{\xi/a - 1}\right)^{1/2}. \tag{5.5.69}$$

When $a \to 0$ and $b \to 0$, the inhomogeneous elastic half-space degenerates into a homogeneous elastic half-space. Equation (5.5.63) shows that torsional surface waves do not exist.

When $a/\xi = 0.01$ and $b/\xi = 0.001$, the dimensionless wave velocity c/c_0 has been computed numerically for different values of dimensionless wavenumber ξh. Figure 5.30 shows the dispersion curves of torsional surface waves when the density of elastic half-space is constant (i.e. $b = 0$). Curves 1 and 2 show that the variation of the density in the elastic half-space leads to evident influence on the wave velocity of torsional surface waves. In order to compare the propagation of torsional surface waves with that of Love waves, curve 3 in Fig. 5.29 shows the wave velocity of Love wave calculated from Eq. (5.5.62). It is observed that the wave velocity of torsional surface waves in the inhomogeneous medium is larger than that of Love waves in the homogeneous medium. In the absence of the covering layer over the inhomogeneous half-space, Eqs. (5.5.64) and (5.5.65) give the wave velocity equation torsional surface waves. Equation (5.5.64) gives constant velocity whose amplitude

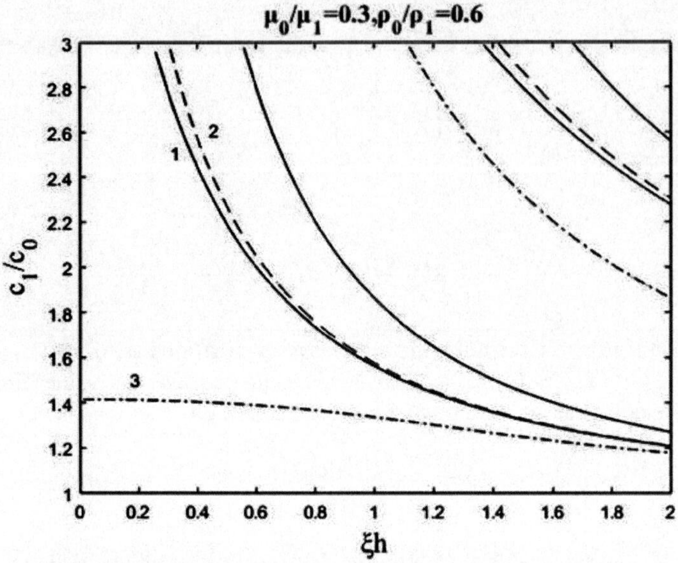

Fig. 5.30 Dispersion curves of torsional surface waves in the inhomogeneous elastic half-space with a homogeneous covering layer (— : $\boldsymbol{a/k} = 0.01, \boldsymbol{b/k} = 0.001$; – – – : $\boldsymbol{a/k} = 0.01, \boldsymbol{b/k} = 0$; –. – .– : $\boldsymbol{a/k} = \boldsymbol{b/k} = 0$)

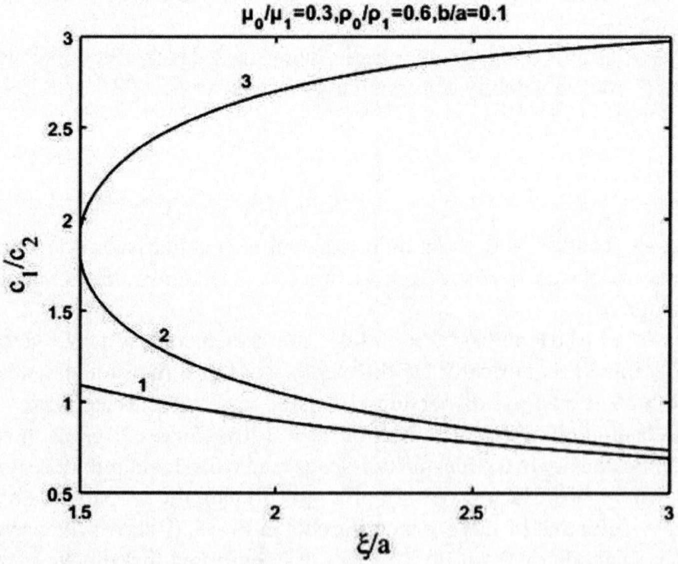

Fig. 5.31 Dispersion curves of torsional surface waves in the vertically inhomogeneous elastic half-space without a homogeneous covering layer

is dependent on the ratio of shear modulus to mass density. It is observed that Eq. (5.5.66) gives two real roots for $(c/c_1)^2$ and for small values of ξ/a. The curves 1 and 2 are shown in Fig. 5.31. Curve 3 is obtained by Eq. (5.5.64). It is observed that there are three torsional surface waves propagating in the inhomogeneous half-space for small values of ξ/a.

Chapter 6
Guided Waves

A variety of structural, such as rods, beams, plates, tubes and spherical shells, are often encountered in engineering application. A common feature of this type of structure is that there are two parallel surfaces or a cross section with a closed boundary. Since there are two surfaces, the waves located between the two surfaces will undergo multiple reflections at two parallel surfaces, and these round-trip waves, sometimes, also called forward and backward waves, will produce complex interference phenomenon. Due to the existence of characteristic geometric dimensions, such as "layer thickness" or "wall thickness", the velocity of the waves propagating between two surfaces will depend on the frequency of the wave, resulting in so-called geometric dispersion of elastic waves. An infinite plate has two parallel interfaces. The P wave, SV wave and SH wave in the plate will reflect back and forth on the two parallel surfaces and finally travel in a direction parallel to the plate surface. The parallel boundaries restrict the propagation of waves in the plate. Such a system is often called an elastic plate waveguide. In addition, cylindrical rods, cylindrical or spherical shells and layered structure are also typical elastic waveguides. Some literatures call such a waveguide with two parallel surfaces or a waveguide with a cross section of closed boundary as a closed waveguide and call a half-space with a free surface or an infinite space with a cylindrical hole as an open waveguide. This chapter mainly introduces several common closed waveguides, including slab waveguides, cylindrical waveguides, tube waveguides and spherical shell waveguides, establishes the dispersion equations of guided waves in these waveguides and discusses the propagation characteristics of guided waves.

6.1 Flexural Waves in Beams

There are a variety of elastic theories of beams, wherein Euler–Bernoulli beam theory and Timoshenko beam theory are often used. Both of these two beam theories assume that the cross section of the beam remains flat during bending (called the assumption

© Science Press 2022

P. Wei, *Theory of Elastic Waves*,
https://doi.org/10.1007/978-981-19-5662-1_6

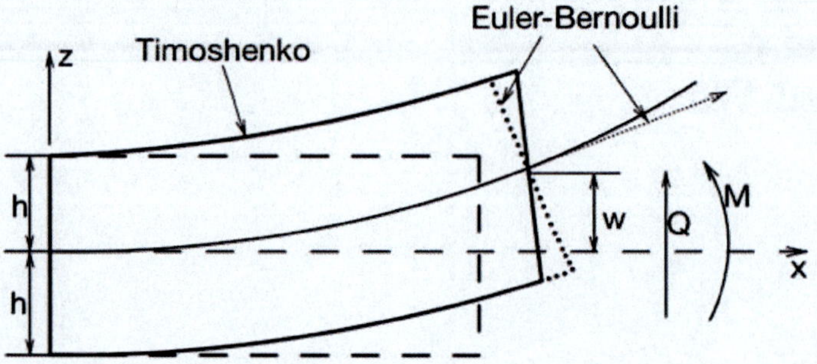

Fig. 6.1 Sketch of the cross-section deformation assumption of Timoshenko beam $\left(\varphi(x) \neq \frac{\partial w}{\partial x}\right)$ and Euler–Bernoulli beam $\left(\varphi(x) = \frac{\partial w}{\partial x}\right)$

of a flat cross section). The difference is that the Euler–Bernoulli beam theory believes that the rotation of the cross section is only caused by the bending deformation, and the cross section always remains perpendicular to the axis of the beam during bending deformation process. The Timoshenko beam theory believes that the rotation of the cross section results from not only the bending deformation but also the shear deformation. Since the shear deformation generates additional rotation (i.e. the shear angle), thus, cross section is no longer perpendicular to the axis of the beam. The differences of two kinds of beam theories on the assumption of the cross-section deformation are shown in Fig. 6.1.

Establish a coordinate system as shown in Fig. 6.1. The axis of the beam is along the x-axis, the height of the beam is along the z-axis, and the thickness of the beam is along the y-axis. Under the assumption of a flat cross section, the cross section of the beam only rotates around the neutral axis after deformation, and the rotation angle is represented by $\varphi(x)$. The axis of the beam becomes a curve from a straight line after bending deformation and can be represented by $w(x)$(referred to as the deflection curve). The displacement at any point (x, y, z) on the cross section of the beam can be expressed as

$$u_x(x, y, z) = -z\varphi(x), u_y(x, y, z) = 0, u_z(x, y, z) = w(x), \tag{6.1.1}$$

where u_x, u_y and u_z are the three-dimensional coordinate components of the displacement vector, φ is the rotation angle of the cross section, and $w(x)$ is the deflection of a point on the axis of the beam.

If the shear deformation of the beam is not considered, and only the bending deformation of the beam is taken into considered, then

$$\varphi = \frac{\partial w}{\partial x}. \tag{6.1.2}$$

Such a beam is generally called Euler–Bernoulli beam. If the bending deformation and shear deformation of the beam are both considered at the same time, the total rotation angle can be expressed as

$$\varphi = \frac{\partial w}{\partial x} - \gamma. \tag{6.1.3}$$

Among them, $\frac{\partial w}{\partial x}$ is the rotation angle caused by bending deformation, and γ is the rotation angle caused by shear deformation. The relationship between the three rotation angles is shown in Fig. 6.2. Such a beam is called Timoshenko beam. It can be seen that the fundamental difference between the two beam theories lies in the assumption of beam deformation. The Euler–Bernoulli beam ignores the shear deformation of the beam and is suitable for slender beams where the bending deformation is absolutely dominant. The Timoshenko beam considers the bending deformation and shear deformation of the beam at the same time and is suitable for short and thick beams that the shear deformation is not too small.

Under the framework of Timoshenko beam theory, the normal strain, normal stress, shear strain and shear stress at any point on the cross section of the beam are

$$\varepsilon_x = -z\frac{\partial \varphi}{\partial x}, \quad \sigma_x = E\varepsilon_x, \tag{6.1.4a}$$

$$\gamma = \frac{\partial w}{\partial x} - \varphi, \quad \tau = G\gamma. \tag{6.1.4b}$$

The internal forces on the cross section, namely the bending moment and the shear force, are, respectively,

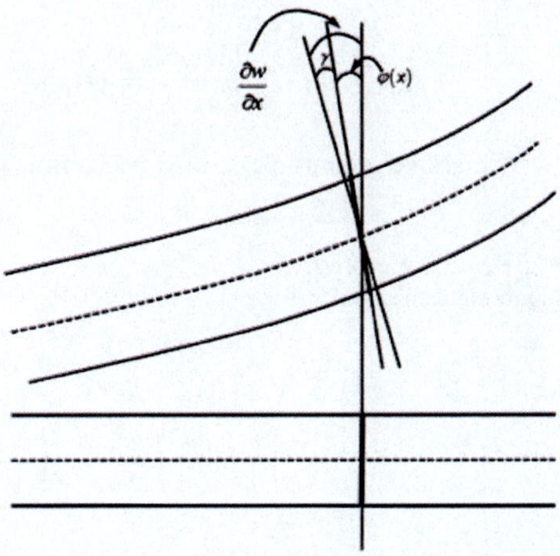

Fig. 6.2 Sketch of decomposition of rotation of cross section of Timoshenko beam

$$M = \int_A \sigma_x z \mathrm{d}A = -EI\frac{\partial \varphi}{\partial x}, \quad Q = \int_A \tau \mathrm{d}A = GA\left(\frac{\partial w}{\partial x} - \varphi\right). \tag{6.1.5}$$

The cross section of the beam can be various, and the distribution of the shear stress on the cross section is generally not uniform. In order to consider the influence of this inhomogeneity, suppose $Q = \int_A \hat{\tau}(z)\mathrm{d}A = A\kappa\tau(x)$, among them, $\tau(x)$ is the shear stress on the cross section calculated by Timoshenko beam theory, and it is uniformly distributed across the entire cross section. $\hat{\tau}(z)$ is the true shear stress on the cross section of the beam, and it is generally non-uniformly distributed on the cross section of the beam. κ is called the shear correction coefficient of cross section, and $\kappa\tau(x)$ is the average shear stress on the actual cross section. The value of the shear correction coefficient of cross section is as follows

$\kappa = \frac{10(1+v)}{12+11v}$ for rectangular cross sections;

$\kappa = \frac{6(1+v)}{7+6v}$ for circular cross sections,

where v is the Poisson's ratio of material.

Take a micro-element of the beam to perform force analysis, as shown in Fig. 6.3. Here, the internal forces of the beam only consider the bending moment and shear force. There is a distributed load $q(x)$ on the upper surface of the beam. We will discuss the governing equation of deformation, which is also the balance equation of the beam. Refer to Fig. 6.3, the vertical force balance condition and the moment balance condition around the neutral axis of the micro-element are

$$-M(x) + [M(x) + \mathrm{d}M(x)] - [Q(x) + \mathrm{d}Q(x)]\mathrm{d}x - \frac{1}{2}q(x)^2\mathrm{d}x^2 = 0, \tag{6.1.6a}$$

$$[Q(x) + \mathrm{d}Q(x)] - Q(x) - q(x)\mathrm{d}x = 0. \tag{6.1.6b}$$

Ignoring the high-order infinitesimal quantities, we get

$$\frac{\partial M(x)}{\partial x} = Q(x), \quad \frac{\partial Q(x)}{\partial x} = q(x). \tag{6.1.7}$$

The above equation can also be combined and written as

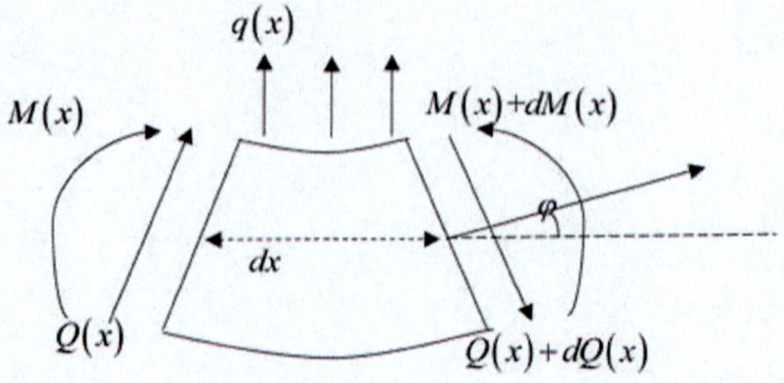

Fig. 6.3 Force analysis of micro-element of beam

$$\frac{\partial^2 M(x)}{\partial x^2} = q(x). \tag{6.1.8}$$

Therefore, the deformation governing equation can be expressed by the following ordinary differential equations

$$\frac{\mathrm{d}^2}{\mathrm{d}x^2}\left(EI\frac{\mathrm{d}\varphi}{\mathrm{d}x}\right) = q(x), \tag{6.1.9a}$$

$$\frac{\mathrm{d}w}{\mathrm{d}x} = \varphi - \frac{1}{\kappa AG}\frac{\mathrm{d}}{\mathrm{d}x}\left(EI\frac{\mathrm{d}\varphi}{\mathrm{d}x}\right). \tag{6.1.9b}$$

The above equations are a set of coupled equations about the unknown quantities $\varphi(x)$ and $w(x)$. Combine the above two equations and eliminate the unknown function $\varphi(x)$, we get

$$EI\frac{\mathrm{d}^4 w}{\mathrm{d}x^4} = q(x) - \frac{EI}{\kappa AG}\frac{\mathrm{d}^2 q}{\mathrm{d}x^2}. \tag{6.1.10}$$

The above equation is exactly the deformation governing equation of Timoshenko beam with a distributed load on the surface.

When ignoring the shear deformation, namely $\varphi = \frac{\partial w}{\partial x}$, the last term of the above governing equation disappears, thus, the beam deformation governing equation is simplified to

$$EI\frac{\mathrm{d}^4 w}{\mathrm{d}x^4} = q(x). \tag{6.1.11}$$

This equation is exactly the deformation governing equation of Euler–Bernoulli beam. It can be seen that the deformation governing equation of Euler–Bernoulli beam is the degradation of the deformation governing equation of Timoshenko beam.

Under the action of dynamic load, the deformation of a beam is not only a function of coordinates, but also a function of time. The displacement of any point on the cross section can be expressed as

$$u_x(x, y, z, t) = -z\varphi(x, t), \quad u_y(x, y, z, t) = 0, \quad u_z(x, y, z, t) = w(x, t). \tag{6.1.12}$$

Under the dynamic deformation of the beam, there is not only a vertical inertia force existing in the cross section, but also an inertia moment around the neutral axis existing. Therefore, the governing equation can be expressed as

$$\rho A\frac{\partial^2 w}{\partial t^2} - q(x, t) = \frac{\partial}{\partial x}\left[\kappa AG\left(\frac{\partial w}{\partial x} - \varphi\right)\right], \tag{6.1.13a}$$

$$\rho I\frac{\partial^2\varphi}{\partial t^2}=\frac{\partial}{\partial x}\left(EI\frac{\partial\varphi}{\partial x}\right)+\kappa AG\left(\frac{\partial w}{\partial x}-\varphi\right). \tag{6.1.13b}$$

Among them, ρ is the mass bulk density of the beam, A is the cross-sectional area, E is the modulus of elasticity, G is the shear modulus and I is the geometrical moment of inertia of the cross section of the beam.

For beams of uniform cross section made of the same material, the above two equations can be combined into

$$\begin{aligned}&EI\frac{\partial^4 w}{\partial x^4}+\rho A\frac{\partial^2 w}{\partial t^2}-\left(\rho I+\frac{\rho EI}{\kappa G}\right)\frac{\partial^4 w}{\partial x^2\partial t^2}+\frac{\rho^2 I}{\kappa G}\frac{\partial^4 w}{\partial t^4}\\&=q(x,t)+\frac{\rho I}{\kappa AG}\frac{\partial^2 q}{\partial t^2}-\frac{EI}{\kappa AG}\frac{\partial^2 q}{\partial x^2}.\end{aligned} \tag{6.1.14}$$

In the above equation, if $G\to\infty$, it indicates that the shear stiffness of the material is very large, and thus, the shear deformation can be ignored. If $I\to 0$, it indicates that the geometric moment of inertia of the cross section is very small, and the inertia effect of the rotation can be ignored. If $G\to\infty$, and $I\to 0$, the above equation degenerates to the deformation governing equation of Euler–Bernoulli beam, namely

$$EI\frac{\partial^4 w}{\partial x^4}+\rho A\frac{\partial^2 w}{\partial t^2}=q(x,t). \tag{6.1.15}$$

If there is an axial force on the cross section of the beam, the axis of the beam will elongate or shorten, so that the displacement of any point on the cross section of the beam can be expressed as

$$u_x(x,y,z,t)=u_0(x,t)-z\varphi(x,t), \tag{6.1.16a}$$

$$u_y(x,y,z,t)=0, \tag{6.1.16b}$$

$$u_z(x,y,z,t)=\omega(x,t), \tag{6.1.16c}$$

where u_0 is the additional displacement of the beam along the x direction. The force analysis of the micro-element of beam considering the axis forces is shown in Fig. 6.4. The vertical component of the axial force N on the left side of the beam micro-element is expressed as

$$N_y=N\sin\varphi\approx N\varphi. \tag{6.1.17}$$

The vertical component of the axial force $(N+dN)$ on the right side of the beam element is expressed as

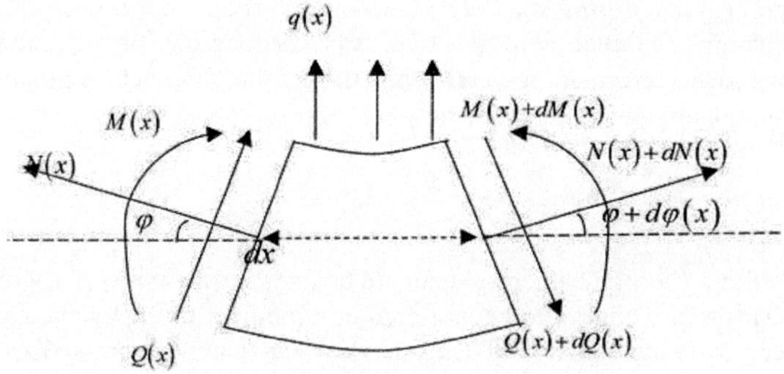

Fig. 6.4 Force analysis of micro-element of beam considering axial force

$$N_y = (N + \mathrm{d}N)\sin(\varphi + \mathrm{d}\varphi) \approx (N + \mathrm{d}N)(\varphi + \mathrm{d}\varphi)$$
$$= N\varphi + N\mathrm{d}\varphi + \varphi\mathrm{d}N + dNd\varphi. \tag{6.1.18}$$

When the axial force is constant along the axial direction,

$$N_y = N\ \sin(\varphi + \mathrm{d}\varphi) \approx N(\varphi + \mathrm{d}\varphi) = N\varphi + N\mathrm{d}\varphi, \tag{6.1.19}$$

then the governing equation of Timoshenko beam becomes

$$\rho A\frac{\partial^2 w}{\partial t^2} = N\frac{\partial^2 w}{\partial x^2} + \frac{\partial}{\partial x}\left[\kappa AG\left(\frac{\partial w}{\partial x} - \varphi\right)\right] + q(x,t), \tag{6.1.20a}$$

$$\rho I\frac{\partial^2 \varphi}{\partial t^2} = N\frac{\partial^2 w}{\partial x^2} + \frac{\partial}{\partial x}\left(EI\frac{\partial \varphi}{\partial x}\right) + \kappa AG\left(\frac{\partial w}{\partial x} - \varphi\right). \tag{6.1.20b}$$

Among them, $N(x,t)$ is the axial force, and its relationship with the axial stress σ_{xx} is

$$N(x,t) = \int_A \sigma_{xx}\mathrm{d}A. \tag{6.1.21}$$

Combine the two equations and eliminate the unknown $\varphi(x)$ to obtain the deformation governing equation of Timoshenko beam considering the axial force

$$EI\frac{\partial^4 w}{\partial x^4} + N\frac{\partial^2 w}{\partial x^2} + \rho A\frac{\partial^2 w}{\partial t^2} - \left(\rho I + \frac{\rho EI}{\kappa G}\right)\frac{\partial^4 w}{\partial x^2 \partial t^2} + \frac{\rho^2 I}{\kappa G}\frac{\partial^4 w}{\partial t^4}$$
$$= q + \frac{\rho I}{\kappa AG}\frac{\partial^2 q}{\partial t^2} - \frac{EI}{\kappa AG}\frac{\partial^2 q}{\partial x^2}. \tag{6.1.22}$$

In the process of deformation, due to the internal friction of the material, damping force and damping moment will generate. It is assumed that the damping force is proportional to the deformation speed, while the damping moment is proportional to the deformation angular velocity, namely

$$Q^c = -\eta_1 \frac{\partial w}{\partial t}, \quad M^c = -\eta_2 \frac{\partial \varphi}{\partial t}. \tag{6.1.23}$$

The force analysis of the micro-element of beam is shown in Fig. 6.5. Considering the contribution of damping force and damping moment to the balance equations, the coupled governing equation of Timoshenko beam can be expressed as

$$\rho A \frac{\partial^2 w}{\partial t^2} + N \frac{\partial w}{\partial x} - \eta_1 \frac{\partial w}{\partial t} = \frac{\partial}{\partial x}\left[\kappa AG\left(\frac{\partial w}{\partial x} - \varphi\right)\right] + q(x, t), \tag{6.1.24a}$$

$$\rho I \frac{\partial^2 \varphi}{\partial t^2} = N \frac{\partial w}{\partial x} - \eta_2 \frac{\partial \varphi}{\partial t} + \frac{\partial}{\partial x}\left(EI \frac{\partial \varphi}{\partial x}\right) + \left[\kappa AG\left(\frac{\partial w}{\partial x} - \varphi\right)\right]. \tag{6.1.24b}$$

The Timoshenko beam model is a mechanical model proposed and developed by the Russian-American scientist and engineer Timoshenko in the early twentieth century. The model considers shear stress and rotational inertia, making it suitable for describing the mechanical behavior of short beams, laminated beams and high-frequency excitation with wavelengths close to the thickness. Different from Euler–Bernoulli beam theory, the deformation governing equation of Timoshenko beam is a fourth-order differential equation, and there is a second-order spatial derivative present. Due to the consideration of additional shear deformation, the stiffness of the beam has been reduced, which makes the deflection greater under a static load than that of Euler–Bernoulli beam. Under the given boundary conditions, the natural frequency of the beam is also reduced. Euler–Bernoulli beam theory believes that the cross section is perpendicular to the axis of the beam before and after deformation. In other words, the influence of transverse shear deformation and transverse normal

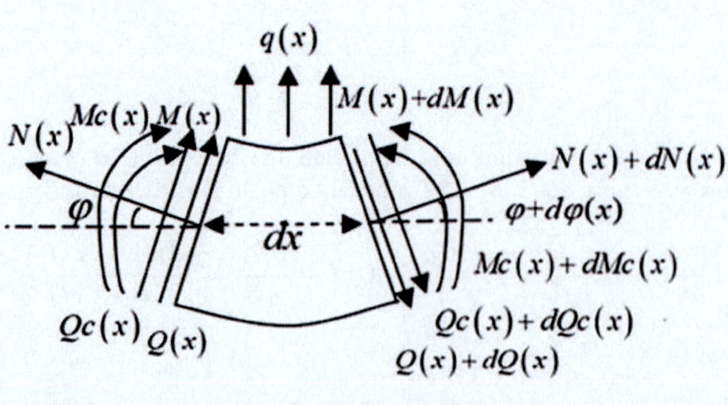

Fig. 6.5 Force analysis of micro-element of beam considering damping force and damping moment

strain is very small, so they can be ignored. These assumptions are valid for slender beams. No transverse shear means that the rotation of the cross section is only caused by bending. For thick beams, high-frequency modal excitation, composite beams, and transverse shear cannot be ignored. Adding transverse shear deformation to the Euler–Bernoulli beam model gives the Timoshenko beam theory. In the Timoshenko beam theory, in order to simplify the derivation of the equation of motion, the shear strain is assumed to be a constant value on a given cross section. In fact, the shear stress is always unevenly distributed across the cross section. For this reason, a shear correction factor is introduced to explain this simplification. Usually, when the slenderness is relatively large (>100), the Euler–Bernoulli theory can be used, and when the slenderness is relatively small, the Timoshenko beam theory is more suitable [25].

Let us now discuss the flexural wave modes that may exist in a beam of infinite length. Suppose the vertical displacement (deflection) of the beam is

$$w(x,t) = Ae^{i(kx-\omega t)}. \tag{6.1.25}$$

Substituting the wave solution Eq. (6.1.25) into the Eq. (6.1.15) and letting the distributed load be zero, we get

$$\left[k^4 - \frac{\rho A}{EI}\omega^2\right]Ae^{i(kx-\omega t)} = 0. \tag{6.1.26}$$

The condition for the above equation to have a non-zero solution is

$$k^4 - \frac{\rho A}{EI}\omega^2 = 0. \tag{6.1.27}$$

Let $a^4 = \frac{\rho A}{EI} > 0$, then $k^2 = \pm a^2\omega$. So the wave number k has four possible values, namely

$$k_1 = -k_2 = a\sqrt{\omega} = k', \quad k_3 = -k_4 = ia\sqrt{\omega} = ik'. \tag{6.1.28}$$

The wave motion solution can be generally expressed as

$$\begin{aligned} w(x,t) =& A_1e^{i(k_1x-\omega t)} + A_2e^{i(k_2x-\omega t)} + A_3e^{i(k_3x-\omega t)} + A_4e^{i(k_4x-\omega t)} \\ =& A_1e^{i(k'x-\omega t)} + A_2e^{i(-k'x-\omega t)} + A_3e^{-k'x-i\omega t} + A_4e^{k'x-i\omega t}. \end{aligned} \tag{6.1.29}$$

The above equation shows that there are four kinds of flexural waves in the beam. The first and second waves are traveling waves, but the propagation direction is opposite. The third and fourth waves are evanescent waves, but the attenuation directions are opposite. The propagation speeds of the first and second waves are the same, namely

$$C_1 = C_2 = \frac{\omega}{k} = \frac{\omega}{a\sqrt{\omega}} = \frac{\sqrt{\omega}}{a}. \tag{6.1.30}$$

Since the propagation speed depends on the frequency, these two kinds of traveling waves are both dispersive waves.

Considering that the subspace $\text{span}\{e^{ik'x}, e^{-ik'x}\}$ is the same as $\text{span}\{\cos k'x, \sin k'x\}$, and $\text{span}\{e^{-k'x}, e^{k'x}\}$ is the same as $\text{span}\{chk'x, shk'x\}$. So the wave motion solution can also be expressed as

$$w(x,t) = (C_1 \cos k'x + C_2 \sin k'x + C_3 chk'x + C_4 shk'x)e^{-i\omega t}. \tag{6.1.31}$$

Among them, shx and chx are hyperbolic sine function and hyperbolic cosine function, namely

$$shx = \frac{e^{k'x} - e^{-k'x}}{2}, \quad chx = \frac{e^{k'x} + e^{-k'x}}{2}. \tag{6.1.32}$$

For hyperbolic sine function and hyperbolic cosine function, the following relationship exists

$$\begin{aligned} &\frac{\mathrm{d}}{\mathrm{d}x} shx = chx, \quad \frac{\mathrm{d}}{\mathrm{d}x} chx = shx, \quad ch^2x - sh^2x = 1, \\ &sh(x)|_{x=0} = 0, \; ch(x)|_{x=0} = 1. \end{aligned} \tag{6.1.33}$$

For Timoshenko beam, $I \neq 0$, $G < \infty$ in Eq. (6.1.14), let the distributed load $q(x, t) = 0$, and substituting the wave motion solution of Eq. (6.1.24) into the governing Eq. (6.1.14) of Timoshenko beam leads to

$$\left[EIk^4 - \rho A\omega^2 - \left(\rho I + \frac{\rho EI}{\kappa G}\right)k^2\omega^2 + \frac{\rho^2 I}{\kappa G}\omega^4\right]w = 0, \tag{6.1.34a}$$

$$\left[k^4 - \left(\frac{\rho}{E} + \frac{\rho}{\kappa G}\right)k^2\omega^2 + \left(\frac{\rho^2}{E\kappa G}\omega^2 - \frac{\rho A}{EI}\right)\omega^2\right]Ae^{i(kx-\omega t)} = 0. \tag{6.1.34b}$$

The condition for the above equation to have a non-zero solution is

$$k^4 - \left(\frac{\rho}{E} + \frac{\rho}{\kappa G}\right)k^2\omega^2 + \left(\frac{\rho^2}{E\kappa G}\omega^2 - \frac{\rho A}{EI}\right)\omega^2 = 0. \tag{6.1.35}$$

Let $\lambda = k^2$, $a = -(\frac{\rho}{E} + \frac{\rho}{\kappa G})\omega^2$, $b = \left(\frac{\rho^2}{E\kappa G}\omega^2 - \frac{\rho A}{EI}\right)\omega^2$, then Eq. (6.1.35) is simplified to

$$\lambda^2 + a\lambda + b = 0. \tag{6.1.36}$$

The solution of the above equation is $\lambda = k^2 = \frac{-a \pm \sqrt{a^2 - 4b}}{2}$, that is

$$k_1 = -k_2 = \sqrt{\frac{-a + \sqrt{a^2 - 4b}}{2}}, \quad k_3 = -k_4 = \sqrt{\frac{-a - \sqrt{a^2 - 4b}}{2}}. \tag{6.1.37}$$

Therefore, the wave motion solution is

$$W(x, t) = A_1 e^{i(k_1 x - \omega t)} + A_2 e^{i(k_2 x - \omega t)} + A_3 e^{i(k_3 x - \omega t)} + A_4 e^{i(k_4 x - \omega t)}. \tag{6.1.38}$$

It can also be expressed as

$$W(x, t) = (C_1 shk_1 x + C_2 chk_2 x + C_3 \sin k_3 x + C_4 \cos k_4 x) e^{-i\omega t}. \tag{6.1.39}$$

Note that $b > 0$ when $\omega > \omega_{cr} = \sqrt{\frac{AG\kappa}{\rho I}}$. $k_i (i = 1, 2, 3, 4)$ are all real numbers, which indicates that there are two traveling waves along the same propagation direction in the Timoshenko beam. In contrast, there is only one traveling wave along the same direction in the Euler-Bernoulli beam.

Different from the wave motion solution, the vibration solution (also called the standing wave solution) needs to consider the boundary conditions and initial conditions at both ends of the beam. Because the boundary conditions must be considered, the frequency ω can only take discrete values. When the distributed load is not considered, it is called the free vibration solution. When considering the distributed load that changes with time, it is called the forced vibration solution. Only the free vibration solution is discussed here.

First, take Euler–Bernoulli beam as an example, suppose the vibration solution is

$$w(x, t) = W(x) T(t). \tag{6.1.40}$$

Substituting into the Euler–Bernoulli beam governing equation i.e. Eq. (6.1.15) and using the method of separating variables, we get

$$T''(t) + \omega^2 T(t) = 0, \tag{6.1.41a}$$

$$\frac{\mathrm{d}^4 W(x)}{\mathrm{d}x^4} - \frac{\rho A \omega^2}{EI} W(x) = 0. \tag{6.1.41b}$$

The solution of Eq. (6.1.41a) is

$$T(t) = A_1 e^{i\omega t} + A_2 e^{-i\omega t} = C_1' \cos \omega t + C_2' \sin \omega t. \tag{6.1.42}$$

Let $\lambda^4 = \frac{\rho A}{EI} \omega^2 = \frac{\omega^2}{a^2}$, $a^2 = \frac{EI}{\rho A}$, then the characteristic equation of ordinary differential Eq. (6.1.41b) is

$$s^4 - \lambda^4 = 0. \tag{6.1.43}$$

The characteristic roots are

$$s_1 = -s_2 = \lambda, \quad s_3 = -s_4 = i\lambda. \tag{6.1.44}$$

Thus, the solution of Eq. (6.1.41b) is

$$\begin{aligned} W(x) =& D_1 e^{\lambda x} + D_2 e^{-\lambda x} + D_3 e^{i\lambda x} + D_4 e^{-i\lambda x} \\ =& C_1 ch\lambda x + C_2 sh\lambda x + C_3 \cos \lambda x + C_4 \sin \lambda x. \end{aligned} \tag{6.1.45}$$

The boundary conditions of a beam can be expressed in many ways. The following are some of the most common ones:

(1) Fixed-end

The deflection and rotation angle of the fixed end are both zero, which can be expressed as

$$w(x)|_{x=0} = 0, \quad \left.\frac{\mathrm{d}}{\mathrm{d}x} w(x)\right|_{x=0} = 0. \tag{6.1.46}$$

(2) Sliding support end

The shear force and rotation angle at the sliding support end are zero, which can be expressed as

$$\left.\frac{dw(x)}{\mathrm{d}x}\right|_{x=0} = 0, \quad Q(x)|_{x=0} = 0. \tag{6.1.47}$$

Considering that

$$Q(x) = \frac{\mathrm{d}M}{\mathrm{d}x} = \frac{\mathrm{d}}{\mathrm{d}x}\left(EI\frac{\mathrm{d}\varphi}{\mathrm{d}x}\right) = EI\frac{\mathrm{d}^2\varphi(x)}{\mathrm{d}x^2} = EI\frac{\mathrm{d}^3 w(x)}{\mathrm{d}x^3},$$

the condition of zero shear force can also be expressed as

$$\left.\frac{\mathrm{d}^3}{\mathrm{d}x^3} w(x)\right|_{x=0} = 0. \tag{6.1.48}$$

(3) Hinge support end

The deflection and bending moment at the hinge support end are both zero, which can be expressed as

$$w(x)|_{x=0} = 0, \quad M(x)|_{x=0} = 0. \tag{6.1.49}$$

Considering that

$$M(x) = EI\frac{\mathrm{d}\varphi}{\mathrm{d}x} = EI\frac{\mathrm{d}^2 w(x)}{\mathrm{d}x^2},$$

the condition that the bending moment is zero can also be expressed as

$$\left.\frac{\mathrm{d}^2}{\mathrm{d}x^2}w(x)\right|_{x=0} = 0. \tag{6.1.50}$$

(4) Free end

At the free end of the beam, the shear and bending moment on the cross section are both zero, which can be expressed as

$$Q(x)|_{x=0} = 0, \quad M(x)|_{x=0} = 0, \tag{6.1.51}$$

or equivalently expressed as

$$\left.\frac{\mathrm{d}^3}{\mathrm{d}x^3}w(x)\right|_{x=0} = 0, \quad \left.\frac{\mathrm{d}^2}{\mathrm{d}x^2}w(x)\right|_{x=0} = 0. \tag{6.1.52}$$

(5) Spring support end

At the spring support end of the beam, the shear force and the bending moment are not zero, and their magnitude depends on the deflection and the rotation angle. Assuming that the elastic coefficient of the displacement spring is k_w and the torsional elastic coefficient of the torsion spring is k_θ, then the boundary conditions can be expressed as

$$\begin{aligned} Q(x)|_{x=0} &= k_w w(x)|_{x=0}, \\ M(x)|_{x=0} &= k_\theta \left.\frac{\mathrm{d}w(x)}{\mathrm{d}x}\right|_{x=0}, \end{aligned} \tag{6.1.53}$$

or equivalently expressed as

$$\begin{aligned} \left.\frac{\mathrm{d}^3}{\mathrm{d}x^3}w(x)\right|_{x=0} &= k_w w(x)|_{x=0}, \\ \left.\frac{\mathrm{d}^2}{\mathrm{d}x^2}w(x)\right|_{x=0} &= k_\theta \left.\frac{\mathrm{d}w(x)}{\mathrm{d}x}\right|_{x=0}. \end{aligned} \tag{6.1.54}$$

When $k_w \to 0$ and$k_\theta \to 0$, the spring support end degenerates into a free end. When $k_w \to \infty$ and $k_\theta \to \infty$, the spring support end degenerates into a fixed end.

By adjusting the values of k_θ andk_w, the intermediate state of boundary conditions between the free end and the fixed end can be described.

Let us take the beam with two ends fixed as an example to discuss the free vibration solution. From the boundary conditions at the fixed end, we get

$$\begin{aligned}&D_1 + D_2 + D_3 + D_4 = 0,\\&\lambda D_1 - \lambda D_2 + i\lambda D_3 - i\lambda D_4 = 0,\\&D_1 e^{\lambda l} + D_2 e^{-\lambda l} + D_3 e^{i\lambda l} + D_4 e^{-i\lambda l} = 0,\\&\lambda D_1 e^{\lambda l} - \lambda D_2 e^{-\lambda l} + \lambda D_3 i e^{i\lambda l} - \lambda D_4 i e^{-i\lambda l} = 0.\end{aligned} \tag{6.1.55}$$

Written in the matrix form as

$$\begin{bmatrix} 1 & 1 & 1 & 1 \\ \lambda & -\lambda & i\lambda & -i\lambda \\ e^{\lambda l} & e^{-\lambda l} & e^{i\lambda l} & e^{-i\lambda l} \\ \lambda e^{\lambda l} & -\lambda e^{-\lambda l} & i\lambda e^{i\lambda l} & -i\lambda e^{-i\lambda l} \end{bmatrix} \begin{Bmatrix} D_1 \\ D_2 \\ D_3 \\ D_4 \end{Bmatrix} = 0. \tag{6.1.56}$$

The condition for the above equation to have a non-zero solution is that the coefficient determinant is zero, namely

$$|A(\lambda)| = 0, \tag{6.1.57}$$

where λ can only take discrete values. Considering $\lambda^4 = \frac{\omega^2}{a^2}$, in other words, ω can only take discrete values. These discrete values are the natural frequencies of the beams with two fixed ends, denoted as $\omega_i (i = 1, 2, \ldots)$. Substituting ω_i into $A(\omega_i)\{D_k\} = 0$ can obtain the amplitude solution $\{D_k\}$ corresponding to the natural frequencyω_i, denoted as $D_k^{(i)} (i = 1, 2, \ldots; k = 1, 2, 3, 4)$. Substituting these amplitude solutions $D_k^{(i)}$ corresponding to ω_i to the deflection expression

$$W^{(i)}(x) = D_1^{(i)} e^{\lambda_i x} + D_2^{(i)} e^{-\lambda_i x} + D_3^{(i)} e^{i\lambda_i x} + D_4^{(i)} e^{-i\lambda_i x},$$

leads to the deflection curve corresponding to ω_i, which is called the mode shape corresponding to ω_i. There are infinitely many natural frequencies, and there are also infinitely many corresponding modes, and these modes are orthogonal to each other. The vibration solution of a beam fixed at both ends can be expressed in general

$$\begin{aligned}W(x,t) &= \sum_{i=1}^{\infty} \left[D_1^{(i)} e^{\lambda_i x} + D_2^{(i)} e^{-\lambda_i x} + D_3^{(i)} e^{i\lambda_i x} + D_4^{(i)} e^{-i\lambda_i x} \right] T(t)\\&= \sum_{i=1}^{\infty} W^{(i)}(x) T(t).\end{aligned} \tag{6.1.58}$$

Considering

$$
\begin{aligned}
T(t) &= A_1 e^{i\omega t} + A_2 e^{-i\omega t} = A_1 e^{i\omega t} + A_2 e^{(i\omega t + \pi i)} = \left(A_1 + A_2 e^{\pi i}\right) e^{i\omega t} \\
&= A e^{i\omega t},
\end{aligned} \tag{6.1.59}
$$

the general expression of the vibration solution can also be written as

$$
W(x,t) = \sum_{i=1}^{\infty} A_i \left[D_1^{(i)} e^{\lambda_i x} + D_2^{(i)} e^{-\lambda_i x} + D_3^{(i)} e^{i\lambda_i x} + D_4^{(i)} e^{-i\lambda_i x} \right] e^{i\omega t}, \tag{6.1.60}
$$

where $D_k^{(i)}$ is the combination coefficient of the ith mode shape.

For free vibration (disregarding the existence of distributed load $q(x,t)$), the main concerns are the natural frequency of the beam and the corresponding mode curve. For forced vibration (considering the existence of distributed load $q(x,t)$), we also care about the deflection curve of the beam under the action of distributed load $q(x,t)$. Therefore, it is necessary to determine the combination coefficient A_i corresponding to the distributed load $q(x,t)$.

It is should be pointed that there are second frequency spectrum existing for the Timoshenko beam with the simply supported boundary conditions at both ends. However, the second frequency spectrum is a controversial issue. Abbas and Thomas [28] used the finite element model to investigate the frequencies of vibration of Timoshenko beams with various end conditions. A concept of coupled vibration is introduced and is used to explain the behavior of the Timoshenko beams. It was shown that there is no separate second spectrum of frequencies except for the special case of a hinged-hinged Timoshenko beam. Levinson and Cooke [29] thought there is only a single frequency spectrum. The "two frequency spectra" viewpoint may be expedient as a device to compute frequencies but does not serve to explain the complex dynamical behavior of Timoshenko beams. Stephen [30] made a comprehensive review of the contributions and views on the second spectrum of Timoshenko beam over the past two decades and presented some new results. It is pointed that Timoshenko frequency equation factorizes not solely for hinged–hinged end conditions, as is often claimed, but also for guided–guided and guided–hinged. The numerical example is made by using both TBT and exact plane stress elastodynamic theory. Agreement is excellent for the first spectrum. However, the second spectrum predictions are not in consistent with any single mode of vibration. They concluded that the second spectrum is "unphysical" and should be disregarded. Almeida and Ramo [31] assure that the existence of the second spectrum for classical Timoshenko beam model justifies from several results on exponential decay for dissipative Timoshenko systems and believe that results of exponential decay occur because of the second spectrum. Therefore, they think that the second spectrum plays an important role in explaining some results on exponential decay and should be paid more attention.

In order to avoid the appearance of the second frequency spectrum, a small modification can be made on Timoshenko beam. The modified Timoshenko beam ignores the moment of inertia due to shear deformation, and the motion equation can be modified as

$$\kappa AG\left(\frac{\partial w}{\partial x}-\varphi\right)=-EI\frac{\partial^2\varphi}{\partial x^2}+\rho I\frac{\partial^3 w}{\partial x\partial t^2}, \tag{6.1.61a}$$

$$\kappa AG\left(\frac{\partial^2 w}{\partial x^2}-\frac{\partial\varphi}{\partial x}\right)=\rho A\frac{\partial^2 w}{\partial t^2}. \tag{6.1.61b}$$

After elimination of φ, the motion equation with respect to transverse displacement w can be obtained as

$$EI\frac{\partial^4 w}{\partial x^4}-\rho I\frac{\partial^4 w}{\partial x^2\partial t^2}-\frac{EI}{\kappa AG}\frac{\partial^2}{\partial x^2}\left(\rho A\frac{\partial^2 w}{\partial t^2}\right)=\rho A\frac{\partial^2 w}{\partial t^2}, \tag{6.1.62a}$$

$$EI\frac{\partial^4 w}{\partial x^4}-\rho I\frac{\partial^4 w}{\partial x^2\partial t^2}-\frac{EI}{\kappa AG}\frac{\partial^2}{\partial x^2}\left(\rho A\frac{\partial^2 w}{\partial t^2}\right)+\frac{\rho I}{\kappa AG}\frac{\partial^2}{\partial t^2}\left(\rho A\frac{\partial^2 w}{\partial t^2}\right)=\rho A\frac{\partial^2 w}{\partial t^2}. \tag{6.1.62b}$$

Comparing the equation of motion of the classical Timoshenko beam (corresponding to Eq. (6.1.62b)) with that of the modified Timoshenko beam (corresponding to Eq. (6.1.62b)), it is noted that the first three terms on the left of the equals sign are the same. The first term corresponds to the basic theory of Euler beams. The second term is the additional shear stress corresponding to the moment of inertia caused by bending deformation. The third term is the additional shear stress caused by shear deformation. It can be seen that the fundamental difference between the two models is whether the moment of inertia caused by the shear deformation of the beam element is considered. The extra term in Eq. (6.1.62b) is resulted from that replacing term $\rho I\frac{\partial^2\varphi}{\partial t^2}$ in (6.1.13b) by term $\rho I\frac{\partial^3 w}{\partial x\partial t^2}$ in Eq. (6.1.62a).

The free vibration solution for the simply supported boundary conditions at both ends can be expressed as

$$w(x,t)=\sin\left(\frac{n\pi}{l}x\right)e^{-i\omega_n t}. \tag{6.1.63}$$

Inserting Eq. (6.1.63) into Eq. (6.1.62a) leads to the frequency equation

$$\left[-\mathrm{d}\left(\frac{n\pi}{l}\right)^2-s\right]\omega_n^2+\left(\frac{n\pi}{l}\right)^4=0, \tag{6.1.64}$$

where $\mathrm{d}=\frac{\rho}{E}\left(1+\frac{E}{\kappa G}\right), s=\frac{\rho A}{EI}$. From Eq. (6.1.64), we get the natural frequency of modified Timoshenko beam

$$\omega_n=\frac{\left(\frac{n\pi}{l}\right)^2}{\sqrt{\mathrm{d}\left(\frac{n\pi}{l}\right)^2+s}}. \tag{6.1.65}$$

Inserting Eq. (6.1.63) into Eq. (6.1.62b) leads to the frequency equation of classic Timoshenko beam

$$m\omega_n^2 + \left[-\mathrm{d}\left(\frac{n\pi}{l}\right)^2 - s\right]\omega_n + a\left(\frac{n\pi}{l}\right)^4 = 0. \tag{6.1.66}$$

The natural frequency of classic Timoshenko beam can be obtained as

$$\omega_n = \sqrt{\frac{\left[\mathrm{d}\left(\frac{n\pi}{l}\right)^2 + s\right] \pm \sqrt{\left[\mathrm{d}\left(\frac{n\pi}{l}\right)^2 + s\right]^2 - 4m\left(\frac{n\pi}{l}\right)^4}}{2m}}, \tag{6.1.67}$$

where $m = \frac{\rho^2}{\kappa EG}$.

By comparison of the expressions of natural frequency of modified and classic Timoshenko beams, i.e. Eqs. (6.1.65) and (6.1.67), it is noted that the classic Timoshenko beam has the second frequency spectrum, while the modified Timoshenko beam has only one frequency spectrum. Therefore, the appearance of second frequency spectrum is due to the fact that the moment of inertia caused by the shear deformation is taken into account. In fact, there are two kinds of vibration modes, i.e. longitudinal shear mode and bending mode, for the hinged-hinged Timoshenko beam. The two vibration modes interfere and result in two distinct spectrum of frequency, where the bending mode dominates in the first spectrum of frequency while the longitudinal shear mode dominates in second spectrum of frequency. Further, deformations due to shear and bending are of the same phase for the first spectrum but antiphase for the second spectrum. The modified Timoshenko beam model can eliminate the second frequency spectrum.

6.2 Flexural Waves in Plate

The classic plate theory is also called Kirchhoff theory of plate, which is named after Kirchhoff because of the famous Kirchhoff assumption. Considers that a linear elastic and isotropic homogeneous thin plate. The bending of the plate is limited to a small deformation range. During the bending process of the plate, it is necessary to meet:

(1) The normal line of the midplane of the plate remains straight and perpendicular to the midplane before and after bending.
(2) Ignore the normal stress along the direction perpendicular to the midplane.
(3) Only the inertia force of the translation of the mass is included, and its inertia moment of rotation is omitted.
(4) The midplane of the plate does not stretch and contract.

It is assumed that the straight line perpendicular to the midplane of the plate before deformation is still a straight line perpendicular to the midplane of the plate after deformation in Kirchhoff's assumption. Therefore, the rotations of cross section are only due to bending deformation. This results in two angular displacements ψ_x and

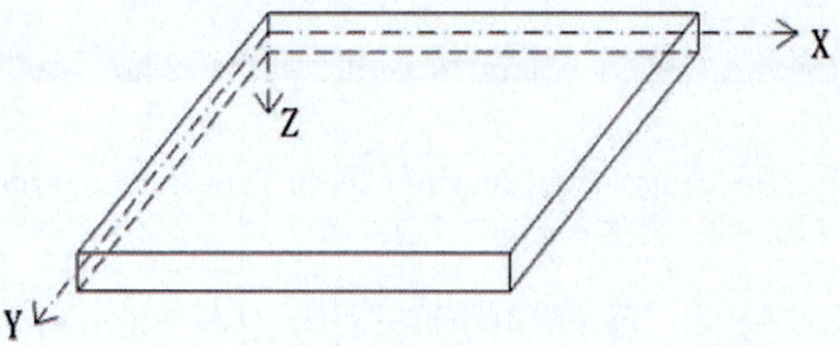

Fig. 6.6 Sketch of isotropic thin plate of equal thickness

ψ_y in the horizontal direction of the plate, which can be directly represented by the first derivative of the midplane deflection $w(x, y, t)$. As a result, the deflection $w(x, y, t)$ is the only independent variable in Kirchhoff plate theory, which greatly simplifies the plate problem.

Consider an isotropic homogeneous thin plate with a rectangular boundary, as shown in Fig. 6.6. Take the midplane of the plate as the xoy plane, the z-axis is perpendicular to the xoy plane, and the plate thickness is h. Introduce the Cartesian coordinate system (x, y, z), and set the deflection of the plate as $w(x, y, t)$. According to the Kirchhoff hypothesis of the plate, the displacement u, v, w at any point on the cross section of the plate can be expressed as

$$u = -\frac{\partial w}{\partial x}z, \quad v = -\frac{\partial w}{\partial y}z, \quad w = w(x, y, t). \tag{6.2.1}$$

The above equation shows that the in-plane displacement of each point on the cross section of the plate is linearly distributed along the thickness direction and is related to the slope of the deflection surface w along the x and y directions. The rotation angle of the cross section caused by the shear deformation should be zero. According to the geometric equation, we get

$$\gamma_{xz} = \frac{\partial u}{\partial z} + \frac{\partial w}{\partial x} = 0, \quad \gamma_{yz} = \frac{\partial v}{\partial z} + \frac{\partial w}{\partial y} = 0, \quad \varepsilon_z = \frac{\partial w}{\partial z} = 0. \tag{6.2.2}$$

The in-plane strain components $\left(\varepsilon_x, \varepsilon_y, \gamma_{xy}\right)$ can be expressed by the deflection w as follows

$$\varepsilon_x = -\frac{\partial^2 w}{\partial x^2}z, \quad \varepsilon_y = -\frac{\partial^2 w}{\partial y^2}z, \quad \gamma_{xy} = -2\frac{\partial^2 w}{\partial x \partial y}z. \tag{6.2.3}$$

The above equation shows that the in-plane strain component of each point on the cross section of the plate is also linearly distributed along the thickness. Combining Eqs. (6.2.2) and (6.2.3), it can be seen that the deformation of the plate is in a plane strain state under the Kirchhoff assumption.

Considering the assumption (2), that is, $\sigma_z = 0$, it is derived from Hooke's law

$$\varepsilon_x = \frac{1}{E}\left(\sigma_x - v\sigma_y\right), \quad \varepsilon_y = \frac{1}{E}\left(\sigma_y - v\sigma_x\right), \quad \gamma_{xy} = \frac{\tau_{xy}}{G}. \tag{6.2.4}$$

The stress components are

$$\sigma_x = \frac{E}{1 - v^2}\left(\varepsilon_x + v\varepsilon_y\right) = -\frac{E}{1 - v^2}z\left(\frac{\partial^2 w}{\partial x^2} + v\frac{\partial^2 w}{\partial y^2}\right), \tag{6.2.5a}$$

$$\sigma_y = \frac{E}{1 - v^2}\left(\varepsilon_y + v\varepsilon_x\right) = -\frac{E}{1 - v^2}z\left(\frac{\partial^2 w}{\partial y^2} + v\frac{\partial^2 w}{\partial x^2}\right), \tag{6.2.5b}$$

$$\tau_{xy} = G\gamma_{xy} = -2Gz\frac{\partial^2 w}{\partial x \partial y}. \tag{6.2.5c}$$

The above equation shows that the stress components at each point of the cross section are also linearly distributed along the thickness.

After obtaining the distribution of normal stress and shear stress along the thickness of the cross section of the plate, we can further get the internal forces by performing an integral operation along the thickness, that is, calculating the force moment with respect to the center plane. The internal forces on the cross section, namely the bending moment and torque moment, can be obtained as

$$M_x = \int_{-\frac{h}{2}}^{\frac{h}{2}} \sigma_x z dz = -D\left(\frac{\partial^2 w}{\partial x^2} + v\frac{\partial^2 w}{\partial y^2}\right), \tag{6.2.6a}$$

$$M_y = \int_{-\frac{h}{2}}^{\frac{h}{2}} \sigma_y z dz = -D\left(\frac{\partial^2 w}{\partial y^2} + v\frac{\partial^2 w}{\partial x^2}\right), \tag{6.2.6b}$$

$$M_{xy} = M_{yx} = \int_{-\frac{h}{2}}^{\frac{h}{2}} \tau_{xy} z dz = -D(1 - v)\frac{\partial^2 w}{\partial x \partial y}, \tag{6.2.6c}$$

where

$$D = \frac{Eh^3}{12\left(1 - v^2\right)}.$$

It is called the bending stiffness of the plate. The bending moment and torque in the plate are related to the bending stiffness of the plate and the curvature and twisting curvature of the deflection surface. Based on the Kirchhoff assumption, the internal force (bending moments and torques) at each point on the cross section of the plate can be represented by the two-dimensional deflection surface function

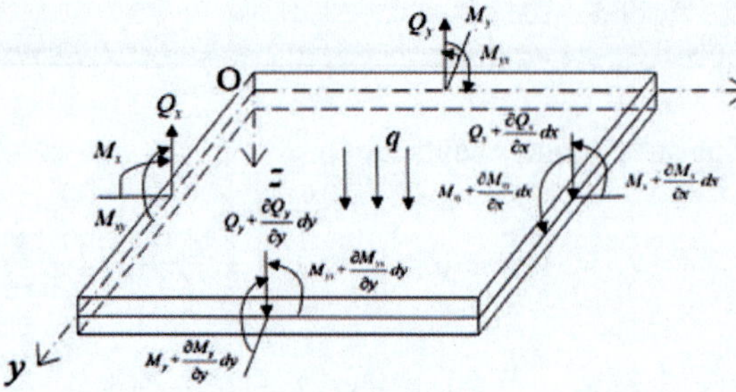

Fig. 6.7 Sketch of force analysis of micro-element body of thin plate

w, and thus, the deflection function w becomes the only unknown quantity to be determined. Compared with the method of three-dimensional elastic mechanics, the thin plate theory based on Kirchhoff's assumption greatly simplifies the problem.

Take a micro-element body of the thin plate and carry out the force analysis, as shown in Fig. 6.7. According to assumption (3), namely ignoring the inertia moment and only considering the inertial force, the moment balance equation in the x and y directions and the motion equation in the z direction are as follows

$$\frac{\partial M_x}{\partial x} + \frac{\partial M_{yx}}{\partial y} - Q_x = 0, \tag{6.2.7a}$$

$$\frac{\partial M_{xy}}{\partial x} + \frac{\partial M_y}{\partial y} - Q_y = 0, \tag{6.2.7b}$$

$$\frac{\partial Q_x}{\partial x} + \frac{\partial Q_y}{\partial y} + q - \rho h \frac{\partial^2 w}{\partial t^2} = 0. \tag{6.2.7c}$$

Substituting the expressions of bending moment and torque (6.2.6) into the moment balance Eqs. (6.2.7a) and (6.2.7b), the shear force expression can be obtained

$$Q_x = -D\frac{\partial}{\partial x}\left(\frac{\partial^2 w}{\partial x^2} + \frac{\partial^2 w}{\partial y^2}\right) = -D\frac{\partial}{\partial x}\nabla^2 w, \tag{6.2.8a}$$

$$Q_y = -D\frac{\partial}{\partial y}\left(\frac{\partial^2 w}{\partial x^2} + \frac{\partial^2 w}{\partial y^2}\right) = -D\frac{\partial}{\partial y}\nabla^2 w, \tag{6.2.8b}$$

where $\nabla^2 = \frac{\partial^2}{\partial x^2} + \frac{\partial^2}{\partial y^2}$ is the Laplace operator. Substituting Eq. (6.2.8) into Eq. (6.2.7c), we finally obtain the equation of motion of the thin plate

$$\frac{\partial^4 w}{\partial x^4} + 2\frac{\partial^4 w}{\partial x^2 \partial y^2} + \frac{\partial^4 w}{\partial y^4} + \frac{\rho h}{D}\frac{\partial^2 w}{\partial t^2} = \frac{q}{D}, \tag{6.2.9}$$

or abbreviated it as

$$\nabla^4 w + \frac{\rho h}{D}\frac{\partial^2 w}{\partial t^2} = \frac{q}{D}. \tag{6.2.10}$$

In the above equation, the operator $\nabla^4 = \frac{\partial^4}{\partial x^4} + 2\frac{\partial^4}{\partial x^2 \partial y^2} + \frac{\partial^4}{\partial y^4} = (\frac{\partial^2}{\partial x^2} + \frac{\partial^2}{\partial y^2})^2$. ρ is the mass bulk density, and q is the transverse load per unit area on the surface of the plate.

Now consider the problem of flexural wave propagation in the plate. Ignore the lateral load, that is, $q = 0$, and set the wave motion solution as

$$w(x, y, t) = A \exp[i(k_1 x + k_2 y - \omega t)], \tag{6.2.11}$$

where A and ω are the amplitude and circular frequency of the wave, respectively, and

$$k_1 = k \sin\alpha, \quad k_2 = k \cos\alpha, \tag{6.2.12}$$

are the component of the bending wave vector $\boldsymbol{k}$ in the x and y directions, and α represents the angle between the propagation direction of the bending wave and the y coordinate axis. Substitute the wave solution (6.2.11) into the governing Eq. (6.2.9), we get

$$\left(k_1^4 + 2k_1^2 k_2^2 + k_2^4 - \frac{\rho h}{D}\omega^2\right) A \exp[i(k_1 x + k_2 y - \omega t)] = 0. \tag{6.2.13}$$

The condition for Eq. (6.2.13) to have a non-zero solution is

$$k_1^4 + 2k_1^2 k_2^2 + k_2^4 - \frac{\rho h}{D}\omega^2 = 0. \tag{6.2.14}$$

Substituting Eq. (6.2.12) into Eq. (6.2.14) leads to

$$k^4\left(\sin^4\alpha + 2\sin^2\alpha\cos^2\alpha + \cos^4\alpha\right) - \frac{\rho h}{D}\omega^2 = 0. \tag{6.2.15}$$

Considering

$$\sin^4\alpha + 2\sin^2\alpha\cos^2\alpha + \cos^4\alpha = \left(\sin^2\alpha + \cos^2\alpha\right)^2 = 1. \tag{6.2.16}$$

Equation (6.2.15) can be further simplified as

$$k^4 - \frac{\rho h}{D}\omega^2 = 0. \tag{6.2.17}$$

Let $\lambda^4 = \frac{\rho h}{D} > 0$, then the characteristic equation is

$$k^4 - \lambda^4\omega^2 = 0. \tag{6.2.18}$$

The solution is $k^2 = \pm\lambda^2\omega$, thus,

$$k_1 = -k_2 = \lambda\sqrt{\omega} = k', \quad k_3 = -k_4 = i\lambda\sqrt{\omega} = ik'. \tag{6.2.19}$$

The wave motion solution can be generally expressed as

$$\begin{aligned} w(x,t) = {} & A_1 e^{ik'(x \sin\alpha + y \cos\alpha) - i\omega t} + A_2 e^{-ik'(x \sin\alpha + y \cos\alpha) - i\omega t} \\ & + A_3 e^{-k'(x \sin\alpha + y \cos\alpha) - i\omega t} + A_4 e^{k'(x \sin\alpha + y \cos\alpha) - i\omega t}. \end{aligned} \tag{6.2.20}$$

The above equation shows that there are two kinds of flexural waves in the plate. The first is a pair of traveling waves with opposite propagation directions, which have the same propagation speed; the second type is a pair of evanescent waves with opposite attenuation directions. They are standing waves without propagating speed. The propagation velocity of a pair of traveling flexural waves with opposite propagation directions is

$$C_1 = C_2 = \frac{\omega}{k} = \frac{\omega}{\lambda\sqrt{\omega}} = \frac{\sqrt{\omega}}{\lambda}. \tag{6.2.21}$$

Since the propagation speed depends on the frequency, the pair of traveling flexural waves are both dispersive waves [26, 27].

In the Kirchhoff plate theory, the rotation angles ψ_x and ψ_y caused by shear deformation are ignored. This makes the deformation problem of the plate ultimately boil down to the solution of the deflection surface of the plate, which greatly simplifies the problem discussed. However, large errors are brought about for plates with relatively large plate thicknesses. Different from the Kirchhoff plate theory, the Mindlin plate theory not only considers the transverse shear deformation but also considers the influence of the inertia moment. In Mindlin plate theory, the straight line perpendicular to the midplane of the plate before deformation is still a straight line after deformation, but not necessarily perpendicular to the midplane of the plate. The rotation angle of the cross section of the plate can no longer be expressed by deflection, that is, $\psi_x \neq \frac{dw}{dx}$ and $\psi_y \neq dw/dy$. Therefore, in addition to the deflection w of the midplane, two rotation angles ψ_x and ψ_y as independent variates are introduced to describe the rotation of the cross section along the x- and y-axes. These three variables are independent variables. If it is assumed that the shear deformations along the *x*- and *y*-axes are γ_x and γ_y, respectively, the relationship between them is

$$\psi_x = dw/dx - \gamma_x, \quad \psi_y = dw/dy - \gamma_y.$$

The terms of (1) and (3) in Kirchhoff's hypothesis are revised in the Mindlin plate theory, however, the terms of (2) and (4) in Kirchhoff's hypothesis are retained, namely they are still valid for the Mindlin plate.

In the Cartesian coordinate system, let $z = 0$ denote the midplane of the plate. The three displacement components u, v, w along the x, y, z coordinate directions at any point on the cross section of the plate can be written as

$$u(x, y, z) = -z\psi_x(x, y), \quad v(x, y, z) = -z\psi_y(x, y), \quad w(x, y, z) = w(x, y). \tag{6.2.22}$$

According to the geometric relationship, the strain component can be further obtained as

$$\begin{aligned} &\varepsilon_x = -z\frac{\partial \psi_x}{\partial x}, \quad \varepsilon_y = -z\frac{\partial \psi_y}{\partial y}, \quad \gamma_{xy} = -z\left(\frac{\partial \psi_x}{\partial y} + \frac{\partial \psi_y}{\partial x}\right), \\ &\gamma_x = \frac{\partial w}{\partial x} - \psi_x, \quad \gamma_y = \frac{\partial w}{\partial y} - \psi_y. \end{aligned} \tag{6.2.23}$$

Considering the assumption (2), that is, $\sigma_z = 0$. According to Hooke's law, we have

$$\begin{aligned} &\varepsilon_x = \frac{1}{E}(\sigma_x - v\sigma_y), \quad \varepsilon_y = \frac{1}{E}(\sigma_y - v\sigma_x), \\ &\gamma_{xy} = \frac{\tau_{xy}}{G}, \quad \gamma_x = \frac{\tau_{xz}}{G}, \quad \gamma_y = \frac{\tau_{yz}}{G}. \end{aligned} \tag{6.2.24}$$

By solving the stress components and substituting them into the strain components Eq. (6.2.23), we get

$$\sigma_x = \frac{E}{1 - v^2}(\varepsilon_x + v\varepsilon_y) = -\frac{E}{1 - v^2}z\left(\frac{\partial \psi_x}{\partial x} + v\frac{\partial \psi_y}{\partial y}\right), \tag{6.2.25a}$$

$$\sigma_y = \frac{E}{1 - v^2}(\varepsilon_y + v\varepsilon_x) = -\frac{E}{1 - v^2}z\left(\frac{\partial \psi_y}{\partial y} + v\frac{\partial \psi_x}{\partial x}\right), \tag{6.2.25b}$$

$$\tau_{xy} = G\gamma_{xy} = -Gz\left(\frac{\partial \psi_x}{\partial y} + \frac{\partial \psi_y}{\partial x}\right), \tag{6.2.25c}$$

$$\tau_{xz} = G\gamma_x = G\left(\frac{\partial w}{\partial x} - \psi_x\right), \tag{6.2.25d}$$

$$\tau_{yz} = G\gamma_y = G\left(\frac{\partial w}{\partial y} - \psi_y\right). \tag{6.2.25e}$$

After obtaining the internal normal stress, shear stress distribution and transverse shear stress distribution on the cross section of the plate, we can perform the integral operation along the thickness to calculate the moment with respect to the midplane. Then, we further obtain the internal force of the cross section, namely bending moment, torque and shear force.

$$M_x = \int_{-\frac{h}{2}}^{\frac{h}{2}} \sigma_x z \mathrm{d}z, \quad M_y = \int_{-\frac{h}{2}}^{\frac{h}{2}} \sigma_y z \mathrm{d}z, \quad M_{xy} = \int_{-\frac{h}{2}}^{\frac{h}{2}} \tau_{xy} z \mathrm{d}z,$$

$$Q_x = \int_{-\frac{h}{2}}^{\frac{h}{2}} \tau_{xz} \mathrm{d}z, \quad Q_y = \int_{-\frac{h}{2}}^{\frac{h}{2}} \tau_{yz} \mathrm{d}z. \tag{6.2.26}$$

Substituting the stress expression (6.2.25) into the formula (6.2.26) leads to the bending moment, torque and shear force

$$M_x = -D\left(\frac{\partial \psi_x}{\partial x} + v\frac{\partial \psi_y}{\partial y}\right), \tag{6.2.27a}$$

$$M_y = -D\left(\frac{\partial \psi_y}{\partial y} + v\frac{\partial \psi_x}{\partial x}\right), \tag{6.2.27b}$$

$$M_{xy} = M_{yx} = -\frac{D(1-v)}{2}\left(\frac{\partial \psi_x}{\partial y} + \frac{\partial \psi_y}{\partial x}\right), \tag{6.2.27c}$$

$$Q_x = C\left(\frac{\partial w}{\partial x} - \psi_x\right), \tag{6.2.27d}$$

$$Q_y = C\left(\frac{\partial w}{\partial y} - \psi_y\right). \tag{6.2.27e}$$

In the above equation, $D = Eh^3/12(1-v^2)$ is the bending stiffness of the plate, $C = \kappa Eh/2(1+v)$ is the shear stiffness of the plate, and κ is the shear correction factor.

In the Mindlin's plate theory, the inertia moment is taken into account. Therefore, the dynamic Eq. (6.2.7) of the micro-element body is modified to,

$$\frac{\partial M_x}{\partial x} + \frac{\partial M_{yx}}{\partial y} - Q_x - \rho I\frac{\partial^2 \psi_x}{\partial t^2} = 0, \tag{6.2.28a}$$

$$\frac{\partial M_{xy}}{\partial x} + \frac{\partial M_y}{\partial y} - Q_y - \rho I\frac{\partial^2 \psi_y}{\partial t^2} = 0, \tag{6.2.28b}$$

$$\frac{\partial Q_x}{\partial x} + \frac{\partial Q_y}{\partial y} + q - \rho h\frac{\partial^2 w}{\partial t^2} = 0. \tag{6.2.28c}$$

Substituting the expressions (6.2.27) of bending moment, torque and shearing force into (6.2.28), we get the differential equation of motion represented by the midplane deflection w and the two rotation angles ψ_x and ψ_y.

$$D\left(\frac{\partial^2\psi_x}{\partial x^2}+\frac{(1-v)}{2}\frac{\partial^2\psi_x}{\partial y^2}+\frac{(1+v)}{2}\frac{\partial^2\psi_y}{\partial x\partial y}\right)+C\left(\frac{\partial w}{\partial x}-\psi_x\right)+\rho I\frac{\partial^2\psi_x}{\partial t^2}=0, \tag{6.2.29a}$$

$$D\left(\frac{(1+v)}{2}\frac{\partial^2\psi_x}{\partial x\partial y}+\frac{(1-v)}{2}\frac{\partial^2\psi_y}{\partial x^2}+\frac{\partial^2\psi_y}{\partial y^2}\right)+C\left(\frac{\partial w}{\partial y}-\psi_y\right)+\rho I\frac{\partial^2\psi_y}{\partial t^2}=0, \tag{6.2.29b}$$

$$C\left(\frac{\partial^2 w}{\partial x^2}+\frac{\partial^2 w}{\partial y^2}-\frac{\partial\psi_x}{\partial x}-\frac{\partial\psi_y}{\partial y}\right)+\rho h\frac{\partial^2 w}{\partial t^2}=q, \tag{6.2.29c}$$

where $q(x, y, t)$ is the distributed load acting on the surface of the plate, and $I = h^3/12$ is the geometric moment of inertia per unit width of the cross section.

Let us now discuss the problem of flexural wave propagation in a plate. Ignore the lateral distributed load, that is, $q = 0$, and set the wave motion solution as

$$\psi_x(x, y, t) = A_1 \exp[i(k_1x + k_2y - \omega t)], \tag{6.2.30a}$$

$$\psi_y(x, y, t) = A_2 \exp[i(k_1x + k_2y - \omega t)], \tag{6.2.30b}$$

$$w(x, y, t) = A_3 \exp[i(k_1x + k_2y - \omega t)], \tag{6.2.30c}$$

where A_i is the amplitude of the wave, ω is the circular frequency of the wave, and

$$k_1 = k\sin\alpha,\quad k_2 = k\cos\alpha, \tag{6.2.31}$$

is the component of the wave vector $\boldsymbol{k}$ in the x and y directions, and α represents the angle between the propagation direction of the flexural wave and the y coordinate axis. Substitute the wave motion solution (6.2.30) into the governing Eq. (6.2.29), we get

$$D\left(k_1^2A_1+\frac{1-v}{2}k_2^2A_1+\frac{1+v}{2}k_1k_2A_2\right)-C(ik_1A_3-A_1)+\rho I\omega^2A_1=0, \tag{6.2.32a}$$

$$D\left(k_2^2A_2+\frac{1-v}{2}k_1^2A_2+\frac{1+v}{2}k_1k_2A_1\right)-C(ik_2A_3-A_2)+\rho I\omega^2A_2=0, \tag{6.2.32b}$$

$$C(k_1^2A_3+k_2^2A_3+ik_1A_1+ik_2A_2)+\rho h\omega^2A_3=0. \tag{6.2.32c}$$

Written in matrix form

$$\begin{bmatrix} D\left(k_1^2+\frac{1-v}{2}k_2^2\right)+C+\rho I\omega^2 & D\frac{1+v}{2}k_1k_2 & -iCk_1 \\ D\frac{1+v}{2}k_1k_2 & D\left(k_2^2+\frac{1-v}{2}k_1^2\right)+C+\rho I\omega^2 & -iCk_2 \\ iCk_1 & iCk_2 & k^2C+\rho h\omega^2 \end{bmatrix} \begin{Bmatrix} A_1 \\ A_2 \\ A_3 \end{Bmatrix} = 0 \tag{6.2.33}$$

The condition for the above equation to have a non-zero solution is that the coefficient determinant is zero, that is,

$$\begin{vmatrix} D\left(k_1^2+\frac{1-v}{2}k_2^2\right)+C+\rho I\omega^2 & D\frac{1+v}{2}k_1k_2 & -Cik_1 \\ D\frac{1+v}{2}k_1k_2 & D\left(k_2^2+\frac{1-v}{2}k_1^2\right)+C+\rho I\omega^2 & -Cik_2 \\ Cik_1 & Cik_2 & Ck^2+\rho h\omega^2 \end{vmatrix} = 0. \tag{6.2.34}$$

The expansion of determinant leads to

$$\begin{aligned} &CD^2\frac{1-v}{2}k^6+\left(C^2D+CD\frac{3-v}{2}\rho I\omega^2+D^2\frac{1-v}{2}\rho h\omega^2-C^2D\frac{v}{2}\sin^2 2\alpha\right)k^4 \\ &+\left(C^2\rho I\omega^2+C\rho^2I^2\omega^4+CD\frac{3-v}{2}\rho h\omega^2+D\frac{3-v}{2}\rho^2Ih\omega^4\right)k^2 \\ &+\left(C^2+2C\rho I\omega^2+\rho^2I^2\omega^4\right)\rho h\omega^2=0. \end{aligned} \tag{6.2.35}$$

The above equation can be simply written as

$$ak^6+bk^4+ck^2+d=0 \tag{6.2.36}$$

where

$$a=CD^2\frac{1-v}{2},$$

$$b=C^2D+CD\frac{3-v}{2}\rho I\omega^2+D^2\frac{1-v}{2}\rho h\omega^2-C^2D\frac{v}{2}\sin^2 2\alpha,$$

$$c=C^2\rho I\omega^2+C\rho^2I^2\omega^4+CD\frac{3-v}{2}\rho h\omega^2+D\frac{3-v}{2}\rho^2Ih\omega^4,$$

$$d=\left(C^2+2C\rho I\omega^2+\rho^2I^2\omega^4\right)\rho h\omega^2.$$

Through the change of function

$$k^2=k'-\frac{b}{3a}. \tag{6.2.37}$$

The Eq. (6.2.36) is transformed into

$$k'^3 + pk' + q = 0, \tag{6.2.38}$$

where $p = \frac{c}{a} - \frac{b^2}{3a^2}$, $q = \frac{2b^3}{27a^3} - \frac{bc}{3a^2} + \frac{d}{a}$. According to the Cardino's equation, the solutions of the equation are

$$k_1^{'} = m + n,\ k_2^{'} = \lambda m + \lambda^2 n,\ k_3^{'} = \lambda^2 m + \lambda n, \tag{6.2.39}$$

where

$$m = \sqrt[3]{-\frac{q}{2} + \sqrt{\left(\frac{q}{2}\right)^2 + \left(\frac{p}{3}\right)^3}},\ n = \sqrt[3]{-\frac{q}{2} - \sqrt{\left(\frac{q}{2}\right)^2 + \left(\frac{p}{3}\right)^3}},\ \lambda = \frac{-1 + \sqrt{3}i}{2}.$$

Then, we have

$$k_1 = -k_2 = \sqrt{k_1^{'} - \frac{b}{3a}},\ k_3 = -k_4 = \sqrt{k_2^{'} - \frac{b}{3a}},\ k_5 = -k_6 = \sqrt{k_3^{'} - \frac{b}{3a}}. \tag{6.2.40}$$

Thus, the wave motion solution is

$$\begin{aligned} w(x, y, t) =& A_1 e^{i(k_1(x \sin\alpha + y \cos\alpha) - \omega t)} + A_2 e^{i(-k_1(x \sin\alpha + y \cos\alpha) - \omega t)} \\ &+ A_3 e^{i(k_3(x \sin\alpha + y \cos\alpha) - \omega t)} + A_4 e^{i(-k_3(x \sin\alpha + y \cos\alpha) - \omega t)} \\ &+ A_5 e^{i(k_5(x \sin\alpha + y \cos\alpha) - \omega t)} + A_6 e^{i(-k_5(x \sin\alpha + y \cos\alpha) - \omega t)}. \end{aligned} \tag{6.2.41}$$

The above formula shows that there are 3 pairs of flexural waves with opposite propagation directions in the Mindlin plate. If the wave number is a real number, the corresponding flexural wave is a traveling wave. If the wave number is a pure imaginary number, the corresponding wave is a standing wave. If the wave speed is a complex number, the corresponding wave is an attenuated traveling wave. For classic elastic plates, this situation is impossible because there is no energy dissipation.

6.3 Guided Waves in Plate (Lamb Wave)

Consider an infinite plate of thickness 2h and a rectangular coordinate system as Fig. 6.8. The plate is symmetric about the middle plane $z = 0$ and extends infinitely in the direction of x and y. Our problem is to solve the wave equation under boundary conditions at $z = \pm h$.

$$\nabla^2 \phi = c_P^{-2} \frac{\partial^2 \phi}{\partial t^2}, \tag{6.3.1a}$$

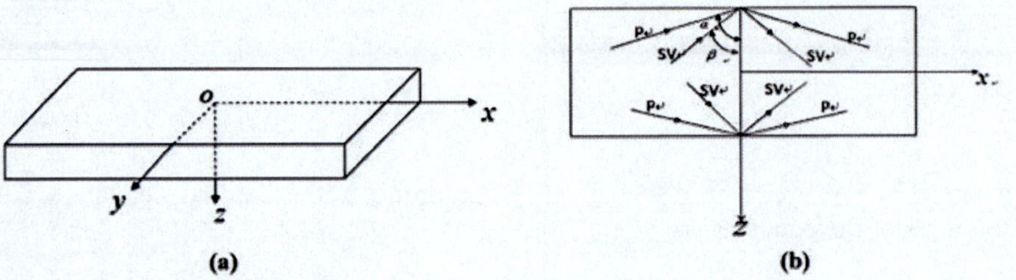

Fig. 6.8 Wave in an infinite plate. **a** Plate waveguide. **b** A series of reflection waves on the upper and lower surfaces of the plate waveguide

$$\nabla^2 \psi = c_S^{-2} \frac{\partial^2 \psi}{\partial t^2}. \tag{6.3.1b}$$

We consider a wave along x direction and assume that the displacements u and w along x and z directions are not zero, while the displacement v along y direction is zero. Thus, $\frac{\partial(\cdot)}{\partial y} = 0$, $\sigma_{xy} = \sigma_{yz} = 0$. Assume the solution

$$\phi = f(z) \exp[ik(x - ct)], \tag{6.3.2a}$$

$$\psi = g(z) \exp[ik(x - ct)], \tag{6.3.2b}$$

where k and c are the wave number and velocity along x direction (often called apparent wave number and apparent wave velocity). These waves are formed by the interference of reflected waves that are repeatedly reflected at the upper and lower surfaces of the plate. A series of reflection waves in the plate are shown in Fig. 6.8. These reflection waves interfere with each other and eventually form plate waves, which propagate in the plate under the guidance of the upper and lower surfaces, as shown in Fig. 6.9. According to the Snell's law that the reflection waves follow, the reflection waves have the same frequency and the same wave number along the x direction. So, the following relationship exists

$$\omega = k_P c_P = k_S c_S = kc, \tag{6.3.3a}$$

$$k = k_P \sin\alpha = k_S \sin\beta, \tag{6.3.3b}$$

$$c = \frac{c_P}{\sin\alpha} = \frac{c_S}{\sin\beta}. \tag{6.3.3c}$$

In Eq. (6.3.3a), k_P and k_S are the wave numbers of reflection P and SV waves, respectively. c_P and c_S are wave velocities of reflection P and SV waves, respectively. k and c are apparent wave number and apparent wave velocity, respectively. α and β are the reflection angle of reflection P and SV waves, respectively.

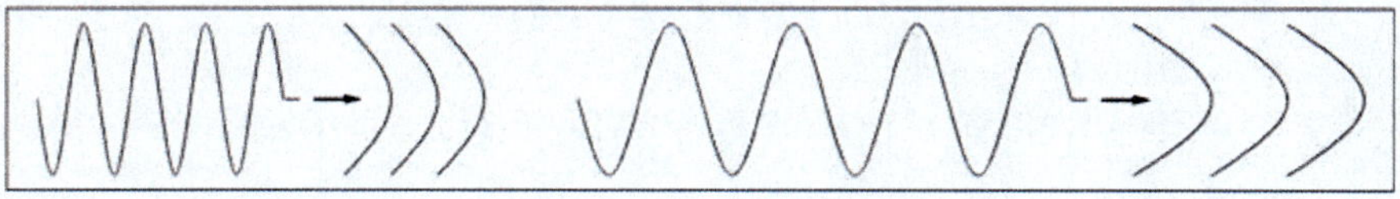

Fig. 6.9 Plate wave (Lamb wave) propagating in the plate waveguide

Inserting the solution, i.e. Eq. (6.3.2), into the wave motion equation Eq. (6.3.1) leads to

$$f'' + \eta_P^2 f = 0, \tag{6.3.4a}$$

$$g'' + \eta_S^2 g = 0, \tag{6.3.4b}$$

where

$$\eta_P^2 = k^2\left[\left(\frac{c}{c_P}\right)^2 - 1\right] = \left(\frac{\omega}{c_P}\right)^2 - k^2 = k_P^2 - k^2 = (k_P \cos\alpha)^2, \tag{6.3.5a}$$

$$\eta_S^2 = k^2\left[\left(\frac{c}{c_S}\right)^2 - 1\right] = \left(\frac{\omega}{c_S}\right)^2 - k^2 = k_S^2 - k^2 = (k_S \cos\beta)^2. \tag{6.3.5b}$$

Obviously, η_P and η_S are the wave numbers of the reflection longitudinal wave and the reflection transverse wave along the plate thickness direction, which are the components of the wave vector k_P and k_S along z direction. The solutions of Eq. (6.3.4) are

$$f(z) = A \sin \eta_P z + B \cos \eta_P z, \tag{6.3.6a}$$

$$g(z) = C \sin \eta_S z + D \cos \eta_S z. \tag{6.3.6b}$$

Thus, the solution of the wave equation can be written as

$$\phi = (A \sin \eta_P z + B \cos \eta_P z) \exp[ik(x - ct)], \tag{6.3.7a}$$

$$\psi = (C \sin \eta_S z + D \cos \eta_S z) \exp[ik(x - ct)]. \tag{6.3.7b}$$

Using the relations of displacement potential and stress potential, we obtain the corresponding expression of displacement and stress,

$$u = \left(ikf(z) - g'(z)\right) \exp[ik(x - ct)], \tag{6.3.8a}$$

$$w = \left(f'(z) + ikg(z)\right) \exp[ik(x - ct)], \tag{6.3.8b}$$

$$\sigma_{xx} = \mu\left[\left(2\eta_P^2 - k_S^2\right) f(z) - i2kg'(z)\right] \exp[ik(x - ct)], \tag{6.3.9a}$$

$$\sigma_{zz} = \mu\left[\left(k^2 - \eta_S^2\right) f(z) + i2kg'(z)\right] \exp[ik(x - ct)], \tag{6.3.9b}$$

$$\sigma_{zx} = \mu\left[i2kf'(z) + \left(\eta_S^2 - k^2\right) g(z)\right] \exp[ik(x - ct)], \tag{6.3.9c}$$

$$\sigma_{yy} = \nu(\sigma_{xx} + \sigma_{zz}), \tag{6.3.9d}$$

where μ is the shear modulus and ν is Poisson's ratio. Several specific boundary conditions will be discussed next.

6.3.1 Mixed Boundary Condition

Mixed boundary conditions can be expressed as

$$w(x, z) = \sigma_{zx}(x, z) = 0, \quad (z = \pm h). \tag{6.3.10}$$

Such boundary conditions physically indicate that the plate is rigidly clamped in the normal direction of the boundary, but the surface is lubricated and can move freely between the plate and the rigid constraint.

Inserting Eq. (6.3.6) into Eqs. (6.3.8b) and (6.3.9c), we obtain

$$\eta_P \cos \eta_P h \cdot A - \eta_P \sin \eta_P h \cdot B + ik \sin \eta_S h \cdot C + ik \cos \eta_S h \cdot D = 0, \tag{6.3.11a}$$

$$\eta_P \cos \eta_P h \cdot A + \eta_P \sin \eta_P h \cdot B - ik \sin \eta_S h \cdot C + ik \cos \eta_S h \cdot D = 0, \tag{6.3.11b}$$

$$\begin{aligned} &2ik\eta_P \cos \eta_P h \cdot A - 2ik\eta_P \sin \eta_P h \cdot B \\ &\quad + \left(\eta_S^2 - k^2\right) \sin \eta_S h \cdot C + \left(\eta_S^2 - k^2\right) \cos \eta_S h \cdot D = 0, \end{aligned} \tag{6.3.11c}$$

$$\begin{aligned} &2ik\eta_P \cos \eta_P h \cdot A + 2ik\eta_P \sin \eta_P h \cdot B \\ &\quad - \left(\eta_S^2 - k^2\right) \sin \eta_S h \cdot C + \left(\eta_S^2 - k^2\right) \cos \eta_S h \cdot D = 0. \end{aligned} \tag{6.3.11d}$$

After adding and subtracting the first two equations as well as adding and subtracting the last two equations, two sets of equations can be obtained by rearranging them

$$\begin{aligned}&\eta_P \cos \eta_P h \cdot A + ik \cos \eta_S h \cdot D = 0,\\&2ik\eta_P \cos \eta_P h \cdot A + \left(\eta_S^2 - k^2\right) \cos \eta_S h \cdot D = 0,\end{aligned} \tag{6.3.12a}$$

and

$$\begin{aligned}&\eta_P \sin \eta_P h \cdot B - ik \sin \eta_S h \cdot C = 0,\\&2ik\eta_P \sin \eta_P h \cdot B - \left(\eta_S^2 - k^2\right) \sin \eta_S h \cdot C = 0.\end{aligned} \tag{6.3.12b}$$

The former is the equation set concerning A and B. The latter is the equation set concerning C and D. This decomposition has obvious physical significance. The displacements given by terms including B and C are symmetric with respect to the x-axis, while the displacements given by terms including A and D are antisymmetric with respect to the x-axis, which can be seen from Eqs. (6.3.6) and (6.3.8). Thus, the deformations caused by the waves are decomposed into symmetric and antisymmetric parts in Eq. (6.3.12).

Consider first the antisymmetric case. According to the existence of a non-trivial solution in Eq. (6.3.12a), we can obtain

$$\eta_P\left(\eta_s^2 + k^2\right) \cos \eta_P h \cdot \cos \eta_S h = 0, \tag{6.3.13}$$

which is the frequency equation for the antisymmetric mode. If Eq. (6.3.13) is valid, wave number should satisfy

$$\eta_P = 0, \quad \text{or } \eta_P = \frac{m\pi}{2h}, \quad \text{or } \eta_S = \frac{n\pi}{2h} (m, n = 1, 3, 5 \ldots). \tag{6.3.14}$$

According to (6.3.12a), when $\eta_P = 0$ or $\eta_P = \frac{m\pi}{2h}$, the amplitudes satisfy $A \neq 0$ and$D = 0$, which indicates that the antisymmetric mode is generated by the reflection of P wave. When $\eta_S = \frac{n\pi}{2h}$, the amplitudes satisfy $A = 0$ and$D \neq 0$, which indicates that the antisymmetric mode is generated by the reflection of SV wave. Therefore, under mixed boundary conditions, P wave and SV wave are uncoupled and can exist separately in the antisymmetric modes. This is the same as the reflection pattern of P wave and SV wave in half-space under the same boundary condition, that is, there is no mode conversion of wave.

Next, consider the symmetric case. According to Eq. (6.3.12b), we can obtain the frequency equation of the symmetric mode,

$$\eta_P\left(\eta_S^2 + k^2\right) \sin \eta_P h \cdot \sin \eta_S h = 0. \tag{6.3.15}$$

If Eq. (6.3.15) is valid, the wave number must satisfy

$$\eta_P = \frac{m\pi}{2h}, \text{ or } \eta_S = \frac{n\pi}{2h} (m = 0, 2, 4 \ldots, n = 2, 4, 6 \ldots). \tag{6.3.16}$$

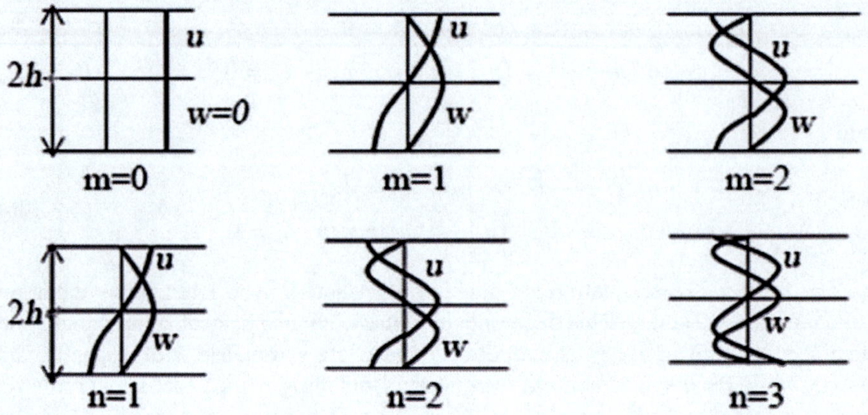

Fig. 6.10 Sketch of displacement distribution of symmetric and antisymmetric modes on the cross section

Obviously, P wave and SV wave are also uncoupled in the symmetric mode, that is, P wave and SV wave can exist separately. In fact, the expansion wave and the equal-volume wave are not coupled together in the reflected or flat waveguide problem, which is an important feature of the mixed boundary conditions.

The modes of the wave propagation are determined by different values of *m* and *n*. When *m* and *n* are odd, the modes of antisymmetric expansion wave and equal-volume wave are, respectively, given. When *m* and *n* are even, the modes of symmetric expansion wave and equal-volume wave are, respectively, given. Inserting Eqs. (6.3.14) and (6.3.16) into Eqs. (6.3.7), the propagation modes corresponding to ϕ and Ψ(expansion wave and equal-volume wave) can be obtained. Similarly, the displacement corresponding to guided waves (composed of expansion wave and equal-volume wave) of different orders on the cross section can be obtained from Eqs. (6.3.8). Figure 6.10 shows the displacements u and w of the first three order modes.

From Eq. (6.3.5a), by using the expressions of η_d in Eqs. (6.3.14) and (6.3.16), we can obtain

$$\left(\frac{\omega}{c_P}\right)^2 = \left(\frac{m\pi}{2h}\right)^2 + k^2. \tag{6.3.17a}$$

The dimensionless form of Eq. (6.3.17a) is

$$\Omega^2 = \left(\frac{\omega}{\omega_S}\right)^2 = D^2\left(m^2 + \zeta^2\right), \tag{6.3.17b}$$

where

$$\omega_S = \frac{\pi c_S}{2h}, \quad \zeta = \frac{2hk}{\pi}, \quad D = \frac{c_P}{c_S}.$$

Equation (6.3.17) is the frequency equation of the expansion wave. It can be seen that for a particular expansion wave mode, there is a frequency-wave number continuous spectrum curve corresponding to it, which is a branch of the dispersion relation. Obviously, for different values of m, the corresponding curves are different, and there are an infinite number of such curves. When we define Ω and ζ, the thickness h is used as a fundamental quantity. The long or short waves mentioned later are relative to h. Similarly, the low or high frequencies are relative to ω_s.

From Eq. (6.3.17a), we can obtain

$$\left(\frac{c}{c_P}\right)^2 = \left(\frac{m\pi}{2hk}\right)^2 + 1. \tag{6.3.18a}$$

The dimensionless form of Eq. (6.3.18a) is

$$C^2 = \left(\frac{c}{c_S}\right)^2 = D^2\left[\left(\frac{m}{\zeta}\right)^2 + 1\right]. \tag{6.3.18b}$$

The relation between phase velocity and wave number is given by Eq. (6.3.18), which is also known as the dispersion relation.

Similarly, the frequency spectrum and phase velocity spectrum of the equal-volume wave can also be obtained:

$$\Omega^2 = n^2 + \zeta^2, \tag{6.3.19}$$

$$C^2 = \left(\frac{n}{\zeta}\right)^2 + 1. \tag{6.3.20}$$

Physically, Ω is required to be a positive real number. ζ can be either a real number γ or a purely imaginary number $i\delta$, but not a complex number, which can be seen from Eqs. (6.3.17b) and (6.3.19). Similarly, C is required to be a positive real number physically. But we do not have to force ζ to be a real number. The imaginary ζ corresponds to the stationary wave decaying in x direction. In fact, from $\zeta = i\delta$, we know

$$\exp[ik(x - ct)] = \exp\left[i\zeta\left(\hat{x} - \hat{c}\hat{t}\right)\right] = \exp\left(-\delta\hat{x}\right)\exp(-i\,\Omega\,\hat{t}),$$

where $\hat{x} = \frac{\pi x}{2h}$, $\hat{t} = \frac{\pi c_S t}{2h}$, $\hat{c} = \frac{c}{c_S}$. The above equation represents a vibration that attenuates but not propagates in x direction.

As an example of finding the roots of the dispersion equation, we have performed a calculation on an aluminum plate with the thick of 20 mm. Figure 6.11 shows the dispersion curve of aluminum plate with mixed boundary conditions. The group of

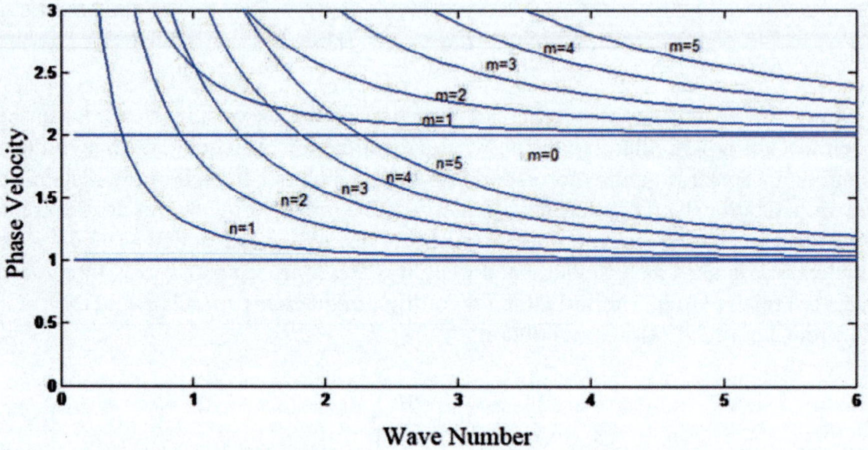

Fig. 6.11 Dispersion curves of guided waves in aluminum plate with mixed boundary conditions

$(m, n = 0, 2, 4...)$ corresponds to the symmetric mode while the group of $(m, n = 1, 3, 5...)$ corresponds to the antisymmetric mode.

6.3.2 Free Boundary Conditions

We have previously discussed the dispersion equation of guided waves in a plate with mixed boundary conditions. Because P and SV waves are always independent or uncoupled there, the dispersion equation can be solved analytically. However, it does not mean that P and SV waves are always independent of each other for any boundary condition. The P and SV waves under free boundary conditions are always coupled together. This is exactly analogous to the previously discussed case of plane strain P and SV waves reflected from an elastic half-space surface. The free boundary condition of a plate can be expressed as

$$\sigma_{zz} = \sigma_{zx} = 0, \quad (z = \pm h). \tag{6.3.21}$$

Consider first the symmetric mode. Assume $A = D = 0$ in Eqs. (6.3.6), and then insert it into Eqs. (6.3.9b) and (6.3.9c). Using the boundary condition Eq. (6.3.21), we can obtain

$$\begin{cases} B\left(k^2 - \eta_S^2\right) \cos \eta_P h + 2iCk\eta_S \cos \eta_S h = 0 \\ 2iBk\eta_P \sin \eta_P h + C\left(k^2 - \eta_S^2\right) \sin \eta_S h = 0 \end{cases}. \tag{6.3.22}$$

Consider the antisymmetry mode next. Assume $B = C = 0$ in Eqs. (6.3.6). Similarly, we can obtain

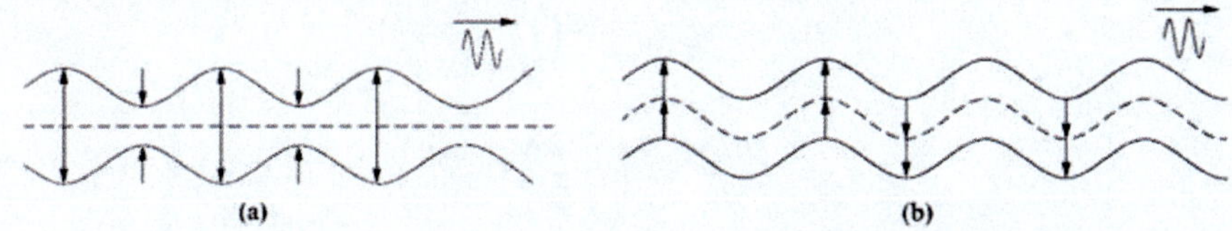

(a) symmetrical mode (no deformation in the midplane) (b) anti-symmetric mode

Fig. 6.12 Surface displacement caused by symmetric mode and antisymmetric mode and the corresponding middle-face displacement. **a** Symmetric mode. **b** Antisymmetric mode

$$\begin{cases} A\left(k^2 - \eta_S^2\right) \sin \eta_P h - 2iDk\eta_S \sin \eta_S h = 0 \\ 2iAk\eta_P \cos \eta_P h - D\left(k^2 - \eta_S^2\right) \cos \eta_S h = 0 \end{cases}. \tag{6.3.23}$$

Equation (6.3.22) is homogeneous linear algebraic equations with undetermined coefficients B and C. Equation (6.3.23) is homogeneous linear algebraic equations with undetermined coefficients A and D. The surface displacement caused by symmetric mode and antisymmetric mode and the corresponding middle-face displacement are shown in Fig. 6.12.

According to the existence of a non-trivial solution in homogeneous linear algebraic equations, we can obtain

$$\frac{2k\eta_S \cos \eta_S h}{\left(\eta_S^2 - k^2\right) \cos \eta_P h} = \frac{\left(k^2 - \eta_S^2\right) \sin \eta_S h}{2k\eta_d \sin \eta_P h} \text{(Symmetric mode)}, \tag{6.3.24}$$

$$\frac{2k\eta_S \sin \eta_S h}{\left(k^2 - \eta_S^2\right) \sin \eta_P h} = \frac{\left(\eta_S^2 - k^2\right) \cos \eta_S h}{2k\eta_d \cos \eta_P h} \text{(Antisymmetric mode)}. \tag{6.3.25}$$

Equations (6.3.24) and (6.3.25) can be combined into an equation, i.e.

$$F(\eta_P(\omega), \eta_S(\omega), k) = \frac{\tan \eta_S h}{\tan \eta_P h} + \left[\frac{4\eta_P \eta_S k^2}{\left(k^2 - \eta_S^2\right)^2}\right]^{\pm 1} = 0. \tag{6.3.26}$$

The dimensionless form of Eq. (6.3.26) is

$$F(\Omega, \zeta) = \frac{\tan\left(\frac{\pi \eta_S'}{2}\right)}{\tan\left(\frac{\pi \eta_P'}{2}\right)} + \left[\frac{4\eta_S' \eta_P' \zeta^2}{\left(\zeta^2 - \eta_S'\right)^2}\right]^{\pm 1} = 0, \tag{6.3.27}$$

where

$$\Omega = \frac{\omega}{\omega_S},\ \omega_S = \frac{\pi c_S}{2h},\ \zeta = \frac{2hk}{\pi},$$

$$\eta_S' = \frac{2h\eta_S}{\pi} = (\Omega^2 - \zeta^2)^{\frac{1}{2}},$$

$$\eta_p' = \frac{2h\eta_P}{\pi} = \left(\left(\frac{\Omega}{D}\right)^2 - \zeta^2\right)^{\frac{1}{2}}.$$

Equation (6.3.26) or Eq. (6.3.27) is called **Rayleigh-Lamb** equation. Since the equation contains the tangent function (multiple-valued function), the solution of the equation is not unique for any given apparent wave number k. It is usually necessary to find all the roots $\omega_i (i = 1,2,\ldots)$ in the frequency range of interest by dichotomy. Different from the mixed boundary conditions, for each root of the equation $\omega_i (i = 1,2,\ldots)$, the corresponding $\eta_P(\omega)$ and $\eta_S(\omega)$ are not zero. Physically, this means that the P wave and SV wave are coupled, that is, for guided waves of any mode, the reflection P wave and SV wave are coexisting. As an example of finding the root of the dispersion equation above, we calculate the steel plate. The thickness of the steel plate is 20 mm, and wave velocities of longitudinal wave and transverse wave are 5790 and 3200 m/s, respectively. The dispersion curves are shown in Fig. 6.13, which describes the dispersion characteristics of the first few order symmetric and antisymmetric modes. Symbol S(blue curve) denotes the symmetric mode, and A(red curve) denotes the antisymmetric mode. The subscript numbers 0,1,2... of A_0, A_1, $A_2\ldots$ and S_0, S_1, $S_2\ldots$ represent the order of the modes.

6.3.3 Fixed Boundary Condition

The fixed boundary conditions can be expressed as

$$u(x, z) = w(x, z) = 0, \quad (z = \pm h). \tag{6.3.28}$$

Inserting Eqs. (6.3.8) into Eq. (6.3.28), we can obtain the homogeneous equations of A, B, C and D

$$ikA \sin \eta_P h + ikB \cos \eta_P h - C\eta_S \cos \eta_S h + D\eta_S \sin \eta_S h = 0, \tag{6.3.29a}$$

$$-ikA \sin \eta_P h + ikB \cos \eta_P h - C\eta_S \cos \eta_S h - D\eta_S \sin \eta_S h = 0, \tag{6.3.29b}$$

$$A\eta_P \cos \eta_P h - B\eta_P \sin \eta_P h + ikC \sin \eta_S h + ikD \cos \eta_S h = 0, \tag{6.3.29c}$$

$$A\eta_P \cos \eta_P h + B\eta_P \sin \eta_P h - ikC \sin \eta_S h + ikD \cos \eta_S h = 0. \tag{6.3.29d}$$

After subtracting the first two equations and adding the last two equations, frequency equation of antisymmetric mode can be obtained,

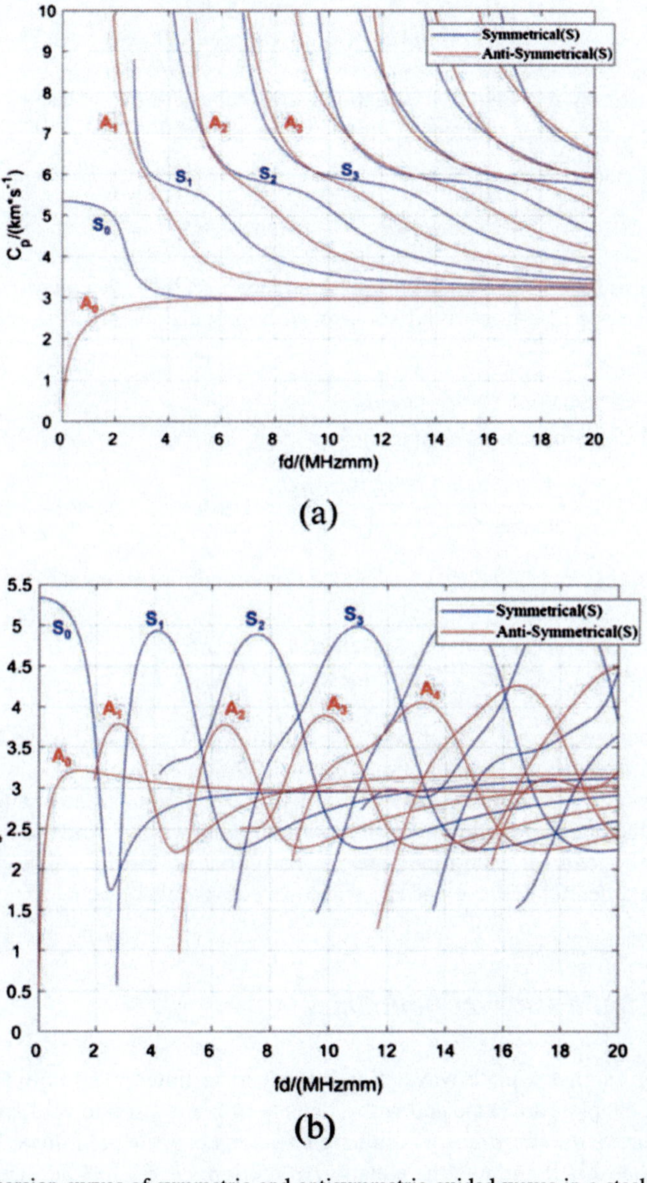

Fig. 6.13 Dispersion curves of symmetric and antisymmetric guided waves in a steel plate with free boundary conditions. **a** Phase velocity. **b** Group velocity

$$\begin{cases} ik \sin \eta_P h \cdot A + \eta_S \sin \eta_S h \cdot D = 0 \\ \eta_P \cos \eta_P h \cdot A + ik \cos \eta_S h \cdot D = 0 \end{cases}. \tag{6.3.30}$$

After adding the first two equations and subtracting the last two equations, frequency equation of symmetric mode can be obtained,

$$\begin{cases} ik \cos \eta_P h \cdot B - \eta_S \cos \eta_S h \cdot C = 0 \\ \eta_P \sin \eta_P h \cdot B - ik \sin \eta_S h \cdot C = 0 \end{cases}. \tag{6.3.31}$$

Equations (6.3.30) and (6.3.31) are homogeneous algebraic equations. According to the existence of a non-trivial solution of homogeneous algebraic equations, we can obtain

$$\frac{k \sin \eta_P h}{\eta_S \sin \eta_S h} = \frac{\eta_P \cos \eta_P h}{-k \cos \eta_S h}, \text{ (antisymmetric mode)}, \tag{6.3.32}$$

$$\frac{k \sin \eta_S h}{\eta_P \sin \eta_P h} = \frac{\eta_S \cos \eta_S h}{-k \cos \eta_P h}, \text{ (symmetric mode)}. \tag{6.3.33}$$

Equations (6.3.32) and (6.3.33) can be combined into an equation, i.e.

$$F(\eta_P, \eta_S, k) = \frac{\tan \eta_S h}{\tan \eta_P h} + \left(\frac{k^2}{\eta_S \eta_P}\right)^{\pm 1} = 0.$$

The above equation is the frequency equation of the guided wave in an infinite plate with fixed boundary conditions. Same as the free boundary condition, for any given apparent wave number k, the solution of the equation is also not unique, which indicates that there are infinitely many modes of plate waves. And the corresponding $\eta_P(\omega)$ and $\eta_S(\omega)$ for each mode are not zero, that is, for the guided wave of any mode, the reflection P wave and SV wave are coexisting.

6.3.4 Liquid Load on Both Sides

Figure 6.14 shows a plate waveguide immersed in liquid. The liquids on top and bottom of the plate are same and with a density of ρ and a sound velocity of c. For an inviscid liquid, the stress tensor is spherical tensor, i.e. $\sigma_{ij} = p\delta_{ij}$. Inserting the stress tensor of inviscid liquid into the equation of motion, i.e. $\sigma_{ij,j} = \rho\ddot{u}_i$, the equation of sound wave in the liquid is obtained,

$$\nabla p(x, z, t) = \rho\ddot{u}_i(x, z, t),$$

or equivalently

$$p(x, z, t) = \rho\ddot{\varphi}(x, z, t).$$

Correspondingly, the constitutive equation is reduced to $p = \lambda\varepsilon_v$ (ε_v is the volume strain of the liquid, and λ is the compression modulus of the liquid). According to the wave equation of fluid, the displacement potential function in the fluid can be written as

$$\begin{aligned}\varphi_1 &= F_1 \exp[i(kx + rz - \omega t)],\\ \varphi_2 &= F_2 \exp[i(kx - rz - \omega t)],\end{aligned} \tag{6.3.34}$$

where $r = \sqrt{\frac{\omega^2}{c^2} - k^2}$. The exponential term of coordinate z in the two potential functions has opposite signs because the projections of propagation directions of sound wave in the upper and lower liquid along coordinate z are opposite. According to the displacement potential and the constitutive equation of liquid, the displacement and sound pressure in the liquid can be obtained (Fig. 6.14),

$$w_1 = \frac{\partial \varphi_1}{\partial z} = irF_1 \exp[i(kx + rz - \omega t)], \tag{6.3.35a}$$

$$p_1 = -\rho\frac{\partial^2 \varphi_1}{\partial t^2} = \rho\omega^2 F_1 \exp[i(kx + rz - \omega t)], \tag{6.3.35b}$$

$$w_2 = \frac{\partial \varphi_2}{\partial z} = -irF_2 \exp[i(kx - rz - \omega t)], \tag{6.3.35c}$$

$$p_2 = -\rho\frac{\partial^2 \varphi_2}{\partial t^2} = \rho\omega^2 F_2 \exp[i(kx - rz - \omega t)]. \tag{6.3.35d}$$

The boundary conditions of the plate with the liquid load on both sides are that the shear stress at the upper and lower boundaries are zero, while the normal stress and the displacement in z direction are continuous, namely

$$\sigma_{zx}|_{z=\pm h} = 0, \tag{6.3.36a}$$

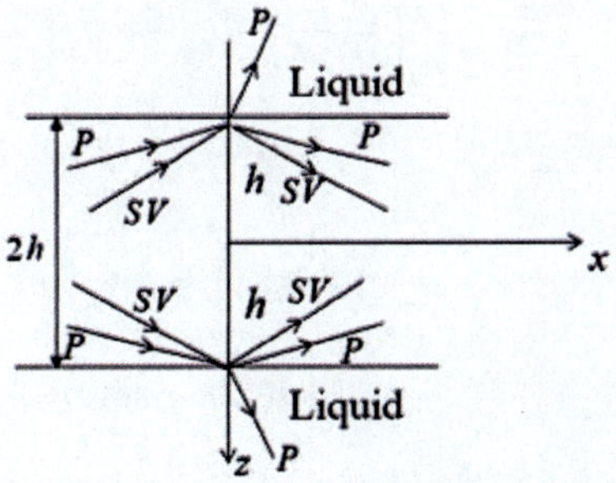

Fig. 6.14 Guided waves in an infinite plate with liquid loads on both sides

$$w_{|z=h} = w_{1|z=h}\,,\; w_{|z=-h} = w_{2|z=-h}\,, \tag{6.3.36b}$$

$$\sigma_{zz} = -p_{1|z=h},\; \sigma_{zz} = -p_{2|z=-h}. \tag{6.3.36c}$$

The condition that shear stress is zero can be expressed as

$$\begin{aligned} &2ik\eta_P \cos(\eta_P h)\cdot A - 2ik\eta_P \sin(\eta_P h)\cdot B \\ &+(\eta_S^2 - k^2)\sin(\eta_S h)\cdot C + (\eta_S^2 - k^2)\cos(\eta_S h)\cdot D = 0, \end{aligned} \tag{6.3.37a}$$

$$\begin{aligned} &2ik\eta_P \cos(\eta_d h)\cdot A + 2ik\eta_P \sin(\eta_P h)\cdot B \\ &-(\eta_S^2 - k^2)\sin(\eta_S h)\cdot C + (\eta_S^2 - k^2)\cos(\eta_S h)\cdot D = 0. \end{aligned} \tag{6.3.37b}$$

The condition that the normal displacement is continuous can be expressed as

$$\begin{aligned} &\eta_P \cos\eta_P h\cdot A - \eta_P \sin\eta_P h\cdot B \\ &+ik\sin\eta_S h\cdot C + ik\cos\eta_S h\cdot D = irF_1\exp(irh), \end{aligned} \tag{6.3.38a}$$

$$\begin{aligned} &\eta_P \cos\eta_P h\cdot A + \eta_P \sin\eta_P h\cdot B \\ &-ik\sin\eta_S h\cdot C + ik\cos\eta_{S_s} h\cdot D = -irF_2\exp(irh). \end{aligned} \tag{6.3.38b}$$

The condition that the normal stress is continuous can be expressed as

$$\begin{aligned} &-\mu(k^2 - \eta_S^2)\sin\eta_P h\cdot A - \mu(k^2 - \eta_S^2)\cos\eta_P h\cdot B \\ &-2i\mu k\eta_S\cos\eta_S h\cdot C + 2i\mu k\eta_S\sin\eta_S h\cdot D = \rho\omega^2 F_1\exp(irh), \end{aligned} \tag{6.3.39a}$$

$$\begin{aligned} &\mu(k^2 - \eta_S^2)\sin\eta_P h\cdot A - \mu(k^2 - \eta_S^2)\cos\eta_P h\cdot B \\ &-2i\mu k\eta_S\cos\eta_S h\cdot C + 2i\mu k\eta_S\sin\eta_S h\cdot D = \rho\omega^2 F_2\exp(irh). \end{aligned} \tag{6.3.39b}$$

Inserting Eq. (6.3.38) into Eq. (6.3.39) and eliminating F_1 and F_2, we get

$$\begin{aligned} &\left[\mu(k^2 - \eta_S^2)\sin\eta_P h + \frac{\rho\omega^2\eta_P}{ir}\cos\eta_P h\right]A \\ &+\left[\mu(k^2 - \eta_S^2)\cos\eta_P h - \frac{\rho\omega^2\eta_P}{ir}\sin\eta_P h\right]B \\ &+\left[2i\mu k\eta_S\cos\eta_S h + \frac{\rho\omega^2 k}{r}\sin\eta_S h\right]C \\ &-\left[2i\mu k\eta_S\sin\eta_S h - \frac{\rho\omega^2 k}{r}\cos\eta_S h\right]D = 0, \end{aligned} \tag{6.3.40a}$$

$$\left[\mu\left(k^2-\eta_S^2\right)\sin\eta_P h+\frac{\rho\omega^2\eta_P}{ir}\cos\eta_P h\right]A$$
$$-\left[\mu\left(k^2-\eta_S^2\right)\cos\eta_P h-\frac{\rho\omega^2\eta_P}{ir}\sin\eta_P h\right]B$$
$$-\left[2i\mu k\eta_S\cos\eta_S h+\frac{\rho\omega^2 k}{r}\sin\eta_S h\right]C$$
$$-\left[2i\mu k\eta_S\sin\eta_S h-\frac{\rho\omega^2 k}{r}\cos\eta_S h\right]D=0. \tag{6.3.40b}$$

After subtracting the two equations in Eq. (6.3.40), we get

$$2i\mu k\eta_P\sin\eta_P hB+\mu\left(k^2-\eta_S^2\right)\sin\eta_S h\cdot C=0, \tag{6.3.41a}$$

$$\left[\mu\left(k^2-\eta_S^2\right)\cos\eta_P h-\frac{\rho\omega^2\eta_P}{ir}\sin\eta_P h\right]\cdot B$$
$$+\left[2i\mu k\eta_S\cos\eta_S h+\frac{\rho\omega^2 k}{r}\sin\eta_S h\right]\cdot C=0. \tag{6.3.41b}$$

The non-zero solution condition requires the determinant of coefficient to be zero, i.e.

$$\frac{\left(k^2-\eta_S^2\right)^2}{\tan\eta_P h}+\frac{4k^2\eta_S\eta_P}{\tan\eta_S h}-i\frac{\rho\omega^2\eta_P}{\mu r}\left(k^2+\eta_S^2\right)=0. \tag{6.3.42}$$

The above equation is the dispersion equation of the symmetric mode of Lamb wave in a plate with liquid loads on both sides. Similarly, by adding Eqs. (6.3.40a) and (6.3.40b), the dispersion equation of antisymmetric mode of Lamb wave can be written as

$$\left(k^2-\eta_S^2\right)^2\tan\eta_P h+4k^2\eta_S\eta_P\tan\eta_S h+i\frac{\rho\omega^2\eta_P}{\mu r}\left(k^2+\eta_S^2\right)=0. \tag{6.3.43}$$

For a thin plate immersed in liquid, due to the presence of surrounding liquid, the energy of carried Lamb waves propagated in the plate will leak to the surrounding medium, and its vibration amplitude will decay with the increase of propagation distance. In this case, the wave number k is a complex number. Mathematically, it implies that Eqs. (6.3.42) and (6.3.43) have solutions only in complex number domain. In other words, solutions in the real number domain do not exist. We assume that the wave number has the form

$$k=k_r+ik_i,$$

where the real part k_r is the wave number in the general sense, i.e. $k_r = \frac{\omega}{c_P}$, while the imaginary part k_i corresponds to the attenuation rate of the amplitude of Lamb wave. Therefore, in order to obtain the solutions of Eqs. (6.3.42) and (6.3.43), we should to scan the complex plane. That is, for a given frequencyω, we should find the corresponding complex wave number $k = k_r + ik_i$ so that Eqs. (6.3.42) and (6.3.43) are satisfied. Meanwhile, the dispersion relation ($k_r \sim \omega$) and the attenuation relation ($k_r \sim \omega$) are both obtained. For this situation, we cannot use the dichotomy which is suitable to find the root in the real number domain, in general, instead, we should use the contour integral method of complex function to find the root of these two transcendental equations.

6.4 Guided Waves in Cylindrical Rod

The gradient operator and Laplace operator in cylindrical coordinate system are, respectively,

$$\nabla = \frac{\partial}{\partial r}\boldsymbol{e}_r + \frac{1}{r}\frac{\partial}{\partial \theta}\boldsymbol{e}_\theta + \frac{\partial}{\partial z}\boldsymbol{e}_z, \tag{6.4.1}$$

$$\nabla^2 = \frac{\partial^2}{\partial r^2} + \frac{1}{r}\frac{\partial}{\partial r} + \frac{1}{r^2}\frac{\partial^2}{\partial \theta^2} + \frac{\partial^2}{\partial z^2}. \tag{6.4.2}$$

The wave equation can still be expressed as

$$\nabla^2 \phi = c_d^{-2}\frac{\partial^2 \phi}{\partial t^2}, \tag{6.4.3a}$$

$$\nabla^2 \chi = c_s^{-2}\frac{\partial^2 \chi}{\partial t^2}, \tag{6.4.3b}$$

$$\nabla^2 \eta = c_s^{-2}\frac{\partial^2 \eta}{\partial t^2}. \tag{6.4.3c}$$

The displacement field can be expressed as

$$\mathbf{u} = \nabla\phi + \nabla \times \boldsymbol{\psi} = \nabla\phi + \nabla \times (\chi \mathbf{e}_z) + \nabla \times \nabla \times (\eta \mathbf{e}_z). \tag{6.4.4}$$

In this way, we get the displacement-potential relationship as follows

$$u_r = \frac{\partial \phi}{\partial r} + \frac{\partial \chi}{r\partial \theta} + \frac{\partial^2 \eta}{\partial z \partial r}, \tag{6.4.5a}$$

$$u_\theta = \frac{\partial \phi}{r\partial \theta} - \frac{\partial \chi}{\partial r} + \frac{\partial^2 \eta}{r\partial z \partial \theta}, \tag{6.4.5b}$$

$$u_z = \frac{\partial \phi}{\partial z} - \frac{\partial}{r\partial r}\left(r\frac{\partial \eta}{\partial r}\right) - \frac{\partial^2 \eta}{r^2 \partial^2 \theta}. \tag{6.4.5c}$$

By substituting them into the expressions for strain and stress

$$\varepsilon_{rr} = \frac{\partial u_r}{\partial r}, \quad \varepsilon_{\theta\theta} = \frac{1}{r}\frac{\partial u_\theta}{\partial \theta} + \frac{u_r}{r}, \quad \varepsilon_{zz} = \frac{\partial u_z}{\partial z}, \tag{6.4.6a,b,c}$$

$$\varepsilon_{\theta z} = \frac{1}{2}\left(\frac{\partial u_\theta}{\partial z} + \frac{1}{r}\frac{\partial u_z}{\partial \theta}\right), \tag{6.4.6d}$$

$$\varepsilon_{zr} = \frac{1}{2}\left(\frac{\partial u_r}{\partial z} + \frac{\partial u_z}{\partial r}\right), \tag{6.4.6e}$$

$$\varepsilon_{r\theta} = \frac{1}{2}\left(\frac{1}{r}\frac{\partial u_r}{\partial \theta} + \frac{\partial u_\theta}{\partial r} - \frac{u_\theta}{r}\right), \tag{6.4.6f}$$

$$\sigma_{rr} = \lambda\varepsilon_{kk} + 2\mu\varepsilon_{rr}, \quad (\varepsilon_{kk} = \varepsilon_{rr} + \varepsilon_{\theta\theta} + \varepsilon_{zz}), \tag{6.4.7a}$$

$$\sigma_{\theta\theta} = \lambda\varepsilon_{kk} + 2\mu\varepsilon_{\theta\theta}, \tag{6.4.7b}$$

$$\sigma_{zz} = \lambda\varepsilon_{kk} + 2\mu\varepsilon_{zz}, \tag{6.4.7c}$$

$$\sigma_{r\theta} = 2\mu\varepsilon_{r\theta}, \tag{6.4.7d}$$

$$\sigma_{\theta z} = 2\mu\varepsilon_{\theta z}, \tag{6.4.7e}$$

$$\sigma_{rz} = 2\mu\varepsilon_{rz}, \tag{6.4.7f}$$

the relations of strain—potential and stress—potential are obtained.

Guided waves propagating in an infinitely long cylindrical rod of radius a, as shown in Fig. 6.15, are discussed below. Assume that the surface of the cylindrical rod is free. Thus, the boundary conditions can be expressed as

$$\sigma_{rr} = \sigma_{r\theta} = \sigma_{rz} = 0 \quad (\text{at}\, r = a). \tag{6.4.8}$$

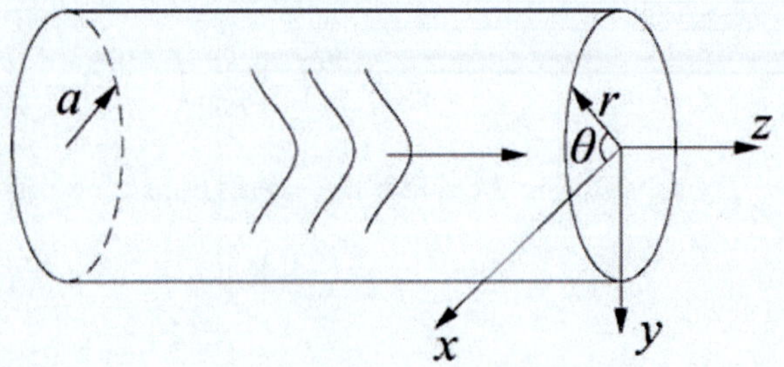

Fig. 6.15 Guided waves in an infinitely long cylindrical rod

6.4.1 Axisymmetric Torsional Waves

Cylindrical rods can undergo various deformation modes, such as stretching, twisting, bending and their combination. The torsional deformation satisfies the following conditions

$$u_r(r, z, t) = u_z(r, z, t) = 0, \quad u_\theta(r, z, t) \neq 0, \quad \frac{\partial(\cdot)}{\partial\theta} = 0. \tag{6.4.9a,b,c}$$

The guided waves in a cylindrical bar in the torsional deformation mode are called axisymmetric torsional waves. In this situation, the non-zero strain components are

$$\varepsilon_{r\theta} = \frac{1}{2} r \frac{\partial}{\partial r}\left(\frac{u_\theta}{r}\right), \quad \varepsilon_{z\theta} = \frac{1}{2}\frac{\partial u_\theta}{\partial z}. \tag{6.4.10a,b}$$

For axisymmetric torsional waves, it is more convenient to use the circumferential displacement u_θ as the basic unknown quantity directly. The wave equation in term of the circumferential displacement can be expressed as

$$\frac{\partial^2 u_\theta}{\partial r^2} + \frac{1}{r}\frac{\partial u_\theta}{\partial r} + \frac{\partial^2 u_\theta}{\partial z^2} - \frac{u_\theta}{r^2} = c_s^{-2}\frac{\partial^2 u_\theta}{\partial t^2}. \tag{6.4.11}$$

Considering the traveling wave propagating along the z axis, we can assume the solution of the above equation

$$u_\theta = U(r)\exp[ik(z - ct)]. \tag{6.4.12}$$

Substituting it into Eq. (6.4.11), we get

$$\frac{\mathrm{d}^2 U}{\mathrm{d}r^2} + \frac{1}{r}\frac{\mathrm{d}U}{\mathrm{d}r} + \left(\eta_s^2 - \frac{1}{r^2}\right)U = 0, \tag{6.4.13}$$

where $\eta_s = \left[\left(\frac{\omega}{c_s}\right)^2 - k^2\right]^{\frac{1}{2}}$, $\omega = kc$.

When $\eta_s = 0$, this equation is Euler's equation, and its general solution is

$$U(r) = C_1 r + C_2 \frac{1}{r}. \tag{6.4.14}$$

When $\eta_s \neq 0$, this equation is the first-order Bessel equation, and its general solution is

$$U(r) = C_3 J_1(\eta_s r) + C_4 Y_1(\eta_s r). \tag{6.4.15}$$

In order to ensure u_θ is bounded when $r = 0$, we must let $C_2 = C_4$=0. Then, the solution becomes

$$u_\theta(r, z) = \begin{cases} C_1 r \ \exp[ik(z - ct)], \ \eta_s = 0 \\ C_3 J_1(\eta_s r) \exp[ik(z - ct)], \ \eta_s \neq 0. \end{cases} \tag{6.4.16}$$

By using the free surface conditions of the cylindrical rod, we get

$$\sigma_{r\theta}|_{r=a} = 2\mu\varepsilon_{r\theta}|_{r=a} = \mu\left(\frac{\partial u_\theta}{\partial r} - \frac{u_\theta}{r}\right)_{r=a} = 0,$$

where (exp$[ik(z - ct)]$ is omitted here)

$$\begin{aligned} \frac{\partial u_\theta}{\partial r} - \frac{u_\theta}{r} &= C_3 \frac{\mathrm{d}J_1(\eta_s r)}{\mathrm{d}r} - \frac{1}{r} C_3 J_1(\eta_s r) \\ &= C_3 \left[\frac{\eta_s}{2}(J_0(\eta_s r) - J_2(\eta_s r)) - \frac{1}{r} J_1(\eta_s r)\right] \\ &= -\, C_3 \eta_s J_2(\eta_s r). \end{aligned} \tag{6.4.17}$$

In the derivation of above formula, we have used the recursive relation of Bessel function

$$J_v'(x) = \frac{1}{2}(J_{v-1}(x) - J_{v+1}(x)),$$

$$J_{v-1}(x) + J_{v+1}(x) = \frac{2v}{x} J_v(x).$$

For the stress-free condition to be true, it requires $\eta_s = 0$ or $J_2(\eta_s a) = 0$. Let $\alpha_1, \alpha_2 \ldots$ is the zero roots of $J_2(x)$. Here, $\alpha_1 = 5.136$, $\alpha_2 = 8.417$, $\alpha_3 = 11.620\ldots$, If the solution corresponding to $\eta_s = 0$ is also included, and let $\alpha_0 = 0$, then, the requirement can be expressed as $\eta_s a = \alpha_n (n = 0, 1, 2, 3 \ldots)$, in general. The requirement implies that η_s can only take a series of discrete values corresponding to α_n. According to the definition of η_s above, the frequency equation or dispersion relation of the axisymmetric torsion waves in a cylinder rod can be written as

$$\omega^2 = c_s^2\left(\eta_s^2 + k^2\right) = c_s^2\left(\frac{\alpha_n^2}{a^2} + k^2\right). \tag{6.4.18}$$

For any given n, a corresponding mode of axisymmetric torsion waves can be obtained from above formula. Let

$$\omega_s = \frac{c_s}{a},\ \xi = ak,\ \Omega = \frac{\omega}{\omega_s}.$$

The regularization form, or the dimensionless form, of the dispersion relation can be expressed as

$$\Omega^2 = \alpha_n^2 + \xi^2. \tag{6.4.19}$$

By using $\omega = kc$ and let $C^2 = \left(\frac{c}{c_s}\right)^2$, the regularization form of the dispersion relation can also expressed in terms of phase velocity as

$$C^2 = \frac{\alpha_n^2}{\xi^2} + 1. \tag{6.4.20}$$

The corresponding dispersion curves are shown in Fig. 6.16. As Ω and C are required to be positive real numbers physically, however, ξ can be real number ν or purely imaginary number $i\delta$ but not complex number. Since η_s is a function of frequency, as a result, α_n is also a function of frequency, so all torsional guided waves (except zero order) are dispersive.

6.4.2 *Axisymmetric Compression Waves*

In the tension–compression deformation mode, the displacement field in the cylindrical rod satisfies

$$u_r(r,z,t) \neq 0,\quad u_z = u_z(r,z,t) \neq 0,\quad u_\theta(r,z,t) = 0,\quad \frac{\partial(\cdot)}{\partial\theta} = 0. \tag{6.4.21}$$

Thus, the guided wave propagating in the cylindrical bar is called the axisymmetric compression wave.

The potential function $\chi = 0$ is known from Eq. (6.4.5b), so the displacement potential relation becomes

$$\begin{cases} u_r = \frac{\partial\phi}{\partial r} + \frac{\partial^2\eta}{\partial z\partial r} \\ u_z = \frac{\partial\phi}{\partial z} - \frac{\partial}{r\partial r}\left(r\frac{\partial\eta}{\partial r}\right) \end{cases}. \tag{6.4.22}$$

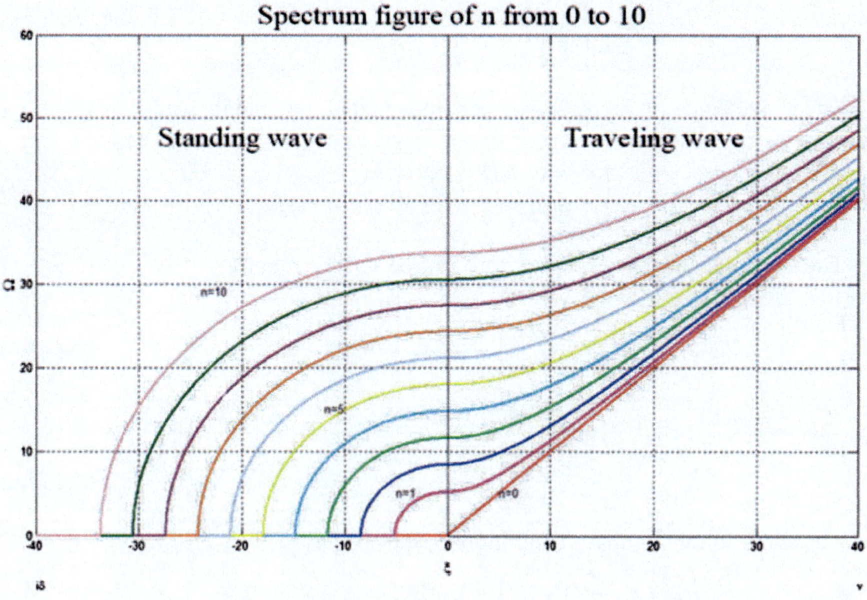

Fig. 6.16 Dispersion curve of axisymmetric torsional wave in a circular rod

Correspondingly, the relationship between stress and potential is simplified as

$$\begin{cases} \sigma_{rr} = \frac{\lambda}{c_d^2}\frac{\partial^2\phi}{\partial t^2} + 2\mu\left(\frac{\partial^2\phi}{\partial r^2} + \frac{\partial^3\eta}{\partial r^2\partial\eta}\right) \\ \sigma_{rz} = \mu\left\{2\frac{\partial^2\phi}{\partial r\partial z} + \frac{\partial^3\eta}{\partial r\partial z^2} - \frac{\partial}{\partial r}\left[\frac{1}{r}\frac{\partial}{\partial r}\left(r\frac{\partial\eta}{\partial r}\right)\right]\right\} \end{cases}. \tag{6.4.23}$$

The wave Eqs. (6.4.3a) and (6.4.3c) become

$$\begin{cases} \frac{\partial^2\phi}{\partial r^2} + \frac{1}{r}\frac{\partial\phi}{\partial r} + \frac{\partial^2\phi}{\partial z^2} = \frac{1}{c_d^2}\frac{\partial^2\phi}{\partial t^2} \\ \frac{\partial^2\eta}{\partial r^2} + \frac{1}{r}\frac{\partial\eta}{\partial r} + \frac{\partial^2\eta}{\partial z^2} = \frac{1}{c_s^2}\frac{\partial^2\eta}{\partial t^2} \end{cases}. \tag{6.4.24}$$

Consider that the traveling wave propagates along the axis z, the trial solution is taken as

$$\begin{cases} \phi = f(r)\exp[ik(z - ct)] \\ \eta = g(r)\exp[ik(z - ct)] \end{cases}. \tag{6.4.25}$$

Substitute it into the wave Eq. (6.4.24) leads to

$$\begin{cases} \frac{d^2 f}{dr^2} + \frac{1}{r}\frac{df}{dr} + \eta_d^2 f = 0 \\ \frac{d^2 g}{dr^2} + \frac{1}{r}\frac{dg}{dr} + \eta_s^2 g = 0 \end{cases}. \tag{6.4.26}$$

The above equation is the zero-order Bessel equation, and its solution is

$$\begin{cases} \phi = A J_0(\eta_d r) \exp[ik(z - ct)] \\ \eta = B J_0(\eta_s r) \exp[ik(z - ct)] \end{cases}. \tag{6.4.27}$$

The conditions that stresses are zero in free surface require

$$\left[-\frac{\lambda\omega^2}{c_d^2} J_0(\eta_d r) + 2\mu \frac{d^2 J_0(\eta_d r)}{dr^2} \right]_{r=a} A + 2i\mu k \frac{d^2 J_0(\eta_s r)}{dr^2}\bigg|_{r=a} B = 0, \tag{6.4.28a}$$

$$2ik \frac{dJ_0(\eta_d r)}{dr}\bigg|_{r=a} A - \left\{ k^2 \frac{dJ_0(\eta_s r)}{dr} + \frac{d}{dr}\left[\frac{1}{r}\frac{d}{dr}\left(r \frac{dJ_0(\eta_s r)}{dr} \right) \right] \right\}_{r=a} B = 0. \tag{6.4.28b}$$

The condition that non-zero solution exists leads to

$$\left(k^2 - \eta_s^2\right)^2 \left[\frac{\eta_d a J_0(\eta_d a)}{J_1(\eta_d a)} \right] + 4k^2 \eta_d^2 \left[\frac{\eta_s a J_0(\eta_s a)}{J_1(\eta_s a)} \right] = 2\eta_d^2\left(\eta_s^2 + k^2\right). \tag{6.4.29}$$

This is the frequency equation of axisymmetric compression waves (also called Pochhammer equation).

When the radius is very small compared to the wavelength, the Bessel function can be approximated as a power series

$$J_0(z) = 1 - \frac{1}{4}z^2 + \frac{1}{64}z^4 + \ldots, \quad J_1(z) = \frac{1}{2}z - \frac{1}{16}z^3 + \ldots \tag{6.4.30}$$

Substitute it into the frequency equation and retain to a^2 term, you get

$$c = \frac{\omega}{k} = \left(\frac{E}{\rho}\right)^{\frac{1}{2}} \left(1 - \frac{1}{4}v^2 k^2 a^2\right), \tag{6.4.31}$$

where $E = \frac{(3\lambda+2\mu)\mu}{(\lambda+\mu)}$ is Young's modulus of elasticity and v is Poisson's ratio. The first term is the velocity of compression wave in the one-dimensional deformation problem, denoted as $c_0 = \left(\frac{E}{\rho}\right)^{\frac{1}{2}}$. The term, i.e. $\left(1 - \frac{1}{4}v^2k^2a^2\right)$, is called Rayleigh correction term, which is produced in the three-dimensional elastic theory. It can be seen that the propagation velocity of longitudinal wave propagating in a cylindrical rod is different from that in an infinite medium. Compared with the velocity of one-dimensional compression waves, the longitudinal wave velocity propagated in the cylindrical rod is somewhat reduced. The larger the dimensionless wave numberka, the greater the decrease in wave velocity. The reasons for the decrease of wave speed can be explained as follows: Axisymmetric guided waves are formed by the superposition of expansion waves and equal-volume waves propagating in a cylindrical

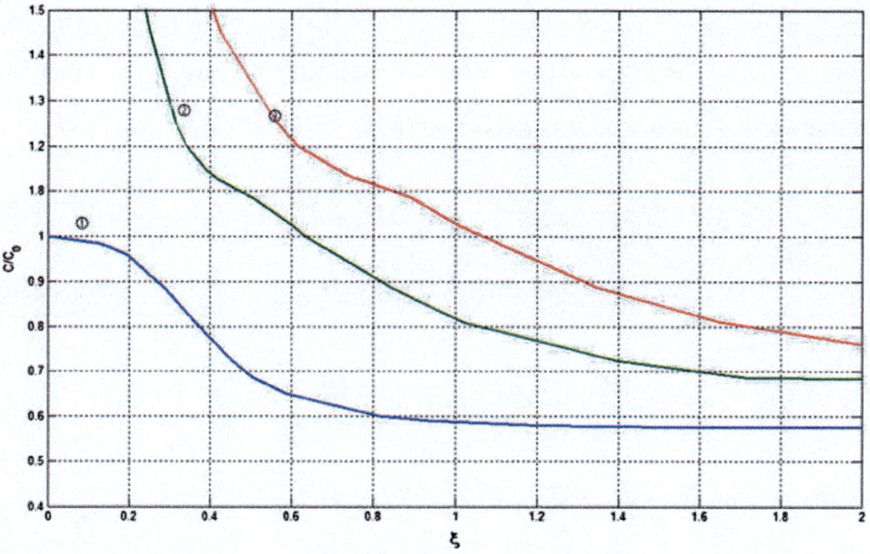

Fig. 6.17 Dispersion curves of the first three order modes of axisymmetric compression of a circular rod

rod. Due to the multiple reflection of the expansion wave and the equal-volume wave propagating in the cylindrical rod on the cylindrical surface, their propagation path is actually repeatedly broken line. Its projection along the axis of the cylindrical rod is greatly reduced than the actual distance traveled. The macro behavior is reflected in the propagation speed along the axis direction, which is smaller than the speed of the one-dimension compressional wave. Moreover, the larger the reflection angle of the expansion wave and the equal-volume wave propagating in the cylindrical rod on the cylinder surface, the larger the apparent wave number along the axis direction, thus, the more the propagation velocity of the guided wave decreases. Figure 6.17 gives the phase velocity curves of the lower three modes, where the dimensionless speed and the dimensionless wave number are defined as

$$\overline{c} = \frac{c}{c_0}, \; \xi = \frac{ak}{2\pi} = \frac{a}{\lambda}.$$

6.4.3 *Non-axisymmetric Guided Waves (Bending Waves)*

This is the most general case of guided wave in a cylindrical rod, in which the three displacement components u_r, u_θ, u_z and the three potential functions ϕ, χ, η are all non-zero, and they are all functions of coordinates and time, i.e. (r, θ, z, t). Since the potential function is decoupled, it is convenient to take the potential function as the basic unknown quantity. Take the trial solution as

$$f(\theta, r, z, t) = R(r)H(\theta)\exp[ik(z - ct)]. \tag{6.4.32}$$

Substitute it into the wave equation leads to

$$\frac{\mathrm{d}^2 R(r)}{\mathrm{d}r^2} + \frac{1}{r}\frac{\mathrm{d}R(r)}{\mathrm{d}r} + \left(\eta_\alpha^2 - \frac{\lambda^2}{r^2}\right)R(r) = 0, \tag{6.4.33a}$$

$$H''(\theta) + \lambda^2 H(\theta) = 0, \tag{6.4.33b}$$

where

$$\eta_\alpha^2 = \frac{\omega^2}{c_\alpha^2} - k^2.$$

The solution of the equation about θ can be expressed as

$$H(\theta) = A_1 \sin\lambda\theta + B_1 \cos\lambda\theta, \quad (\lambda = 1, 2, \ldots). \tag{6.4.34}$$

The equation about r becomes Bessel equation of order n, and its solution is

$$R(r) = C_n J_n(\eta_\alpha r) + D_n Y_n(\eta_\alpha r). \tag{6.4.35}$$

Thus, the solution of guided bending wave is

$$f_n = C_n J_n(\eta_\alpha r)[A_n \sin n\theta + B_n \cos n\theta]\exp[ik(z - ct)]. \tag{6.4.36}$$

In the above formula, the term including D_n is discarded to ensure that the problem is bounded at $r = 0$. η_α is either η_d or η_s depending on the different situations. Considering the symmetry of the displacement triple (u_r, u_θ, u_z) with respect to θ in the case of guided bending waves, we discard either $\sin(n\theta)$ or $\cos(n\theta)$ in the expression so that the expressions of (u_r, u_θ, u_z) contain only $\sin(n\theta)$ (which satisfies symmetry with respect to θ) or $\cos(n\theta)$ (which satisfies antisymmetry with respect to θ). Then, we get an expression of the potential function.

$$\phi = AJ_n(\eta_d r)\cos(n\theta)\exp[ik(z - ct)], \tag{6.4.37a}$$

$$\chi = BJ_n(\eta_s r)\sin(n\theta)\exp[ik(z - ct)], \tag{6.4.37b}$$

$$\eta = CJ_n(\eta_s r)\cos(n\theta)\exp[ik(z - ct)]. \tag{6.4.37c}$$

The above equation indicates that there are infinitely many possible distributions of potential or displacement in both radial and circumferential directions. For a particular value of n, there will be a family of modes of the bending wave. Substitute (6.4.37) into the stress potential relationship, namely

$$\sigma_{rr} = \frac{\lambda}{c_d^2}\frac{\partial^2\phi}{\partial t^2} + 2\mu\left(\frac{\partial^2\phi}{\partial r^2} + \frac{\partial^2\chi}{r\partial\theta\partial r} + \frac{\partial^3\eta}{\partial z\partial r^2}\right), \tag{6.4.38a}$$

$$\sigma_{r\theta} = \mu\left(2\frac{\partial^2\phi}{r\partial r\partial\theta} - \frac{\partial\phi}{r^2\partial\theta} + \frac{\partial^2\chi}{r^2\partial\theta^2} - \frac{\partial^2\chi}{\partial r^2} + \frac{1}{r}\frac{\partial\chi}{\partial r} + 2\frac{\partial^3\eta}{r\partial r\partial\theta\partial z} - \frac{1}{r^2}\frac{\partial^2\eta}{\partial\theta\partial z}\right), \tag{6.4.38b}$$

$$\sigma_{rz} = \mu\left(2\frac{\partial^2\phi}{\partial r\partial z} + \frac{\partial^2\chi}{r\partial\theta\partial z} + \frac{\partial^3\eta}{\partial r\partial z^2} - \frac{\partial^3\eta}{r^2\partial\theta^2\partial z} - \frac{\partial^3\eta}{\partial r^3} - \frac{1}{r}\frac{\partial^2\eta}{\partial r^2} + \frac{1}{r^2}\frac{\partial\eta}{\partial r}\right), \tag{6.4.38c}$$

then, the stress expressions are obtained. Further, by using the free surface condition, i.e.

$$(\sigma_{rr}, \sigma_{r\theta}, \sigma_{rz})|_{r=a} = 0, \tag{6.4.39}$$

we can obtain the ternary equations set for coefficients A, B and C. The dispersion equations of the guided bending waves are obtained by the requirement that the determinant of the coefficients is zero, namely

$$|D_{ij}| = 0, \ (i, j = 1, 2, 3) \tag{6.4.40}$$

where

$D_{11} = \left[2n(n-1) - (\eta_s^2 - k^2)a^2\right]J_n(\eta_d a) + 2\eta_d a J_{n+1}(\eta_d a)$,
$D_{12} = 2k\eta_s a^2 J_n(\eta_s a) - 2ka(n+1)J_{n+1}(\eta_s a)$,
$D_{13} = 2n(n-1)J_n(\eta_s a) - 2n\eta_s a J_{n+1}(\eta_s a)$,
$D_{21} = -2n(n-1)J_n(\eta_d a) + 2n\eta_d a J_{n+1}(\eta_d a)$,
$D_{22} = -k\eta_s a^2 J_n(\eta_s a) + 2n(n+1)ka J_{n+1}(\eta_s a)$,
$D_{23} = -\left[2n(n-1) - \eta_s^2 a^2\right]J_n(\eta_s a) - 2\eta_s a J_{n+1}(\eta_s a)$,
$D_{31} = 2nka J_n(\eta_d a) - 2k\eta_d a^2 J_{n+1}(\eta_d a)$,
$D_{32} = n\eta_s a J_n(\eta_s a) + (k^2 - \eta_s^2)a^2 J_{n+1}(\eta_s a)$,
$D_{33} = nka J_n(\eta_s a)$.

The three possible guided wave modes in the infinite length cylindrical rod discussed above, namely the axisymmetric compression wave, the axisymmetric torsional wave and the bending wave, are shown in Fig. 6.18. Among them, the compression and tensile deformation are distributed periodically along the axis of cylindrical rod for the axisymmetric compression wave. The clockwise and counterclockwise torsional deformations are distributed periodically along the axis of cylindrical rod for the torsional wave. The concave-convex bending deformations are distributed periodically along the axis of cylindrical rod for the bending wave. In order to better reflect the deformation characteristics corresponding to the three guided wave modes and the differences between them, the deformations of the three guided wave modes simulated by the finite element method are given in Fig. 6.19.

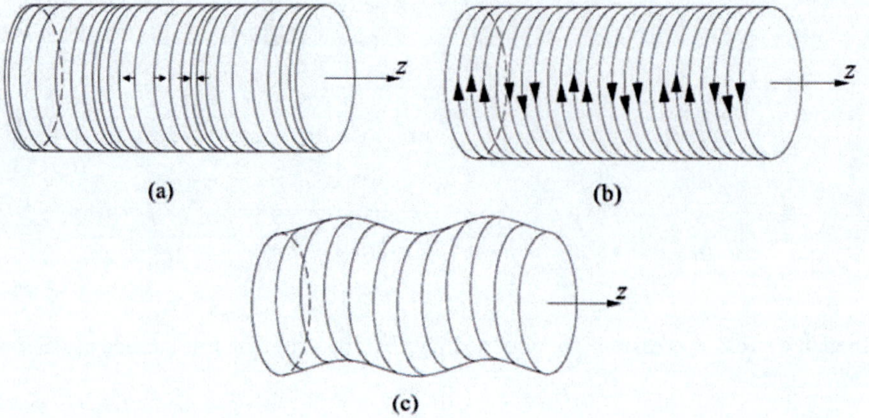

Fig. 6.18 Three guided wave modes in infinite cylindrical rod. **a** Compression wave (period distribution of tension and compression deformation). **b** Torsional wave (period distribution of clockwise and counterclockwise torsional deformation). **c** bending wave (periodic distribution of concave-convex bending deformation)

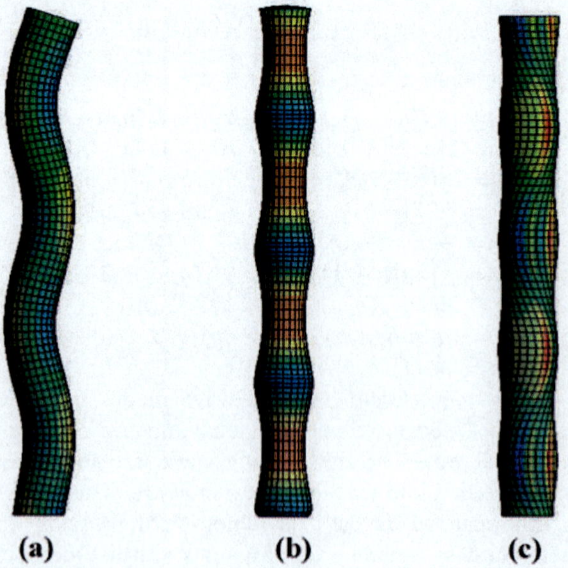

Fig. 6.19 Finite element simulation of three guided wave modes in an infinitely long cylindrical rod. **a** Bending wave (periodic distribution of concave-convex bending deformation). **b** compression wave (period distribution of tension and compression deformation). **c** torsional wave (period distribution of clockwise/counterclockwise torsional deformation)

6.4.4 Surface with Liquid Load

The mathematical treatment of this problem is essentially the same as that of the dispersion equation for a cylinder without a liquid load, but there are two points to be noticed: (1) the potential function ϕ_w can be used to represent the longitudinal wave in water when considering the longitudinal wave propagating in water; (2) the wave number of the guided wave should be complex number rather than real number because the amplitudes of the guided waves of the various modes in the circular rod decay gradually due to the leakage of energy into the water during the propagation process. In the cylindrical coordinates system, the wave equation in water can be expressed as

$$\nabla^2 \phi = c_d^{-2} \frac{\partial^2 \phi}{\partial t^2}. \tag{6.4.41}$$

The solution of (6.4.41) can be expressed by Hankel function as

$$\phi_w = \left[C H_0^{(2)}(k_w r) + D H_0^{(1)}(k_w r) \right] e^{ik(z-ct)}, \tag{6.4.42}$$

where

$$H_0^{(1,2)}(z) = J_0(z) \pm i Y_0(z), \tag{6.4.43}$$

represent the first and second kinds of Hankel functions. The Hankel function is actually a combination of the Bessel function and the Neumann function. Hankel functions of the first kind represent cylindrical waves converging from infinity toward the surface of a cylinder, while Hankel functions of the second kind represent cylindrical waves diffusing outward from the surface of a cylinder. By means of the Hankel function of the second kind, the longitudinal wave diffusing outward from the cylinder surface in water can be expressed as

$$\phi_w = C H_0^{(2)}(k_w r) e^{ik(z-ct)}, \tag{6.4.44}$$

where

$$k_w^2 = \left(\frac{\omega}{c_w} \right)^2 - k^2, \quad H_0^{(2)}(k_w r) = J_0(k_w r) - i Y_0(k_w r).$$

Substituting the potential function into $u = \nabla \phi_w$, and using the property of Hankel function, i.e. $H_0'(x) = -H_1(x)$, the radial displacement of the longitudinal wave in water is obtained as

$$u_r' = -k_w C H_1^{(2)}(k_w r) e^{ik(z-ct)}. \tag{6.4.45}$$

Further, the pressure of water is obtained as

$$\sigma_{rr}^{'} = \lambda_w \nabla \cdot \mathbf{u} = \lambda_w \Big[-(k_w^2 + k^2) H_0^{(2)}(k_w r) \Big] C e^{ik(z-ct)}. \tag{6.4.46}$$

The continuous conditions of normal stress and normal displacement on the surface of cylinder and the condition that the surface shear stress is zero can be expressed as

$$\sigma_{rr}|_{r=a} = \sigma_{rr}^{'}\big|_{r=a}, \quad \sigma_{rz}|_{r=a} = 0, \quad u_r|_{r=a} = u_r^{'}\big|_{r=a}. \tag{6.4.47a,b,c}$$

By application of Eq. (6.4.47), the following three equations can be obtained

$$\left[-\frac{\lambda \omega^2}{c_{\mathrm{d}}^2} J_0(\eta_{\mathrm{d}} r) + 2\mu \frac{\mathrm{d}^2 J_0(\eta_{\mathrm{d}} r)}{\mathrm{d}r^2} \right]_{r=a} A + 2i\mu k \frac{\mathrm{d}^2 J_0(\eta_s r)}{\mathrm{d}r^2} \bigg|_{r=a} B \\ = -\lambda_w \left(k_w^2 + k^2 \right) H_0^{(2)}(k_w a) C, \tag{6.4.48a}$$

$$2ik \frac{\mathrm{d}J_0(\eta_{\mathrm{d}} r)}{\mathrm{d}r}\bigg|_{r=a} A - \left\{ k^2 \frac{\mathrm{d}J_0(\eta_s r)}{\mathrm{d}r} + \frac{\mathrm{d}}{\mathrm{d}r}\left[\frac{1}{r} \frac{\mathrm{d}}{\mathrm{d}r} \left(r \frac{\mathrm{d}J_0(\eta_s r)}{\mathrm{d}r} \right) \right] \right\}_{r=a} B = 0, \tag{6.4.48b}$$

$$-A\eta_{\mathrm{d}} J_1(\eta_{\mathrm{d}} a) - ikB\eta_s J_1(\eta_s a) + k_w C H_1^{(2)}(k_w a) = 0. \tag{6.4.48c}$$

According to the condition that the equation has a non-zero solution, we get the dispersion equation

$$|D_{mn}| = 0, \quad (m, n = 1, 2, 3) \tag{6.4.49}$$

where

$D_{11} = -\left(\frac{\lambda \omega^2}{c_d^2} + \mu \eta_d^2 \right) J_0(\eta_d a) + \mu \eta_d^2 J_2(\eta_d a),$

$D_{12} = i\mu k \eta_s^2 (J_0(\eta_s a) - J_2(\eta_s a))$, $D_{13} = \lambda_w \left(k_w^2 + k^2 \right) H_0^{(2)}(k_w a)$,
$D_{21} = -2ik\eta_d J_1(\eta_d a)$, $D_{22} = \eta_s \left(k^2 - \eta_s^2 \right) J_1(\eta_s a)$, $D_{23} = 0$,
$D_{31} = -\eta_d J_1(\eta_d a)$, $D_{32} = -ik\eta_s J_1(\eta_s a)$, $D_{33} = k_w H_1^{(2)}(k_w a)$.

Equation (6.4.49) is a non-linear function about the wave number k and the angular frequency ω. When finding the corresponding wave number for a given angular frequency, it is important to note that there are roots only in the complex number domain, and no roots exist in the real number domain. It physically embodies the energy leakage characteristics in the process of guided wave propagation.

6.5 Waves in Cylindrical Tube

The water pipes and oil pipelines buried in the ground will appear cracking, corrosion and other defects in the long service process. The health monitoring of pipeline structure has thus great significance in engineering practice. Therefore, it is very necessary to study the propagation mode and propagation characteristics of guided waves in pipeline structure. Consider an infinite long circular tube with an inside diameter of $2a$ and an outside diameter of $2b$, as shown in Fig. 6.20. Compared with the infinitely long cylindrical rod, the circular tube has both inner and outer boundaries. The boundary conditions of the inner and outer boundaries can be varied depending on the purpose of the tube and the environment in which it is used. In this section, we will discuss the mode classification and dispersion characteristics of guided waves in a circular tube based on several common boundary conditions [4,8].

6.5.1 *Axisymmetric Torsional Waves*

When the displacement field satisfies the following conditions

$$u_r(r, z, t) = u_z(r, z, t) = 0, \tag{6.5.1a}$$

$$u_\theta(r, z, t) \neq 0, \tag{6.5.1b}$$

$$\frac{\partial(\cdot)}{\partial\theta} = 0, \tag{6.5.1c}$$

the guided waves in a circular tube are called axisymmetric torsional waves. It is similar to the derivation of axisymmetric compression wave in a solid circular rod. The non-zero displacement solution of a circular tube is

$$u_\theta = \begin{cases} \left(C_1 r + C_2 \frac{1}{r}\right) \exp[ik(z - ct)], \ \eta_s = 0 \\ [C_3 J_1(\eta_s r) + C_4 Y_1(\eta_s r)] \exp[ik(z - ct)], \ \eta_s \neq 0. \end{cases} \tag{6.5.2a}$$

The above equation can also be rewritten as

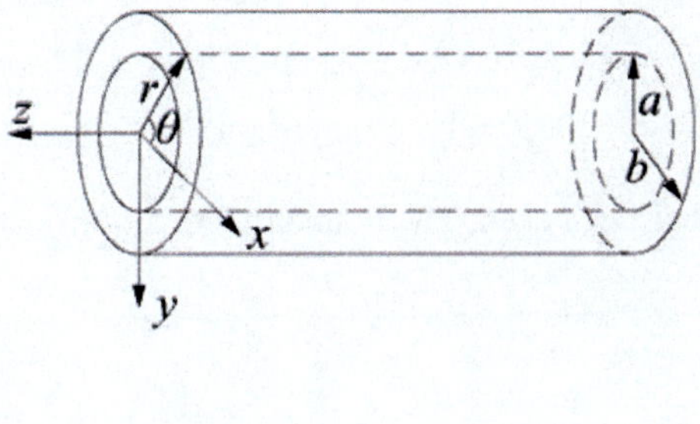

Fig. 6.20 Guided waves in an infinite long tube

$$u_\theta = \begin{cases} \left(C_1 r + C_2 \frac{1}{r}\right) \exp[ik(z - ct)], \ \eta_s = 0 \\ \left[C_3 H_1^{(1)}(\eta_s r) + C_4 H_1^{(2)}(\eta_s r)\right] \exp[ik(z - ct)], \ \eta_s \neq 0. \end{cases} \tag{6.5.2b}$$

If the form of Eq. (6.5.2b) is adopted, the first and second kind of Bessel functions in the following discussion can be replaced by the first and second kind of Hankel functions, respectively. The two group of functions, i.e. Bessel functions and Hankel functions are essentially the same.

For the case that the internal and external boundaries are both free, the boundary conditions can be expressed as following.

When $\eta_s = 0$,

$$\sigma_{r\theta} = 2\mu\varepsilon_{r\theta} = \mu\left(\frac{\partial u_\theta}{\partial r} - \frac{u_\theta}{r}\right) = \mu\frac{1}{r^2}C_2.$$

Consider that the inner and outer boundaries are both free, $\sigma_{r\theta}$ must be zero at $r = a$ and $r = b$. So C_2 must be zero.

When $\eta_s \neq 0$, the boundary conditions are

$$\sigma_{r\theta}|_{r=a} = 2\mu\varepsilon_{r\theta}|_{r=a} = \mu\left(\frac{\partial u_\theta}{\partial r} - \frac{u_\theta}{r}\right)_{r=a} = 0, \tag{6.5.3a}$$

$$\sigma_{r\theta}|_{r=b} = 2\mu\varepsilon_{r\theta}|_{r=b} = \mu\left(\frac{\partial u_\theta}{\partial r} - \frac{u_\theta}{r}\right)_{r=b} = 0. \tag{6.5.3b}$$

Consider (exp$[ik(z - ct)]$ is omitted here)

$$\begin{aligned} \frac{\partial u_\theta}{\partial r} - \frac{u_\theta}{r} &= C_3\frac{\mathrm{d}J_1(\eta_s r)}{\mathrm{d}r} + C_4\frac{\mathrm{d}Y_1(\eta_s r)}{\mathrm{d}r} - \frac{1}{r}[C_3 J_1(\eta_s r) + C_4 Y_1(\eta_s r)] \\ &= C_3\left[\frac{\eta_s}{2}(J_0(\eta_s r) - J_2(\eta_s r)) - \frac{1}{r}J_1(\eta_s r)\right] \\ &\quad + C_4\left[\frac{\eta_s}{2}(Y_0(\eta_s r) - Y_2(\eta_s r)) - \frac{1}{r}Y_1(\eta_s r)\right] \\ &= -C_3\eta_s J_2(\eta_s r) - C_4\eta_s Y_2(\eta_s r), \end{aligned} \tag{6.5.4}$$

where the recursive relation of Bessel function

$$J_v'(x) = \frac{1}{2}(J_{v-1}(x) - J_{v+1}(x)), \tag{6.5.5a}$$

$$J_{v-1}(x) + J_{v+1}(x) = \frac{2v}{x}J_v(x), \tag{6.5.5b}$$

has been used. The boundary conditions can be explicitly expressed as

$$\begin{cases} -C_3\eta_s J_2(\eta_s a) - C_4\eta_s Y_2(\eta_s a) = 0 \\ -C_3\eta_s J_2(\eta_s b) - C_4\eta_s Y_2(\eta_s b) = 0 \end{cases}. \tag{6.5.6}$$

The conditions that the non-zero solutions exist require the determinant of coefficients equal to zero, i.e.

$$\begin{vmatrix} \eta_s J_2(\eta_s a) & \eta_s Y_2(\eta_s a) \\ \eta_s J_2(\eta_s b) & \eta_s Y_2(\eta_s b) \end{vmatrix} = \eta_s^2[J_2(\eta_s a)Y_2(\eta_s b) - Y_2(\eta_s a)J_2(\eta_s b)] = 0. \tag{6.5.7}$$

This is the dispersion equation of torsional guided waves. Since $\eta_s = \left[\left(\frac{\omega}{c_s}\right)^2 - k^2\right]^{\frac{1}{2}}$, $\omega = kc$, the Eq. (6.5.7) is a transcendental equation about the wave number k and angular frequency ω.

6.5.2 Axisymmetric Compression Waves

In the following cases,

$$u_r(r, z, t) \neq 0, \quad u_z(r, z, t) \neq 0, \quad u_\theta(r, z, t) = 0, \quad \frac{\partial(\cdot)}{\partial\theta} = 0, \tag{6.5.8}$$

guided waves propagating in a circular tube are called axisymmetric compression waves. Similar to the axisymmetric compression wave in a solid circular rod, the solutions of the potential function are

$$\begin{cases} \phi = [A_1 J_0(\eta_d r) + A_2 Y_0(\eta_d r)]\exp[ik(z - ct)] \\ \eta = [A_3 J_0(\eta_s r) + A_4 Y_0(\eta_s r)]\exp[ik(z - ct)] \end{cases}. \tag{6.5.9a}$$

The above equation can also be rewritten as

$$\begin{cases} \phi = \left[A_1 H_0^{(1)}(\eta_d r) + A_2 H_0^{(2)}(\eta_d r)\right]\exp[ik(z - ct)] \\ \eta = \left[A_3 H_0^{(1)}(\eta_s r) + A_4 H_0^{(2)}(\eta_s r)\right]\exp[ik(z - ct)] \end{cases}. \tag{6.5.9b}$$

(1) For the case that the internal and external boundaries are both free, the boundary conditions can be expressed as

$$\sigma_{rr}|_{r=a} = 0, \quad \sigma_{rz}|_{r=a} = 0, \tag{6.5.10a}$$

$$\sigma_{rr}|_{r=b} = 0, \quad \sigma_{rz}|_{r=b} = 0. \tag{6.5.10b}$$

Substituting the expression of potential function into the above boundary conditions, we can get

$$\begin{aligned}&\left[-\frac{\lambda\omega^2}{c_{\mathrm{d}}^2}J_0(\eta_{\mathrm{d}}r)+2\mu\frac{\mathrm{d}^2J_0(\eta_{\mathrm{d}}r)}{\mathrm{d}r^2}\right]_{r=a}A_1\\&+\left[-\frac{\lambda\omega^2}{c_{\mathrm{d}}^2}Y_0(\eta_{\mathrm{d}}r)+2\mu\frac{\mathrm{d}^2Y_0(\eta_{\mathrm{d}}r)}{\mathrm{d}r^2}\right]_{r=a}A_2\\&+2i\mu k\frac{\mathrm{d}^2J_0(\eta_sr)}{\mathrm{d}r^2}\bigg|_{r=a}A_3+2i\mu k\frac{\mathrm{d}^2Y_0(\eta_sr)}{\mathrm{d}r^2}\bigg|_{r=a}A_4=0\end{aligned}\quad, \tag{6.5.11a}$$

$$\begin{aligned}&2ik\frac{\mathrm{d}J_0(\eta_{\mathrm{d}}r)}{\mathrm{d}r}\bigg|_{r=a}A_1+2ik\frac{\mathrm{d}Y_0(\eta_{\mathrm{d}}r)}{\mathrm{d}r}\bigg|_{r=a}A_2-\\&\left\{k^2\frac{\mathrm{d}J_0(\eta_sr)}{\mathrm{d}r}+\frac{\mathrm{d}}{\mathrm{d}r}\left[\frac{1}{r}\frac{\mathrm{d}}{\mathrm{d}r}\left(r\frac{\mathrm{d}J_0(\eta_sr)}{\mathrm{d}r}\right)\right]\right\}_{r=a}A_3\\&-\left\{k^2\frac{\mathrm{d}Y_0(\eta_sr)}{\mathrm{d}r}+\frac{\mathrm{d}}{\mathrm{d}r}\left[\frac{1}{r}\frac{\mathrm{d}}{\mathrm{d}r}\left(r\frac{\mathrm{d}Y_0(\eta_sr)}{\mathrm{d}r}\right)\right]\right\}_{r=a}A_4=0\end{aligned}\quad, \tag{6.5.11b}$$

$$\begin{aligned}&\left[-\frac{\lambda\omega^2}{c_{\mathrm{d}}^2}J_0(\eta_{\mathrm{d}}r)+2\mu\frac{\mathrm{d}^2J_0(\eta_{\mathrm{d}}r)}{\mathrm{d}r^2}\right]_{r=b}A_1\\&+\left[-\frac{\lambda\omega^2}{c_{\mathrm{d}}^2}Y_0(\eta_{\mathrm{d}}r)+2\mu\frac{\mathrm{d}^2Y_0(\eta_{\mathrm{d}}r)}{\mathrm{d}r^2}\right]_{r=b}A_2\\&+2i\mu k\frac{\mathrm{d}^2J_0(\eta_sr)}{\mathrm{d}r^2}\bigg|_{r=b}A_3+2i\mu k\frac{\mathrm{d}^2Y_0(\eta_sr)}{\mathrm{d}r^2}\bigg|_{r=b}A_4=0\end{aligned}\quad, \tag{6.5.11c}$$

$$\begin{aligned}&2ik\frac{\mathrm{d}J_0(\eta_{\mathrm{d}}r)}{\mathrm{d}r}\bigg|_{r=b}A_1+2ik\frac{\mathrm{d}Y_0(\eta_{\mathrm{d}}r)}{\mathrm{d}r}\bigg|_{r=b}A_2-\\&\left\{k^2\frac{\mathrm{d}J_0(\eta_sr)}{\mathrm{d}r}+\frac{\mathrm{d}}{\mathrm{d}r}\left[\frac{1}{r}\frac{\mathrm{d}}{\mathrm{d}r}\left(r\frac{\mathrm{d}J_0(\eta_sr)}{\mathrm{d}r}\right)\right]\right\}_{r=b}A_3\\&-\left\{k^2\frac{\mathrm{d}Y_0(\eta_sr)}{\mathrm{d}r}+\frac{\mathrm{d}}{\mathrm{d}r}\left[\frac{1}{r}\frac{\mathrm{d}}{\mathrm{d}r}\left(r\frac{\mathrm{d}Y_0(\eta_sr)}{\mathrm{d}r}\right)\right]\right\}_{r=b}A_4=0\end{aligned}\quad. \tag{6.5.11d}$$

The dispersion equation is given by the condition that the equation has a non-zero solution, i.e.

$$|D_{mn}|=0,\ (m,n=1,2,3,4), \tag{6.5.12}$$

where

$$\begin{aligned}D_{11}&=-\left(\tfrac{\lambda\omega^2}{c_d^2}+\mu\eta_d^2\right)J_0(\eta_da)+\mu\eta_d^2J_2(\eta_da),\\D_{12}&=-\left(\tfrac{\lambda\omega^2}{c_d^2}+\mu\eta_d^2\right)Y_0(\eta_da)+\mu\eta_d^2Y_2(\eta_da),\\D_{13}&=i\mu k\eta_s^2(J_0(\eta_sa)-J_2(\eta_sa)),\\D_{14}&=i\mu k\eta_s^2(Y_0(\eta_sa)-Y_2(\eta_sa)),\end{aligned}$$

$$
\begin{aligned}
&D_{21} = -2ik\eta_d J_1(\eta_d a), \quad D_{22} = -2ik\eta_d Y_1(\eta_d a),\\
&D_{23} = \eta_s(k^2 - \eta_s^2)J_1(\eta_s a), \quad D_{24} = \eta_s(k^2 - \eta_s^2)Y_1(\eta_s a),\\
&D_{31} = -\left(\tfrac{\lambda\omega^2}{c_d^2} + \mu\eta_d^2\right)J_0(\eta_d b) + \mu\eta_d^2 J_2(\eta_d b),\\
&D_{32} = -\left(\tfrac{\lambda\omega^2}{c_d^2} + \mu\eta_d^2\right)Y_0(\eta_d b) + \mu\eta_d^2 Y_2(\eta_d b),\\
&D_{33} = i\mu k\eta_s^2(J_0(\eta_s b) - J_2(\eta_s b)), D_{34} = i\mu k\eta_s^2(Y_0(\eta_s b) - Y_2(\eta_s b)),\\
&D_{41} = -2ik\eta_d J_1(\eta_d b), D_{42} = -2ik\eta_d Y_1(\eta_d b),\\
&D_{43} = \eta_s(k^2 - \eta_s^2)J_1(\eta_s b), D_{44} = \eta_s(k^2 - \eta_s^2)Y_1(\eta_s b).
\end{aligned}
$$

(2) For the case that the outer surface of the circular tube is fixed while the inner surface is free, the boundary conditions can be expressed as

$$
u_r|_{r=b} = u_z|_{r=b} = 0, \tag{6.5.13a}
$$

$$
\sigma_{rr}|_{r=a} = \sigma_{rz}|_{r=a} = 0. \tag{6.5.13b}
$$

Using the potential function expression (6.5.9), the above boundary conditions can be explicitly expressed as

$$
\begin{aligned}
&\left[-\frac{\lambda\omega^2}{c_\mathrm{d}^2}J_0(\eta_\mathrm{d} r) + 2\mu\frac{\mathrm{d}^2 J_0(\eta_\mathrm{d} r)}{\mathrm{d}r^2}\right]_{r=a} A_1\\
&+\left[-\frac{\lambda\omega^2}{c_\mathrm{d}^2}Y_0(\eta_\mathrm{d} r) + 2\mu\frac{\mathrm{d}^2 Y_0(\eta_\mathrm{d} r)}{\mathrm{d}r^2}\right]_{r=a} A_2\\
&+2i\mu k\left.\frac{\mathrm{d}^2 J_0(\eta_s r)}{\mathrm{d}r^2}\right|_{r=a} A_3 + 2i\mu k\left.\frac{\mathrm{d}^2 Y_0(\eta_s r)}{\mathrm{d}r^2}\right|_{r=a} A_4 = 0,
\end{aligned}
$$

$$
\begin{aligned}
&2ik\left.\frac{\mathrm{d}J_0(\eta_\mathrm{d} r)}{\mathrm{d}r}\right|_{r=a} A_1 + 2ik\left.\frac{\mathrm{d}Y_0(\eta_\mathrm{d} r)}{\mathrm{d}r}\right|_{r=a} A_2\\
&-\left\{k^2\frac{\mathrm{d}J_0(\eta_s r)}{\mathrm{d}r} + \frac{\mathrm{d}}{\mathrm{d}r}\left[\frac{1}{r}\frac{\mathrm{d}}{\mathrm{d}r}\left(r\frac{\mathrm{d}J_0(\eta_s r)}{\mathrm{d}r}\right)\right]\right\}_{r=a} A_3\\
&-\left\{k^2\frac{\mathrm{d}Y_0(\eta_s r)}{\mathrm{d}r} + \frac{\mathrm{d}}{\mathrm{d}r}\left[\frac{1}{r}\frac{\mathrm{d}}{\mathrm{d}r}\left(r\frac{\mathrm{d}Y_0(\eta_s r)}{\mathrm{d}r}\right)\right]\right\}_{r=a} A_4 = 0,
\end{aligned}
$$

$$
A_1\eta_\mathrm{d} J_1(\eta_\mathrm{d} b) + A_2\eta_\mathrm{d} Y_1(\eta_\mathrm{d} b) + ikA_3\eta_s J_1(\eta_s b) + ikA_4\eta_s Y_1(\eta_s b) = 0,
$$

$$
ikA_1 J_0(\eta_\mathrm{d} b) + ikA_2 Y_0(\eta_\mathrm{d} b) + A_3\eta_s^2 J_2(\eta_s b) + A_4\eta_s^2 Y_2(\eta_s b) = 0.
$$

Correspondingly, the dispersion equation is

$$
|D_{mn}| = 0, \quad (m, n = 1, 2, 3, 4) \tag{6.5.14}
$$

where

$$
\begin{aligned}
D_{11} &= -\left(\tfrac{\lambda\omega^2}{c_d^2} + \mu\eta_d^2\right) J_0(\eta_d a) + \mu\eta_d^2 J_2(\eta_d a),\\
D_{12} &= -\left(\tfrac{\lambda\omega^2}{c_d^2} + \mu\eta_d^2\right) Y_0(\eta_d a) + \mu\eta_d^2 Y_2(\eta_d a),\\
D_{13} &= i\mu k\eta_s^2 (J_0(\eta_s a) - J_2(\eta_s a)),\\
D_{14} &= i\mu k\eta_s^2 (Y_0(\eta_s a) - Y_2(\eta_s a)),\\
D_{21} &= -2ik\eta_d J_1(\eta_d a),\ D_{22} = -2ik\eta_d Y_1(\eta_d a),\\
D_{23} &= \eta_s\left(k^2 - \eta_s^2\right) J_1(\eta_s a),\ D_{24} = \eta_s\left(k^2 - \eta_s^2\right) Y_1(\eta_s a),\\
D_{31} &= -\eta_d J_1(\eta_d b),\ D_{32} = -\eta_d Y_1(\eta_d b),\\
D_{33} &= ik\eta_s J_1(\eta_s b),\ D_{34} = ik\eta_s Y_1(\eta_s b),\\
D_{41} &= ikJ_0(\eta_d b),\ D_{42} = ikY_0(\eta_d b),\\
D_{43} &= \eta_s^2 J_2(\eta_s b),\ D_{44} = \eta_s^2 Y_2(\eta_s b).
\end{aligned}
$$

6.5.3 Non-axisymmetric Waves (Bending Waves)

Similar to the case of a solid circular rod, the three displacement components u_r, u_θ, u_z and the three potential functions ϕ, χ, η of the asymmetric bending wave are not zero. Moreover, they are all functions of coordinates and time (r, θ, z, t). The three potential functions can be expressed as

$$f_n = [C_n J_n(\eta_\alpha r) + D_n Y_n(\eta_\alpha r)][A_n \sin n\theta + B_n \cos n\theta] \exp[ik(z - ct)], \tag{6.5.15a}$$

or

$$f_n = \left[C_n H_n^{(1)}(\eta_\alpha r) + D_n H_n^{(2)}(\eta_\alpha r)\right][A_n \sin n\theta + B_n \cos n\theta] \exp[ik(z - ct)]. \tag{6.5.15b}$$

η_α is either η_d or η_s depending on the different situations. Consider that the displacement mode (u_r, u_θ, u_z) is symmetry with respect to θ in the case of bending guided waves. We discard either $\sin(n\theta)$ or $\cos(n\theta)$ in the expression so that the expressions of u_r, u_θ, u_z contain only $\sin(n\theta)$ (which satisfies symmetry with respect to θ) or $\cos(n\theta)$ (which satisfies antisymmetry with respect to θ). Then, we get an expression of the potential function.

$$\phi = [AJ_n(\eta_d r) + BY_n(\eta_d r)] \cos(n\theta) \exp[ik(z - ct)], \tag{6.5.16a}$$

$$\chi = [CJ_n(\eta_s r) + DY_n(\eta_s r)] \sin(n\theta) \exp[ik(z - ct)], \tag{6.5.16b}$$

$$\eta = [EJ_n(\eta_s r) + FY_n(\eta_s r)] \cos(n\theta) \exp[ik(z - ct)]. \tag{6.5.16c}$$

The above equations indicate that there are infinitely possible distributions of potential or displacement in both radial and circumferential directions. For a particular value n, there is a family of modes of bending waves corresponding with it. If the three expressions of (6.5.16) are substituted into the stress potential relation, the explicit expression of the stress is obtained

$$\sigma_{rr} = \frac{\lambda}{c_d^2}\frac{\partial^2\phi}{\partial t^2} + 2\mu\left(\frac{\partial^2\phi}{\partial r^2} + \frac{\partial^2\chi}{r\partial\theta\partial r} + \frac{\partial^3\eta}{\partial z\partial r^2}\right), \tag{6.5.17a}$$

$$\sigma_{r\theta} = \mu\left(2\frac{\partial^2\phi}{r\partial r\partial\theta} - \frac{\partial\phi}{r^2\partial\theta} + \frac{\partial^2\chi}{r^2\partial\theta^2} - \frac{\partial^2\chi}{\partial r^2} + \frac{1}{r}\frac{\partial\chi}{\partial r} + 2\frac{\partial^3\eta}{r\partial r\partial\theta\partial z} - \frac{1}{r^2}\frac{\partial^2\eta}{\partial\theta\partial z}\right), \tag{6.5.17b}$$

$$\sigma_{rz} = \mu\left(2\frac{\partial^2\phi}{\partial r\partial z} + \frac{\partial^2\chi}{r\partial\theta\partial z} + \frac{\partial^3\eta}{\partial r\partial z^2} - \frac{\partial^3\eta}{r^2\partial\theta^2\partial z} - \frac{\partial^3\eta}{\partial r^3} - \frac{1}{r}\frac{\partial^2\eta}{\partial r^2} + \frac{1}{r^2}\frac{\partial\eta}{\partial r}\right). \tag{6.5.17c}$$

Then, we use the boundary condition that the inner and outer surfaces of the circular pipe are free, i.e.

$$(\sigma_{rr}, \sigma_{r\theta}, \sigma_{rz})|_{r=a,b} = 0, \tag{6.5.18}$$

and a six-element equation set with respect to the coefficients (A, B, C, D, E, F) in the expression of the potential function can be obtained. Further, the dispersion equation is obtained by using the condition that non-zero solutions exists,

$$\left|D_{ij}\right| = 0, \quad (i, j = 1, 2\ldots 6), \tag{6.5.19}$$

where

$$\begin{aligned}
D_{11} &= \left[2n(n-1) - (\eta_s^2 - k^2)a^2\right]J_n(\eta_d a) + 2\eta_d a J_{n+1}(\eta_d a),\\
D_{12} &= 2k\eta_s a^2 J_n(\eta_s a) - 2ka(n+1)J_{n+1}(\eta_s a),\\
D_{13} &= -2n(n-1)J_n(\eta_s a) + 2n\eta_s a J_{n+1}(\eta_s a),\\
D_{14} &= \left[2n(n-1) - (\eta_s^2 - k^2)a^2\right]Y_n(\eta_d a) + 2\eta_d a Y_{n+1}(\eta_d a),\\
D_{15} &= 2k\eta_s a^2 Y_n(\eta_s a) - 2ka(n+1)Y_{n+1}(\eta_s a),\\
D_{16} &= -2n(n-1)Y_n(\eta_s a) + 2n\eta_s a Y_{n+1}(\eta_s a),\\
D_{21} &= 2n(n-1)J_n(\eta_d a) - 2n\eta_d a J_{n+1}(\eta_d a),\\
D_{22} &= -k\eta_s a^2 J_n(\eta_s a) + 2ka(n+1)J_{n+1}(\eta_s a),\\
D_{23} &= -\left[2n(n-1) - \eta_s^2 a^2\right]J_n(\eta_s a) - 2\eta_s a J_{n+1}(\eta_s a),\\
D_{24} &= 2n(n-1)Y_n(\eta_d a) - 2n\eta_d a Y_{n+1}(\eta_d a),\\
D_{25} &= -k\eta_s a^2 Y_n(\eta_s a) + 2ka(n+1)Y_{n+1}(\eta_s a),\\
D_{26} &= -\left[2n(n-1) - \eta_s^2 a^2\right]Y_n(\eta_s a) - 2\eta_s a Y_{n+1}(\eta_s a),\\
D_{31} &= 2nka J_n(\eta_d a) + 2k\eta_d a^2 J_{n+1}(\eta_d a),\\
D_{32} &= -n\eta_s a J_n(\eta_s a) + (\eta_s^2 - k^2)a^2 J_{n+1}(\eta_s a),\\
D_{33} &= nka J_n(\eta_s a),\\
D_{34} &= -2nka Y_n(\eta_d a) + 2k\eta_d a^2 Y_{n+1}(\eta_d a),
\end{aligned}$$

$D_{35} = -n\eta_s a Y_n(\eta_s a) + (\eta_s^2 - k^2)a^2 Y_{n+1}(\eta_s a),$
$D_{36} = nkaY_n(\eta_s a).$

The remaining elements D_{41} to D_{66} are the same as the elements D_{11} to D_{36}, but replace a by b.

6.5.4 Inner and Outer Surfaces with Liquid Load

For subsea pipelines, there are liquid loads on both inner and outer surfaces. Therefore, it is of practical significance to study the guided waves in the pipe with liquid load. In the following, we will discuss the effects of liquid loads on axisymmetric compression waves as an example. Similar discussions can be made for other types of guided waves.

There are longitudinal waves which diffuse outward in the liquid outside the pipe. Its potential function is assumed as

$$\phi_w^{'} = A_5 H_0^{(2)}(k_w r)e^{ik(z-ct)}, \tag{6.5.20}$$

where $k_w^2 = \left(\frac{\omega}{c_w}\right)^2 - k^2$. The displacement and pressure generated by the longitudinal wave in the liquid are, respectively,

$$u_r^{'} = \nabla\phi_w = -k_w A_5 H_1^{(2)}(k_w r)e^{ik(z-ct)}, \tag{6.5.21}$$

$$\sigma_{rr}^{'} = \lambda_w \nabla\cdot\mathbf{u}' = \lambda_w\left[-(k_w^2 + k^2)H_0^{(2)}(k_w r)\right]A_5 e^{ik(z-ct)}. \tag{6.5.22}$$

There is also a compressional wave in the liquid inside the tube. However, due to the boundedness of the displacement corresponding to the longitudinal wave at r = 0, the potential function of the longitudinal wave can be expressed as

$$\phi_w^{''} = A_6 J_0(k_w r)e^{ik(z-ct)}. \tag{6.5.23}$$

The corresponding displacement and pressure are

$$u_r^{''} = -k_w A_6 J_1(k_w r)e^{ik(z-ct)}, \tag{6.5.24}$$

$$\sigma_{rr}^{''} = \lambda_w\left[-(k_w^2 + k^2)J_0(k_w r)\right]A_6 e^{ik(z-ct)}. \tag{6.5.25}$$

The potential function inside the cylindrical tube is the same as Eq. (6.5.9).

The continuous condition of radial pressure and radial displacement on the inner and outer surface of the pipe and the conditions that the shear stress on the inner and outer surface is zero can be expressed as

$$u_r|_{r=a} = u''_r|_{r=a}, \quad u_r|_{r=b} = u'_r\Big|_{r=b}, \tag{6.5.26a}$$

$$\sigma_{rr}|_{r=a} = \sigma''_{rr}|_{r=a}, \quad \sigma_{rr}|_{r=b} = \sigma'_{rr}\Big|_{r=b}, \tag{6.5.26b}$$

$$\sigma_{rz}|_{r=a} = 0, \ \sigma_{rz}|_{r=b} = 0. \tag{6.5.26c}$$

From Eq. (6.5.26), the following linear algebraic equations can be obtained

$$(D_{mn})\{A_n\} = 0, \quad \{A_n\} = (A_1, A_1, \ldots A_6)^T. \tag{6.5.27}$$

According to the condition that the equation has a non-zero solution, the dispersion equation is obtained as

$$\left|D_{ij}\right| = 0, \ (i, j = 1, 2 \ldots 6) \tag{6.5.28}$$

where

$D_{11} = -\left(\frac{\lambda\omega^2}{c_d^2} + \mu\eta_d^2\right) J_0(\eta_d a) + \mu\eta_d^2 J_2(\eta_d a)$,
$D_{12} = -\left(\frac{\lambda\omega^2}{c_d^2} + \mu\eta_d^2\right) Y_0(\eta_d a) + \mu\eta_d^2 Y_2(\eta_d a)$,
$D_{13} = i\mu k\eta_s^2 (J_0(\eta_s a) - J_2(\eta_s a))$, $D_{14} = i\mu k\eta_s^2 (Y_0(\eta_s a) - Y_2(\eta_s a))$,
$D_{15} = 0$, $D_{16} = -\lambda_w\left[-\left(k_w^2 + k^2\right) J_0(k_w a)\right]$,
$D_{21} = -2ik\eta_d J_1(\eta_d a)$, $D_{22} = -2ik\eta_d Y_1(\eta_d a)$,
$D_{23} = \eta_s\left(k^2 - \eta_s^2\right) J_1(\eta_s a)$, $D_{24} = \eta_s\left(k^2 - \eta_s^2\right) Y_1(\eta_s a)$,
$D_{25} = 0$, $D_{26} = 0$,
$D_{31} = -\eta_d J_1(\eta_d a)$, $D_{32} = -\eta_d Y_1(\eta_d a)$, $D_{33} = ik\eta_s J_1(\eta_s a)$,
$D_{34} = ik\eta_s Y_1(\eta_s a)$, $D_{35} = 0$, $D_{36} = k_w J_1(k_w a)$,
$D_{41} = -\eta_d J_1(\eta_d b)$, $D_{42} = -\eta_d Y_1(\eta_d b)$, $D_{43} = ik\eta_s J_1(\eta_s b)$,
$D_{44} = ik\eta_s Y_1(\eta_s b)$, $D_{45} = k_w H_1^{(2)}(k_w b)$, $D_{46} = 0$,
$D_{51} = -2ik\eta_d J_1(\eta_d b)$, $D_{52} = -2ik\eta_d Y_1(\eta_d b)$, $D_{53} = \eta_s\left(k^2 - \eta_s^2\right) J_1(\eta_s b)$,
$D_{54} = \eta_s\left(k^2 - \eta_s^2\right) Y_1(\eta_s b)$, $D_{55} = 0$, $D_{56} = 0$,
$D_{61} = -\left(\frac{\lambda\omega^2}{c_d^2} + \mu\eta_d^2\right) J_0(\eta_d b) + \mu\eta_d^2 J_2(\eta_d b)$,
$D_{62} = -\left(\frac{\lambda\omega^2}{c_d^2} + \mu\eta_d^2\right) Y_0(\eta_d b) + \mu\eta_d^2 Y_2(\eta_d b)$,
$D_{63} = i\mu k\eta_s^2 (J_0(\eta_s b) - J_2(\eta_s b))$, $D_{64} = i\mu k\eta_s^2 (Y_0(\eta_s b) - Y_2(\eta_s b))$,
$D_{65} = -\lambda_w\left[-\left(k_w^2 + k^2\right) H_0^{(2)}(k_w b)\right]$, $D_{66} = 0$.

Due to the existence of liquid outside the pipe, the longitudinal wave in the liquid is a cylindrical wave diffusing outwards, which will carry a part of energy, and this will make the energy of the guided wave in the cylindrical pipe gradually reduce and become a leaky wave. Different from the cylindrical waves diffusing outwards in the liquid outside pipe, the cylindrical waves in the liquid inside pipe are standing waves along radius direction, which cannot dissipate the energy of guided waves.

For torsional guided waves in the cylindrical pipe, the liquids inside and outside the pipe have no influences on the guided waves of torsional mode because the liquid cannot bear shear stress. The dispersion curves of each mode of torsional guided waves are not any different from those of a pipe in air.

6.6 Guided Waves in Spherical Shell

The gradient operator and Laplace operator in spherical coordinate system are [3,4]

$$\nabla = \frac{\partial}{\partial r}\mathbf{e}_r + \frac{1}{r}\frac{\partial}{\partial \theta}\mathbf{e}_\theta + \frac{1}{r\sin\theta}\frac{\partial}{\partial \varphi}\mathbf{e}_\varphi, \tag{6.6.1}$$

$$\nabla^2 = \frac{1}{r^2}\frac{\partial}{\partial r}\left(r^2\frac{\partial}{\partial r}\right) + \frac{1}{r^2\sin\theta}\frac{\partial}{\partial \theta}\left(\sin\theta\frac{\partial}{\partial \theta}\right) + \frac{1}{r^2\sin^2\theta}\frac{\partial^2}{\partial \varphi^2}. \tag{6.6.2}$$

The displacement field decomposition in spherical coordinate system can be expressed as

$$\mathbf{u} = \nabla\Phi + \nabla\times\mathbf{F} = \nabla\Phi + \nabla\times(r\Psi\mathbf{e}_r) + \nabla\times\nabla\times(r\Pi\mathbf{e}_r) = \mathbf{L}+\mathbf{M}+\mathbf{N}, \tag{6.6.3}$$

where **L, M** and **N** are the contribution of the displacement potential function (Φ, Ψ, Π) to the displacement field, respectively. Similar to the P, SV and SH waves in rectangular coordinates system, the displacement potential function (Φ, Ψ, Π) represents the elastic waves of three polarization modes in spherical coordinates system. Substituting (6.6.3) into the governing equation of the displacement field, the wave equations satisfied by the displacement potential function can be obtained as follows

$$\nabla^2\Phi = c_d^{-2}\frac{\partial^2\Phi}{\partial t^2}, \tag{6.6.4a}$$

$$\nabla^2\Psi = c_s^{-2}\frac{\partial^2\Psi}{\partial t^2}, \tag{6.6.4b}$$

$$\nabla^2\Pi = c_s^{-2}\frac{\partial^2\Pi}{\partial t^2}, \tag{6.6.4c}$$

where $c_d^2 = \frac{\lambda+2\mu}{\rho}$ and $c_s^2 = \frac{\mu}{\rho}$ represent the wave velocities of expansion wave and shear wave, respectively. The relation between displacement and potential functions in spherical coordinate system is as follows

$$u_r = \frac{\partial\Phi}{\partial r} - \frac{1}{r}\left[\frac{\cos\theta}{\sin\theta}\frac{\partial\Pi}{\partial\theta} + \frac{\partial^2\Pi}{\partial\theta^2} + \frac{1}{\sin^2\theta}\frac{\partial^2\Pi}{\partial\varphi^2}\right], \tag{6.6.5}$$

$$u_\theta = \frac{\partial \Phi}{r\partial \theta} + \frac{1}{\sin\theta}\frac{\partial \Psi}{\partial \varphi} + \frac{\partial \Pi}{r\partial \theta} + \frac{\partial^2 \Pi}{\partial r \partial \theta}, \tag{6.6.6}$$

$$u_\varphi = \frac{1}{r\sin\theta}\frac{\partial \Phi}{\partial \varphi} - \frac{\partial \Psi}{\partial \theta} + \frac{1}{r}\left[\frac{1}{\sin\theta}\frac{\partial \Pi}{\partial \varphi} + \frac{r}{\sin\theta}\frac{\partial^2 \Pi}{\partial \varphi\, \partial r}\right]. \tag{6.6.7}$$

Substitute the above equations into the expression of strain and stress as follows

$$\varepsilon_{rr} = \frac{\partial u_r}{\partial r},\ \varepsilon_{\theta\theta} = \frac{1}{r}\frac{\partial u_\theta}{\partial \theta} + \frac{u_r}{r},\ \varepsilon_{\varphi\varphi} = \frac{1}{r\sin\theta}\frac{\partial u_\varphi}{\partial \varphi} + \frac{\cot\theta}{r}u_\theta + \frac{u_r}{r},$$

$$\varepsilon_{r\theta} = \varepsilon_{\theta r} = \frac{1}{2}\left(\frac{1}{r}\frac{\partial u_r}{\partial \theta} + \frac{\partial u_\theta}{\partial r} - \frac{u_\theta}{r}\right),\ \varepsilon_{r\varphi} = \varepsilon_{\varphi r} = \frac{1}{2}\left(\frac{1}{r\sin\theta}\frac{\partial u_r}{\partial \varphi} + \frac{\partial u_\varphi}{\partial r} - \frac{u_\varphi}{r}\right),$$

$$\varepsilon_{\theta\varphi} = \varepsilon_{\varphi\theta} = \frac{1}{2}\left(\frac{1}{r\sin\theta}\frac{\partial u_\theta}{\partial \varphi} + \frac{1}{r}\frac{\partial u_\varphi}{\partial \theta} - \frac{\cot\theta}{r}u_\varphi\right),$$

$$\sigma_{rr} = \lambda(\nabla\cdot\mathbf{u}) + 2\mu\frac{\partial u_r}{\partial r},\quad \sigma_{\theta\theta} = \lambda(\nabla\cdot\mathbf{u}) + 2\mu\left(\frac{1}{r}\frac{\partial u_\theta}{\partial \theta} + \frac{u_r}{r}\right),$$

$$\sigma_{\varphi\varphi} = \lambda(\nabla\cdot\mathbf{u}) + 2\mu\left(\frac{1}{r\sin\theta}\frac{\partial u_\varphi}{\partial \varphi} + \frac{\cot\theta}{r}u_\theta + \frac{u_r}{r}\right),$$

$$\sigma_{r\theta} = \sigma_{\theta r} = \mu\left(\frac{1}{r}\frac{\partial u_r}{\partial \theta} + \frac{\partial u_\theta}{\partial r} - \frac{u_\theta}{r}\right),$$

$$\sigma_{r\varphi} = \sigma_{\varphi r} = \mu\left(\frac{1}{r\sin\theta}\frac{\partial u_r}{\partial \varphi} + \frac{\partial u_\varphi}{\partial r} - \frac{u_\varphi}{r}\right),$$

$$\sigma_{\theta\varphi} = \sigma_{\varphi\theta} = \mu\left(\frac{1}{r\sin\theta}\frac{\partial u_\theta}{\partial \varphi} + \frac{1}{r}\frac{\partial u_\varphi}{\partial \theta} - \frac{\cot\theta}{r}u_\varphi\right). \tag{6.6.8}$$

The relation between strain, stress and potential functions can be obtained

$$\begin{aligned}\varepsilon_{rr} =& \frac{\partial^2 \Phi}{\partial r^2} + \frac{1}{r^2}\left(\frac{\cos\theta}{\sin\theta}\frac{\partial \Pi}{\partial \theta} + \frac{\partial^2 \Pi}{\partial \theta^2} + \frac{1}{\sin^2\theta}\frac{\partial^2 \Pi}{\partial \varphi^2}\right)\\ &- \frac{1}{r}\left(\frac{\cos\theta}{\sin\theta}\frac{\partial^2 \Pi}{\partial r\partial \theta} + \frac{\partial^3 \Pi}{\partial r\partial \theta^2} + \frac{1}{\sin^2\theta}\frac{\partial^3 \Pi}{\partial r\partial \varphi^2}\right),\end{aligned}$$

$$\begin{aligned}\varepsilon_{\theta\theta} =& \frac{1}{r}\left(\frac{\partial^2 \Phi}{r\partial \theta^2} - \frac{\cos\theta}{\sin^2\theta}\frac{\partial \Psi}{\partial \varphi} + \frac{1}{\sin\theta}\frac{\partial^2 \Psi}{\partial \theta\partial \varphi} + \frac{\partial^2 \Pi}{r\partial \theta^2} + \frac{\partial^3 \Pi}{\partial r\partial \theta^2}\right)\\ &+ \frac{1}{r}\left[\frac{\partial \Phi}{\partial r} - \frac{1}{r}\left(\frac{\cos\theta}{\sin\theta}\frac{\partial \Pi}{\partial \theta} + \frac{\partial^2 \Pi}{\partial \theta^2} + \frac{1}{\sin^2\theta}\frac{\partial^2 \Pi}{\partial \varphi^2}\right)\right],\end{aligned}$$

$$\varepsilon_{\varphi\varphi} = \frac{1}{r\sin\theta}\left[\frac{1}{r\sin\theta}\frac{\partial^2 \Phi}{\partial \varphi^2} - \frac{\partial^2 \Psi}{\partial \varphi\partial \theta} + \frac{1}{r}\left(\frac{1}{\sin\theta}\frac{\partial^2 \Pi}{\partial \varphi^2} + \frac{r}{\sin\theta}\frac{\partial^3 \Pi}{\partial \varphi^2\partial r}\right)\right]$$

$$+\frac{\cot\theta}{r}\left(\frac{\partial\Phi}{r\partial\theta}+\frac{1}{\sin\theta}\frac{\partial\Psi}{\partial\varphi}+\frac{\partial\Pi}{r\partial\theta}+\frac{\partial^2\Pi}{\partial r\partial\theta}\right)$$

$$+\frac{1}{r}\left[\frac{\partial\Phi}{\partial r}-\frac{1}{r}\left(\frac{\cos\theta}{\sin\theta}\frac{\partial\Pi}{\partial\theta}+\frac{\partial^2\Pi}{\partial\theta^2}+\frac{1}{\sin^2\theta}\frac{\partial^2\Pi}{\partial\varphi^2}\right)\right],$$

$$\varepsilon_{r\theta}=\frac{1}{2}\{\frac{1}{r}[\frac{\partial^2\Phi}{\partial\theta\partial r}-\frac{1}{r}(-\frac{1}{\sin^2\theta}\frac{\partial\Pi}{\partial\theta}+\frac{\cos\theta}{\sin\theta}\frac{\partial^2\Pi}{\partial\theta^2}+\frac{\partial^3\Pi}{\partial\theta^3}$$

$$-\frac{2\cos\theta}{\sin^3\theta}\frac{\partial^2\Pi}{\partial\varphi^2}+\frac{1}{\sin^2\theta}\frac{\partial^3\Pi}{\partial\theta\partial\varphi^2})]-\frac{1}{r^2}\frac{\partial\Phi}{\partial\theta}+\frac{1}{r}\frac{\partial^2\Phi}{\partial r\partial\theta}$$

$$+\frac{1}{\sin\theta}\frac{\partial^2\Psi}{\partial r\partial\varphi}-\frac{1}{r^2}\frac{\partial\Pi}{\partial\theta}+\frac{1}{r}\frac{\partial^2\Pi}{\partial r\partial\theta}$$

$$+\frac{\partial^3\Pi}{\partial r^2\partial\theta}-\frac{1}{r}\left(\frac{\partial\Phi}{r\partial\theta}+\frac{1}{\sin\theta}\frac{\partial\Psi}{\partial\varphi}+\frac{\partial\Pi}{r\partial\theta}+\frac{\partial^2\Pi}{\partial r\partial\theta}\right)\},$$

$$\varepsilon_{r\varphi}=\frac{1}{2}\{\frac{1}{r\sin\theta}\left[\frac{\partial^2\Phi}{\partial\varphi\partial r}-\frac{1}{r}\left(\frac{\cos\theta}{\sin\theta}\frac{\partial^2\Pi}{\partial\varphi\partial\theta}+\frac{\partial^3\Pi}{\partial\varphi\partial\theta^2}+\frac{1}{\sin^2\theta}\frac{\partial^3\Pi}{\partial\varphi^3}\right)\right]$$

$$-\frac{1}{r^2\sin\theta}\frac{\partial\Phi}{\partial\varphi}+\frac{1}{r\sin\theta}\frac{\partial^2\Phi}{\partial r\partial\varphi}-\frac{\partial^2\Psi}{\partial r\partial\theta}-\frac{1}{r^2}\left(\frac{1}{\sin\theta}\frac{\partial\Pi}{\partial\varphi}+\frac{r}{\sin\theta}\frac{\partial^2\Pi}{\partial\varphi\partial r}\right)$$

$$+\frac{1}{r}\left(\frac{2}{\sin\theta}\frac{\partial^2\Pi}{\partial r\partial\varphi}+\frac{r}{\sin\theta}\frac{\partial^3\Pi}{\partial r^2\partial\varphi}\right)$$

$$-\frac{1}{r}\left[\frac{1}{r\sin\theta}\frac{\partial\Phi}{\partial\varphi}-\frac{\partial\Psi}{\partial\theta}+\frac{1}{r}\left(\frac{1}{\sin\theta}\frac{\partial\Pi}{\partial\varphi}+\frac{r}{\sin\theta}\frac{\partial^2\Pi}{\partial\varphi\partial r}\right)\right]\},$$

$$\varepsilon_{\varphi\theta}=\frac{1}{2}\{\frac{1}{r\sin\theta}\left(\frac{\partial^2\Phi}{r\partial\varphi\partial\theta}+\frac{1}{\sin\theta}\frac{\partial^2\Psi}{\partial\varphi^2}+\frac{\partial^2\Pi}{r\partial\varphi\partial\theta}+\frac{\partial^3\Pi}{\partial r\partial\varphi\partial\theta}\right)$$

$$+\frac{1}{r}[-\frac{\cos\theta}{r\sin^2\theta}\frac{\partial\Phi}{\partial\varphi}+\frac{1}{r\sin\theta}\frac{\partial^2\Phi}{\partial\theta\partial\varphi}-\frac{\partial^2\Psi}{\partial\theta^2}$$

$$+\frac{1}{r}\left(-\frac{\cos\theta}{\sin^2\theta}\frac{\partial\Pi}{\partial\varphi}+\frac{1}{\sin\theta}\frac{\partial^2\Pi}{\partial\theta\partial\varphi}-\frac{r\cos\theta}{\sin^2\theta}\frac{\partial^2\Pi}{\partial r\partial\varphi}+\frac{r}{\sin\theta}\frac{\partial^3\Pi}{\partial r\partial\varphi\partial\theta}\right)]$$

$$-\frac{\cot\theta}{r}\left[\frac{1}{r\sin\theta}\frac{\partial\Phi}{\partial\varphi}-\frac{\partial\Psi}{\partial\theta}+\frac{1}{r}\left(\frac{1}{\sin\theta}\frac{\partial\Pi}{\partial\varphi}+\frac{r}{\sin\theta}\frac{\partial^2\Pi}{\partial\varphi\partial r}\right)\right],$$

$$\sigma_{rr}=\lambda\nabla^2\Phi+2\mu\frac{\partial^2\Phi}{\partial r^2}+2\mu l\left[\frac{\partial^2(r\Pi)}{\partial r^2}-r\nabla^2\Pi\right],$$

$$\sigma_{\theta\theta}=\lambda\nabla^2\Phi+2\mu\left(\frac{1}{r}\frac{\partial\Phi}{\partial r}+\frac{1}{r^2}\frac{\partial^2\Phi}{\partial\theta^2}\right)+2\mu\frac{1}{r}\frac{\partial}{\partial\theta}\left[\frac{1}{r\sin\theta}\frac{\partial^2(r\Psi)}{\partial\varphi\partial r}\right]$$

$$+2\mu l\left[\frac{1}{r^2}\frac{\partial^3(r\Pi)}{\partial\theta^2\partial r}+\frac{1}{r}\left(\frac{\partial^2(r\Pi)}{\partial r^2}-r\nabla^2\Pi\right)\right],$$

$$\sigma_{\varphi\varphi}=\lambda\nabla^2\Phi+2\mu\left[\frac{1}{r\sin\theta}\frac{\partial}{\partial\varphi}\left(\frac{1}{r\sin\theta}\frac{\partial\Phi}{\partial\varphi}\right)+\frac{1}{r}\frac{\partial\Phi}{\partial r}+\frac{1}{r^2}\cot\theta\frac{\partial\Phi}{\partial\theta}\right]$$

$$+2\mu\left[\frac{\cot\theta}{r^2\sin\theta}\frac{\partial(r\Psi)}{\partial\varphi}-\frac{1}{r^2\sin\theta}\frac{\partial^2(r\Psi)}{\partial\varphi\partial\theta}\right]$$

$$+2\mu l\left[\frac{1}{r^2\sin^2\theta}\frac{\partial^3(r\Pi)}{\partial\varphi^2\partial r}+\frac{1}{r}\left(\frac{\partial^2(r\Pi)}{\partial r^2}-r\nabla^2\Pi\right)+\frac{\cot\theta}{r^2}\frac{\partial^2(r\Psi)}{\partial\theta\partial r}\right],$$

$$\sigma_{r\theta}=\frac{2\mu}{r}\left[\frac{\partial^2\Phi}{\partial r\partial\theta}-\frac{1}{r}\frac{\partial\Phi}{\partial\theta}\right]-\frac{\mu}{r}\left[\frac{1}{r\sin\theta}\frac{\partial(r\Psi)}{\partial\varphi}-r\frac{\partial}{\partial r}\left(\frac{1}{r\sin\theta}\frac{\partial(r\Psi)}{\partial\varphi}\right)\right]$$

$$+\frac{\mu l}{r}\left[\frac{\partial}{\partial\theta}\left(\frac{\partial^2(r\Pi)}{\partial r^2}-r\nabla^2\Pi\right)-\frac{1}{r}\frac{\partial^2(r\Pi)}{\partial\theta\partial r}+r\frac{\partial}{\partial r}\left(\frac{1}{r}\frac{\partial^2(r\Pi)}{\partial\theta\partial r}\right)\right],$$

$$\sigma_{r\varphi}=2\mu\left[\frac{1}{r\sin\theta}\frac{\partial^2\Phi}{\partial r\partial\varphi}-\frac{1}{r^2\sin\theta}\frac{\partial\Phi}{\partial\varphi}\right]+\mu\left[\frac{2}{r^2}\frac{\partial(r\Psi)}{\partial\theta}-\frac{1}{r}\frac{\partial^2(r\Psi)}{\partial\theta\partial r}\right]$$

$$+\mu l\left[\frac{1}{r\sin\theta}\frac{\partial}{\partial\varphi}\left(\frac{2\partial^2(r\Pi)}{\partial r^2}-r\nabla^2\Pi\right)-\frac{2}{r^2\sin\theta}\frac{\partial^2(r\Pi)}{\partial\varphi\partial r}\right],$$

$$\sigma_{\theta\varphi}=2\mu\left[\frac{1}{r^2\sin\theta}\frac{\partial^2\Phi}{\partial\theta\partial\varphi}-\frac{\cot\theta}{r^2\sin\theta}\frac{\partial\Phi}{\partial\varphi}\right]$$

$$+2\mu l\left[\frac{1}{r^2\sin\theta}\frac{\partial^3(r\Pi)}{\partial r\partial\theta\partial\varphi}-\frac{\cot\theta}{r^2\sin\theta}\frac{\partial^2(r\Pi)}{\partial r\partial\varphi}\right]$$

$$+\mu\left[\frac{\cot\theta}{r^2}\frac{\partial(r\Psi)}{\partial\theta}-\frac{1}{r^2}\frac{\partial^2(r\Psi)}{\partial\theta^2}+\frac{1}{r^2\sin^2\theta}\frac{\partial^2(r\Psi)}{\partial\varphi^2}\right]. \quad (6.6.9)$$

The guided waves in a sphere with an inner diameter of a and an outer diameter of b are discussed below. In spherical coordinates system (r,θ,φ), the solution of the displacement potential function of Eq. (6.6.4) can be expressed in spherical wave function form as follows

$$\Phi=[A_{mn}j_n(\alpha r)+B_{mn}y_n(\alpha r)]P_n^m(\cos\theta)e^{(im\varphi-i\omega t)}, \quad (6.6.10a)$$

$$\Psi=\left[A'_{mn}j_n(\beta r)+B'_{mn}y_n(\beta r)\right]P_n^m(\cos\theta)e^{(im\varphi-i\omega t)}, \quad (6.6.10b)$$

$$\Pi=\left[A''_{mn}j_n(\beta r)+B''_{mn}y_n(\beta r)\right]P_n^m(\cos\theta)e^{(im\varphi-i\omega t)}. \quad (6.6.10c)$$

where j_n and y_n represent spherical Bessel functions of the first and second kinds, respectively. $P_n^m(\cos\theta)$ is associated Legendre polynomial. A_{mn}, B_{mn}, A'_{mn}, B'_{mn}, A''_{mn} and B''_{mn} are arbitrary constants. α and β represent the wave number of compression wave and shear wave, respectively, i.e.

$$\alpha=\frac{\omega}{c_d},\quad \beta=\frac{\omega}{c_s}. \quad (6.6.11)$$

If let

$$v_{mn}(r) = A_{mn} j_n(\alpha r) + B_{mn} y_n(\alpha r), \tag{6.6.12a}$$

$$g_{mn}(r) = A'_{mn} j_n(\beta r) + B'_{mn} y_n(\beta r), \tag{6.6.12b}$$

$$f_{mn}(r) = A''_{mn} j_n(\beta r) + B''_{mn} y_n(\beta r), \tag{6.6.12c}$$

then $v_{mn}(r)$, $f_{mn}(r)$ and $g_{mn}(r)$ represent the radial displacement mode of guided waves, while $P_n^m(\cos\theta)$ represents the displacement mode of guided wave along the longitude direction. They are the fundamental to distinguish guided waves of different orders. $e^{(im\varphi - i\omega t)}$ represents the guided wave propagating along the latitude direction, where m is the angular wave number, and the relation between it and the linear wave number $k(= \frac{2\pi}{\Lambda})$ is $m = kb$ (b is the outer diameter of the hollow sphere). The relation between the wave length Λ and the angular wave number m is $\Lambda = \frac{2\pi b}{m}$, or $m = \frac{2\pi b}{\Lambda}$, that is, the number of complete waveforms corresponding to one period contained on the equator or meridian. When m is an integer, it means that the guided wave which propagates along latitude direction degenerates into a standing wave. Considering that $P_n^m(\cos\theta)$ should be bounded at the poles, where θ=0, π. m can only be an integer. When m and n are both integers, Eq. (6.6.10) actually represents the natural vibration (standing wave) of order (n, m) of the sphere, and the corresponding frequencies ω_{nm} are the ones that correspond to the natural vibrations of order (n, m). When the guided waves on the sphere are discussed, the conditions of boundedness at the north and south poles and periodicity in the latitudinal direction can be relaxed, as long as the free boundary conditions on the sphere are retained. So m and n are not limited to integers, they can be any real number. This is the differences between the vibrational solution and the wave solution.

For guided waves of order (n, m), according to the relationship between potential function and displacement, the displacement components corresponding to the guided wave can be expressed as

$$u_r = U_r(r) P_n^m(\cos\theta) e^{im\varphi - imt}, \tag{6.6.13a}$$

$$u_\theta = \left(U_\theta^1(r) \frac{\mathrm{d}}{\mathrm{d}\theta} + U_\theta^2(r) \frac{im}{\sin\theta} \right) P_n^m(\cos\theta) e^{im\varphi - imt}, \tag{6.6.13b}$$

$$u_\varphi = -\left(U_\varphi^2(r) \frac{\mathrm{d}}{\mathrm{d}\theta} + U_\varphi^1(r) \frac{im}{\sin\theta} \right) P_n^m(\cos\theta) e^{im\varphi - imt}, \tag{6.6.13c}$$

where

$$U_r(r) = \frac{\mathrm{d}}{\mathrm{d}r} v_{mn}(r) + n(n+1) \frac{f_{mn}(r)}{r},$$

$$U_\theta^1(r) = U_\varphi^1(r) = \frac{v_{mn}(r)}{r} + \frac{1}{r} \frac{\mathrm{d}(f_{mn}(r) r)}{\mathrm{d}r},$$

$$U_\theta^2(r) = U_\varphi^2(r) = g_{mn}(r).$$

Meanwhile, according to the relationship between potential function and stresses, the stress components corresponding to the guided waves of order (n, m) are

$$\sigma_{rr} = T_{rr}(r)P_n^m(\cos\theta)e^{(im\varphi - i\omega t)}, \tag{6.6.14a}$$

$$\sigma_{r\theta} = \left(T_{r\theta}^1(r)\frac{\mathrm{d}}{\mathrm{d}\theta} + T_{r\theta}^2(r)\frac{im}{\sin\theta}\right)P_n^m(\cos\theta)e^{(im\varphi - i\omega t)}, \tag{6.6.14b}$$

$$\sigma_{r\varphi} = -\left(T_{r\varphi}^2(r)\frac{\mathrm{d}}{\mathrm{d}\theta} + T_{r\varphi}^1(r)\frac{im}{\sin\theta}\right)P_n^m(\cos\theta)e^{(im\varphi - i\omega t)}, \tag{6.6.14c}$$

where

$$T_{rr}(r) = \left(-\lambda\alpha^2 + 2\mu\frac{\mathrm{d}^2}{\mathrm{d}r^2}\right)v_{mn}(r) + 2\mu n(n+1)\frac{\mathrm{d}}{\mathrm{d}r}\left(\frac{f_{mn}(r)}{r}\right), \tag{6.6.15a}$$

$$T_{r\theta}^1(r) = T_{r\varphi}^1(r) = 2\mu\frac{\mathrm{d}}{\mathrm{d}r}\left(\frac{v_{mn}(r)}{r}\right) + \mu\left[\frac{1}{r^2}n(n+1) + \frac{\mathrm{d}^2}{\mathrm{d}r^2} - \frac{2}{r^2}\right]f_{mn}(r), \tag{6.6.15b}$$

$$T_{r\theta}^2(r) = T_{r\varphi}^2(r) = \mu\left(\frac{\mathrm{d}}{\mathrm{d}r} - \frac{1}{r}\right)g_{mn}(r). \tag{6.6.15c}$$

6.6.1 Inner and Outer Free Surfaces

If the inner and outer surfaces of the sphere are free, the boundary conditions can be expressed as

$$\sigma_{rr}|_{r=a} = \sigma_{r\theta}|_{r=a} = \sigma_{r\varphi}\big|_{r=a} = 0, \tag{6.6.16a}$$

$$\sigma_{rr}|_{r=b} = \sigma_{r\theta}|_{r=b} = \sigma_{r\varphi}\big|_{r=b} = 0. \tag{6.6.16b}$$

Substitute Eq. (6.6.14) into Eq. (6.6.16), and make use of $T_{r\theta}^1(r) = T_{r\varphi}^1(r)$ and $T_{r\theta}^2(r) = T_{r\varphi}^2(r)$, we can obtain on the inner boundary

$$T_{rr}(a)P_n^m(\cos\theta)e^{(im\varphi - i\omega t)} = 0, \tag{6.6.17a}$$

$$\left(T_{r\theta}^1(a)\frac{\mathrm{d}}{\mathrm{d}\theta} + T_{r\theta}^2(a)\frac{im}{\sin\theta}\right)P_n^m(\cos\theta)e^{(im\varphi - i\omega t)} = 0, \tag{6.6.17b}$$

$$-\left(T^2_{r\varphi}(a)\frac{\mathrm{d}}{\mathrm{d}\theta}+T^1_{r\varphi}(a)\frac{im}{\sin\theta}\right)P^m_n(\cos\theta)e^{(im\varphi-i\omega t)}=0. \tag{6.6.17c}$$

By application of the property of Legendre functions

$$(2n+1)\left(1-x^2\right)\frac{\mathrm{d}}{\mathrm{d}x}\left(P^m_n(x)\right)=(n+1)(n+m)P^m_{n-1}(x)-n(n+m-1)P^m_{n+1}(x), \tag{6.6.18}$$

we get

$$\frac{\mathrm{d}}{\mathrm{d}\theta}P^m_n(\cos\theta)=\frac{(n+1)(n+m)}{(2n+1)\sin\theta}P^m_{n-1}(\cos\theta)-\frac{n(n+m-1)}{(2n+1)\sin\theta}P^m_{n+1}(\cos\theta). \tag{6.6.19}$$

The orthogonality condition of the spherical harmonic functions on the spherical surface $0\le\theta\le\pi, 0\le\varphi\le 2\pi$ can be expressed as

$$\int\limits_0^\pi\int\limits_0^{2\pi}P^m_n(\cos\theta)e^{im\varphi}P^k_l(\cos\theta)e^{ik\varphi}\sin\theta\,\mathrm{d}\theta\,\mathrm{d}\varphi=\begin{cases}0, & n\ne l\ or\ m\ne k\\ \left(N^m_n\right)^2, & n=l\ and\ m=k\end{cases}. \tag{6.6.20}$$

where

$$\left(N^m_n\right)^2=\frac{(n+m)!}{(n-m)!}\frac{2}{2n+1}\pi\delta_m,$$

$$\delta_m=\begin{cases}2, & m=0\\ 1, & m\ne 0\end{cases}.$$

The application of Eqs. (6.6.19) and (6.6.20) onto the boundary conditions (6.6.17) leads to the boundary conditions at the inner boundary where $r=a$,

$$T_{rr}(a)=0, \tag{6.6.21a}$$

$$T^1_{r\theta}(a)=0,\quad \text{or}\quad T^2_{r\theta}(a)=0, \tag{6.6.21b}$$

$$T^1_{r\varphi}(a)=0,\quad \text{or}\quad T^2_{r\varphi}(a)=0. \tag{6.6.21c}$$

Consider that Eq. (6.6.15) is valid, the boundary condition at $r=a$ can also be expressed as

$$\begin{Bmatrix} T_{rr}(a) \\ T^1_{r\theta}(a) \\ T^2_{r\theta}(a) \end{Bmatrix} = 0. \quad (6.6.22)$$

Similarly, the boundary conditions at the outer boundary $r = b$ can be expressed as

$$\begin{Bmatrix} T_{rr}(b) \\ T^1_{r\theta}(b) \\ T^2_{r\theta}(b) \end{Bmatrix} = 0. \quad (6.6.23)$$

Equations (6.6.22) and (6.6.23) can be combined into the following linear algebraic equations

$$(D)_{6\times6}\{A\}_{6\times1} = 0, \quad (6.6.24)$$

where

$$(D) = \begin{pmatrix} D_1 & 0 \\ 0 & D_2 \end{pmatrix} = \begin{pmatrix} \mathrm{d}_{11} & \mathrm{d}_{12} & \mathrm{d}_{13} & \mathrm{d}_{14} & 0 & 0 \\ \mathrm{d}_{21} & \mathrm{d}_{22} & \mathrm{d}_{23} & \mathrm{d}_{24} & 0 & 0 \\ \mathrm{d}_{31} & \mathrm{d}_{32} & \mathrm{d}_{33} & \mathrm{d}_{34} & 0 & 0 \\ \mathrm{d}_{41} & \mathrm{d}_{42} & \mathrm{d}_{43} & \mathrm{d}_{44} & 0 & 0 \\ 0 & 0 & 0 & 0 & \mathrm{d}_{55} & \mathrm{d}_{56} \\ 0 & 0 & 0 & 0 & \mathrm{d}_{65} & \mathrm{d}_{66} \end{pmatrix},$$

$$\{A\} = \left(A_{mn}, B_{mn}, A''_{mn}, B''_{mn}, A'_{mn}, B'_{mn}\right)^T,$$

where D_1 is a square matrix of 4×4 order and represents the dispersion equation of Lamb-like waves coupled by compressional wave Φ and shear wave Π. Lamb-like waves have not only in-plane displacements u_φ and u_θ, but also the radial displacement u_r. By comparison with the guided waves in cylindrical shells, it is noted that Lamb-like waves in spherical shells correspond to axisymmetric compression waves and bending waves in cylindrical shells. Lamb-like waves can be understood as spherical guided waves propagating from one pole to the other. D_2 is a square matrix of 2×2 order and represents the guided wave generated by the shear wave Ψ. Because the potential function Ψ only produces displacements along the spherical surface, i.e. u_φ and u_θ, no radial displacement u_r exists, So, it can be called SH-like waves. The SH-like waves in the sphere shell correspond to the torsional waves in the cylindrical shell. It can be understood as a spherical torsional guided wave traveling from one pole to the other.

According to the condition that the equations have non-zero solutions, the dispersion equations of Lamb-like waves and SH-like waves are, respectively,

$$\Omega_1(n, \omega) = \det(D_1) = 0, \quad (6.6.25)$$

$$\Omega_2(n,\omega) = \det(D_2) = 0, \tag{6.6.26}$$

where

$$
\begin{aligned}
&d_{11} = -\tfrac{\lambda}{2\mu}\alpha^2 j_n(\alpha a) + \tfrac{2\alpha}{a} j_{n+1}(\alpha a) + \tfrac{n^2-n}{a^2} j_n(\alpha a) - \alpha^2 j_n(\alpha a),\\
&d_{12} = -\tfrac{\lambda}{2\mu}\alpha^2 y_n(\alpha a) + \tfrac{2\alpha}{a} y_{n+1}(\alpha a) + \tfrac{n^2-n}{a^2} y_n(\alpha a) - \alpha^2 y_n(\alpha a),\\
&d_{13} = n(n+1)\Big(\tfrac{n-1}{a^2} j_n(\beta a) - \tfrac{\beta}{a} j_{n+1}(\beta a)\Big),\\
&d_{14} = n(n+1)\Big(\tfrac{n-1}{a^2} y_n(\beta a) - \tfrac{\beta}{a} y_{n+1}(\beta a)\Big),\ d_{15} = d_{16} = 0,\\
&d_{21} = 2\big(\tfrac{n-1}{a^2} j_n(\alpha a) - \tfrac{\alpha}{a} j_{n+1}(\alpha a)\big),\ \ d_{22} = 2\big(\tfrac{n-1}{a^2} y_n(\alpha a) - \tfrac{\alpha}{a} y_{n+1}(\alpha a)\big),\\
&d_{23} = \tfrac{1}{a^2} n(n+1) j_n(\beta a) + \tfrac{2\beta}{a} j_{n+1}(\beta a) + \tfrac{n^2-n}{a^2} j_n(\beta a) - \beta^2 j_n(\beta a) - \tfrac{2}{a^2} j_n(\beta a),\\
&d_{24} = \tfrac{1}{a^2} n(n+1) y_n(\beta a) + \tfrac{2\beta}{a} y_{n+1}(\beta a) + \tfrac{n^2-n}{a^2} y_n(\beta a) - \beta^2 y_n(\beta a) - \tfrac{2}{a^2} y_n(\beta a),\\
&d_{25} = d_{26} = 0,\\
&d_{31} = -\tfrac{\lambda}{2\mu}\alpha^2 j_n(\alpha b) + \tfrac{2\alpha}{b} j_{n+1}(\alpha b) + \tfrac{n^2-n}{b^2} j_n(\alpha b) - \alpha^2 j_n(\alpha b),\\
&d_{32} = -\tfrac{\lambda}{2\mu}\alpha^2 y_n(\alpha b) + \tfrac{2\alpha}{b} y_{n+1}(\alpha b) + \tfrac{n^2-n}{b^2} y_n(\alpha b) - \alpha^2 y_n(\alpha b),\\
&d_{33} = n(n+1)\Big(\tfrac{n-1}{b^2} j_n(\beta b) - \tfrac{\beta}{b} j_{n+1}(\beta b)\Big),\\
&d_{34} = n(n+1)\Big(\tfrac{n-1}{b^2} y_n(\beta b) - \tfrac{\beta}{b} y_{n+1}(\beta b)\Big),\ \ d_{35} = d_{36} = 0,\\
&d_{41} = 2\big(\tfrac{n-1}{b^2} j_n(\alpha b) - \tfrac{\alpha}{b} j_{n+1}(\alpha b)\big),\ \ d_{42} = 2\big(\tfrac{n-1}{b^2} y_n(\alpha b) - \tfrac{\alpha}{b} y_{n+1}(\alpha b)\big),\\
&d_{43} = \tfrac{1}{b^2} n(n+1) j_n(\beta b) + \tfrac{2\beta}{b} j_{n+1}(\beta b) + \tfrac{n^2-n}{b^2} j_n(\beta b) - \beta^2 j_n(\beta b) - \tfrac{2}{b^2} j_n(\beta b),\\
&d_{44} = \tfrac{1}{b^2} n(n+1) y_n(\beta b) + \tfrac{2\beta}{b} y_{n+1}(\beta b) + \tfrac{n^2-n}{b^2} y_n(\beta b) - \beta^2 y_n(\beta b) - \tfrac{2}{b^2} y_n(\beta b),\\
&d_{45} = d_{46} = 0,\ d_{51} = d_{52} = d_{53} = d_{54} = 0,\\
&d_{55} = \tfrac{n-1}{a} j_n(\beta a) - \beta j_{n+1}(\beta a),\ \ d_{56} = \tfrac{n-1}{a} y_n(\beta a) - \beta y_{n+1}(\beta a),\\
&d_{61} = d_{62} = d_{63} = d_{64} = 0,\\
&d_{65} = \tfrac{n-1}{b} j_n(\beta b) - \beta j_{n+1}(\beta b),\ \ d_{66} = \tfrac{n-1}{b} y_n(\beta b) - \beta y_{n+1}(\beta b).
\end{aligned}
$$

It can be seen from the expressions above, d_{ij} does not contain m, that is to say, the dispersion equation of guided waves of order (n, m) does not depend on m. This is a natural consequence of the spherical symmetry feature. In fact, the frequency of the natural vibrations of order (n, m) of the spherical shell is also independent of m.This feature is also known as the degenerate natural vibrations of the spherical shell of order (n, m), that is, for any integer n, the natural vibrations corresponding $m = 0, \pm 1, \ldots \pm n$, total $2n + 1$ order, have the same vibration frequency. Since the dispersion equation is independent of m, we can only discuss the case of $m = n$, whether it is Lamb-like wave or SH-like wave. In this case, n and m have the same meaning, that is, n represents the angular wave number of the guided wave. n can be any real number, and when n is an integer, the spherical guided wave degenerates into a spherical standing wave (Fig. 6.21).

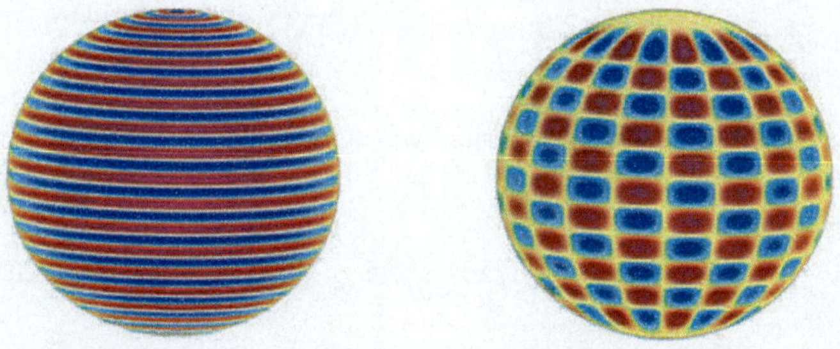

(a) Spherical guided waves (b) Spherical standing waves

Fig. 6.21 Sketch of spherical guided waves and spherical standing waves

6.6.2 Inner and Outer Surfaces with Liquid Loads

This problem is treated mathematically in much the same way as a sphere without a liquid load. But there are two points necessary to be considered: (1) There are longitudinal waves which propagate in the liquid and can be represented by a potential function ϕ_w; (2) the guided waves of various modes in the sphere shell will leak energy into the liquid during their propagation process, and thus, the amplitude of the guided waves attenuates gradually. The effect of the liquid must be considered in terms of the complex wave number, i.e.$k = k^r + ik^m$, and the imaginary part of k corresponds to an attenuation factor.

In spherical coordinate system, the wave equation in the liquid can be expressed as follow in terms of the potential function ϕ_w

$$\nabla^2 \phi_w = c_d^{-2} \frac{\partial^2 \phi_w}{\partial t^2}. \tag{6.6.27}$$

There are spherical waves diffusing outwards in the liquid surrounding the sphere. According to Eq. (6.6.27), the potential function of the liquid outside the spherical shell is as follows

$$\phi'_w = \sum_{n=0}^{+\infty} \sum_{m=-n}^{+n} C_{mn} h_n^{(2)}(k_w r) P_n^m(\cos\theta) e^{im\varphi - i\omega t}. \tag{6.6.28}$$

where $k_w^2 = \frac{\omega^2}{c_w^2}$, c_w stands for the speed of sound in the liquid. Due to $\boldsymbol{u}'_w = \nabla \phi'_w$, the normal displacement is further obtained as

$$u'_r = \sum_{n=0}^{+\infty} \sum_{m=-n}^{+n} C_{mn} \frac{\mathrm{d}}{\mathrm{d}r} h_n^{(2)}(k_w r) P_n^m(\cos\theta) e^{im\varphi - i\omega t}$$

$$= \sum_{n=0}^{+\infty} \sum_{m=-n}^{+n} C_{mn} \left(\frac{n}{r} h_n^{(2)}(k_w r) - k_w h_{n+1}^{(2)}(k_w r) \right) P_n^m(\cos\theta) e^{im\varphi - i\omega t}. \quad (6.6.29)$$

At the interface between the spherical waveguide and the liquid, the pressure of the liquid is

$$\sigma'_{rr} = \lambda_w \nabla \cdot \boldsymbol{u}'_w = -\sum_{n=0}^{+\infty} \sum_{m=-n}^{+n} \lambda_w k_w^2 C_{mn} H_n^{(2)}(k_w r) P_n^m(\cos\theta) e^{(im\varphi - i\omega t)}. \quad (6.6.30)$$

Consider the bounded nature of the displacement at $r = 0$, there is a spherical standing wave in the liquid inside the sphere. The potential function of the spherical standing wave can be expressed as

$$\phi''_w = \sum_{n=0}^{+\infty} \sum_{m=-n}^{+n} E_{mn} j_n(k_w r) P_n^m(\cos\theta) e^{im\varphi - i}. \quad (6.6.31)$$

The corresponding displacement and pressure are

$$u''_r = \sum_{n=0}^{+\infty} \sum_{m=-n}^{+n} E_{mn} \frac{\mathrm{d}}{\mathrm{d}r} j_n(k_w r) P_n^m(\cos\theta) e^{im\varphi - i\omega t}$$
$$= \sum_{n=0}^{+\infty} \sum_{m=-n}^{+n} E_{mn} \left(\frac{n}{r} j_n(k_w r) - k_w j_{n+1}(k_w r) \right) P_n^m(\cos\theta) e^{im\varphi - i\omega t}. \quad (6.6.32)$$

$$\sigma''_{rr} = \lambda_w \nabla \cdot \boldsymbol{u}''_w = -\sum_{n=0}^{+\infty} \sum_{m=-n}^{+n} \lambda_w k_w^2 E_{mn} j_n(k_w r) P_n^m(\cos\theta) e^{(im\varphi - i\omega t)}. \quad (6.6.33)$$

The continuity condition of radial pressure and radial displacement on the inner and outer surfaces of the sphere and the condition that the shear stresses on the inner and outer surfaces are zero corresponding to the guided waves of order (n, m) can be expressed as

$$u_r|_{r=a} = u''_r\big|_{r=a}, \quad u_r|_{r=b} = u'_r\Big|_{r=b}, \quad (6.6.34a)$$

$$\sigma_{rr}|_{r=a} = \sigma''_{rr}\big|_{r=a}, \quad \sigma_{rr}|_{r=b} = \sigma'_{rr}\Big|_{r=b}, \quad (6.6.34b)$$

$$\sigma_{r\theta}|_{r=a} = \sigma_{r\varphi}\big|_{r=a} = 0, \quad (6.6.34c)$$

$$\sigma_{r\theta}|_{r=b} = \sigma_{r\varphi}\big|_{r=b} = 0. \quad (6.6.34d)$$

Similar to the treatment of the boundary conditions that the inner and outer surfaces are both free, the application of the properties of Legendre polynomials and the orthogonality of spherical harmonic functions leads to the linear algebraic equations

$$(D)_{8\times 8}\{A\}_{8\times 1}=0, \tag{6.6.35}$$

where

$$(D)=\begin{pmatrix} D_1 & 0 \\ 0 & D_2 \end{pmatrix}=\begin{pmatrix} d_{11} & d_{12} & d_{13} & d_{14} & d_{15} & 0 & 0 & 0 \\ d_{21} & d_{22} & d_{23} & d_{24} & 0 & d_{26} & 0 & 0 \\ d_{31} & d_{32} & d_{33} & d_{34} & d_{35} & 0 & 0 & 0 \\ d_{41} & d_{42} & d_{43} & d_{44} & 0 & d_{46} & 0 & 0 \\ d_{51} & d_{52} & d_{53} & d_{54} & 0 & 0 & 0 & 0 \\ d_{61} & d_{62} & d_{63} & d_{64} & 0 & 0 & 0 & 0 \\ 0 & 0 & 0 & 0 & 0 & 0 & d_{77} & d_{78} \\ 0 & 0 & 0 & 0 & 0 & 0 & d_{87} & d_{88} \end{pmatrix},$$

$$\{A\}=\left(A_{mn}, B_{mn}, A''_{mn}, B''_{mn}, C_{mn}, E_{mn}, A'_{mn}, B'_{mn}\right)^T.$$

Here, D_1 is a square matrix of 6×6 order and represents leaky Lamb-like waves coupled by compressional Φ and shear waves Π in spherical shell and the compressional waves in the liquid, i.e. ϕ'_w *and* ϕ''_w. The leaky Lamb-like waves generate not only the radial displacement u_r of spherical shell, the radial displacement u'_r *and* u''_r of liquid, but also the displacements u_θ and u_φ. D_2 is a square matrix of 2×2 order, and it represents SH-like waves controlled by a potential function Ψ. The SH-like waves only have in-plane displacement in the spherical shell, i.e. u_θ and u_φ. Since there is no shear stress existing in the liquid inside and outside the spherical shell, the liquid load does not have any effect on the SH-like waves. According to the condition that the equations have non-zero solutions, the dispersion equations of Lamb-like waves and SH-like waves are, respectively,

$$\Omega_1(n,\omega)=det(D_1)=0, \tag{6.6.36a}$$

$$\Omega_2(n,\omega)=det(D_2)=0, \tag{6.6.36b}$$

where

$$d_{11}=-\tfrac{\lambda}{2\mu}\alpha^2 j_n(\alpha a)+\tfrac{2\alpha}{a}j_{n+1}(\alpha a)+\tfrac{n^2-n}{a^2}j_n(\alpha a)-\alpha^2 j_n(\alpha a),$$
$$d_{12}=-\tfrac{\lambda}{2\mu}\alpha^2 y_n(\alpha a)+\tfrac{2\alpha}{a}y_{n+1}(\alpha a)+\tfrac{n^2-n}{a^2}y_n(\alpha a)-\alpha^2 y_n(\alpha a),$$
$$d_{13}=n(n+1)\left(\tfrac{n-1}{a^2}j_n(\beta a)-\tfrac{\beta}{a}j_{n+1}(\beta a)\right),$$
$$d_{14}=n(n+1)\left(\tfrac{n-1}{a^2}y_n(\beta a)-\tfrac{\beta}{a}y_{n+1}(\beta a)\right),$$
$$d_{15}=\lambda_w k_w^2 j_n(k_w a),\ d_{16}=d_{17}=d_{18}=0,$$

$$
\begin{aligned}
&d_{21}=-\tfrac{\lambda}{2\mu}\alpha^2 j_n(\alpha b)+\tfrac{2\alpha}{b}j_{n+1}(\alpha b)+\tfrac{n^2-n}{b^2}j_n(\alpha b)-\alpha^2 j_n(\alpha b),\\
&d_{22}=-\tfrac{\lambda}{2\mu}\alpha^2 y_n(\alpha b)+\tfrac{2\alpha}{b}y_{n+1}(\alpha b)+\tfrac{n^2-n}{b^2}y_n(\alpha b)-\alpha^2 y_n(\alpha b),\\
&d_{23}=n(n+1)\Big(\tfrac{n-1}{b^2}j_n(\beta b)-\tfrac{\beta}{b}j_{n+1}(\beta b)\Big),\\
&d_{24}=n(n+1)\Big(\tfrac{n-1}{b^2}y_n(\beta b)-\tfrac{\beta}{b}y_{n+1}(\beta b)\Big),\\
&d_{25}=0,\ d_{26}=\lambda_w k_w^2 h_n^{(2)}(k_w b),\ d_{27}=d_{28}=0,\\
&d_{31}=\tfrac{n}{a}j_n(\alpha a)-\alpha j_{n+1}(\alpha a),\ \ d_{32}=\tfrac{n}{a}y_n(\alpha a)-\alpha y_{n+1}(\alpha a),\\
&d_{33}=\tfrac{n(n+1)}{a}j_n(\alpha a),\ \ d_{34}=\tfrac{n(n+1)}{a}y_n(\alpha a),\\
&d_{35}=k_w j_{n+1}(k_w a)-\tfrac{n}{a}j_n(k_w a),\ \ d_{36}=d_{37}=d_{38}=0,\\
&d_{41}=\tfrac{n}{b}j_n(\alpha b)-\alpha j_{n+1}(\alpha b),\ \ d_{42}=\tfrac{n}{b}y_n(\alpha b)-\alpha y_{n+1}(\alpha b),\\
&d_{43}=\tfrac{n(n+1)}{b}j_n(\alpha b),\ d_{44}=\tfrac{n(n+1)}{b}y_n(\alpha b),\\
&d_{45}=0,\ d_{46}=k_w h_{n+1}^{(2)}(k_w b)-\tfrac{n}{b}h_n^{(2)}(k_w b),\ \ d_{47}=d_{48}=0,\\
&d_{51}=2\big(\tfrac{n-1}{a^2}j_n(\alpha a)-\tfrac{\alpha}{a}j_{n+1}(\alpha a)\big),\ d_{52}=2\big(\tfrac{n-1}{a^2}y_n(\alpha a)-\tfrac{\alpha}{a}y_{n+1}(\alpha a)\big),\\
&d_{53}=\tfrac{1}{a^2}n(n+1)j_n(\beta a)+\tfrac{2\beta}{a}j_{n+1}(\beta a)+\tfrac{n^2-n}{a^2}j_n(\beta a)-\beta^2 j_n(\beta a)-\tfrac{2}{a^2}j_n(\beta a),\\
&d_{54}=\tfrac{1}{a^2}n(n+1)y_n(\beta a)+\tfrac{2\beta}{a}y_{n+1}(\beta a)+\tfrac{n^2-n}{a^2}y_n(\beta a)-\beta^2 y_n(\beta a)-\tfrac{2}{a^2}y_n(\beta a),\\
&d_{55}=d_{56}=d_{57}=d_{58}=0,\\
&d_{61}=2\big(\tfrac{n-1}{b^2}j_n(\alpha b)-\tfrac{\alpha}{b}j_{n+1}(\alpha b)\big),\ \ d_{62}=2\big(\tfrac{n-1}{b^2}y_n(\alpha b)-\tfrac{\alpha}{b}y_{n+1}(\alpha b)\big),\\
&d_{63}=\tfrac{1}{b^2}n(n+1)j_n(\beta b)+\tfrac{2\beta}{b}j_{n+1}(\beta b)+\tfrac{n^2-n}{b^2}j_n(\beta b)-\beta^2 j_n(\beta b)-\tfrac{2}{b^2}j_n(\beta b),\\
&d_{64}=\tfrac{1}{b^2}n(n+1)y_n(\beta b)+\tfrac{2\beta}{b}y_{n+1}(\beta b)+\tfrac{n^2-n}{b^2}y_n(\beta b)-\beta^2 y_n(\beta b)-\tfrac{2}{b^2}y_n(\beta b),\\
&d_{65}=d_{66}=d_{67}=d_{68}=0,\ d_{71}=d_{72}=d_{73}=d_{74}=d_{75}=d_{76}=0,\\
&d_{77}=\tfrac{n-1}{a}j_n(\beta a)-\beta j_{n+1}(\beta a),\ \ d_{78}=\tfrac{n-1}{a}y_n(\beta a)-\beta y_{n+1}(\beta a),\\
&d_{81}=d_{82}=d_{83}=d_{84}=d_{85}=d_{86}=0,\\
&d_{87}=\tfrac{n-1}{b}j_n(\beta b)-\beta j_{n+1}(\beta b),\ \ d_{88}=\tfrac{n-1}{b}y_n(\beta b)-\beta y_{n+1}(\beta b).
\end{aligned}
$$

It is noted that from the expression above, d_{ij} does not contain m, that is to say, the dispersion equation of guided waves of order (n, m) does not depend on m. m only affects the circumferential displacement distribution of guided waves, but does not affect the propagation velocity of guided waves. When calculating the dispersion curves of spherical guided waves, only the case of $m = n$ can be considered. In this case, n and m have the same meaning, namely n represents the angular wave number of the guided wave. n can be any real number, and when n is an integer, a spherical guided wave (traveling wave) degenerates into a spherical standing wave. Considering the existence of liquid loads, the roots of determinant corresponding to Lamb-like waves should be scaned in the complex domain. That is to say, for any given frequency ω, the roots $n = kb$, where $k = k^r + ik^m$, should be sought in the complex domain. The imaginary part of the complex wave number represents the energy leakage caused by the liquid loads. Since the liquid loads have no effect on the interface conditions of SH-like waves, the roots of determinant corresponding to SH-like wave can still be sought in the real number domain.

References

1. Guitong Y, Shanyuan Z (1988) Elastic dynamics (Chinese). China Railway Publishing House, Beijing
2. Hongliang D (2014) Elastic dynamics (Chinese). Hunan University Press, Changsha
3. Peijun W (2012) Equations of mathematical physics (Chinese). Metallurgieal Industry Press, Beijing
4. Yih-Hsing P, Chao-chow M (1973) Diffraction of elastic wave and dynamic stress concentration. Crane Russak & Company Inc. USA. Chinese version: (trans: Diankui L, Xianyue S) (1993). Science Press, Beijing
5. Caviglia G, Morro A (1992) Inhomogeneous waves in solids and fluids. World Scientific
6. Auld BA (1982) Acoustic fields and waves in solids. Chinese version: (trans: Chengping S). Science Press, Beijing
7. Brekhovsky LM (1960) Waves in layered media. Academic Press. Chinese version: (trans: Xunren Y) (1985). Science Press, Beijing
8. Rose JL (1999) Ultrasonic waves in solid media. Cambridge University Press. Chinese version: (trans: Cunfu H, Bing W, Xiuyan W) (2004). Science Press, Beijing
9. Shaohua G (2014) Introduction of anisotropic elastic wave (Chinese). Science Press, Beijing
10. Achenbach JD (1973) Wave propagation in elastic solids. Elsevier Science Publishers, Amsterdam
11. Abubakar I (1962) Scattering of plane elastic waves at rough surfaces. Math Proc Camb Phil Soc 58:136–157
12. Asano S (1966) Reflection and refraction of elastic waves at a corrugated interface. Bull Seismol Soc Am 56:201–221
13. Yonggang K, Peijun W, Yueqiu L, Peng Z (2021) Modeling elastic wave propagation through a partially saturated poroviscoelastic interlayer by fractional order derivatives. Appl Math Model 100:612–631
14. Peng Z, Peijun W, Yueqiu Li (2018) The elastic wave propagation through the finite and infinite periodic laminated structure of micropolar elasticity. Compos Struct 200:358–370
15. Fengyu J, Peijun W, Li L (2017) Wave propagation through an inhomogeneous slab sandwiched by the piezoelectric and the piezomagnetic half spaces. Ultrasonics 73:22–33
16. Li L, Wei PJ, Guo X (2016) Rayleigh wave on the half-space with a gradient piezoelectric layer and imperfect interface. Appl Math Model 40:8326–8337
17. Peng Z, Peijun W, Yueqiu L (2016) Wave propagation through a micropolar slab sandwiched by two elastic half-spaces. J Vibr Acoust 138(4):1–17, 041008
18. Xiao G, Peijun W, Man L, Li L (2016) Dispersive relations of elastic waves in one-dimensional piezoelectric/piezomagnetic phononic crystal with functionally graded interlayers. Ultrasonics 70:158–171
19. Xiao G, Peijun W, Li L (2016) Dispersion relations of elastic waves in one-dimensional piezoelectric phononic crystal with mechanically and dielectrically imperfect interfaces. Mech Mater 93:168–183
20. Yueqiu L, Peijun W (2016) Reflection and transmission through a microstructured slab sandwiched by two half-spaces. Eur J Mech A/Solids 57:1–17
21. Yueqiu L, Peijun W, Yahong Z (2016) Band gaps of elastic waves in 1-D phononic crystal with dipolar gradient elasticity. Acta Mechanica 227(4):1005–1023

© Science Press 2022

P. Wei, *Theory of Elastic Waves*,
https://doi.org/10.1007/978-981-19-5662-1

22. Zhao YP, Wei PJ (2009) The band gap of 1D viscoelastic phononic crystal. Comput Mater Sci 46(3):603–606 (SCI, EI)
23. Zheng M, Wei PJ (2009) Band gaps of elastic waves in 1D phononic crystal with imperfect interfaces. Int J Miner Metall Mater 16(5):608–614
24. Dey S, Gupta AK, Gupta S (1996) Propagation of torsional surface waves in a homogeneous substratum over a heterogeneous half-space. Int J Numer Anal Meth Geomech 20:287–294
25. Mengsi H, Peijun W, Lina Z, Yueqiu L (2021) Multiple fields coupled elastic flexural waves in the thermoelastic semiconductor microbeam with consideration of small scale effects. Composite Struct 270:114104
26. Yishuang H, Peijun W (2021) Modelling the flexural waves in a nanoplate based on the fractional order nonlocal strain gradient elasticity and thermoelasticity. Composite Struct 266:113793
27. Yishuang H, Peijun W (2021) Modelling flexural wave propagation by the nonlocal strain gradient elasticity with fractional derivatives. Math Mech Solid
28. Abbas BA, Thomas J (1977) The second frequency spectrum of Timoshenko beams. J Sound Vib 51:123–137
29. Levinson M, Cooke DW (1982) on the two frequency spectra of Timoshenko beams. J Sound Vib 84(3):319–326
30. Stephen NG (2006) The second spectrum of Timoshenko beam theory—further assessment. J Sound Vib 292:372–389
31. Almeida DSJ, Ramo AJA (2017) On the nature of dissipative Timoshenko systems at light of the second spectrum of frequency. Z Angew Math Phys 68:145–154
32. David E., Muller (1956) A Method for Solving Algebraic Equations Using an Automatic Computer. Mathematical Tables and Other Aids to Computation 10(56) 208-215
33. Lily, Yau Adi, Ben-Israel (1998) The Newton and Halley Methods for Complex Roots. The American Mathematical Monthly 105(9) 806-818
34. Zhang Kai. Modeling and dispersion curve inversion of Rayleigh waves in viscoelastic media[D].Wuhan: China University of Geo-sciences, 2011.
35. C. G., Lai (2002) Solution of the Rayleigh Eigenproblem in Viscoelastic Media. Bulletin of the Seismological Society of America 92(6) 2297-2309
36. Fornberg B., A Practical Guide to Pseudo spectral Methods,Cambridge University Press, Cambridge UK, 1996.
37. Trefethen L. N., Spectral Methods in MATLAB ~SIAM, Philadelphia, 2000.
38. Boyd J. P., Chebyshev and Fourier Spectral Methods, 2nd ed. Dover, New York, 2001.
39. Shen J., et al., Spectral Methods: Algorithms, Analysis and Applications, Springer-Verlag Berlin Heidelberg 2011.
40. J. A., Weideman S. C., Reddy (2000) A MATLAB differentiation matrix suite. ACM Transactions on Mathematical Software 26(4) 465-519
41. Guoli, Liu Jianmin, Qu (1998) Guided Circumferential Waves in a Circular Annulus. Journal of Applied Mechanics 65(2) 424-430
42. Christine, Valle Jianmin, Qu Laurence J., Jacobs (1999) Guided circumferential waves in layered cylinders. International Journal of Engineering Science 37(11) 1369-1387
43. Doyle J. F., Wave propagation in structures: an FFT-based spectral analysis methodology, Springer-Verlag, New York, 1989.
44. A. T. I., Adamou R. V., Craster (2004) Spectral methods for modelling guided waves in elastic media. The Journal of the Acoustical Society of America 116(3) 1524-1535
45. Florian, Karpfinger Boris, Gurevich Andrey, Bakulin (2008) Modeling of wave dispersion along cylindrical structures using the spectral method. The Journal of the Acoustical Society of America 124(2) 859-865
46. F., Hernando Quintanilla M.J.S., Lowe R.V., Craster (2016) Full 3D dispersion curve solutions for guided waves in generally anisotropic media. Journal of Sound and Vibration 363545-559
47. ŁQuintanilla F.H., Lowe M., Craster R.V., Modelling guided elastic wave singe nerally anisotropic media using a spectral collocation method, Journal of the Acoustical Society of America, 2015,137(3):1180-1194

48. ŁFlorian Karpfinger, Boris Gurevicha and Andrey Bąkulin, Modeling of axi symmetric wave modes in a poroelastic cylinder using spectral method, The Jo urnal of the Acoustical Society of America, 2008, 124, EL230